ENVIRONMENTAL AND HEALTH MANAGEMENT OF NOVEL CORONAVIRUS DISEASE (COVID-19)

ENVIRONMENTAL AND HEALTH MANAGEMENT OF NOVEL CORONAVIRUS DISEASE (COVID-19)

Edited by

MOHAMMAD HADI DEHGHANI

Department of Environmental Health Engineering,
School of Public Health,
Tehran University of Medical Sciences, Tehran, Iran;
Institute for Environmental Research,
Center for Solid Waste Research,
Tehran University of Medical Sciences, Tehran, Iran

RAMA RAO KARRI

Petroleum and Chemical Engineering,
Faculty of Engineering, Universiti Teknologi Brunei (UTB),
Gadong, Brunei Darussalam

SHARMILI ROY

Division of Oncology, School of Medicine,
Stanford University, Palo Alto, CA, United States

ELSEVIER

ACADEMIC PRESS

An imprint of Elsevier

Academic Press is an imprint of Elsevier
125 London Wall, London EC2Y 5AS, United Kingdom
525 B Street, Suite 1650, San Diego, CA 92101, United States
50 Hampshire Street, 5th Floor, Cambridge, MA 02139, United States
The Boulevard, Langford Lane, Kidlington, Oxford OX5 1GB, United Kingdom

Library of Congress Cataloging-in-Publication Data
A catalog record for this book is available from the Library of Congress

British Library Cataloguing-in-Publication Data
A catalogue record for this book is available from the British Library

ISBN: 978-0-323-85780-2

For information on all Academic Press publications visit our website at https://www.elsevier.com/books-and-journals

Publisher: Andre Gerhard Wolff
Acquisitions Editor: Kattie Washington
Editorial Project Manager: Kristi Anderson
Production Project Manager: Selvaraj Raviraj
Cover Designer: Mark Rogers

Typeset by TNQ Technologies

In the Name of God, the Most Gracious, the Most Merciful

I am thankful to God Almighty that I succeeded in writing this book with the help of my colleagues.

I dedicate this book to my parents, my brothers and my sister who are always praying for me.

I especially appreciate my lovely wife and children who contributed to my progress and success with their patience and forbearance.

I dedicate this book to the defenders and martyrs of healthcare in the Iran, Islamic Republic of.

I also dedicate this book to healthcare professionals around the world who are at the forefront of the fight against COVID-19.

Prof. Mohammad Hadi Dehghani

I dedicate this book to the memory of my Father

Karri Sri Ramulu (10/3/1939—10/12/1989)—Ex. Army

I also dedicate this to my mother Karri Kannathalli, who protected, guided, and supported me in all these years. She was my inspiration for driving me to achieve my best. You are my superwoman and constant inspiration.

I also thank my lovely wife, Soni. Without her support, this book as well as my research achievements would not be possible.

I also dedicate this book to all the frontline warriors all over the world.

Dr. Rama Rao Karri

I dedicate this book to my father Mr. Pranab Jyoti Roy and my mother Mrs. Kalyani Roy. Without their continuous support and motivation, it would not be possible for me to achieve my best. I also thank my husband Mr. Vivek Majumdar. Without his support, inspiration, and continuous motivation, this book as well as my research achievements would not be possible. Lastly, I want to thank my PhD supervisor Dr. Minhaz Uddin Ahmed for his proper guidance and supervision, without which my research achievements would not be feasible.

Dr. Sharmili Roy

Contents

Contributors

Bashir Adelodun
Department of Agricultural Civil Engineering, Kyungpook National University, Daegu, Korea; Department of Agricultural and Biosystems Engineering, University of Ilorin, Ilorin, Kwara State, Nigeria

Kamoru Akanni Adeniran
Department of Agricultural and Biosystems Engineering, University of Ilorin, Ilorin, Kwara State, Nigeria

Jamiu Adetayo Adeniran
Environmental Engineering Research Laboratory, Department of Chemical Engineering, University of Ilorin, Ilorin, Kwara State, Nigeria; Atmospheric Chemistry and Modeling Group, Department of Atmospheric and Oceanic Sciences, Peking University, Beijing, China

Fidelis Odedishemi Ajibade
Department of Civil and Environmental Engineering, Federal University of Technology, Akure, Ondo State, Nigeria; Key Laboratory of Environmental Biotechnology, Research Centre for Eco-Environmental Sciences, Chinese Academy of Sciences, Beijing, PR China; University of Chinese Academy of Sciences, Beijing, PR China

Temitope F. Ajibade
Department of Civil and Environmental Engineering, Federal University of Technology, Akure, Ondo State, Nigeria; University of Chinese Academy of Sciences, Beijing, PR China; Key Laboratory of Urban Pollutant Conversion, Institute of Urban Environment, Chinese Academy of Sciences, Xiamen, PR China

Zakaria Al-Qodah
Al-Balqa Applied University, Faculty of Engineering Technology, Department of Chemical Engineering, Amman, Jordan

Aldo Alvarez-Risco
Universidad de Lima, Facultad de Ciencias Empresariales y Económicas. Carrera de Negocios Internacionales. Lima, Peru

Hashim Olalekan Bakare
Department of Chemical Engineering, University of Ilorin, Ilorin, Kwara State, Nigeria

Moumita Bishai
Department of Botany, Gurudas College, Kolkata, West Bengal, India

Maria Cadonna
ADEP – Agenzia per la Depurazione, Autonomous Province of Trento, Trento, Italy

Turgay Cetinkaya
Food Processing Department, Armutlu Vocational School, Yalova University, Yalova, Turkey

Zafer Ceylan
Van Yüzüncü Yıl University, Faculty of Tourism, Department of Gastronomy and Culinary Arts, Van, Turkey

Prasanna Raja Chandrasekaran
Department of In Vitro Pharmacology, Natural Remedies Private Limited, Bangalore, Karnataka, India

Kyung Sook Choi
Department of Agricultural Civil Engineering, Kyungpook National University, Daegu, Korea; Institute of Agricultural Science & Technology, Kyungpook National University, Daegu, Korea

Tanima Chowdhury
Quality Department, Regional Head Office, ALPLA India Pvt. Ltd., Hyderabad, Telangana, India

Sun Ah Chung
Universidad Cristiana de Bolivia. Comunidad Científica de Estudiantes de Medicina, Santa Cruz, Bolivia

Francesca Cutrupi
Department of Civil, Environmental and Mechanical Engineering (DICAM), University of Trento, Trento, Italy

Monalisha Ghosh Dastidar
Research School of Electrical, Energy, and Materials Engineering, College of Engineering and Computer Science, Australian National University, ACT, Canberra, Australia

Mohammad Hadi Dehghani
Department of Environmental Health Engineering, School of Public Health, Tehran University of Medical Sciences, Tehran, Iran; Institute for Environmental Research, Center for Solid Waste Research, Tehran University of Medical Sciences, Tehran, Iran

Shyla Del-Aguila-Arcentales
Escuela Nacional de Marina Mercante "Almirante Miguel Grau", Callao, Peru

Alaa El Din Mahmoud
Environmental Sciences Department, Faculty of Science, Alexandria University, Alexandria, Egypt; Green Technology Group, Faculty of Science, Alexandria University, Alexandria, Egypt

Mohammad Mahdi Emamjomeh
Social Determinants of Health Research Center, Research Institute for Prevention of Non-Communicable Diseases, Qazvin University of Medical Sciences, Qazvin, Iran

Paola Foladori
Department of Civil, Environmental and Mechanical Engineering (DICAM), University of Trento, Trento, Italy

Franko O. Garcia-Solorzano
Universidad Científica del Sur. Facultad de Ciencias de la Salud, Carrera de Medicina Humana, Lima, Peru

Madhumita Goala
Nehru College, Pailapool, Affiliated Assam University, Silchar, Cachar, Assam, India

Zahra Mohammadi Goldar
Student Research Committee, Qazvin University of Medical Sciences, Qazvin, Iran

Chloé Guitart
Infection Control Programme, University of Geneva Hospitals and Faculty of Medicine, Geneva, Switzerland

Hajar Haghighi
Department of Health Management, Policy & Economics, School of Public Health, Tehran University of Medical Sciences (TUMS), Tehran, Iran

Marjan Hashemi
Environmental and Occupational Hazards Control Research Center, Shahid Beheshti University of Medical Sciences, Tehran, Iran

Khalid S. Hashim
Department of Civil Engineering, Liverpool John Moores University, Liverpool, United Kingdom

Sara Hemati
Department of Environmental Health Engineering, School of Health, Shahrekord University of Medical Sciences, Shahrekord, Iran

Palmer J. Hernández-Yépez
Universidad Privada Norbert Wiener. Grupo de Investigación en Gestión y Salud Pública, Lima, Peru

Chaudhery Mustansar Hussain
Department of Chemistry and Environmental Science, New Jersey Institute of Technology, Newark, NJ, United States

Rahmat Gbemisola Ibrahim
Kwara State Ministry of Health, Ilorin, Kwara State, Nigeria

Ramanaiah Illuri
Department of Pharmacology, Vidya Herbs Pvt. Ltd, Bangalore, Karnataka, India

Fiorella Inga-Berrospi
Universidad Privada Norbert Wiener. Grupo de Investigación en Gestión y Salud Pública, Lima, Peru

Samuel Jacob
Department of Biotechnology, School of Bioengineering, College of Engineering and Technology, Faculty of Engineering and Technology, SRM Institute of Science and Technology, Kattankulathur, Tamil Nadu, India

Hojatollah Kakaei
Department of Occupational Health Engineering, School of Health, Ilam University of Medical Sciences, Ilam, Iran

Kadir Karakus
Department of Animal Science, Faculty of Agriculture, Malatya Turgut Özal University, Malatya, Turkey

Rama Rao Karri
Petroleum and Chemical Engineering, Faculty of Engineering, Universiti Teknologi Brunei (UTB), Gadong, Brunei Darussalam

Lokeshwaran Kirubananthan
Research and Development, Bhat Biotech India (P) Limited, Bangalore, Karnataka, India

Federica Maestrini
Department of Civil, Environmental and Mechanical Engineering (DICAM), University of Trento, Trento, Italy

Alaa El Din Mahmoud
Environmental Sciences Department, Faculty of Science, Alexandria University Alexandria, Egypt

Serena Manara
Department of Cellular Computational and Integrative Biology (CIBIO), University of Trento, Trento, Italy

Christian R. Mejia
Universidad Continental, Lima, Peru

Fazel Mohammadi-Moghadam
Department of Environmental Health Engineering, School of Health, Shahrekord University of Medical Sciences, Shahrekord, Iran

Sneha Mohapatra
Department of Biotechnology, School of Bioengineering, College of Engineering and Technology, Faculty of Engineering and Technology, SRM Institute of Science and Technology, Kattankulathur, Tamil Nadu, India

Milad Mousazadeh
Student Research Committee, Qazvin University of Medical Sciences, Qazvin, Iran; Department of Environmental Health Engineering, School of Health, Qazvin University of Medical Sciences, Qazvin, Iran

Ahmad Mukhtar
Department of Chemical Engineering, Universiti Teknologi PETRONAS (UTP), Seri Iskandar, Perak, Malaysia

Sukanya Nag
Department of Biotechnology, School of Bioengineering, College of Engineering and Technology, Faculty of Engineering and Technology, SRM Institute of Science and Technology, Kattankulathur, Tamil Nadu, India

Zohreh Naghdali
Student Research Committee, Qazvin University of Medical Sciences, Qazvin, Iran; Department of Environmental Health Engineering, School of Health, Qazvin University of Medical Sciences, Qazvin, Iran

Shirsendu Nandi
Operations Management and Quantitative Techniques, FORE School of Management, New Delhi, India

Sajesh Nithianandam
Department of Biotechnology, School of Bioengineering, College of Engineering and Technology, Faculty of Engineering and Technology, SRM Institute of Science and Technology, Kattankulathur, Tamil Nadu, India

Heshmatollah Nourmoradi
Biotechnology and Medicinal Plant Research Center, Ilam University of Medical Sciences, Ilam, Iran; Department of Environmental Health Engineering, School of Health, Ilam University of Medical Sciences, Ilam, Iran

Elvan Ocak
Van Yüzüncü Yıl University, Engineering Faculty, Food Engineering Department, Van, Turkey

Golden Odey
Department of Agricultural Civil Engineering, Kyungpook National University, Daegu, Korea

Biswaranjan Paital
Redox Regulation Laboratory, Odisha University of Agriculture and Technology, College of Basic Science and Humanities, Bhubaneswar, Odisha, India

Alexandra Peters
Infection Control Programme, University of Geneva Hospitals and Faculty of Medicine, Geneva, Switzerland

Didier Pittet
Infection Control Programme, University of Geneva Hospitals and Faculty of Medicine, Geneva, Switzerland

Neda Rahimian
Endocrine Research Center, Institute of Endocrinology and Metabolism, Iran University of Medical Sciences (IUMS), Tehran, Iran

Ramesh Rajendran
Centre for Bioseparation Technology, Vellore Institute of Technology, Vellore, Tamil Nadu, India

Gunasekaran Rajeswari
Department of Biotechnology, School of Bioengineering, College of Engineering and Technology, Faculty of Engineering and Technology, SRM Institute of Science and Technology, Kattankulathur, Tamil Nadu, India

Archana Ramadoss
Department of Research and Development, Nanolane, Le Mans, France

Azam Raoofi
Department of Health Management, Policy & Economics, School of Public Health, Tehran University of Medical Sciences (TUMS), Tehran, Iran

Shrestha Rastogi
Department of Biotechnology, School of Bioengineering, College of Engineering and Technology, Faculty of Engineering and Technology, SRM Institute of Science and Technology, Kattankulathur, Tamil Nadu, India

Brenda Rojas Román
Universidad Cristiana de Bolivia. Comunidad Científica de Estudiantes de Medicina, Santa Cruz, Bolivia

Sharmili Roy
Division of Oncology, School of Medicine, Stanford University, Palo Alto, CA, United States

Simar Sakhuja
Department of Biotechnology, School of Bioengineering, College of Engineering and Technology, Faculty of Engineering and Technology, SRM Institute of Science and Technology, Kattankulathur, Tamil Nadu, India

Miguel A. Sandoval
Universidad de Santiago de Chile USACH, Facultad de Química y Biología, Departamento de Química de los Materiales, Laboratorio de Electroquímica Medio Ambiental, LEQMA, Santiago, Chile; Universidad de Guanajuato, División de Ciencias Naturales y Exactas, Departamento de Ingeniería Química, Guanajuato, Mexico

Satish Kumar Santhosh Venkat
Department of Biotechnology, School of Bioengineering, College of Engineering and Technology, Faculty of Engineering and Technology, SRM Institute of Science and Technology, Kattankulathur, Tamil Nadu, India

Ravi Siddharth
Department of Biotechnology, School of Bioengineering, College of Engineering and Technology, Faculty of Engineering and Technology, SRM Institute of Science and Technology, Kattankulathur, Tamil Nadu, India

Mika Sillanpää
Chemistry Department, College of Science, King Saud University, Riyadh, Saudi Arabia; Faculty of Science and Technology, School of Applied Physics, University Kebangsaan Malaysia, Bangi, Selangor, Malaysia; Department of Biological and Chemical Engineering, Aarhus University, Nørrebrogade 44, Aarhus C, Denmark

Senthil Nathan Sri Laxma Alankar
Department of Biotechnology, School of Bioengineering, College of Engineering and Technology, Faculty of Engineering and Technology, SRM Institute of Science and Technology, Kattankulathur, Tamil Nadu, India

Amirhossein Takian
Department of Health Management, Policy & Economics, School of Public Health, Tehran University of Medical Sciences (TUMS), Tehran, Iran; Department of Global Health & Public Policy, School of Public Health, Tehran University of Medical Sciences (TUMS), Tehran, Iran; Health Equity Research Center (HERC), Tehran University of Medical Sciences (TUMS), Tehran, Iran

AbdulGafar Olatunji Tiamiyu
Department of Chemical Engineering, University of Ilorin, Ilorin, Kwara State, Nigeria

Yılmaz Uçar
Fatsa Faculty of Marine Sciences, Ordu University, Ordu, Turkey

Jaime A. Yáñez
Universidad Peruana de Ciencias Aplicadas. Facultad de Educación, Carrera de Educación y Gestión Del Aprendizaje, Lima, Peru; Teoma Global. Gerencia Corporativa de Asuntos Científicos y Regulatorios, Lima, Peru

Foreword

The global impacts of COVID-19 in health care settings and environmental elements such as food processing and distributing systems, and air quality, as well as health care and municipal solid and liquid wastes management are considered in this book. Moreover, different remediation technologies, especially the measures that can be used by individuals and environmental health managers at the time of an outbreak, are included and discussed in detail.

The lack of improved conceptual edited books is visibly felt in the present pandemic. Environmental and health manager's, microbiologists and virologists facing difficulties in getting the latest measures, which is scattered in public domain. This is the driving force to prepare this edited book.

Alireza Mesdaghinia
Emeritus Professor of Environmental Health,
Tehran University of Medical Sciences, Iran
President, Iranian Association of Environmental Health

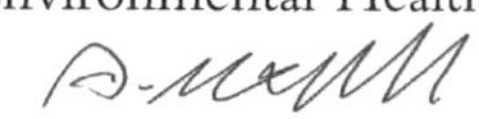

Preface

Today, the rapid and growing prevalence of COVID-19 has led experts to accurately assess the behavior and prevention of this contagious virus in the environment. Environmental and health managers, microbiologists, and virologists face difficulties in getting information on the latest measures, which are scattered in the public domain. This is the driving force behind this book. An effort has been made in this book to present the global impacts of COVID-19 in health care settings and the environment, including waste, wastewater, food products, and air quality. Moreover, different remediation technologies, especially the measures that individuals and environmental health managers can adopt at the time of an outbreak, might be considered as an important source of information in the prevention of COVID-19 epidemics are provided in detail.

The scope of outbreak of COVID-19 and preventive measures makes the publication of this book needed at this time.

Because health guidelines and protocols are updated regularly, we refer readers to the latest guidelines released by the World Health Organization.

Editors,
Mohammad Hadi Dehghani
Department of Environmental Health Engineering,
School of Public Health,
Tehran University of Medical Sciences, Tehran, Iran;
Institute for Environmental Research,
Center for Solid Waste Research,
Tehran University of Medical Sciences, Tehran, Iran

Rama Rao Karri
Petroleum and Chemical Engineering,
Faculty of Engineering, Universiti Teknologi Brunei (UTB),
Gadong, Brunei Darussalam

Sharmili Roy
Division of Oncology, School of Medicine,
Stanford University, Palo Alto, CA, United States

Acknowledgments

This edited book has been supported by the Tehran University of Medical Sciences. I also thank my coeditors, especially Dr. Rama Rao Karri, for without his support and cooperation this book would not have been possible. I also thank my colleagues in department of Environmental Health Engineering for their valuable supports. I also thank all the authors who contributed chapters consisting of their valuable research.

Prof. Mohammad Hadi Dehghani

I thank Prof. Zohrah, Vice Chancellor of the University of Technology Brunei, and the higher management for the support. I also thank my coeditors, for without their support and cooperation, this book would not have been possible. I also thank all the authors who contributed chapters consisting of their valuable research.

Dr. Rama Rao Karri

I thank Stanford University for encouragement and providing the necessary facilities for completion of this work. I also thank my coeditors. I appreciate all the authors who contributed chapters with full dedications and efforts to complete this book.

Dr. Sharmili Roy

Introduction

Coronaviruses (CoV) are a large family of viruses that can infect animals and humans. Many known CoVs cause a range of respiratory infections in humans, such as MERS-CoV and SARS-CoV. In rare cases, animal CoVs infect humans and then spread between them. SARS-CoV-2 is a new virus that has led to the outbreak of a respiratory disease known as novel coronavirus 2019 (COVID-19). This virus is spreading rapidly around the world. Evidence shows that COVID-19 transmission occurs primarily between individuals through direct contact and then indirectly or in close contact with infected persons through contaminated secretions or respiratory droplets from the infected person when coughing, sneezing, or talking.

Health care centers, especially hospitals, can be sites where outbreaks of infectious diseases such as COVID-19 occur. Therefore, if environmental and public health are not well observed in these centers, staff and patients will face severe risks of viral and bacterial infections.

Although health care professionals and other health care workers are on the line to combat COVID-19, public participation is also essential to rapidly control the pandemic worldwide. Therefore, informing the public about health protocols to control this epidemic is very important.

The rapid spread of COVID-19 and protocol awareness requirements highlight the significance of the publication of this book. We believe that this book can be very effective in raising awareness to prevent and control the COVID-19 through various environments such as air, wastewater, medical wastes, food products, and different surfaces. This book focuses on several aspects of the novel coronavirus 2019 (COVID-19). The book consists of four sections: (1) introduction, transmission routes, and sampling technologies; (2) remediation measures in waste and wastewater environments; (3) environmental and health management aspects; and (4) challenges and opportunities faced due to COVID-19.

This book caters widely to students and academic communities who are working in the fields of environmental health science, microbiology, and virology, health policy and public health, medicine, and bioengineering. Also, this book can be extensively useful for researchers, professional, policymakers, and industrial practitioners who are working in the environmental science field, water and wastewater industry, food industry, air quality and air pollution, waste management, virology and microbiology, health care settings, infectious diseases and epidemiology, clinical nursing, and oncology medicine.

Even though there have been a few recent books on CoVs, so far, there have been no books published that focus exclusively on environmental health managers' measures for the prevention of COVID-19 outbreak.

Editors,
Mohammad Hadi Dehghani
Department of Environmental Health Engineering, School of Public Health, Tehran University of Medical Sciences, Tehran, Iran; Institute for Environmental Research, Center for Solid Waste Research, Tehran University of Medical Sciences, Tehran, Iran

Rama Rao Karri
Petroleum and Chemical Engineering, Faculty of Engineering, Universiti Teknologi Brunei (UTB), Gadong, Brunei Darussalam

Sharmili Roy
Division of Oncology, School of Medicine, Stanford University, Palo Alto, CA, United States

SECTION I

Introduction, transmission routes and sampling technologies

CHAPTER 1

Updated insight into COVID-19 disease and health management to combat the pandemic

Sharmili Roy[1], Archana Ramadoss[2]
[1]Division of Oncology, School of Medicine, Stanford University, Palo Alto, CA, United States;
[2]Department of Research and Development, Nanolane, Le Mans, France

1.1 Introduction to SARS-CoV-2

Severe acute respiratory syndrome coronavirus 2 (SARS-CoV-2) is a pandemic potential zoonotic virus that belongs to the Coronaviridae family.[1–3] As of January 2021, more than 92 million people around the globe are actively battling the virus and over 2 million people have already succumbed to the disease so far (https://covid19.who.int/). Although the origin of this virus remains speculative, earliest reports of patient cases were linked to a live sea food market in Wuhan in the Hubei province of China.[3] The zoonotic virus is thought to originate in bats and has been transmitted to humans through an intermediate host animal such as palm civet cats or pangolins.[3] However, the intermediate animal host is yet to be established.

The contagious nature and the enhanced transmission rate of the virus are the primary causes for such rapid spread of the disease globally.[4,5] The human-to-human transmission of the virus primarily occurs through direct or indirect contact with infected respiratory water droplets.[6] Such contact is established either by inhalation or direct contact with a contaminated surface or infected body fluid such as saliva and urine.[6–8] The ability of the virus to remain active in suspension of infected respiratory droplets over a long period of time and to be carried over to long distances makes this pathogen airborne.[9,10]

Public health management strategies directly contribute to suppression of human-to-human disease spread. For instance, Taiwan's success in battling the first wave of viral infection is principally credited for the quick and strict observance of public safety measures such as use of facial masks, social distancing between individuals, and strict travel restrictions.[11] Another example is that of Hong Kong, home to a population of 7.5 million, which effectively curbed the spread of SARS-CoV-2 using strategies such as

Environmental Management of COVID-19
ISBN 978-0-323-85780-2
https://doi.org/10.1016/B978-0-323-85780-2.00017-2

prompt and swift surveillance of the disease progression, quarantine and social-distancing measures, closure of schools, and obligatory use of face masks.[12] Another example is that of Vietnam, home to 96 million people, which has surprised the world with less than 2000 COVID-19 cases in the country.[13] The success is primarily attributed to early risk assessment of the situation and placement of strict travel and border restrictions with foreign countries including quarantines apart from the standard health safety procedures.[14] Such successful examples have inspired many nations around the globe with hope that are actively battling the virus outbreak. However, attempts to contain the spread of the virus worldwide has not been uniform among all the nations globally since not all nations are in the same phase of infection, and other reasons, including less stringent public safety measures, delayed governmental policies, and/or even absence of prior experience with such a coronavirus epidemic.[15] For instance, in Italy, the country was taken by surprise when the initial outbreak of the virus was reported. The level of governmental unpreparedness, a decentralized policy of following different health safety measures in different regions of the country, in addition to not-so-sophisticated health care facilities caused significant economic and human loss in the country.[16] The world bank forecasts a great economic dip in the world economy since the world war era as the aftermath of this pandemic.[17] Given the deep impact of the pandemic as the world enters into a great socioeconomic dip, the importance of health management strategies cannot be stressed enough.

In humans, COVID-19 presents with clinical symptoms ranging from mild in healthy adults to severe in individuals with immune-compromised or weakened immune system and the elderly.[18–20] Some of the clinical symptoms include fever, sore throat, fatigue, loss of smell and sense of taste for milder cases, while acute-to-severe respiratory distress syndrome (ARDS) is among the major complications that require medical attention.[18] The rapid global spread of the disease warranting increased hospitalization of COVID-19 patients and constant monitoring of the disease progression in patients has pressured the scientific community worldwide. Scientists and health care professionals around the globe are working at record speeds to understand the origin and nature of SARS-CoV-2 to search for suitable novel technologies for rapid virus detection and vaccines to contain the spread of the virus.[21,22] Currently, over 10 vaccines against SARS-CoV-2 are already available for public use, and many more are expected to be commercialized in the first quarter of 2021.[23–25] Albeit the success and enthusiasm regarding the commercial availability of COVID-19 vaccines, concerns such as efficiency of the vaccine over time and its effects on pregnant women are currently being evaluated.[26]

Early detection of the virus and isolation and further quarantine are important steps of actions in the health care management strategies to control the COVID-19 pandemic.[27,28] Health care diagnostics and advances in biotechnology play an instrumental role in successful implementation of the health care management strategies.[29] Molecular biotechnological tools such polymerase chain reactions (PCRs) are the current global standard for detection of the virus in suspected individuals. Rapid detection of the virus in suspected individuals is of prime importance for efficient implementation of health management strategies aimed at containing the spread of the virus.

Keeping in mind all of the aforementioned facts, this chapter provides an overview of the epidemiology of coronavirus and the current classical health management strategies and issues to tackle this pandemic. The chapter particularly highlights the role of standard as well as novel biomolecular diagnostic techniques (with capacity to offer rapid and robust detection of SARS-CoV-2) as tools for successful implementation of such public safety measures issued by medical policy makers and governing authorities.

1.2 Overview of SARS-CoV-2 virus

SARS-CoV-2 is classified as a member of the Coronaviridae family.[30] The detailed classification of the virus is shown in Fig. 1.1. The newly identified SARS-CoV-2 is a bona fide human pathogen like some other famous members of this family, such as HCoV-229E, HCoV-NL63, HCoV-OC43, HCoV-HKU1, SARS-CoV, and MERS-CoV.[31,32] In the past, members of the betacoronavirus genus have caused human life-threatening respiratory diseases.[33] Examples include the SARS outbreak in 2003 and the MERS outbreak in 2012 with a mortality rate of 10% and 35%, respectively.[34,35] Based on the mortality analyses of COVID-19 cases, the mortality rate is estimated to be between 0.4% and 9%.[36]

The members of coronaviruses are a group of large enveloped viruses that carry positive-sense single-stranded RNA (+ssRNA) as their genomes.[32,37] They are known to infect a wide range of host organisms from chicken

Category	Coronaviruses
Realm	*Riboviria*
Order	*Nidovirales*
Suborder	*Cornidovirineae*
Family	*Coronaviridae*
Subfamily	*Orthocoronavirinae*
Genus	*Betacoronavirus*
Subgenus	*Sarbecovirus*
Species	*Severe acute respiratory syndrome-related coronavirus*
Individuum	SARS-CoVUrbani, SARS-CoVGZ-02, Bat SARS CoVRf1/2004, Civet SARS CoVSZ3/2003, SARS-CoVPC4-227, SARSr-CoVBtKY72, SARS-CoV-2 Wuhan-Hu-1, SARSr-CoVRatG13, and so on.

Figure 1.1 A detailed taxonomical classification of SARS-CoV-2 in the Coronaviridae family. *(The image is adopted from Gorbalenya AE, Baker SC, Baric RS, et al. The species severe acute respiratory syndrome-related coronavirus: classifying 2019-nCoV and naming it SARS-CoV-2.* Nat Microbiol *2020;**5**(4):536–544. 2020/04/01. https://10.1038/s41564-020-0695-z with permission.)*

to humans.[37] Although the origin of SARS-CoV-2 remains unclear, SARS-CoV-2 shares over ~96% genetic similarity with RaTG13, a coronavirus strain originally thought to be found among the bats that are trapped in the Yunan caves of People's Republic of China (PRC).[38–41]

Detailed structural studies using electron microscopy reveals that SARS-CoV-2 contains an icosahedral viral head structure and appears spherical with diameter in the range of ~100–120 nm.[42,43] The virus possesses numerous envelope proteins (E) such as spike (S), the transmembrane glycoprotein (M and E), and the nucleocapsid protein (N), as shown in Fig. 1.2.[44] The uncanny resemblance of S protein on the virus envelope to

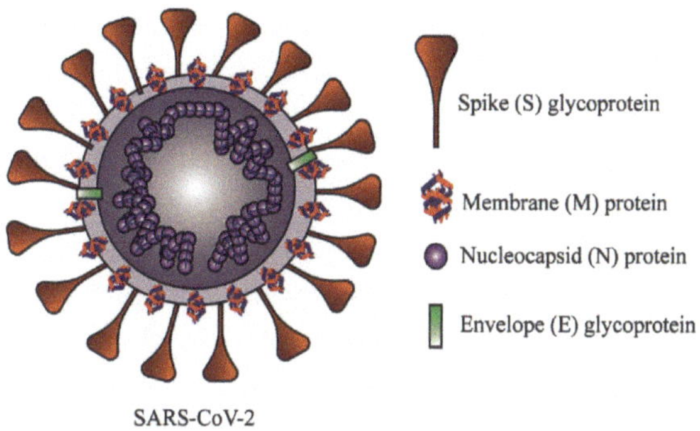

Figure 1.2 Virion morphology and RNA genomic structure of SARS-CoV-2. This structure of the virions of the members of betacoronavirus genus shows the arrangement of spike (S) glycoproteins, membrane proteins, nucleocapsid (M and N), and envelope glycoprotein (E). *(Image adopted from Sarkar C, Mondal M, Torequl Islam M, et al. Potential therapeutic options for COVID-19: current status, challenges, and future perspectives.* Front Pharmacol *2020;**11**(1428). 2020-09-15. http://10.3389/fphar.2020.572870 with permission.)*

that of the flares/corona of the sun or the crown of a queen, gives the virus its classical name of coronavirus. SARS-CoV-2 actively uses the S protein to engage with the host cell.[45] The S glycoprotein mediates viral anchoring on the host cell and fusion of the membranes for viral entry into the host cell. The virus uses human angiotensin converting enzyme 2 (ACE-2) to facilitate this viral entry.[46] Thus, current efforts to develop vaccines and antibodies to arrest the spread of the virus largely target the trimeric S protein.[43] Accumulation of point mutations in the S protein of the virus leading to the genomic evolution of SARS-CoV-2 has been elaborated in the following section.

1.2.1 Genomic evolution of SARS-CoV-2

SARS-CoV-2 carries +ssRNA with a genomic size of ~30 Kb, one of the largest genomic size among the known RNA viruses.[43] A key feature of this family of viruses is that they possess large open reading frame (ORF1a and ORF1b) that occupies nearly two-thirds of the genome (from the 3′ proximal end of the genome) and encodes for nearly 16 nonstructural proteins while the rest of the genome (toward the 5′ proximal end) encodes for all the known structural proteins of the virus.[47–49]

SARS-CoV-2 appears to have a remarkable adaptation to its host organism.[50,51] One of the 16 nonstructural proteins is nsp 13 that encodes for an *exo-ribonuclease* that provides a proofreading activity for the virus, thereby efficiently maintaining the infectivity and virulence of the virus.[52] Recently, comparative studies on SARS-CoV-2 strains isolated from patients in Wuhan (in January 2020) and that isolated from patients in Western countries (predominantly in the US, Spain, and Italy in July 2020) suggests that the virus has accumulated point mutations on its S protein, i.e., D614G, and is currently the predominant strain in the Western world, as shown in Fig. 1.3.[53] The study also demonstrated clinical evidence that the newly accumulated D614G mutation renders the virus more highly infectious than its original strain; however, the effect of this mutation on severity of the disease is not yet known.[53]

Currently, not much has been understood regarding the immunity achieved upon infection, although it appears that the severity of the infection is linked to the larger amount of the antibodies produced.[54] However, current scientific reports support the possibility of SARS-CoV-2 reinfection in individuals who have already had COVID-19, i.e., the antibodies produced by the body against this SARS-CoV-2 could last for a

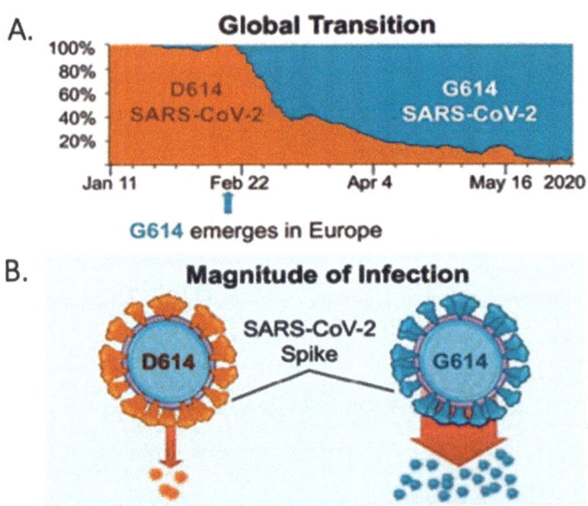

Figure 1.3 (A) Evolution of the virus: A mutation in spike protein of SARS-CoV-2 was first observed in Europe around mid-February and as of July 2020, we observe a global dominance of the mutated virus strain with a magnitude of infection over 9 times more efficient than its D614 strain. (B) Structural mutation in spike protein (from D614 to G614) showing the impact of the mutation in the viral replicability. *(Image reproduced with permission from Korber B, Fischer WM, Gnanakaran S, et al. Tracking changes in SARS-CoV-2 spike: evidence that D614G increases infectivity of the COVID-19 virus.* Cell *2020;**182**(4):812–827. e19. 2020/08/20. https://doi.org/10.1016/j.cell.2020.06.043.)*

month or less, thus leaving a previously affected individual once again vulnerable to the disease.[55,56] This appears to be a unique feature of SARS-CoV-2 compared to infections from other known infectious coronaviruses. In the past, researchers have observed that the antibodies created in the infected individuals protected them for a maximum period of three years.[57] Thus, a reinfection of masses and multiple waves of coronavirus infection is a great possibility until mass vaccination is initiated. A deeper understanding of the viral transmission, and the advantages and limitations of the current classical diagnostic tools and novel testing approaches are important to highlight the concern and importance of practicing public health safety measures during the current pandemic.

1.2.2 Transmission of coronaviruses

The rapid spread of SARS-CoV-2 across the globe has put the spotlight on the role of SARS-CoV-2 transmission dynamics.[19] Reproduction number (or R_0) is a basic parameter that allows us to estimate a disease outbreak and intensity of the infection. R_0 is defined as an average number of secondary infections caused by an affected individual when the individual is introduced to a susceptible population.[58] The knowledge of R_0 of SARS-CoV-2 is a principle tool that aids to assess the current trends and to predict the future trends of viral infectivity, i.e., disease-spreading potential of the virus. The transmission rate of SARS-CoV-2 ranges between $R_0 = 2.2$ and $R_0 = 3.9$, i.e., 1 infected person could infect ~ 4 people in the vicinity, while the median R_0 of SARS-CoV-1 and the mean R_0 of MERS were between 0.58 and 0.69 respectively.[53] The higher the R_0 the stronger the human-to-human transmission.

Among the various viral human-to-human transmission routes, three principle routes of transmission have been identified by WHO:

1. Direct or close contact with diseased individuals is a major mode of viral transmission for COVID-19 disease spreading.[59] Presence of susceptible individuals within <1 m distance from the infected person facilitates close contact viral transmission via infected respiratory droplets.
2. Contact with contaminated surfaces: Studies show that SARS-CoV-2 are highly stable over plastic and stainless surfaces and remain viable for a period of at least 72 h.[60] Thus, direct contact with such contaminated surfaces carrying sufficient concentration of the virus induces the disease in susceptible individuals.

3. Airborne transmission occurs when a susceptible individual comes in contact with a suspended contaminated respiratory droplet. Liquid droplets of diameter <5 μm can remain suspended in air and are also capable of traveling up to six feet of distance in the course of air. Such suspended droplets are also called as aerosols.[10,61] An infected individual upon sneezing generates millions of respiratory droplets that could remain suspended in the air for a long period of time and in turn contaminate other individuals nearby. This kind of viral transmission is often observed in mass public gatherings or in crowded places, and in hospital settings where medical procedures were carried on COVID-19-positive patients. Scientists have demonstrated that SARS-CoV-2 remains stable and viable in the air for at least 3 h and over surfaces such as plastic/stainless steel for at least 72 h.[60]

Multiple other modes of human-to-human SARS-CoV-2 transmission such as mother-to-child and fecal route transmission have been documented. Overall, the probability of human-to-human transmission of this virus is very high and can occur upon direct or indirect contact with infected respiratory droplets or with contaminated surfaces.[62] An important point to note in this context is that the viral transmission risk is also dependent on other key factors apart from the R_0 factor such as identifying high-risk environments, effective contact tracing for constant monitoring of disease progression, and careful implementation of government health care policies.[63]

Worldwide, many nations have adopted different policies to implement classic health care and safety management measures such as use of facial masks, implementing social distancing, etc., to curb the spread of COVID-19. Most common national level health crisis management strategies include imposing nationwide lockdown, curfews that ban outside travels at certain hours, effective isolation and contact tracing, implementing social distancing, and bans on public gatherings and transportation.[15] Strict implementation of such public safety measures has reduced the infection rate by over 50%—60% in many Asian and European countries.[15] For instance, South Korea's success in COVID-19 management is mainly attributed to rapid implementation of strict government policies drafted in collaboration with its scientific community. The country was quick to issue stay-at-home orders, pursued extensive and efficient contact tracing to identify potential cases, and most importantly, provided rapid testing and secured required numbers of personnel to implement strict social crisis management strategies.[64] For a detailed review on all the SARS-CoV-2 transmission modes and updates on detailed government policies to tackle the viral transmission, readers are directed to other excellent reviews.[65]

1.2.3 Clinical characteristics for COVID-19

Clinically, SARS-CoV-2 is infectious to human beings irrespective of age.[66] Younger adults often develop milder symptoms compared to older adults, particularly if the older adult has other existing medical conditions.[67] The median age of infection in adults is ~47 years. People with COVID-19 most often develop mild-to-moderate symptoms without the need for medical assistance. Most often over 80% of the COVID-19 patients develop mild respiratory disease in the form of dyspnea (i.e., breathing difficulties). Based on the statements furnished by WHO, most common symptoms reported by COVID-19 patients include fever (i.e., over ~99% of the patients develop fever at some stage during their infection period), dry cough, and tiredness. Classical symptoms exhibited by patients include fever above 38°C, fatigue, dry cough, sore throat, diarrhea, and the characteristic shortness of breath that results in hypoxia.[68] Individuals with weakened immune systems or with other chronic illnesses (~10%–20% of the patients) experience severe hypoxia and often require the support of ventilators.[69–71] Body aches, sore throat, diarrhea, loss of taste and smell, skin rashes or finger/toe discoloration, and headache are among less common symptoms.[72] More serious symptoms include acute to severe breathing difficulties including shortness of breath, chest pain, and/or loss of speech or movement that require immediate medical attention.

Presentation of clinical symptoms occur on an average of 5–6 days postinitial viral exposure, while it could also take up to 11 days for the onset of the symptoms. In some cases, in spite of the infection, the individuals remain asymptomatic.[73] The viral transmission dynamics by such asymptomatic individuals is not very well understood currently.

SARS-CoV-2 primarily infects the human respiratory system and with potential to infect other organs of the body such as brain, liver, stomach, kidneys, etc.[74] The pulmonary infection causes diffuse alveolar damage, which is considered as the first sign of damage. Specific host immune response results in cytokine dysregulation causing massive infiltration of large macrophages and T-cells on the respiratory tract parenchyma and induces pneumocytic proliferation (https://www.ncbi.nlm.nih.gov/pmc/articles/PMC7152395/). Such massive infiltration of the immune cells can be observed as patches on chest X-ray scans.[75] SARS-CoV-2 can further proceed and infect the gastrointestinal tract particularly targeting the enterocytes where the virus enters and replicates causing diarrhea.[71]

Currently, suspected individuals are recommended to undergo serological testing that checks for the presence of antibodies to the virus and molecular-based diagnostic testing such as polymerase chain reaction (PCR) for pathogen detection. SARS-CoV-2 positive individuals are advised to undergo imaging-based diagnosis to assess the lung infection.[76–78] In the following sections, various challenges and new age technologies for pathogen detection and monitoring of disease progression are elaborately discussed.

1.2.4 Diagnostic detection of COVID-19

In addition to the ongoing struggle to contain the pandemic, the countries face major socioeconomic impact.[79,80] In an attempt to prevent further damage, many governmental authorities have strongly recommended coordination between the medical community and local public health authorities for developing COVID-19 health care management strategies including the use of novel diagnostic testing approaches for rapid pathogen detection.[81] Currently WHO recommended primary confirmatory diagnostic technique includes Nucleic Acid Amplification Tests (NAAT) such as real-time reverse transcriptase PCR (RT-PCR) for SARS-CoV-2 detection[82] (https://www.who.int/publications/i/item/10665-331501). A detailed overview on the principle and method of diagnosis using this technology can be found in Ref.[83] This section provides an overview of analytical issues at various stages of sample collection and processing. It further provides an overview of the primary COVID-19 diagnostic techniques highlighting their advantages and limitations.

1.2.4.1 Preanalytical issues impacting the diagnosis

As of January 2021, a typical COVID-19 diagnosis relies on genomic detection of SARS-CoV-2 from the respiratory tract samples such as nasopharyngeal (NP) swabs and/or an oropharyngeal (OP) swabs collected of suspected individuals.[84,85] The collected sample is placed into a viral transportation medium and is transported to nearby clinical laboratories for analysis.[86,87] Molecular diagnostic analysis based on nucleic acid amplification such as RT-PCR is performed to detect the pathogen in the collected sample.[88]

One of the most common issues regarding RT-PCR is the obtention of false-negative results, i.e., pathogens could go undetected in the patient sample. The rate of obtention of false-negative results are in the range of ~2%–29%.[89] The false negative results pose a greater risk to society, since such individuals could further spread the disease adding to the difficulty in containing the virus outbreak.[90] To reduce such false negatives, technical issues must be well considered and addressed. For instance, technical issues such as stability of the RNA sample collected, the time of swab collection (i.e., the NP/OP swabs should be taken at the onset of the symptoms as viral loads tend to be higher in the region around this time), and swift swab collection procedures (i.e., collection of respiratory samples particularly NP/OP swabs requires proper prior training).[91]

Mishandling of such airborne pathogens during sample collection, handling, and transportation could be dangerous and would lead to a new outbreak of the disease among health care professionals. Such hazardous situations could be avoided by providing adequate formal training and sufficient supply of personal protective equipment (PPE) to all health care professionals dealing with COVID-19 samples and patients. The reuse of the sample processing kits should be avoided under all circumstances.[92] Thus, practicing safe and recommended methods of sample collection and maintenance of high hygienic standards in places of sample collection, manipulation, and storage are important steps toward ensuring safe diagnosis of COVID-19 in suspected individuals.

1.2.4.2 Primary diagnostic techniques for COVID-19 detection

Clinical diagnosis of COVID-19 disease is based on observation of symptoms, epidemiological history, and testing by standard molecular testing methods. Currently, the three most commonly used as molecular diagnosis for SARS-CoV-2 are RT-PCR, Loop-Mediated Isothermal Amplification (LAMP) and high-throughput Next Generation Sequencing (NGS) of the whole genome.[93] However, exploitation of NGS technology is limited due to its dependency on many instrumentations and the expenses incurred for such analyses. On the other hand, RT-PCR and RT-LAMP are cost-effective and straightforward technologies that allow the detection of pathogenic diseases. In addition, these techniques are portable, rapid, and robust in nature, requiring less instrumentation.[94] Currently, many new biosensors and nanomaterials-based techniques are being developed for the detection of infectious diseases particularly useful in the current scenario.[95–97] In the following sections, a brief overview of the three classical techniques listed above is presented.

1.2.4.2.1 RT-PCR

RT-PCR is one of the current standard method for COVID-19 detection. Respiratory samples such as NP/OP swabs collected from suspected individuals are processed to obtain the pathogenic genomic material, i.e., RNA as the source material, which is then subjected to nucleic acid amplification techniques such as RT-PCR. RT-PCR uses the reverse transcriptase enzyme to convert the RNA material obtained from the sample into complementary DNA (cDNA), which is further amplified using a DNA polymerase and primers to specific biomarker genes.[98] The technology uses fluorescent primers to amplify the genes under investigation. This facilitates real-time quantification of the relative expression of the biomarker's genes in the sample. Thus, higher expression of viral genes compared to various control samples is considered as a positive outcome for the assay. For SARS-CoV-2 diagnosis, viral genes such as N, E, S, and RdRp are being used as biomarkers.[99] SARS-CoV-2 positive samples are further confirmed by whole genome sequencing such as NGS technology.[82]

In spite of having multiple advantages (listed in the introduction of this section), there are some limitations in using the RT-PCR-based diagnosis.[100] As suspected cases of COVID-19 are surging, necessitating the rapid detection of the virus, limitations of RT-PCR based detection techniques for COVID-19 disease is more profoundly visible now than before.[101] Limitations in the detection technique include unreliable stability of the results, i.e., the efficiency of target gene and control gene do not remain identical at all times, obtention of false positive and negative results, and the unverified clinical significance of positive PCR amplicons. These are among the key issues that need to be addressed.[102,103] Multiple factors influence the molecular test, such as condition of the sample obtained, anatomical location from where it was extracted, the time of disease, i.e., the stage or the phase of the disease in the individual, and time spent for sample transportation.[104] In addition, the RT-PCR is time and labor consuming. Other methods of isothermal amplification of nucleic acids such LAMPs, antigen-based serological tests, and serum antibody tests could provide solutions to these issues.

1.2.4.2.2 RT-LAMP assays

LAMP assays are a robust, highly sensitive, high accuracy, and rapid diagnostic tool for SARS-CoV-2 detection.[105–109] It is a popular choice point-of-care (POC) technique employed for rapid COVID-19 testing.[110] This promising nucleic acid LAMP amplification technique uses six different target sequences for the same viral gene, which increases the sensitivity and specificity of the process.[111] Moreover, as LAMP does not require expensive instruments and reagents, it is highly cost-effective. The technique uses one single heat block for nucleic acid amplification process, and results can be obtained in less than 60 min.[111] These are the critical advantages of the technique that improve cost reduction and assist in rapid COVID-19 detection at this crucial time.[112]

Recently, researchers have developed colorimetry-based reverse transcriptase LAMP (RT-LAMP) technique for further rapid detection of COVID-19 that allows one-step reverse transcription, which can be detected by a color change and is visible to the naked eyes.[110] Yu et al. developed the RT-LAMP technique for the detection of genomic RNA from SARS-CoV-2 strain and showed that technique has a limit of detection (LOD) up to 100 copies/reaction, thus demonstrating that the technique has very high sensitivity for pathogen identification. They also experimented on various other strains of coronaviruses such as human coronaviruses (hCoV-229E, hCoV-OC43), Middle East respiratory syndrome coronavirus (MERS-CoV), and SARS-CoV2 (Fig. 1.4). They claimed that this RT-LAMP process is highly specific for detection of SARS-CoV-2, and due to the colorimetric method, this tool can be exploited as a potential POC test for COVID-19 detection in the near future.[113]

This technique, however, has a few limitations, such as requiring high sample purity, sophisticated sample processing steps, issues regarding result reproducibility, and in some cases, sample stability issues have also been reported.[114,115] Currently, research is aimed at developing the techniques suitable for POC testing.

1.2.4.2.3 NGS techniques

NGS is an emerging technology that can potentially overcome some of the limitations mentioned earlier and can efficiently cater to the growing diagnostic demand. For quite some time, NGS has been used as tool to study and understand the genomic details of a pathogen. For instance, at the initial stage of SARS-CoV-2 outbreak NGS technology was used for identification and discovery of the viral strain.[116–118] Early identification of this virus has helped to conduct rapid and promising advanced research on the virus, which has led to the rapid production of medicines and vaccines against viral infection.[119,120] Later, RT-PCR techniques helped to detect the specific sequences of the SARS-CoV-2 genome. Moreover, NGS tools

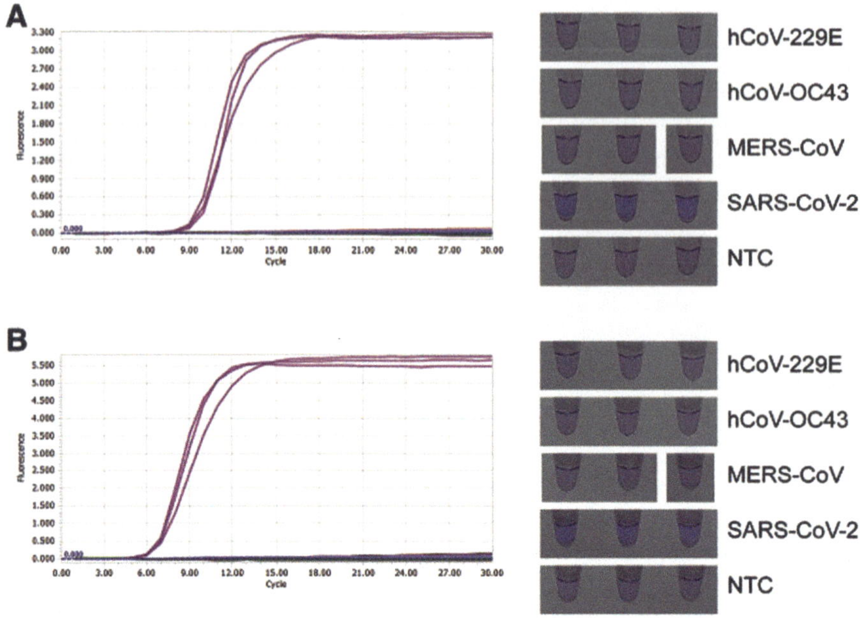

Figure 1.4 RT-LAMP assay for detection of SARS-CoV-2. Series of selectivity tests are conducted among human coronaviruses, MESR-CoV, SARS-CoV, SARS-CoV-2, and negative templates by colorimetric RT-LAMP process. (A) and (B) represents different sets of newly designs SARS-CoV-2 LAMP primers. *(This figure is reproduced with permission from Park G-S, Ku K, Baek S-H, et al. Development of reverse transcription loop-mediated isothermal amplification assays targeting severe acute respiratory syndrome coronavirus 2.* J Mol Diagn *2020. 2020/04/07. https://doi.org/10.1016/j.jmoldx.2020.03.006.)*

have been instrumental for identification and detection of the mutation frequency in the virus and thus providing a detailed overview of the evolution of the virus over time.[121,122]

Nanopore sequencing is a third-generation sequencing technique that has been used for detailed viral genome analysis from various clinical specimens.[123,124] This technology also measures RNA molecules from the viral genomic sample without the necessity of generating cDNA for analysis. Such techniques offer complete genome information of the pathogen, thus providing novel research insights that help in understanding the nature of the pathogen.[117]

The use of computational tools for analyzing the sequencing data including the genome analyses using NCBI's Basic Local Alignment Search Tool (BLAST) is often time and labor consuming to generate whole outcomes.[125,126] Thus, such genome analysis techniques are highly expensive and time-consuming compared to simple RT-PCR, RT-LAMP, or other molecular diagnostic assays.

1.2.4.2.4 Serum-based antigen/antibody-based detection techniques

Antigen-based detection techniques employ patient serum and detect antigens present in patient samples. Many antigen detection techniques are highly sensitive, robust and rapid. Antigen-based detection techniques are currently available in the market. Many serology-based tests are highly compatible with high-throughput analysis, i.e., thousands of samples could be processed in very less time compared to conventional techniques.[127] The success of such serology-based tests and their commercialization will provide rapid answers to suspected COVID-19 patients, and subsequently contribute to increased efficiency in contact tracing and precautionary isolation approaches that will truly contribute to efficient containment of the disease.[128–130]

Although such techniques are highly specific, they are very sensitive to the mutations accumulated by the RNA, i.e., they cannot successfully identify the pathogen, should they evolve over time. In spite of the aforementioned issues, the capacity of the technology for rapid COVID-19 detection testing has powered the advancement of various serological approaches.[131] Many nations worldwide have exploited such serological methods for initiating surveillance of suspected individuals. For example, as of January 31, 2020, Singapore has employed more serological based SARS-CoV-2 testing than the traditional RT-PCR technique for mass COVID-19 testing in the country. They have been successful in identifying many positive cases using this approach.[132]

A major advantage of this technique is that many asymptomatic patients have also been diagnosed as COVID-19 positive since this technique involves detection of antigens and antibodies in the patient samples.[133] Hence, alongside molecular testing, application of serological testing is also an important assay.

In spite of all these advantages, these antigen-based detections have few limitations, such as the importance of time of sample collection, i.e., often antigen productions are low at the initial phase of the symptom onset, low sensitivity of the technique, potential false-negative results, chances of cross-contamination with closely similar antigens, requirement of trained personnel for sample obtention, requirement of a preenrichment processes.[134,135] Further development on ELISA-based detection is continuing and near future these problems will be resolved.

1.2.4.3 Novel Point-of-Care approaches and their limitations

Various biomolecular and immunoassays have been developed for the rapid detection of SARS-CoV-2. Assay selection is an important issue for efficient detection of SARS-CoV2. Few important POC immunoassays that are currently employed for the detection of the virus are lateral flow assays (LFAs), serological assays, and high-throughput immunoanalyzer assays.[136] The LFA process identifies antigens of SARS-CoV-2 by detecting antibodies such as IgM and IgG against COVID-19 disease.[137] This immunoassay, similar to qRT-PCR process, offers rapid and cost-effective detection of the virus. However, this assay is not as sensitive as other biomolecular assays (Table 1.1).[138–140] Few studies have showed nonspecific detection of IgM against COVID-19 in the immunoassays.[141,142] Thus, the ongoing research, on the one hand, is looking for specific monoclonal antibodies against SARS-CoV-2, and on the other, is searching for other speedy antigen detection assays that can be reliably employed for the diagnosis of COVID-19.[143]

Assay selection for molecular detection of SARS-CoV-2 is also a current issue. Rapid amplification-based sequencing methods play a key role in the detection of the disease. The preliminary quick and robust detection is performed using molecular techniques such as qRT-PCR, RT-PCR, and RT-LAMP process.[149] A complete genome analysis is performed using NGS platforms in the second stage of further confirmation of the virus in positive samples.[150]

RT-PCR assay is the most commonly employed molecular technique in the first stage of the virus detection in patient samples.[6] Other molecular methods, such as LAMP, CRISPR-Cas9, and microarray-based techniques have shown great potential for the detection of the disease.[107,151–153] One of the major advantages of these molecular techniques is that the sample is amplified before analysis, and such amplification procedures significantly reduce the false-positive results. This helps to avoid amplicon contamination. These techniques are highly specific, and since only a few strains of coronaviruses that cause respiratory disorders are circulating at the moment, such high specificity is another added advantage for these techniques.[154]

The aforementioned advantages could also prove to be a limitation since both SARS-CoV and SARS-CoV-2 are a group of SARS-like bat origin coronaviruses.[155] Therefore, there is a possibility and risk for making nonspecific molecular observations in some samples. To overcome such issues regarding nonspecificity, the WHO and CDC recommends the use

Table 1.1 Advantages and disadvantages of different techniques to detect SARS-CoV-2.

Diagnostic technology	Detection limit (LOD)	Specificity	Response time	Specimen	Advantages	Disadvantages	Identification	References
RT-PCR	3.9 copies/ reaction	99%	~190 min	Samples from fecal	High specificity and sensitivity, used as gold standards procedure	Thermal cycler is needed, costly reagents and fully trained laboratory skilled personnel is needed and time consuming	RNA	94
Real-time RT-PCR	11.2 genomic RNA copies/ reaction	Good specificity, tested with different other coronaviruses	Not mentioned	Nasal, oral and throat swabs	Good sensitivity and specificity	High-end sophisticated instrument is needed	Genomic RNA	144
One-step qRT-PCR	10 copies/ reaction	Good specificity, checked with other strain of viruses	Not specifically mentioned	Nasal swabs and stool samples	Very specific and sensitivity wise is high with the vast capacity of samples experiments	Costly, laborious, and false-negative chances are there	RNA	145

Continued

Table 1.1 Advantages and disadvantages of different techniques to detect SARS-CoV-2.—cont'd

Diagnostic technology	Detection limit (LOD)	Specificity	Response time	Specimen	Advantages	Disadvantages	Identification	References
RT-LAMP	10 copies/reaction	As good as RT-PCR process	~40 min	Nasal, oral and lower respiratory swabs	No high-end instruments are needed, perfect for on spot examination, affordable, rapid and good sensitivity	False-negative chances and cross-contamination issues may occur	RNA	146
RCA	100 copies/reaction	Good specificity	~90 min	Respiratory swabs, and urine samples	Good specificity, rapid	Not so good sensitivity, difficult to prepare circular probe for the experiment	Genomic DNA	147
ELISA	Not specific	95%	240 min	Nasal, oral and lower respiratory swabs	Good for antibody testing	Costly, time consuming and specific instruments are needed and as compatible as PCR process	Antigen of SARS-CoV-2	148

of at least two molecular assays for testing for instance, the first-line screening with E gene assay followed by RdRp gene assay as confirmatory examination.[94]

1.2.5 Late detection issues

Generally, late detection of patients or asymptomatic individuals are problematic in terms of isolating infected patients from healthy populations. Many studies indicated that the viral loads of SARS-CoV-2 RNA is immensely high in the bronchoalveolar specimens, i.e., deep respiratory samples than Nasal/oral swabs.[156] Thus, the typical samples such as the sputum or bronchoalveolar specimens collected from lower respiratory tract contain the highest SARS-CoV-2 viral loads.[157] A recent study reported that the positive samples collected from bronchoalveolar lavage fluid showed the highest number of viral loads present in the sample.[156] The classic symptoms of COVID-19 can be two types, such as mild and severe. Mild symptoms are generally cough, fever, mild headache, sneezing, and fatigue.[158] However, these mild symptoms are similar symptoms of flu/influenza.[159] Sometimes, it may be confusing to differentiate between normal flu and COVID-19 because of such mild symptoms until severe symptoms such as lung congestion and shortness of breath appear.[160] By this time, it is often too late to treat the infected patient. Such patients must be admitted to a hospital, particularly to the intensive care unit (ICU) and may probably need life support. Alternatively, some studies have also reported patients with high viral loads of SARS-CoV-2 in their fecal material.[20,161] Therefore, for COVID-19 testing with swabs from respiratory tracts are not sufficient. Other studies have also reported SARS-CoV-2 viral loads from enterocytes in the digestive tract by means of electron microscopy.[81] Thus, for efficient sample collection and diagnosis, multiple swab samples should be obtained from the same suspected patients.[162,163]

1.2.6 Monitoring of severely affected COVID-19 patients

Continuous monitoring of critically ill COVID-19 patients is a great challenge. Critical patients have a high amount of SARS-CoV-2 RNA load in their body organs. Since higher viral loads have also been found in fecal material and in the respiratory tract of the infected patients, analyzing at least two samples from patients at the beginning of hospital admission is very crucial for patient monitoring.[156,162,164,165] Practicing standard hygienic practices such as thorough cleaning is mandatory, especially in the clinical houses or hospitals, objects/articles that are related to COVID-19 patients, and the patient-frequented washrooms. Listed below are the CDC guidelines for maintaining high hygienic standards while handling COVID-19 patients, their clinical samples, and other related articles[166,167]:

(1) At the preliminary stage, the container containing the obtained patient sample should be sealed with a screw cap. It must inform the patient on the spot of examination centers.

(2) The container must have low risk of breakage and low chances of contamination, so it is suggested to carry a cotton swab inside a conical tube.

(3) The details of the individuals should be noted and recorded officially. The information should be written on the specimen container. For example, containers should have their name, number, and a specific barcode with date of collection.

(4) The outer surface of the container should be wiped with 70% ethanol and the container should be packed in a zipper bag before transportation.

(5) While transferring the specimens to the testing laboratories or hospitals, clinical staffs should wear PPE. The personnel handling the patient sample should have proper training. And a separate route should be used for such transportation.

(6) After transferring the specimen to the biosafety level 2 (BL2) laboratory, the personnel should change PPE, masks, and gloves. These articles should be disposed in a closed container after properly disinfecting.

(7) The laboratory bench should be cleaned with bleach or other strong disinfectants.

(8) The extraction of nucleic acids should be performed inside the higher biosafety cabinet (BSC) with standard protocol and precautions.

(9) A standard operating process should be used in all clinical laboratories. Currently real-time RT-PCR is being used for COVID-19 diagnosis.

(10) While using commercial kits, the instructions and all other manufacture information should be followed properly.

(11) To prevent the chances of cross-contamination during RNA extractions and pre-PCR process personnel need to take care of extra precautions and should follow the guidelines of biosafety rules. Personnel should wear gloves, masks, gown, and eye protection in a BL2 laboratory.

(12) At the end, all infectious waste such as residual specimens, contaminated reagents, PPE, tips, etc., should be properly disposed by following the institute recommended guidelines of handling infectious waste samples.

Further, the contaminated patient samples like swabs, stool cultures, surfaces, and instruments such as PCR machine, electron microscopy, and CT scanners have to be handled with extra attention, and proper sanitization protocols should be followed.[168] Some studies claim that handling respiratory samples are harder than handling stool samples from the patients because of the presence of higher viral loads in respiratory samples than in stools samples.[169] Currently, the preferred method for detecting SARS-CoV-2 is real-time PCR technique from a respiratory swab sample.[162,163,170]

1.2.7 Other health aspects due to COVID-19 and related health management factors

In a very broad spectrum, handling with proper management of COVID-19 patients and associated factors are very challenging and hectic. Implementation of good health management policies and recruitment of sufficient amount of manpower for implementing those measures together with the appropriate use of advanced technologies is crucial for efficient containment of the virus.

This pandemic situation has resulted in a high level of work stress among health care professionals who face the danger of being directly exposed to SARS-CoV-2 on a daily basis. The high-level work stress, long periods of confinement, and self-isolation approaches have negatively impacted mental health, giving rise to problems/issues such as depression and anxiety.[171] Pressure of loss of jobs due to economic recession and loss of loved ones due to COVID-19 bring greater risk of psychological stress in such individuals.[172] To reduce such mental pressures among health care professionals, equal work distribution should be made practical. The governing authorities and health care policy makers should undertake strong management factors with good leadership to support health care professionals and families. Fig. 1.5 shows few health managements factors that are necessary for maximal effective containment of virus. This section of the chapter also discusses other health management factors for other health issues caused by COVID-19.

1.2.7.1 Direct COVID-19 risk management

As mentioned earlier, early detection of COVID-19 with gold standard technologies are necessary alongside a few other issues that should be considered, for instance, postvaccination surveillance,[173] and the timely evaluation of preexisting medical risks that could trigger another disease outbreak causing further damage.[174] In general, the requirement of the

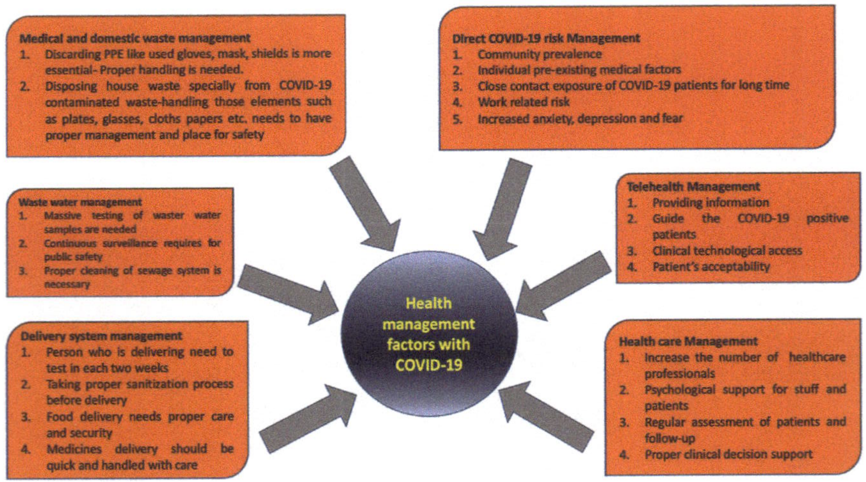

Figure 1.5 The framework of health management factors for taking good care of COVID-19 patients and environment at the same time during the pandemic.

physical presence of health care professionals in close proximity to the COVID-19 patients for treatments and other purposes poses a great risk and increases the chance of direct contamination with the virus manifold times. Thus, proper health care management is required to reduce such risks. For instance, the increase in the number of health care professionals dealing with COVID-19 patients and samples, and adequate production and supply of PPE are essential in addition to rapid COVID-19 testing on a daily basis.

1.2.7.2 Medical and domestic waste management

The requirement of proper environmental waste management strategies is a highly significant, growing concern as each day passes during this pandemic. Due to large-scale disposing of huge amounts of medical and domestic waste on a daily basis, environmental pollution is becoming a more serious problem. Medical and domestic waste in many countries such as China, USA, Spain, and Italy has increased in excessive quantities.[175] Reports indicate that in Wuhan, over 240 metric tons of contaminated medical waste are being generated on a daily basis.[176] Household waste has also increased due to online shopping and home delivery options, which are not getting recycled.[177,178] Because of the fear of further spreading of the virus, used and contaminated articles such as gloves, masks, expired medications, and other contaminated PPE cannot be recycled.[178] In future, researchers need to focus on more recycling and disposing processes of medical and domestic waste.

1.2.7.3 Telehealth management

Telehealth management is one of the best approaches to guide and advise patients. This approach optimizes in mitigating the risk of exposure, and thus is an excellent initiative toward reducing the spread of the virus. In the past, telehealth had been suggested due to a huge number of patients queries and the phone consulting facilitated appropriate medical care without the need for huge patient lines and/or waiting time.[179] The main aim for implementing telehealth management is to provide access to health care systems from rural areas.[180] However, in the recent scenario of COVID-19, telehealth management is proving to be a great boon. This management is used across all the medical specialties for maintaining patient access to care while successfully maintaining physical distance from the patients.[181]

For example, hospital admission formalities of a COVID-19 patients such as information about the patient history and the current status of COVID-19 disease can be done using telehealth systems. In general, through this management process, consultants will directly discuss with patients and provide them the required services during the telehealth consulting sessions. This way, telehealth care reduces by half the number of patients visiting the medical center and successfully avoids the exposure of low-risk patients to the virus in hospital settings.[182] The study by Aziz et al. provides a new norm for handling medical visits for pregnant women under the current COVID-19 circumstances. The authors demonstrated that medical consultations of pregnant women who are in <11 weeks of gestation, i.e., the first doctor visit could also be done virtually where patient's clinical history can be taken. Genetic screening and clinical requirements could be reviewed. At a later stage, around 11−14 weeks of gestation, physical examination and prenatal blood examinations could be performed. The following 4 weeks, patient could continue to meet their doctors virtually depending upon the examination reports. Further medical visits of the patient up until the time of delivery could be virtual or on site depending on the patient's condition, while physical presence would be required for an ultrasound examination.[182] By following such practices, the exposure of high-risk and low-risk patients can be significantly reduced, which will play a major role in exposure prevention.[183]

1.2.7.4 Wastewater management

Many findings have reported that traces of SARS-CoV-2 is found in wastewater along with particles of disinfecting elements.[184] The byproducts of disinfectants cause toxicity in wastewater and discharges into rivers and oceans, which is spoiling the environment and public health indirectly.[185] The record from UN's world water development showed the 80% of wastewater is polluting the environment without enough treatment process.[186] One of the studies by Wang et al. stated that SARS-CoV strain was found in medical waste, domestic sewage, and tap water, which has led to fast spread of the virus.[187] Hence, the cleaning and treatment process and proper waste management is highly necessary for containment of the COVID-19 situation. To reduce the chances of virus and the quantity of disinfectant in wastewater, large-scale waste management strategies and treatment plants could be set up.[188] In such situations, the industrial partners and governmental agencies should come forward and work together for the establishment of a central wastewater treatment setup that will benefit everyone.

1.2.7.5 Other health management factors

Besides the aforementioned health management issues and strategies, there are a few other factors on which we must focus. For instance, increasing home confinements have drastically increased the concept of online food and article shopping. With increased home delivery of food, groceries, and other products, proper sanitization protocols and minimal contact with people should be strictly followed by the delivery and transportation system to minimize the risk for further virus spread. The CDC has laid proper guidelines that should be followed to minimize such risks. In addition, as side effects to some health management strategies, other mental health issues such depression, anxiety, and other disparities have been reported in some people due to long periods of nationwide lockdowns, self-isolation, and quarantine rules.[189,190] Therefore, a good balance in implementing the health management strategies considering all of these factors is necessary to successfully overcome this pandemic.

1.3 Conclusions

This chapter provides an overview of the COVID-19 disease caused by SARS-CoV-2, the responsible viral agent for the currently ongoing pandemic and world health crisis that has killed millions of people worldwide. A brief history, evolution, transmission of this virus, and comparisons of various diagnostic technologies are reported in this chapter. In addition to this, preventive measures by CDC and WHO are listed along with many health management factors that are discussed in this chapter. SARS-CoV-2 is a zoonotic virus and is thought to have spread from the Wuhan's live seafood markets in the Hubei province of China. The salient features of this virus such as the high replication rate, very high human-to-human transmission capacity, the potential to infect most body parts, and the persistent virulence of the virus over contaminated surfaces has made it one of the deadliest pathogens known to mankind. Therefore, understanding the origin and nature of the virus, the route of transmission, mode of interaction with host, and pathogenesis are essential for developing better medications, effective therapeutics, and vaccines to overcome this pandemic.

Among the various health management strategies implemented by various nations across the globe, lessons should be learned from nations that have been successful in implementing the public health management policies. Few health care management factors listed in this chapter such as adequate support to health care professionals and provisions for health care facilities, appropriate use of diagnostic tools and advances in technologies for monitoring the disease spread, and proper establishment of waste management are essential to contain the virus spread. Thus, ensuring a proper health care risk assessment, rapid establishment of the governmental health care management strategies upon careful consideration of a wide range of health care factors mentioned in this chapter could help to achieve success in containing the current pandemic, further preventing more human loss.

Since COVID-19 first emerged in December 2019, the scientific communities, public authorities, private enterprises, and people around the globe have relentlessly contributed to the efforts initiated by their countries during the pandemic. We hope that vaccines and better medications should soon reach all the people and that the world returns to normalcy.

References

1. Fields BN, Knipe DM, Howley PM. *Fields virology*. Wolters Kluwer Health/Lippincott Williams & Wilkins; 2013.
2. Mackenzie JS, Smith DW. COVID-19: a novel zoonotic disease caused by a coronavirus from China: what we know and what we don't. *Microbiol Aust* 2020;**41**(1):45–50.
3. Banerjee A, Doxey AC, Mossman K, Irving AT. Unraveling the zoonotic origin and transmission of SARS-CoV-2. *Trends Ecol Evol* 2020;**36**(3):180–4. https://doi.org/10.1016/j.tree.2020.12.002. 2020/12/13.
4. Petersen E, Koopmans M, Go U, et al. Comparing SARS-CoV-2 with SARS-CoV and influenza pandemics. *Lancet Infect Dis* 2020;**20**(9):e238–44. https://doi.org/10.1016/S1473-3099(20)30484-9. 2020/09/01.
5. Sanche S, Lin YT, Xu C, Romero-Severson E, Hengartner N, Ke R. High contagiousness and rapid spread of severe acute respiratory syndrome coronavirus 2. *Emerg Infect Dis* 2020;**26**(7):1470–7. Journal Article.
6. Chan JF-W, Yuan S, Kok K-H, et al. A familial cluster of pneumonia associated with the 2019 novel coronavirus indicating person-to-person transmission: a study of a family cluster. *Lancet* 2020;**395**(10223):514–23. https://doi.org/10.1016/S0140-6736(20)30154-9. 2020/02/15.
7. Liu J, Liao X, Qian S, et al. Community transmission of severe acute respiratory syndrome coronavirus 2, Shenzhen, China, 2020. *Emerg Infect Dis* 2020;**26**(6):1320. https://doi.org/10.3201/eid2606.200239. Journal.
8. Jones DL, Baluja MQ, Graham DW, et al. Shedding of SARS-CoV-2 in feces and urine and its potential role in person-to-person transmission and the environment-based spread of COVID-19. *Sci Total Environ* 2020;**749**:141364. https://doi.org/10.1016/j.scitotenv.2020.141364. 2020/12/20.
9. Feng Y, Marchal T, Sperry T, Yi H. Influence of wind and relative humidity on the social distancing effectiveness to prevent COVID-19 airborne transmission: a numerical study. *J Aerosol Sci* 2020;**147**:105585. https://doi.org/10.1016/j.jaerosci.2020.105585. 2020/09/01.

10. Klompas M, Baker MA, Rhee C. Airborne transmission of SARS-CoV-2: theoretical considerations and available evidence. *J Am Med Assoc* 2020;**324**(5):441–2. https://doi.org/10.1001/jama.2020.12458.
11. Chang H-H, Meyerhoefer CD. COVID-19 and the demand for online food shopping services: empirical evidence from Taiwan. *Am J Agric Econ* 2020. https://doi.org/10.1111/ajae.12170. 2020/11/05.
12. Gibney E. Whose coronavirus strategy worked best? Scientists hunt most effective policies. *Nature* 2020:15–6.
13. Ray R, Rojas F. *Covid-19 impact on Asia and beyond.*
14. Goniewicz K, Khorram-Manesh A, Hertelendy AJ, Goniewicz M, Naylor K, Burkle FM. Current response and management decisions of the European union to the COVID-19 outbreak: a review. *Sustainability* 2020;**12**(9). https://doi.org/10.3390/su12093838.
15. Han E, Tan MMJ, Turk E, et al. Lessons learnt from easing COVID-19 restrictions: an analysis of countries and regions in Asia Pacific and Europe. *Lancet* 2020;**396**(10261):1525–34. https://doi.org/10.1016/S0140-6736(20)32007-9. 2020/11/07.
16. La Regina M, Tanzini M, Fineschi V, et al. Responding to COVID-19: the experience from Italy and recommendations for management and prevention. *Int J Qual Health Care* 2020. https://doi.org/10.1093/intqhc/mzaa057.
17. World B. *COVID-19 to plunge global economy into worst recession since World War II.* Washington, DC: Press Release; 2020.
18. Manoharan L, Cattrall JWS, Harris C, et al. *Early clinical characteristics of Covid-19: scoping review.* medRxiv; 2020. https://doi.org/10.1101/2020.07.31.20165738. 2020.07.31.20165738.
19. Cevik M, Kuppalli K, Kindrachuk J, Peiris M. Virology, transmission, and pathogenesis of SARS-CoV-2. *BMJ* 2020;**371**:m3862. https://doi.org/10.1136/bmj.m3862.
20. Paules CI, Marston HD, Fauci AS. Coronavirus infections—more than just the common cold. *J Am Med Assoc* 2020;**323**(8):707–8. https://doi.org/10.1001/jama.2020.0757.
21. Graham BS, Mascola JR, Fauci AS. Novel vaccine technologies: essential components of an adequate response to emerging viral diseases. *J Am Med Assoc* 2018;**319**(14):1431–2. https://doi.org/10.1001/jama.2018.0345.
22. Sheahan TP, Sims AC, Leist SR, et al. Comparative therapeutic efficacy of remdesivir and combination lopinavir, ritonavir, and interferon beta against MERS-CoV. *Nat Commun* January 10, 2020;**11**(1):222. https://doi.org/10.1038/s41467-019-13940-6.
23. Dai L, Gao GF. Viral targets for vaccines against COVID-19. *Nat Rev Immunol* 2020;**21**:73–82. https://doi.org/10.1038/s41577-020-00480-0. 2020/12/18.
24. Islam KU, Iqbal J. An update on molecular diagnostics for COVID-19. *Front Cell Infect Microbiol* 2020;**10**:694. https://doi.org/10.3389/fcimb.2020.560616.
25. Koirala A, Joo YJ, Khatami A, Chiu C, Britton PN. Vaccines for COVID-19: the current state of play. *Paediatr Respir Rev* 2020;**35**:43–9. https://doi.org/10.1016/j.prrv.2020.06.010. 2020/09/01.
26. Dodd RH, Pickles K, Nickel B, et al. Concerns and motivations about COVID-19 vaccination. *Lancet Infect Dis* 2020.
27. Guan W-j, Chen R-c, Zhong N-s. Strategies for the prevention and management of coronavirus disease 2019. *Eur Respir J* 2020:2000597. https://doi.org/10.1183/13993003.00597-2020.
28. Bhuiyan MN, Ganesh R, Ghosh AK. COVID-19: a 2020 update. *Indian J Med Sci* 2020;**72**(2):88.
29. Home O. *Testing for COVID-19: a way to lift confinement restrictions.*

30. Gorbalenya AE, Baker SC, Baric RS, et al. The species severe acute respiratory syndrome-related coronavirus: classifying 2019-nCoV and naming it SARS-CoV-2. *Nat Microbiol* 2020;**5**(4):536–44. https://doi.org/10.1038/s41564-020-0695-z. 2020/04/01.
31. Gussow AB, Auslander N, Faure G, Wolf YI, Zhang F, Koonin EV. Genomic determinants of pathogenicity in SARS-CoV-2 and other human coronaviruses. *Proc Natl Acad Sci U S A* 2020;**117**(26):15193. https://doi.org/10.1073/pnas.2008176117.
32. Adachi S, Koma T, Doi N, Nomaguchi M, Adachi A. Commentary: origin and evolution of pathogenic coronaviruses. *Front Immunol* 2020;**11**:811. https://doi.org/10.3389/fimmu.2020.00811.
33. Fung TS, Liu DX. Human coronavirus: host-pathogen interaction. *Annu Rev Microbiol* 2019;**73**(1):529–57. https://doi.org/10.1146/annurev-micro-020518-115759. 2019/09/08.
34. Giannis D, Ziogas IA, Gianni P. Coagulation disorders in coronavirus infected patients: COVID-19, SARS-CoV-1, MERS-CoV and lessons from the past. *J Clin Virol* 2020;**127**:104362. https://doi.org/10.1016/j.jcv.2020.104362. 2020/06/01.
35. Meo S, Alhowikan A, Al-Khlaiwi T, et al. Novel coronavirus 2019-nCoV: prevalence, biological and clinical characteristics comparison with SARS-CoV and MERS-CoV. *Eur Rev Med Pharmacol Sci* 2020;**24**(4):2012–9.
36. Korber B, Fischer WM, Gnanakaran S, et al. *Spike mutation pipeline reveals the emergence of a more transmissible form of SARS-CoV-2*. bioRxiv; 2020. https://doi.org/10.1101/2020.04.29.069054. 2020.04.29.069054.
37. Buonaguro L, Tagliamonte M, Tornesello ML, Buonaguro FM. SARS-CoV-2 RNA polymerase as target for antiviral therapy. *J Transl Med* 2020;**18**:1–8.
38. Burki T. The origin of SARS-CoV-2. *Lancet Infect Dis* 2020;**20**(9):1018–9.
39. Andersen KG, Rambaut A, Lipkin WI, Holmes EC, Garry RF. The proximal origin of SARS-CoV-2. *Nat Med* 2020;**26**(4):450–2. https://doi.org/10.1038/s41591-020-0820-9. 2020/04/01.
40. Gallaher WR. A palindromic RNA sequence as a common breakpoint contributor to copy-choice recombination in SARS-COV-2. *Arch Virol* 2020;**165**(10):2341–8. https://doi.org/10.1007/s00705-020-04750-z. 2020/10/01.
41. Baldwin H, Corman V, Nkrumah E, et al. *Widespread coronavirus infection in Ghanaian bats: seasonal, demographic and ecological factors influence infection risk. Epidemiology and ecology of virus and host: bats and coronaviruses in Ghana, West Africa*. 2015. p. 30.
42. Bar-On YM, Flamholz A, Phillips R, Milo R. SARS-CoV-2 (COVID-19) by the numbers. *Elife* 2020;**9**:e57309. https://doi.org/10.7554/eLife.57309.
43. Brahim Belhaouari D, Fontanini A, Baudoin J-P, et al. The strengths of scanning electron microscopy in deciphering SARS-CoV-2 infectious cycle. Original research. *Front Microbiol* 2014;**11**. https://doi.org/10.3389/fmicb.2020.02014. 2020-August-19 2020.
44. Sarkar C, Mondal M, Torequl Islam M, et al. Potential therapeutic options for COVID-19: current status, challenges, and future perspectives. *Front Pharmacol* 2020;**11**(1428). https://doi.org/10.3389/fphar.2020.572870. 2020-September-15.
45. Turoňová B, Sikora M, Schürmann C, et al. In situ structural analysis of SARS-CoV-2 spike reveals flexibility mediated by three hinges. *Science* 2020;**370**(6513):203. https://doi.org/10.1126/science.abd5223.
46. Yan R, Zhang Y, Li Y, Xia L, Guo Y, Zhou Q. Structural basis for the recognition of SARS-CoV-2 by full-length human ACE2. *Science* 2020;**367**(6485):1444. https://doi.org/10.1126/science.abb2762.
47. Yap YL, Zhang XW, Danchin A. Relationship of SARS-CoV to other pathogenic RNA viruses explored by tetranucleotide usage profiling. *BMC Bioinf* 2003;**4**(1):43. https://doi.org/10.1186/1471-2105-4-43. 2003/09/20.

48. Hoque MN, Chaudhury A, Akanda MAM, Hossain MA, Islam MT. Genomic diversity and evolution, diagnosis, prevention, and therapeutics of the pandemic COVID-19 disease. *PeerJ* 2020;**8**:e9689.
49. Chen S, Luo H, Chen L, et al. An overall picture of SARS coronavirus (SARS-CoV) genome-encoded major proteins: structures, functions and drug development. *Curr Pharmaceut Des* 2006;**12**(35):4539–53.
50. Welkers MRA, Han AX, Reusken CBEM, Eggink D. Possible host-adaptation of SARS-CoV-2 due to improved ACE2 receptor binding in mink. *Virus Evol* 2021;**7**(1). https://doi.org/10.1093/ve/veaa094.
51. Wu K, Peng G, Wilken M, Geraghty RJ, Li F. Mechanisms of host receptor adaptation by severe acute respiratory syndrome coronavirus. *J Biol Chem* 2012;**287**(12):8904–11. https://doi.org/10.1074/jbc.M111.325803.
52. Jain J, Gaur S, Chaudhary Y, Kaul R. The molecular biology of intracellular events during Coronavirus infection cycle. *VirusDisease* 2020:1–5.
53. Korber B, Fischer WM, Gnanakaran S, et al. Tracking changes in SARS-CoV-2 spike: evidence that D614G increases infectivity of the COVID-19 virus. *Cell* 2020;**182**(4):812–27. https://doi.org/10.1016/j.cell.2020.06.043. e19. 2020/08/20.
54. Cavanagh D. *Coronaviruses and toroviruses. Principles and practice of clinical virology*. 2004. p. 379–97.
55. Buchrieser J, Dufloo J, Hubert M, et al. Syncytia formation by SARS-CoV-2-infected cells. *EMBO J* 2020;**39**(23):e106267. https://doi.org/10.15252/embj.2020106267. 12/01/2020.
56. Röltgen K, Wirz OF, Stevens BA, et al. *SARS-CoV-2 antibody responses correlate with resolution of RNAemia but are short-lived in patients with mild illness*. medRxiv; 2020. https://doi.org/10.1101/2020.08.15.20175794. 2020.08.15.20175794.
57. Callaway E, Ledford H, Mallapaty S. Six months of coronavirus: the mysteries scientists are still racing to solve. *Nature* July 2020;**583**(7815):178–9. https://doi.org/10.1038/d41586-020-01989-z.
58. Sanche S, Lin YT, Xu C, Romero-Severson E, Hengartner N, Ke R. *Early release-high contagiousness and rapid spread of severe acute respiratory syndrome coronavirus 2*. 2020.
59. World Health O. *Transmission of SARS-CoV-2: implications for infection prevention precautions: scientific brief.* 09 July 2020.
60. Aboubakr HA, Sharafeldin TA, Goyal SM. Stability of SARS-CoV-2 and other coronaviruses in the environment and on common touch surfaces and the influence of climatic conditions: a review. *Transbound Emerg Dis* 2020. https://doi.org/10.1111/tbed.13707. 10.1111/tbed.13707.
61. Setti L, Passarini F, De Gennaro G, et al. Airborne transmission route of COVID-19: why 2 meters/6 feet of inter-personal distance could not Be enough. *Int J Environ Res Publ Health* 2020;**17**(8):2932. https://doi.org/10.3390/ijerph17082932.
62. Dhand R, Li J. Coughs and sneezes: their role in transmission of respiratory viral infections, including SARS-CoV-2. *Am J Respir Crit Care Med* 2020;**202**(5):651–9.
63. Cevik M, Marcus JL, Buckee C, Smith TC. SARS-CoV-2 transmission dynamics should inform policy. *Clin Infect Dis* 2020:ciaa1442. https://doi.org/10.1093/cid/ciaa1442.
64. Lee S, Hwang C, Moon MJ. Policy learning and crisis policy-making: quadruple-loop learning and COVID-19 responses in South Korea. *Policy Soc* 2020;**39**(3):363–81. https://doi.org/10.1080/14494035.2020.1785195. 2020/07/02.
65. Kennedy D, Seale A, Bausch D, Ritchie H, Roser M. *How experts use data to identify emerging COVID-19 success stories*. Our World in Data; 2020.

66. Wu Z, McGoogan JM. Characteristics of and important lessons from the coronavirus disease 2019 (COVID-19) outbreak in China: summary of a report of 72 314 cases from the Chinese center for disease control and prevention. *J Am Med Assoc* 2020;**323**(13):1239–42. https://doi.org/10.1001/jama.2020.2648.
67. Wang D, Hu B, Hu C, et al. Clinical characteristics of 138 hospitalized patients with 2019 novel coronavirus–infected pneumonia in Wuhan, China. *J Am Med Assoc* 2020;**323**(11):1061–9. https://doi.org/10.1001/jama.2020.1585.
68. Wang L, Wang Y, Ye D. Review of the 2019 novel coronavirus (SARS-CoV-2) based on current. *Int J Antimicrob Agents* 2019;**55**(6):105948. https://doi.org/10.1016/j.ijantimicag.2020.105948.
69. Desforges M, Le Coupanec A, Dubeau P, et al. Human coronaviruses and other respiratory viruses: underestimated opportunistic pathogens of the central nervous system? *Viruses* 2020;**12**(1):14.
70. Battaglini D, Brunetti I, Anania P, et al. Neurological manifestations of severe SARS-CoV-2 infection: potential mechanisms and implications of individualized mechanical ventilation settings. *Front Neurol* 2020;**11**:845.
71. Korsman SNJ, van Zyl GU, Nutt L, Andersson MI, Preiser W. Human coronaviruses. In: Korsman SNJ, van Zyl GU, Nutt L, Andersson MI, Preiser W, editors. *Virology*. Churchill Livingstone; 2012. p. 94–5.
72. Thevarajan I, Buising KL, Cowie BC. Clinical presentation and management of COVID-19. *Med J Aust* 2020;**213**(3):134–9. https://doi.org/10.5694/mja2.50698.
73. Pollock AM, Lancaster J. Asymptomatic transmission of covid-19. *BMJ* 2020;**371**:m4851. https://doi.org/10.1136/bmj.m4851.
74. Liya G, Yuguang W, Jian L, et al. Studies on viral pneumonia related to novel coronavirus SARS-CoV-2, SARS-CoV, and MERS-CoV: a literature review. *APMIS* 2020;**128**(6):423–32. https://doi.org/10.1111/apm.13047. 2020/06/01.
75. Wang L, Wang Y, Ye D, Liu Q. Review of the 2019 novel coronavirus (SARS-CoV-2) based on current evidence. *Int J Antimicrob Agents* 2020;**55**(6):105948. https://doi.org/10.1016/j.ijantimicag.2020.105948. 2020/06/01.
76. Hope MD, Raptis CA, Shah A, Hammer MM, Henry TS. A role for CT in COVID-19? What data really tell us so far. *Lancet (London, England)* 2020;**395**(10231):1189–90.
77. Agricola E, Beneduce A, Esposito A, et al. Heart and lung multimodality imaging in COVID-19. *JACC Cardiovasc Imaging* 2020;**13**(8):1792–808. https://doi.org/10.1016/j.jcmg.2020.05.017. 2020/08/01.
78. Shi F, Wang J, Shi J, et al. Review of artificial intelligence techniques in imaging data acquisition, segmentation and diagnosis for COVID-19. *IEEE Rev BiomedEng* 2020. https://doi.org/10.1109/RBME.2020.2987975. 1–1.
79. Fernandes N. *Economic effects of coronavirus outbreak (COVID-19) on the world economy. Available at SSRN 3557504.* 2020.
80. Lange SJ, Ritchey MD, Goodman AB, et al. Potential indirect effects of the COVID-19 pandemic on use of emergency departments for acute life-threatening conditions — United States. *Am J Transplant* 2020;**20**(9):2612–7. https://doi.org/10.1111/ajt.16239. January–May 2020. 2020/09/01.
81. Tang Y-W, Schmitz JE, Persing DH, Stratton CW. Laboratory diagnosis of COVID-19: current issues and challenges. *J Clin Microbiol* 2020;**58**(6):e00512–20. https://doi.org/10.1128/JCM.00512-20.
82. World Health O. *Laboratory testing of 2019 novel coronavirus (2019-nCoV) in suspected human cases: interim guidance*. 2020. 17 January 2020.
83. Tahamtan A, Ardebili A. Real-time RT-PCR in COVID-19 detection: issues affecting the results. *Expert Rev Mol Diagn* 2020;**20**(5):453–4. https://doi.org/10.1080/14737159.2020.1757437.

84. Roy S, Baranwal A. Diverse molecular techniques for early diagnosis of COVID-19 and other coronaviruses. In: Chandra P, Roy S, editors. *Diagnostic strategies for COVID-19 and other coronaviruses*. Springer Singapore; 2020. p. 135–59.
85. Coden E, Russo F, Arosio AD, Castelnuovo P, Karligkiotis A, Volpi L. *Optimum naso-oropharyngeal swab procedure for COVID-19: step by step preparation and technical hints*. Laryngoscope; 2020.
86. Zou L, Ruan F, Huang M, et al. SARS-CoV-2 viral load in upper respiratory specimens of infected patients. *N Engl J Med* 2020;**382**(12):1177–9. https://doi.org/10.1056/NEJMc2001737. 2020/03/19.
87. Druce J, Garcia K, Tran T, Papadakis G, Birch C. Evaluation of swabs, transport media, and specimen transport conditions for optimal detection of viruses by PCR. *J Clin Microbiol* 2012;**50**(3):1064. https://doi.org/10.1128/JCM.06551-11.
88. Patel R, Babady E, Theel ES, et al. Report from the American society for microbiology COVID-19 international summit, 23 March 2020: value of diagnostic testing for SARS–CoV-2/COVID-19. *Am Soc Microbiol* 2020;**11**(2):e00722–20. https://doi.org/10.1128/mBio.00722-20.
89. Arevalo-Rodriguez I, Buitrago-Garcia D, Simancas-Racines D, et al. False-negative results of initial RT-PCR assays for COVID-19: a systematic review. *PLoS One* 2020;**15**(12):e0242958. https://doi.org/10.1371/journal.pone.0242958.
90. Baranwal A, Mahapatra S, Purohit B, Roy S, Chandra P. Insights into novel coronavirus and COVID-19 outbreak. In: Chandra P, Roy S, editors. *Diagnostic strategies for COVID-19 and other coronaviruses*. Springer Singapore; 2020. p. 1–17.
91. Pan Y, Zhang D, Yang P, Poon LLM, Wang Q. Viral load of SARS-CoV-2 in clinical samples. *Lancet Infect Dis* 2020;**20**(4):411–2. https://doi.org/10.1016/S1473-3099(20)30113-4. 2020/04/01.
92. Coroiu A, Moran C, Campbell T, Geller AC. Barriers and facilitators of adherence to social distancing recommendations during COVID-19 among a large international sample of adults. *PloS One* 2020;**15**(10):e0239795. https://doi.org/10.1371/journal.pone.0239795.
93. Li X, Geng M, Peng Y, Meng L, Lu S. Molecular immune pathogenesis and diagnosis of COVID-19. *J Pharm Anal* 2020;**10**(2):102–8. https://doi.org/10.1016/j.jpha.2020.03.001. 2020/04/01.
94. Corman VM, Landt O, Kaiser M, et al. Detection of 2019 novel coronavirus (2019-nCoV) by real-time RT-PCR. *Euro Surveill* 2020;**25**(3):2000045. https://doi.org/10.2807/1560-7917.ES.2020.25.3.2000045.
95. Kumar A, Roy S, Srivastava A, et al. Chapter 10 - nanotherapeutics: a novel and powerful approach in modern healthcare system. In: Maurya PK, Singh S, editors. *Nanotechnology in modern animal biotechnology*. Elsevier; 2019. p. 149–61.
96. Purohit B, Kumar A, Mahato K, Roy S, Chandra P. Chapter 9 - cancer cytosensing approaches in miniaturized settings based on advanced nanomaterials and biosensors. In: Maurya PK, Singh S, editors. *Nanotechnology in modern animal biotechnology*. Elsevier; 2019. p. 133–47.
97. Roy S, Malode SJ, Shetti NP, Chandra P. Modernization of biosensing strategies for the development of lab-on-chip integrated systems. Bioelectrochemical interface engineering. *Wiley Online Books* 2019:325–42. https://doi.org/10.1002/9781119611103.ch17. 2020/06/03.
98. Overbergh L, Giulietti A, Valckx D, Decallonne R, Bouillon R, Mathieu C. The use of real-time reverse transcriptase PCR for the quantification of cytokine gene expression. *J Biomol Tech* 2003;**14**(1):33–43.
99. Vashist SK. In vitro diagnostic assays for COVID-19: recent advances and emerging trends. *Diagnostics* 2020;**10**(4). https://doi.org/10.3390/diagnostics10040202.

100. Godlee F. Covid 19: where's the strategy for testing? *BMJ* 2020;**369**:m2518. https://doi.org/10.1136/bmj.m2518.
101. Ai T, Yang Z, Hou H, et al. Correlation of chest CT and RT-PCR testing for coronavirus disease 2019 (COVID-19) in China: a report of 1014 cases. *Radiology* 2020;**296**(2):E32−40. https://doi.org/10.1148/radiol.2020200642. 2020/08/01.
102. Yang S, Rothman RE. PCR-based diagnostics for infectious diseases: uses, limitations, and future applications in acute-care settings. *Lancet Infect Dis* 2004;**4**(6):337−48. https://doi.org/10.1016/S1473-3099(04)01044-8. 2004/06/01.
103. Klein D. Quantification using real-time PCR technology: applications and limitations. *Trends Mol Med* 2002;**8**(6):257−60. https://doi.org/10.1016/S1471-4914(02)02355-9. 2002/06/01.
104. Karthik K, Aravindh Babu RP, Dhama K, et al. Biosafety concerns during the collection, transportation, and processing of COVID-19 samples for diagnosis. *Arch Med Res* 2020;**51**(7):623−30. https://doi.org/10.1016/j.arcmed.2020.08.007. 2020/10/01.
105. Roy S, Rahman IA, Santos JH, Ahmed MU. Meat species identification using DNA-redox electrostatic interactions and non-specific adsorption on graphene biochips. *Food Control* 2016;**61**:70−8. https://doi.org/10.1016/j.foodcont.2015.09.029. 2016/03/01.
106. Roy S, Wei SX, Ying JLZ, Safavieh M, Ahmed MU. A novel, sensitive and label-free loop-mediated isothermal amplification detection method for nucleic acids using luminophore dyes. *Biosens Bioelectron* 2016;**86**:346−52. https://doi.org/10.1016/j.bios.2016.06.065. 2016/12/15.
107. Roy S, Rahman IA, Ahmed MU. Paper-based rapid detection of pork and chicken using LAMP−magnetic bead aggregates. *Anal Methods* 2016;**8**(11):2391−9. https://doi.org/10.1039/C6AY00274A.
108. Azam NFN, Roy S, Lim SA, Uddin Ahmed M. Meat species identification using DNA-luminol interaction and their slow diffusion onto the biochip surface. *Food Chem* 2018;**248**:29−36. https://doi.org/10.1016/j.foodchem.2017.12.046. 2018/05/15.
109. Roy S, Mohd-Naim NF, Safavieh M, Ahmed MU. Colorimetric nucleic acid detection on paper microchip using loop mediated isothermal amplification and crystal violet dye. *ACS Sens* 2017;**2**(11):1713−20. https://doi.org/10.1021/acssensors.7b00671. 2017/11/22.
110. Yu L, Wu S, Hao X, et al. *Rapid colorimetric detection of COVID-19 coronavirus using a reverse tran-scriptional loop-mediated isothermal amplification (RT-LAMP) diagnostic plat-form: iLACO*. medRxiv; 2020. https://doi.org/10.1101/2020.02.20.20025874. 2020.02.20.20025874.
111. Notomi T, Okayama H, Masubuchi H, et al. Loop-mediated isothermal amplification of DNA. *Nucleic Acids Res* 2000;**28**(12). https://doi.org/10.1093/nar/28.12.e63. e63−e63.
112. Kashir J, Yaqinuddin A. Loop mediated isothermal amplification (LAMP) assays as a rapid diagnostic for COVID-19. *Med Hypotheses* 2020;**141**:109786. https://doi.org/10.1016/j.mehy.2020.109786. 2020/08/01.
113. Park G-S, Ku K, Baek S-H, et al. Development of reverse transcription loop-mediated isothermal amplification assays targeting severe acute respiratory syndrome coronavirus 2. *J Mol Diagn* 2020;**22**(6):729−35. https://doi.org/10.1016/j.jmoldx.2020.03.006. 2020/04/07.
114. Imai M, Ninomiya A, Minekawa H, et al. Rapid diagnosis of H5N1 avian influenza virus infection by newly developed influenza H5 hemagglutinin gene-specific loop-mediated isothermal amplification method. *J Virol Methods* 2007;**141**(2):173−80. https://doi.org/10.1016/j.jviromet.2006.12.004. 2007/05/01.
115. Deguo W, Guicheng H, Fugui W, Yonggang L, Daxi R. Drawback of loop-mediated isothermal amplification. *Afr J Food Sci* 2008;**1**(7):083−6.

116. Ozma MA, Maroufi P, Khodadadi E, et al. Clinical manifestation, diagnosis, prevention and control of SARS-CoV-2 (COVID-19) during the outbreak period. *Le infezioni in medicina* June 1, 2020;**28**(2):153–65. https://infezmed.it/media/journal/Vol_28_2_2020_5.pdf.
117. Feng W, Newbigging AM, Le C, et al. Molecular diagnosis of COVID-19: challenges and research needs. *Anal Chem* 2020. https://doi.org/10.1021/acs.analchem.0c02060. 2020/06/23.
118. Narayanan K, Frost I, Heidarzadeh A, et al. *Pooling RT-PCR or NGS samples has the potential to cost-effectively generate estimates of COVID-19 prevalence in resource limited environments.* medRxiv; 2020. https://doi.org/10.1101/2020.04.03.20051995. 2020.04.03.20051995.
119. Wu F, Zhao S, Yu B, et al. A new coronavirus associated with human respiratory disease in China. *Nature* 2020;**579**(7798):265–9. https://doi.org/10.1038/s41586-020-2008-3. 2020/03/01.
120. Zhou P, Yang X-L, Wang X-G, et al. *Discovery of a novel coronavirus associated with the recent pneumonia outbreak in humans and its potential bat origin.* bioRxiv; 2020. https://doi.org/10.1101/2020.01.22.914952. 2020.01.22.914952.
121. Chen G-Q, Zhuang Q-Y, Wang K-C, et al. Identification and survey of a novel avian coronavirus in ducks. *PLOS One* 2013;**8**(8):e72918. https://doi.org/10.1371/journal.pone.0072918.
122. Wang K-C, Chen G-Q, Jiang W-M, et al. Complete genome sequence of a hemagglutination-negative avian paramyxovirus type 4 isolated from China. *Genome Announc* 2013;**1**(2). https://doi.org/10.1128/genomeA.00045-13. e00045-13.
123. Cozzuto L, Liu H, Pryszcz LP, et al. MasterOfPores: a workflow for the analysis of oxford nanopore direct RNA sequencing datasets. Technology and code. *Front Genet* 2020;**11**(211). https://doi.org/10.3389/fgene.2020.00211. 2020-March-17.
124. Kim D, Lee J-Y, Yang J-S, Kim JW, Kim VN, Chang H. The architecture of SARS-CoV-2 transcriptome. *Cell* 2020;**181**(4):914–21. https://doi.org/10.1016/j.cell.2020.04.011. e10.
125. Zhao M, Wang Q, Wang Q, Jia P, Zhao Z. Computational tools for copy number variation (CNV) detection using next-generation sequencing data: features and perspectives. *BMC Bioinf* 2013;**14**(11):S1. https://doi.org/10.1186/1471-2105-14-S11-S1. 2013/09/13.
126. Liu L, Li Y, Li S, et al. Comparison of next-generation sequencing systems. *J Biomed Biotechnol* 2012:251364. https://doi.org/10.1155/2012/251364. 2012/07/05 2012.
127. Al Dahouk S, Tomaso H, Nöckler K, Neubauer H, Frangoulidis D. Laboratory-based diagnosis of brucellosis—a review of the literature. Part I: techniques for direct detection and identification of Brucella spp. *Clin Lab* 2003;**49**(9–10):487–505.
128. Mondal KK, Shanmugam V. Advancements in the diagnosis of bacterial plant pathogens: an overview. *Biotechnol Mol Biol Rev* 2013;**8**(1):1–11.
129. Guido S, Katzer F, Nanjiani I, Milne E, Innes EA. Serology-based diagnostics for the control of bovine neosporosis. *Trends Parasitol* 2016;**32**(2):131–43. https://doi.org/10.1016/j.pt.2015.11.014. 2016/02/01.
130. Zheng T, Finn C, Parrett CJ, et al. A rapid blood test to determine the active status and duration of acute viral infection. *ACS Infect Dis* 2017;**3**(11):866–73. https://accpjournals.onlinelibrary.wiley.com/doi/abs/10.1002/phar.2439.
131. Chau CH, Strope JD, Figg WD. COVID-19 clinical diagnostics and testing technology. *Pharmacotherapy* 2020;**40**(8):857–68. https://doi.org/10.1002/phar.2439. The Journal of Human Pharmacology and Drug Therapy. 2020/08/01.
132. Yong SEF, Anderson DE, Wei WE, et al. Connecting clusters of COVID-19: an epidemiological and serological investigation. *Lancet Infect Dis* 2020;**20**(7):809–15. https://doi.org/10.1016/S1473-3099(20)30273-5. 2020/07/01.

133. Clarke C, Prendecki M, Dhutia A, et al. High prevalence of asymptomatic COVID-19 infection in hemodialysis patients detected using serologic screening. *J Am Soc Nephrol* 2020;**31**(9):1969. https://doi.org/10.1681/ASN.2020060827.
134. Law JW-F, Ab Mutalib N-S, Chan K-G, Lee L-H. Rapid methods for the detection of foodborne bacterial pathogens: principles, applications, advantages and limitations. *Front Microbiol* 2015;**5**:770. https://doi.org/10.3389/fmicb.2014.00770.
135. Garg SK, Tiwari RP, Tiwari D, et al. Diagnosis of tuberculosis: available technologies, limitations, and possibilities. *J Clin Lab Anal* 2003;**17**(5):155–63. https://doi.org/10.1002/jcla.10086. 2003/01/01.
136. Lerner AM, Eisinger RW, Lowy DR, et al. The COVID-19 serology studies workshop: recommendations and challenges. *Immunity* 2020;**53**(1):1–5. https://doi.org/10.1016/j.immuni.2020.06.012. 2020/07/14.
137. Wu J-L, Tseng W-P, Lin C-H, et al. Four point-of-care lateral flow immunoassays for diagnosis of COVID-19 and for assessing dynamics of antibody responses to SARS-CoV-2. *J Infect* 2020;**81**(3):435–42. https://doi.org/10.1016/j.jinf.2020.06.023. 2020/09/01.
138. Chen Y, Chan K-H, Hong C, et al. A highly specific rapid antigen detection assay for on-site diagnosis of MERS. *J Infect* 2016;**73**(1):82–4. https://doi.org/10.1016/j.jinf.2016.04.014.
139. Sizun J, Arbour N, Talbot PJ. Comparison of immunofluorescence with monoclonal antibodies and RT-PCR for the detection of human coronaviruses 229E and OC43 in cell culture. *J Virol Methods* 1998;**72**(2):145–52. https://doi.org/10.1016/S0166-0934(98)00013-5. 1998/06/01.
140. Liu IJ, Chen P-J, Yeh S-H, et al. Immunofluorescence assay for detection of the nucleocapsid antigen of the severe acute respiratory syndrome (SARS)-Associated coronavirus in cells derived from throat wash samples of patients with SARS. *J Clin Microbiol* 2005;**43**(5):2444. https://doi.org/10.1128/JCM.43.5.2444-2448.2005.
141. Tan W, Lu Y, Zhang J, et al. *Viral kinetics and antibody responses in patients with COVID-19*. medRxiv; 2020. https://doi.org/10.1101/2020.03.24.20042382. 2020.03.24.20042382.
142. Arun Krishnan R, Elizabeth Thomas R, Sukumaran A, Paul JK, Vasudevan DM. COVID-19: current trends in invitro diagnostics. *Indian J Clin Biochem* 2020;**35**(3):285–9. https://doi.org/10.1007/s12291-020-00906-5. 2020/07/01.
143. Diao B, Wen K, Chen J, et al. *Diagnosis of acute respiratory syndrome coronavirus 2 infection by detection of nucleocapsid protein*. medRxiv; 2020. https://doi.org/10.1101/2020.03.07.20032524. 2020.03.07.20032524.
144. Chan JF-W, Yip CC-Y, To KK-W, et al. Improved molecular diagnosis of COVID-19 by the novel, highly sensitive and specific COVID-19-RdRp/Hel real-time reverse transcription-PCR assay validated in vitro and with clinical specimens. *J Clin Microbiol* 2020;**58**(5). https://doi.org/10.1128/JCM.00310-20. e00310-20.
145. Wang J, Cai K, Zhang R, et al. Novel one-step single-tube nested quantitative real-time PCR assay for highly sensitive detection of SARS-CoV-2. *Anal Chem* 2020;**92**(13):9399–404. https://doi.org/10.1021/acs.analchem.0c01884. 2020/07/07.
146. Dhamad AE, Rhida MAA. COVID-19: molecular and serological detection methods. *PeerJ* 2020;**8**:e10180.
147. Tian B, Gao F, Fock J, Dufva M, Hansen MF. Homogeneous circle-to-circle amplification for real-time optomagnetic detection of SARS-CoV-2 RdRp coding sequence. *Biosens Bioelectron* 2020;**165**:112356. https://doi.org/10.1016/j.bios.2020.112356. 2020/10/01.

148. Russo A, Minichini C, Starace M, et al. Current status of laboratory diagnosis for COVID-19: a narrative review. *Infect Drug Resist* 2020;**13**:2657–65. https://doi.org/10.2147/IDR.S264020.
149. Nagura-Ikeda M, Imai K, Tabata S, et al. Clinical evaluation of self-collected saliva by quantitative reverse transcription-PCR (RT-qPCR), direct RT-qPCR, reverse transcription–loop-mediated isothermal amplification, and a rapid antigen test to diagnose COVID-19. *J Clin Microbiol* 2020;**58**(9). https://doi.org/10.1128/JCM.01438-20. e01438-20.
150. Zhou P, Yang X-L, Wang X-G, et al. A pneumonia outbreak associated with a new coronavirus of probable bat origin. *Nature* 2020;**579**(7798):270–3. https://doi.org/10.1038/s41586-020-2012-7. 2020/03/01.
151. Ai J-W, Zhang Y, Zhang H-C, Xu T, Zhang W-H. Era of molecular diagnosis for pathogen identification of unexplained pneumonia, lessons to be learned. *Emerg Microb Infect* 2020;**9**(1):597–600. https://doi.org/10.1080/22221751.2020.1738905. 2020/01/01.
152. Roy SXWS, Abd Rahman I, Ahmed MU. Based visual detection of Salmonella bacteria using isothermal DNA amplification and magnetic bead aggregation. *Malays J Microbiol* 2016;**12**(5):332–8.
153. Roy S, Hossain MM, Safavieh M, Lubis HN, Zourob M, Ahmed MU. Chapter 16 isothermal dna amplification strategies for food biosensors. *Food Biosens R Soc Chem* 2017:367–92.
154. Cui J, Li F, Shi Z-L. Origin and evolution of pathogenic coronaviruses. *Nat Rev Microbiol* 2019;**17**(3):181–92. https://doi.org/10.1038/s41579-018-0118-9. 2019/03/01.
155. Lu R, Zhao X, Li J, et al. Genomic characterisation and epidemiology of 2019 novel coronavirus: implications for virus origins and receptor binding. *Lancet* 2020;**395**(10224):565–74. https://doi.org/10.1016/S0140-6736(20)30251-8. 2020/02/22.
156. Zhang W, Du R-H, Li B, et al. Molecular and serological investigation of 2019-nCoV infected patients: implication of multiple shedding routes. *Emerg Microb Infect* 2020;**9**(1):386–9. https://doi.org/10.1080/22221751.2020.1729071. 2020/01/01.
157. Woo PCY, Lau SKP, Wong BHL, et al. Differential sensitivities of severe acute respiratory syndrome (SARS) coronavirus spike polypeptide enzyme-linked immunosorbent assay (ELISA) and SARS coronavirus nucleocapsid protein ELISA for serodiagnosis of SARS coronavirus pneumonia. *J Clin Microbiol* 2005;**43**(7):3054. https://doi.org/10.1128/JCM.43.7.3054-3058.2005.
158. Gandhi RT, Lynch JB, del Rio C. Mild or moderate covid-19. *N Engl J Med* 2020;**383**(18):1757–66. https://doi.org/10.1056/NEJMcp2009249. 2020/10/29.
159. Hsih W-H, Cheng M-Y, Ho M-W, et al. Featuring COVID-19 cases via screening symptomatic patients with epidemiologic link during flu season in a medical center of central Taiwan. *J Microbiol Immunol Infect* 2020;**53**(3):459–66. https://doi.org/10.1016/j.jmii.2020.03.008. 2020/06/01.
160. Yu C, Zhang Z, Guo Y, et al. Lopinavir/ritonavir is associated with pneumonia resolution in COVID-19 patients with influenza coinfection: a retrospective matched-pair cohort study. *J Med Virol* 2021;**93**(1):472–80. https://doi.org/10.1002/jmv.26260. 2021/01/01.
161. Loeffelholz MJ, Tang Y-W. Laboratory diagnosis of emerging human coronavirus infections – the state of the art. *Emerg Microb Infect* 2020;**9**(1):747–56. https://doi.org/10.1080/22221751.2020.1745095. 2020/01/01.
162. Shi X, Gong E, Gao D, et al. Severe acute respiratory syndrome associated coronavirus is detected in intestinal tissues of fatal cases. *Off J Am Coll Gastroenterol (ACG)* 2005;**100**(1).

163. Yeo C, Kaushal S, Yeo D. Enteric involvement of coronaviruses: is faecal—oral transmission of SARS-CoV-2 possible? *Lancet Gastroenterol Hepatol* 2020;**5**(4):335—7.
164. Young BE, Ong SWX, Kalimuddin S, et al. Epidemiologic features and clinical course of patients infected with SARS-CoV-2 in Singapore. *J Am Med Assoc* 2020;**323**(15):1488—94. https://doi.org/10.1001/jama.2020.3204.
165. Cheng PKC, Wong DA, Tong LKL, et al. Viral shedding patterns of coronavirus in patients with probable severe acute respiratory syndrome. *Lancet* 2004;**363**(9422):1699—700. https://doi.org/10.1016/S0140-6736(04)16255-7. 2004/05/22.
166. Hong KH, Lee SW, Kim TS, et al. Guidelines for laboratory diagnosis of coronavirus disease 2019 (COVID-19) in Korea. *Annals of laboratory medicine* 2020;**40**(5):351—60.
167. Jung S, Lee S, Dou X, Kwon EE. Valorization of disposable COVID-19 mask through the thermo-chemical process. *Chem Eng J* 2021;**405**:126658. https://doi.org/10.1016/j.cej.2020.126658. 2021/02/01.
168. Liu Y, Ning Z, Chen Y, et al. Aerodynamic analysis of SARS-CoV-2 in two Wuhan hospitals. *Nature* 2020;**582**(7813):557—60. https://doi.org/10.1038/s41586-020-2271-3. 2020/06/01.
169. Li H, Wang Y, Ji M, et al. Transmission routes analysis of SARS-CoV-2: a systematic review and case report. Hypothesis and theory. *Front Cell Dev Biol* 2020;**8**:618. https://doi.org/10.3389/fcell.2020.00618. 2020-July-10.
170. Heffernan R, Mostashari F, Das D, Karpati A, Kulldorff M, Weiss D. Syndromic surveillance in public health practice, New York city. *Emerg Infect Dis J* 2004;**10**(5):858. https://doi.org/10.3201/eid1005.030646.
171. Cullen W, Gulati G, Kelly BD. Mental health in the COVID-19 pandemic. *QJM Int J Med* 2020;**113**(5):311—2. https://doi.org/10.1093/qjmed/hcaa110.
172. Galbraith N, Boyda D, McFeeters D, Hassan T. The mental health of doctors during the COVID-19 pandemic. *BJPsych Bull* 2020:1—4. https://doi.org/10.1192/bjb.2020.44.
173. Sampath Kumar NS, Chintagunta AD, Jeevan Kumar SP, Roy S, Kumar M. Immunotherapeutics for Covid-19 and post vaccination surveillance. *3 Biotech* 2020;**10**(12):527. https://doi.org/10.1007/s13205-020-02522-9. 2020/11/11.
174. Singh S, Prakash C, Ramakrishna S. Three-dimensional printing in the fight against novel virus COVID-19: technology helping society during an infectious disease pandemic. *Technol Soc* 2020;**62**:101305. https://doi.org/10.1016/j.techsoc.2020.101305. 2020/08/01.
175. Shakil MH, Munim ZH, Tasnia M, Sarowar S. COVID-19 and the environment: a critical review and research agenda. *Sci Total Environ* 2020;**745**:141022. https://doi.org/10.1016/j.scitotenv.2020.141022. 2020/11/25.
176. Ramteke S, Sahu BL. Novel coronavirus disease 2019 (COVID-19) pandemic: considerations for the biomedical waste sector in India. *Case Stud Chem Environ Eng* 2020;**2**:100029. https://doi.org/10.1016/j.cscee.2020.100029. 2020/09/01.
177. Sharma HB, Vanapalli KR, Cheela VRS, et al. Challenges, opportunities, and innovations for effective solid waste management during and post COVID-19 pandemic. *Resour Conserv Recycl* 2020;**162**:105052. https://doi.org/10.1016/j.resconrec.2020.105052. 2020/11/01.
178. Zambrano-Monserrate MA, Ruano MA, Sanchez-Alcalde L. Indirect effects of COVID-19 on the environment. *Sci Total Environ* August 1, 2020;**728**:138813. https://doi.org/10.1016/j.scitotenv.2020.138813.
179. Lurie N, Carr BG. The role of telehealth in the medical response to disasters. *JAMA Int Med* 2018;**178**(6):745—6. https://doi.org/10.1001/jamainternmed.2018.1314.

180. Seto E, Smith D, Jacques M, Morita PP. Opportunities and challenges of telehealth in remote communities: case study of the Yukon telehealth system. *JMIR Med Inf* 2019;**7**(4):e11353.
181. Hollander JE, Carr BG. Virtually perfect? Telemedicine for covid-19. *N Engl J Med* 2020;**382**(18):1679−81. https://doi.org/10.1056/NEJMp2003539. 2020/04/30.
182. Aziz A, Zork N, Aubey JJ, et al. Telehealth for high-risk pregnancies in the setting of the COVID-19 pandemic. *Am J Perinatol* 2020;**37**(8):800−8. https://doi.org/10.1055/s-0040-1712121.
183. Hadi Kharrazi KG, Weiner J. A population health segmentation framework for balancing medical need and COVID-19 risk during and after the pandemic. *Health Affairs Blog* 2020. September 4, 2020.
184. Foladori P, Cutrupi F, Segata N, et al. SARS-CoV-2 from faeces to wastewater treatment: what do we know? A review. *Sci Total Environ* 2020;**743**:140444. https://doi.org/10.1016/j.scitotenv.2020.140444. 2020/11/15.
185. Paleologos EK, O'Kelly BC, Chao-Sheng T, et al. Post Covid-19 water and waste water management to protect public health and geoenvironment. *Environ Geotech* 2020;**0**(0):1−15. https://www.icevirtuallibrary.com/doi/full/10.1680/jenge.20.00067.
186. Usman M, Farooq M, Hanna K. Existence of SARS-CoV-2 in wastewater: implications for its environmental transmission in developing communities. *Environ Sci Technol* 2020;**54**(13):7758−9. https://doi.org/10.1021/acs.est.0c02777. 2020/07/07.
187. Wang X-W, Li J-S, Jin M, et al. Study on the resistance of severe acute respiratory syndrome-associated coronavirus. *J Virol Methods* 2005;**126**(1):171−7. https://doi.org/10.1016/j.jviromet.2005.02.005. 2005/06/01.
188. Kataki S, Chatterjee S, Vairale MG, Sharma S, Dwivedi SK. Concerns and strategies for wastewater treatment during COVID-19 pandemic to stop plausible transmission. *Resour Conserv Recycl* 2021;**164**:105156. https://journals.sagepub.com/doi/full/10.1177/0253717620963345.
189. Khadse PA, Gowda GS, Ganjekar S, Desai G, Murthy P. Mental health impact of COVID-19 on police personnel in India. *Indian J Psychol Med* 2020;**42**(6):580−2. https://doi.org/10.1177/0253717620963345. 2020/11/01.
190. Grimshaw B, Chaudhuri E. Mental-health-related admissions to the acute medical unit during COVID-19. *Clin Med* 2021;**21**(1):e77. https://doi.org/10.7861/clinmed.2020-0635.

CHAPTER 2

Virus-sampling technologies in different environments

Heshmatollah Nourmoradi[1,2], Fazel Mohammadi-Moghadam[3], Sara Hemati[3], Hojatollah Kakaei[4]
[1]Biotechnology and Medicinal Plant Research Center, Ilam University of Medical Sciences, Ilam, Iran; [2]Department of Environmental Health Engineering, School of Health, Ilam University of Medical Sciences, Ilam, Iran; [3]Department of Environmental Health Engineering, School of Health, Shahrekord University of Medical Sciences, Shahrekord, Iran; [4]Department of Occupational Health Engineering, School of Health, Ilam University of Medical Sciences, Ilam, Iran

2.1 Introduction

Biological airborne particles such as viruses are commonly found in the environment. Any respiratory virus that is able to survive and transmit in the air has a potential risk of disease. The three ways of direct contact, inhalable droplets, and airborne transmission can spread respiratory viral diseases. Hand washing and face masking can substantially reduce the first two paths of the viral transmission.[1] An infectious sneeze can spread 40,000 aerosolized droplets in the air up to approximately 2 m distance. As a result, the droplet nucleus can suspend in the gases environment for a long time (up to 30 h) and results in respiratory tract infections.[2,3]

Bioaerosol sampling has been researched for over 200 years.[4] Three key points associated to sampling performance include sampling efficiency for inhaled bioaerosol particles, efficiency of particulate capture in filtration or laboratory culture, and biological recovery efficiency for bioaerosol particle viability.[5,6] Various physical mechanisms are applied to remove particles from the gas media by these samplers. The impaction force can simply remove particles with greater aerodynamic diameters from the air. The smaller particles (less than 100 nm) can be enlarged using higher diffusivity or by size density, and finally, can be collected by impaction.[7] There are several methods for sampling airborne viruses. Although the primary studies were performed on cyclone sampling, most studies have been conducted by the liquid impingers.[7]

Most air sampling methods are a function of the aerodynamic diameter, Brownian motion, inertia and adhesion properties of the airborne particles

Environmental Management of COVID-19
ISBN 978-0-323-85780-2
https://doi.org/10.1016/B978-0-323-85780-2.00010-X

and the temperature gradient of the environment. The aerosolized particles adhere to any surface that comes in contacted with it.[7]

The SARS-CoV-2 has a sphere-shaped structure with an aerodynamic diameter of approximately 120 nm.[8] Coronaviruses such as SARS-CoV-2 can be moved via airborne transmission. People become infected when viruses enter their body through the respiratory system, eyes, nose, and mouth. Other media that can transmit SARS-CoV-2 include water and wastewater, urine, stool, contaminated surface, food, solid waste, etc. The SARS-CoV-2 has a sphere-shaped structure with a single-segmented RNA in the 80-nm-diameter lumen.[9] In this study, the important procedures for environmental sampling and analysis of the virus were investigated.

2.2 Methods of SARS-CoV-2 air sampling

Various sampling principles like impaction, impingement, electrostatic precipitation, filtration, and cyclone have been applied to viral bio-aerosol.[7,10,11] In a study, the collection performance of four devices including all-glass impinger (AGI), impinger (AGI-30), one-stage Anderson impactor (1-STG), gelatin filter, and nucleoporous polycarbonate filter were employed for four bacteriophages. The results showed that about 95% of the cultured phages with a particle diameter of less than 2.1 μm were collected by the Anderson 6-STG impactor.[12]

The recovery efficiency of heating, ventilation, and air conditioning (HVAC) filters and two types of impingers (AGI-30 and biosampler) using transmissible gastroenteritis virus (TGEV) were investigated by other research. The researchers found that by increasing the relative humidity, the collection performance of the samplers declined. However, the biosamplers had a higher efficiency than the AGI-30 impinger.[13]

In an experimental project on the survival of the airborne viral organisms HCoV-229E, the researchers investigated the effect of the nebulization phenomenon on the survival of the virus and the lack of viral infection at 20°C and various relative humidity (RH). At 20°C and moderate RH, HCoV-299E indicated the highest virus retrieval (87%—91%), while increasing the RH had a detrimental effect. At 6°C and higher RH, a recovery rate was observed 100%.[14]

2.2.1 Filtration

Filtration procedure is commonly applied to catch airborne bioaerosols. The bioaerosols trapped on the filter can be analyzed directly by an electron microscope or by agar culture. Based on the analytical method types, polytetrafluoroethylene (PTFE, Teflon), mixed cellulose ester, polycarbonate, polyvinyl chloride (PVC), nylon, gelatin, and other filters can be applied for bioaerosol sampling.[15,16] When flat-bed filters such as nucleopores were used, the accumulated particles were specifically suitable for microscopic analysis. The washable filters such as gelatin filters have been extended for the better preservation of the culturable sample (like bacteria and virus) and it can be dissolved in a buffer solution or cultured directly on agar for analysis.[16]

The collection devices such as impactors and impingers have lower collection efficiency for particles smaller than 500 nm. Therefore, various filters have been commonly employed for aerosolized viruses sampling. PTFE and cellulose filters have been used for the sampling of aerosols containing virus.[17,18] Because the filtration process dehydrates the viruses during sampling, it can inactivate a noteworthy part of the viruses.[7] In a sampling procedure, it was found that only 22% and 10% of influenza viruses was recovered alive using Teflon and gelatin filters, respectively, compared to biosamplers.[19]

An approach was performed to compare the extraction efficiency for bacteriophage MS2 phage using alumina nanofiber against fiberglass filters. The efficiency of nanofiber filter was less than 10%, while fiberglass filter ranged from 32.3% to 162%.[20]

Among conventional filters, the gelatin type can be readily dissolved in a liquid for molecular counting or cell culture without significant changes in virus tissue. In Zhao et al. study, four six-stage samplers ACI, AGI-30, OMNI-3000, and MD8 were explored with gelatin filter for airborne Brucella virus infection. The results showed that the gelatin filters had the highest collection efficiency (100%) without significant dehydration effects on the virus.[21] The collection efficiency of gelatin filters toward the nucleoporous polycarbonate filters for hydrophilic viruses was about 10 times.[12]

2.2.2 Impactor

Impactors such as a slotted-vessel sampler, six-step Anderson sampler and cyclones are applied for airborne viruses sampling. These sampling methods require a suction pump to suck air into the device. Then, the aerosols with higher inertia force hit the internal surface of the device.[22]

During the SARS prevalence in 2003, two air sampling procedures, including a high-resolution modified slotted-vessel sampler and PTFE membrane filters (0.3 μm) were applied.[23] The findings showed that all the medium cultures were negative, although 2 of 10 reverse-transcription polymerase chain reaction (RT-PCR) samples were positive in terms of the virus. The practical limitation of the conventional impactors is the smallest cut-off size for 0.2–0.3 μm particles.[23]

2.2.3 Impinger

The air sampling devices such as liquid impingers present a little information about the aerosol particle size as a significant parameter that influences on airborne particles movement.[24] Liquid impingers are the most frequently devices that are applied for the collection of viral aerosols. Because the liquid medium helps to survive the viruses and subsequently it can be directly applied for virus extraction analysis.[25] AGIs and biosamplers are extensively used as liquid impregnators. During the sampling, the produced bubbles into the liquid phase can enhance the collection performance for finer particles via the diffusion process, but it can resuspend the collected viruses.[26]

AGI-4, AGI-30, and biosampler were employed as reference samplers. The liquid medium can be straightly used for molecular analysis techniques such as PCR. In Hogan et al. study, the collection performance of AGI-30, biosamplers, and frit bubblers for MS2 and T3 bacteriophages for particles size of 20–100 nm was found to be less than 10%.[27] The high centrifugal force of the biosampler can inactivate a significant proportion of influenza viruses.[28]

2.2.4 Cyclone

This device has not been designed as a high performance sampler (>95%) for the particles size of greater than 10 nm. The collection performance for a typical single-stage cyclone for PM_{10} and $PM_{2.5}$ is approximately 30%–90% and 0%–40%, respectively.[29] However, viral particles because of their size (>100 nm) are not easily trapped by this procedure.[30]

The National Institute for Occupational Safety and Health (NIOSH) improved a multistage cyclone with air rate of 3.5 L/min. The performance of this sampler for viral aerosols is similar to the standard midget impinger (SKC) biosampler for 15 min of sampling. The results showed that only 34% of the viral aerosols trapped by the NIOSH cyclone, because of the dryness of the viruses, were infectious.[31] The liquid cyclone collectors have been used for the sampling of viral aerosols.[32,33]

The result of a project showed that cyclone performance cannot meet the sampling needs of small particle-containing viruses and the number of live viruses collected by this device is less than the actual ones.[7] However, ACI has a high capability to collect viral particles in various size ranges and it can also be used to trap particles downstream the impactor sampler.[34]

2.2.5 Electrostatic precipitator

Electrostatic precipitator (ESP) is another sampler for the collection of airborne particles. Few studies have been performed on virus behavior in ESPs that focus solely on sampling efficiency.[35] One of the most important characteristics of electrostatic sampler is the lower particle settling velocity rate compared to inertial based methods, which leads to less damage to microorganisms.[36] The collection efficiency of ESP reached to 99.3%–99.8% for polystyrene latex (PSL) particles with a size range of 0.5–0.5 μm.[37]

2.2.6 MD-8 airscan

This simple system consists of MD-8 portable air sampler attached to a gelatin filter (diameter of 80 mm and pore size of 3 μm). MD-8 airscan was applied for the influenza viruses sampling in air.[38] Lower relative humidity can lead to drying the virus and higher relative moisture can cause to the liquidation of gelatin filters.[7] These types of filters should be employed in a short time (less than 15 min), because they are quickly dried and melted at higher temperatures.[19] MD-8 airscan was used for MERS-CoV and SARS-CoV sampling in hospital indoor air.[39]

In Liu et al. study, MD-8 airscan sampler and PTFE filter in a cassette were used for sampling and identification of SARS-CoV-2 virus. The application of gelatin filter in MD-8 airscan showed good performance for coronavirus sampling.[40]

2.3 Literature review for SARS-CoV-2 sampling in air

In a study for aerosolized SARS-CoV-2 in hospital, portable pump (APEX2, Casella) equipped with the sterilized gelatin filters (pore size of 3 μm and diameter of 25 mm) was used at air rate of 5 L/min for 1 h. The results showed that the samples were positive.[40]

Guo et al. research on the SARS-CoV-2 in a hospital air through wetted wall cyclone sampler (SAAA 2300) with flow rate of 300 L/min for 30 min showed that the existence of the virus in the air samples were positive.[41]

Cheng et al. explored SARS-CoV-2 in the indoor air of COVID-19 Hospitals. Sartorius MD-8 airscan with sterile gelatin filters (diameter of 80 mm and a pore size of 3 μm) was applied for sampling. The sampler was vertically installed at 10 cm from the patients' heads. The sampling was performed at flow rate of 50 L/min for 20 min while patients with or without surgical mask. After sampling, each filter was immersed in 5 mL of viral transport medium (VTM) and placed at 37°C for 10 min. The results showed that out of 377 air samples close the patients, 19 samples were positive using RT-PCR assay.[42]

Ong et al. examined the personal protective equipment and air of the symptomatic SARS-CoV-2 patients. The air sampling was performed in the indoor environment for 2 days through SKC Universal pumps (with 37 mm filter cassettes and 0.3 μm polytetrafluoroethylene filters at flow rate of 5 L/min for 4 h). The results showed that none of the samples contained the virus.[43]

In various studies, PTFE, gelatin, and polycarbonate filters have been used to sample viruses, of which PTFE and gelatin filters showed higher collection efficiency.[44] However, the NIOSH has suggested that PTFE can be used for sampling of airborne pathogens for immunological assay and PCR.[45] Some studies have reported that the use of filters for the viral sampling is not suitable.[46]

Cyclone samplers were also employed for sampling of SARS-like viruses. These devices were used to collect viruses at high air flow rate for 8 h. The NIOSH 0600 method has suggested that the cyclone can be used as a prefilter to remove nonrespirable particles from bioaerosols.[45] Chia et al. used a NIOSH BC 251 bioaerosol sampler for SARS-CoV-2 in patient's rooms of hospital air. The samplers were equipped with SKC AirCheck TOUCH pumps or SKC Universal air sampling pumps with air rate of 3.5 L/min for 4 h.[47]

In Faridi et al. study on SARS-CoV-2 in hospital air, a setup consisted of a pump with air rate of 1.5 L/min and SKC was used. The sampling device was installed at a height of 1.5—1.8 m from the floor and about 2—5 m away from the patient's beds. The air was collected by a sterile impregnator containing 20 mL of Dulbecco's Modified Eagle Medium (DMEM), 100 μg/mL streptomycin, 100 U/mL penicillin, and 1% antifoam reagent (isoamyl alcohol) for 1 h. The specimens were placed on ice and instantaneously transferred to laboratory. The finding showed that the samples were not positive in terms of SARS-CoV-2.[48]

Razzini et al. research was performed for the sampling of SARS-CoV-2 in a hospital air. The samples were taken by an MD-8 Airport portable air sampler with gelatin filters. The sampling time was 40 min with air rate of 50 L/min. The positive samples were detected from the ICU ward and corridor.[49]

2.4 Surface detection for SARS-CoV-2

Sterile synthetic fiber swabs with plastic shafts have been mainly used to collect the environmental surface samples in many studies.[41,49,50] The swabs were premoistened with VTM and rubbed over the surfaces of the target for a few seconds. Then, swabs were immediately positioned in the sterile tubes containing 2–3 mL of VTM. Each swab was separately taken to curb cross-contamination. Finally, all the samples were transferred to the laboratory for SARS-CoV-2 within 4 h under cool conditions (2–8°C).[41,49,50]

2.4.1 Practical tips before sampling

- The sealing plastic bags should be cleaned with 70% ethanol solution, 80% isopropyl alcohol, or 5% NaClO solution before inserted to the transport container.[50]
- To achieve the best results, sampling must be carefully duplicated at the same time and treated as separate samples to determine the precision.[50]
- Use only commercially available laboratory fiber swabs with plastic or wire shafts.[51]

2.4.2 Surface sampling area

The swabbed surface area depends on the equipment size. The World Health Organization (WHO) recommended that normal swab surface area is 25 cm^2.[51] In this regard, the area (25 cm^2) was swabbed by applying pressure to the swab and rotating stick it.[52] For the larger surfaces, develop the surface area up to 50 cm^2 and do not allow the swap to dehydrate entirely.[53]

According to the Yang et al. method, if the surface area of the equipment is greater than 100 cm^2, the sampling area of the object surface is limited to 100 cm^2, otherwise all surface area would be sampled. In their study, a standard specification board with a size 5 × 5 cm was positioned on the object surface, afterward a premoistened sterile swab with VTM was applied to wipe the specification disk five times.[54] Also, entire surfaces of small objects like faucets, pipettes, and door handles were rubbed by a sterile swab soaked with VTM.[54]

2.4.3 Swab types

Acceptable swab types for SARS-CoV-2 sampling are flocked tapered swabs,[55] Dacron swabs,[56] spun polyester swabs,[51] and cotton and foam swabs.[57]

In Santarpia et al., surface sampling of small objects was performed by premoistened sterile gauze pads (with a size 3 × 3 cm) with 3 mL of phosphate buffered saline (PBS). The larger area surfaces were sampled by "S" pattern wiping in two paths to cover the accessible surface as much as probable. Furthermore, the smaller objects (e.g., mobile phones, TV remote controls) were swabbed in one path on every accessible surface.[58] For sampling of the facemasks of COVID-19 patients, it should be cut into small pieces and immediately placed in VTM.[59]

2.5 Municipal wastewater for SARS-CoV-2 sampling

Many studies showed that municipal wastewaters contain SARS-CoV-2. The amount of virus in sewage is much lower than feces due to factors such as solution pH, temperature, and the presence of disturbing factors.[60] Different methods including precipitation, ultracentrifugation, electronegative membrane, and ultrafiltration have been used to prepare samples of sewage, feces, and urine for RNA virus extraction, which are listed later in the chapter.

2.5.1 Sampling procedures for SARS-CoV-2 in municipal wastewater

In Randazzo et al. project on municipal wastewater treatment plants, 500–1000 mL of wastewater was collected by a grab sampling method in high-density polyethylene (HDPE) sterile containers. The specimens were transferred to the laboratory on ice and kept at 4°C. Then, 200 mL of the sewage was processed by precipitation-thickening method as follows.[61] For precipitation, 200 mL of the sample was poured into a 250 mL centrifugation tube and the pH was regulated to 6. Then, to produce aluminum hydroxide precipitate, 2 mL of aluminum chloride (0.9 N) was mixed with the sample. The solution pH was readjusted to six and mixing was performed by a rotary shaker (150 rpm for 15 min) at room temperature. The thickening process was then carried out by centrifugation (1700 g for 20 min). The residue was mixed with 10 mL of beef extract (pH = 7.4) into a 50 mL centrifugation tube by an orbital shaker (150 rpm for 10 min) and centrifuged (1900 g for 30 min). The precipitation was finally resuspended into 1 mL of phosphate-buffered saline (PBS) and 150 μL of the suspension was applied for RNA extraction.[61]

In Zhang et al. research on septic tank effluent, 2 L of wastewater was collected using a sterile plastic bag and it was immediately transferred to the laboratory at 4°C. The sample was then mixed with sodium chloride (0.3 M) and PEG-6000 and it was stayed overnight at 4°C for precipitation. Finally, the mixture was centrifuged (10,000 g for 30 min) and the virus RNA was tested from the precipitate.[62]

A composite sampling of raw wastewater for SARS-CoV-2 was taken by Wu et al. and after transfer of the sample to the laboratory, it was pasteurized (90 min at 60°C) to inactivate the viruses. The pasteurized sample was then filtered (0.22 micron) to separate large solids and bacterial cells. The filter was discarded because the analysis showed that the filter lacked RNA virus. Then, 4 g of PEG-8000 (8% by weight/volume) and 0.9 g of sodium chloride (0.3 M) were added to 40 mL of the filtrate and centrifuged at 12,000 g for 2 h (or centrifuged until the pellet became observable). Finally, the pellets were suspended in Trizol (Thermofisher) and prepared for the RT-qPCR analysis.[63]

To detect the viral RNA, raw wastewater and treated samples were gathered by combined technique. The samples were transferred to the laboratory by sterile bottles via cold chain and kept at 4°C. The specimens were first centrifuged (4500 g for 30 min) and then passed through a cellulose ester syringe filter (0.22 micron). The filtrate was then thickened using the polyethylene glycol method as follows: 25 mL of the filtrate, polyethylene glycol-9000 (80 g/L) and sodium chloride (17.5 g/L) were poured into a tube. The mixture was shaken overnight at 17°C at 100 rpm. Finally, the mixture was centrifuged (1000 g for 90 min) and the pellet was suspended again in 300 μL of RNase-free water for PCR analysis.[64]

On wastewater from the municipal treatment plant, a composite sampling was performed. Initially, 0.25−1 L of sewage was centrifuged to separate bulky particles. The centrifuged sample was then treated with polyethylene glycol or aluminum sulfate (20 mg/L) and centrifuged again. The sample was shaken at 100 rpm for 12 h at 4°C and the sample was then centrifuged (14,000 g for 45 min) at 4°C to produce a viral residue. The sediments were eventually suspended in PBS. The viral suspension was finally filtered (0.22 micron) and Ultra-15 centrifugal tubes with cut-off of 30 kDa were applied to thicken the sample volume to 1 mL. The sample was kept at −20°C to −80°C to extract the viral RNA.[65]

In a project on municipal wastewater, the sampling was performed and kept in the laboratory at 4°C. Firstly, 11 mL of the samples were centrifuged (200,000 g for 1 h) at 4°C. Then, the residue was mixed into 400 μL of PBS 1×. Finally, 200 μL of the suspension was used for viral analysis.[66]

For determination of SARS-CoV-2 in domestic wastewater, 1.9 L of wastewater was taken by composite sampling and carried to the laboratory on ice packs and placed at 4°C. After homogenization, 20 mL of the sample was poured into 38.5 mL ultracentrifuge bottles. Then, 12 mL sucrose cushion (50% sucrose in Tris-NACL-EDT buffer (TNE) buffer [20 Mm Tris-HCl (pH 7.0), 100 mM NaCl, 2 mM EDTA]) was sensibly injected beneath the surface of sewage with pipette so that two separate layers were created. Finally, the mixture was centrifuged (50,000 g for 45–90 min) at 4°C and the residue was homogenized into 200 μL of 1× PBS. The suspension was stored at −20°C to extract virus RNA.[67]

In an approach for direct extraction of virus RNA from raw municipal wastewater, firstly, the pH of 100–200 mL of raw wastewater was adjusted to about 3.5–5 using HCl (2N). The sample was then filtered by an electronegative filter (9 cm diameter with a pore size of 0.45 microns) using a glass funnel. The filter was placed in a 5 mL bead tube from the RNeasy PowerWater Kit and homogenized at 3 × 20 s at 8000 rpm at 10 s interval. Finally, the virus RNA was tested based on the manufacturer's direction of the RNeasy Power Microbiome kit.[68]

In an experimental procedure for the extraction of virus RNA from raw municipal wastewater, 100–200 mL of the sample was centrifuged (4750 g for 30 min). The clear solution was removed from the tube without any disturbance for the residuals and centrifuged (3500 g for 15 min) by Centricon Plus-70 centrifugal filter (cut-off of 10 kDa). The concentrate cup was inversely installed on the top of the sample filter cup. The instrument was then centrifuged (1000 g for 2 min). The concentrated sample (~250 μL) was finally gathered from the collection cup for RNA analysis.[68]

Kocamemi et al. survey on wastewater treatment plant, 250 mL of raw wastewater was collected and large particles and bacteria were centrifuged at 3200 g for 45 min. The clear solution was then filtrated by ultrafiltration (3200 g for 25–40 min). Finally, 200–600 μL of the viral particles on the filter were used to extract RNA.[69]

In another exploration on municipal wastewater, wastewater samples were transported to the laboratory along with an ice pack. First, in order to remove coarse materials and bacteria, the samples were centrifuged in funnel containers of 250−50 mL at 4654 rpm for 30 min. Then, the ultrafiltration of 100−200 mL of the clear solution was performed using Centricon Plus-70 centrifugal ultrafilter with a cut-off of 100 kDa using centrifugation (1500 g for 15 min). The centricon concentrate was 0.44−1.79 g which used to PCR analysis.[70]

In an investigation on municipal wastewater, the samples were received by the laboratory at 4°C and it was immediately centrifuged (4500 g for 30 min) at 4°C. Then, suspended particles were removed from the centrifuged sample by filtration (pore size of 40 μm). Finally, the filtrate was frozen at −20°C and stored for PCR. Then, 50 mL of the sample was filtered by a 50 KDa membrane and 500 μL of the residue on the filter was used for RNA analysis.[71]

To extract of the virus, wastewater samples were firstly collected into 500 mL polypropylene containers and immediately transferred to the laboratory under cold chain for filtration. The samples were then filtered by fiberglass filter (0.7 micron with a diameter of 145 mm). Then, 200 μL of the filtrates were employed to extract viral RNA.[72]

In another study, 200 mL and 5 L of raw and treated wastewaters were taken from the treatment plant, respectively. The samples was mixed with 25 mM $MgCl_2$ and filtered by mixed cellulose ester filter (pore size of 0.8 micron, diameter of 90 mm). Then, the virus RNA was straightly collected from one-fourth of the filter that placed into 5-mL of PowerWater Bead Tube of a RNeasy PowerWater Kit. A volume of 50 μL of the concentrate was used for viral RNA analysis.[73]

2.6 SARS-CoV-2 in urine and stool

Some people with COVID-19 have gastrointestinal symptoms such as diarrhea.[60] Cheung et al. examination also expressed that 17.6% of the patients had gastrointestinal symptoms and 48.1% of them had positive stool tests.[74] Fecal-oral spread of SARS-CoV-2, due to gastrointestinal symptoms in some COVID-19 patients, is possible.[75] Previous studies have shown that large amounts of the virus RNA were present in the urine and feces of the infected animals.[76] The findings of Jeong et al. survey showed that the virus RNA level of COVID-19 during the 8−30 days of their clinical period in the urine and feces of patients was equal or even more than nasopharyngeal. The stool test for SARS-CoV-2 in patients with negative respiratory symptoms may be positive.[60] Therefore, the stool tests are very essential to diagnose COVID-19.[60,75]

2.6.1 Sampling method for SARS-CoV-2 in urine and stool

In an analysis to examine the virus in feces, the patient's feces were diluted (10 times) in 2 mL of phosphate saline solution and then the debris was separated by centrifugation (12,000 rpm for 20 min). The clear liquid was employed for culture and qRT-PCR.[75]

2.7 SARS-CoV-2 sampling in food

Because the global attention during the COVID-19 pandemic is paid on human health, the impact of the disease on foodstuffs, food processing, and food packaging is still not recognized. Therefore, it is crucial to develop the detection methods for SARS-CoV-2 in food products. Because of the low viral load, the heterogeneous distribution of the viral particles and nonoptimal tedious isolation, the SARS-CoV-2 detection in foodstuffs remains a challenge.[77]

In this regard, several methods have been suggested for SARS-CoV-2, such as RT-qPCR, enzyme-free immunosorbent assay, and nanoparticles and enzyme-linked immunosorbent assays (Nano-ELISA).[78,79] In spite of the COVID-19 pandemic, according to the FDA, there is no documentation about SARS-CoV-2 transmission via food products to date.[80–82] The existence of SARS-CoV-2 in refrigerated and iced meat and salmon was evaluated in Fisher et al. study. In this study, the individual portions of salmon, chicken, and pork provided from food stores were sliced (500 mm^3), then 200 μL of 3 × 106 TCID50/ml SARS-CoV-2 was mixed with each cube. All specimens were placed at 4, −20, and −80°C. The samples were then harvested for 1, 2, 5, 7, 14, and 21 days after inoculation. After that, an aliquot of 150 μL of the virus inoculum was poured into a new tube and frozen at −80°C until titration. Each condition was conducted in triplicates. The cell-free virus titer (TCID50/mL) for the samples was defined by limited dilution. The limit of detection (LOD) was 5 × 101 TCID50/mL.[83]

2.8 SARS-CoV-2 detection techniques in air and surface samples

Presently, SARS-CoV-2 detection is mainly conducted via RT-qPCR. However, other available methods for the virus assays can be explained briefly as follows.

2.8.1 RT-PCR

Most of the literature employed PCR analysis for SARS-CoV-2 in air. Real-time PCR, due to high sensitivity, ease of use, and concurrent detection of more respiratory agents, have been extensively applied to diagnosis of viral respiratory infections.[84]

Presently, RT-PCR and RT-LAMP are the most used procedures for SARS-CoV-2. Until today, the RT-PCR method has been broadly applied to detect SARS-CoV-2 in various media like air and surfaces and biological samples such as sputum, urine, blood, saliva, and stool.[85,86]

2.8.2 Droplet digital PCR (ddPCR)

The droplet digital PCR (ddPCR) with a high LOD was recently used for SARS-CoV-2 in surface media. However, this approach has not yet been used for SARS-CoV-2 in air, and more research is required to evaluate the method efficiency for the virus detection in such samples.[87,107]

2.8.3 Sensor-based method

The sensor-based method for the fast analysis of SARS-CoV-2 which presented by Seo et al.[32] can be used for air samples. They introduced a field-effect transistor (FET)-based biosensing device to detect the virus. Due to this fact, graphene sheets of the FET were covered with a specific antibody against SARS-CoV-2 spike protein to produce the sensor. The efficiency of the mentioned procedure is not required to sample pretreatment and a high sensitivity compared to the conventional detection methods.[88] Due to its high sensitivity, this recognition procedure has the ability to distinguish the low burden of SARS-CoV-2 in air.[88] To our literature review, no study has been conducted by this technique for SARS-CoV-2 in air. The available primers applied for RT-qPCR of SARS-CoV-2 RNA in the air and surface media are summarized in Table 2.1.

2.9 SARS-CoV-2 recognition in water

The RT-qPCR assay, which shows an ALOD of b10 copies per reaction, can be suitable for wastewaters containing SARS-CoV-2.[92,93] It is extremely recommended that positive RT-qPCR signals should be validated for environmental samples by sequencing analysis, because the RT-qPCR assay was established for clinical diagnosis.[94,95] However, the available primers that were used for the recognition of SARS-CoV-2 RNA by RT-qPCR and nested RT-PCR techniques in the river water and wastewater are demonstrated in Table 2.2.

Table 2.1 The primers used for RT-qPCR assays to detect SARS-CoV-2 in air and surface media.

Study	Target	Primer name	Sequence (5′–3′)	Type of sample
Chia et al.[57]	Envelope gene	E_Sarbeco_F1 E_Sarbeco_R2 E_Sarbeco_P1	ACAGGTACGTTAATAGTTAATAGCGT ATATTGCAGCAGTACGCACACA ACACTAGCCATCCTTACTGCGCTTCGBHQ1	Air and surface
	ORF1ab	Wu-BNI-F Wu-BNI-R WU-BNI-P-FAM	CTAACATGTTTATCACCCGCG CTCTAGTAGCATGACACCCCTC TAAGACATGTACGTGCATGGATTGGCTTBHQ1	
Santarpia et al.[89]	E gene	F1 R2 P1	ATATTGCAGCAGTACGCACACA ACAGGTACGTTAATAGTTAATAGCGT FAM/ACACTAAGCC/ZEN/ ATCCTTACTGCGCTTCG/3AIBkFG	Air and surface
Ryu et ai.[90]	E gene, RdRp gene, and N gene	Based on the producer's protocols		Surface
Kenarkoohi et al.[91]	ORF1ab and N genes	Based on the producer's protocols		Air
Wu et al.[55]	RdRp, N and E gene	Based on the producer's protocols		Air and surface
Peyrony et al.[53]	ORF1a/b and E genes	Based on the producer's protocols		Surface
Tan et al.[59]	ORF1a/b	Based on the producer's protocols		Air and surface
Ye et al.[56] and Lv et al.[54]	ORF1ab and N genes	Based on the producer's protocols		Surface
Guo et al.[41]	ORF1ab and nucleocapsid protein (NP)	Based on the producer's protocols		Air and surface

Table 2.2 The primers used for RT-qPCR assays for SARS-CoV-2 in water and wastewater.

Study	Target	Primer name	Sequence (5′–3′)	Product length (bp)
Corman et al.[92]	N_Sarbeco	N_Sarbeco_F1	CACATTGGCACCCGCAATC	128
		N_Sarbeco_R1	GAGGAACGAGAAGAGGCTTG	
		N_Sarbeco_P1	FAMACTTCCTCAAGGAACAACATTGCCA-BHQ1	
Nao et al.[96]	S	WuhanCoV-spk1-f	TTGGCAAAATTCAAGACTCACTTT	547
		WuhanCoV-spk2-r	TGTGGTTCATAAAAATTCCTTTGTG	
		NIID_WH-1_F24381 (Nested)	TCAAGACTCACTTTCTTCCAC	493
		NIID_WH-1_R24873 (Nested)	ATTTGAAACAAAGACACCTTCAC	
WHO[97]	ORF1b-nonstructural protein 14	HKU-ORF1b-nsp14F	TGGGGYTTTACRGGTAACCT	132
		HKU- ORF1b-nsp14R	AACRCGCTTAACAAAGCACTC	
		HKU-ORF1b-nsp141P	FAM-TAGTTGTGATGCWATCATGACTAG-TAMRA	
CDC[98]	CDC-N1	2019-nCoV_N1-F	GACCCCAAAATCAGCGAAAT	72
		2019-nCoV_N1-R	TCTGGTTACTGCCAGTTGAATCTG	
		2019-nCoV_N1-P	FAM-ACCCCGCATTACGTTTGGTGGACC-BHQ1	
Rosa et al.[99]	ORF1ab	2274 - CO-FW1	GTGCTAAACCACCGCCTG	368
		2275 - CO-REV1	CAGATCATGGTTGCTTTGTAGGT	
		2276 - CO-FW2 (Nested)	CGCCTGGAGATCAATTTAAACAC	332
		2277 - CO-REV2 (Nested)	ACCTGTAAAACCCCATTGTTGA	

Continued

Table 2.2 The primers used for RT-qPCR assays for SARS-CoV-2 in water and wastewater.—cont'd

Study	Target	Primer name	Sequence (5′–3′)	Product length (bp)
Zhou et al.[100]	S protein	RBD-qF1	CAATGGTTTAACAGGCACAGG	121
		RBD-qR1	CTCAAGTGTCTGTGGATCACG	
Shirato et al.[93]	NIID_2019-nCOV_N	NIID_2019-nCOV_N_F2	AAATTTTGGGGACCAGGAAC	158
		NIID_2019-nCOV_N_R2ver3	TGGCACCTGTGTAGGTCAAC	
		NIID_2019-nCOV_N_P2	FAM-ATGTCGCGCATTGGCATGGA-BHQ1	
Haramoto et al.[95]	ORF1a	NIID_WH-1_F501	TTCGGATGCTCGAACTGCACC	413
		NIID_WH-1_R913	CTTTACCAGCACGTGCTAGAAGG	
		NIID_WH-1_F509	CTCGAACTGCACCTCATGG	346
		NIID_WH-1_R854	CAGAAGTTGTTATCGACATAGC	
Medema et al.[70]	(N) gene	IC-F	ATGACAGCCACTCCTCCG	412
		IC-R	GGAACGAACCAAACAGTCTTC	
		IC-P	AGCAGAGACCCATTCCCTCAGAGC	

2.10 SARS-CoV-2 detection in solid wastes

To our knowledge, no experimental research has been previously carried out to detect the SARS-CoV-2 in domestic solid wastes. However, few scientific papers were found about this topic. It can be said that the most infectious wastes are generated in hospitals and other health care facilities, which incinerated in place or autoclaved before landfilling. Therefore, no study has been conducted in this regard so far.[101] However, insufficient waste management, improper use of personal protective equipment (PPE), and other unfavorable conditions, especially in developing countries, could potentially increase the spread of SARS-CoV-2. Therefore, management and monitoring of the solid wastes of the self-quarantined patients is a vital key for controlling the virus. Treating patients at home can produce infected wastes, which possibly discarded along with domestic waste; subsequently, it can potentially result in risks to workers, other people, and the environment, depending on the transportation and disposal conditions.[102]

2.11 Analysis techniques for SARS-CoV-2

The PCR and antibody-based methods are currently applied for SARS-CoV-2. But other processes, including LAMP, reverse transcription-loop-mediated isothermal amplification (RT-LAMP),[84,103] and CRISPR, are under investigation and may be used soon to diagnose the SARS-CoV-2.[104,105]

Current studies are focused on SARS-Cov-2 RNA to improve the detection methods. Target genes are spiked (S) protein genes, nucleocapsid protein (N1,N2,N3) genes, open reading frame protein (ORF1, ORF1ab, ORF1b) genes, envelop protein (E) gene, and RNA-dependent RNA polymerase (RdRp) gene.[106]

The PCR-based detection methods are generally used and research is indicating many developments in this method. RT-PCR or qPCR is a strategy for measuring of gene expression at the transcript level.[19] The strategy is based on the following ways: (1) isolation of virus RNA from the samples; (2) synthesis of cDNA through reverse transcription kit, and (3) amplification and detection of cDNA. The PCR is set by mixing of the buffer, dNTPs, primers of the target gene, Taq polymerase, cDNA template, and SYBR green dye. The PCR mix is incubated in a PCR machine and the fluorescence created by PCR amplification is determined to present Cycle threshold (Ct) values. The Ct values of the controls are compared with the experimental samples and the relative expression is estimated. Also a key point for the detection of SARS-CoV-2 in the RT-PCR method is the specimen type selection and time of taking it.[19,106]

However, all of the following steps are performed on the basis of the manufacturer's protocol of RNA kit to perceive SARS-CoV-2 in different samples: primers and probes, nucleic acid extraction, mix preparation for all separate primer/probe combinations, controls, amplification cycles (light cycler system), sensitivity, specificity, and positive control for SARS-CoV-2 RT-PCR.[19,105,106]

2.12 Conclusion

SARS-CoV-2 has been detected in a various environments including air, sewage, urine, and feces. There are different procedures such as filtration, impactor, impinger, cyclone, electrostatic separator, and MD-8 airscan are applied for sampling and measuring viruses from air. Among filters, the gelatin type can be readily dissolved in a liquid for molecular counting or cell culture without significant changes in virus tissue. Liquid impingers are the most frequently used devices that are applied for the collection of viral aerosols. Sterile synthetic fiber swabs with plastic shafts have been mainly used to collect the environmental surface samples in many studies. The swabs were premoistened with VTM and rubbed over the surfaces of the target for a few seconds. The amount of virus in sewage is much lower than feces due to factors such as solution pH, temperature, and the presence of disturbing factors. Also, many methods including precipitation, ultracentrifugation, electronegative membrane, and ultrafiltration have been used to prepare samples of food, wastewater, feces, urine, and surfaces. In many studies, the aforementioned methods have been employed for the sampling of the coronaviruses such as SARS-CoV-2 in various environments. Also, PCR procedures have been commonly used to identify the virus from the environmental samples.

References

1. Mitchell AB, Tang B, Shojaei M, et al. A novel sampling method to detect airborne influenza and other respiratory viruses in mechanically ventilated patients: a feasibility study. *Ann Intensive Care* 2018;**8**(1):45.
2. Cox CS. *Aerobiological pathway of microorganisms*. Wiley; 1987.
3. Xie X, Li Y, Chwang A, Ho P, Seto W. How far droplets can move in indoor environments-revisiting the Wells evaporation-falling curve. *Indoor Air* 2007;**17**(3):211–25.
4. Gregory PH. *The microbiology of the atmosphere. The microbiology of the atmosphere*. 1961.
5. Macher J. *Bioaerosols: assessment and control*. 1999.

6. Vincent JH. *Aerosol sampling: science, standards, instrumentation and applications*. John Wiley & Sons; 2007.
7. Verreault D, Moineau S, Duchaine C. Methods for sampling of airborne viruses. *Microbiol Mol Biol Rev* 2008;**72**(3):413–44.
8. Qian G-Q, Yang N-B, Ding F, et al. Epidemiologic and clinical characteristics of 91 hospitalized patients with COVID-19 in Zhejiang, China: a retrospective, multi-centre case series. *QJM: Int J Med* 2020;**113**(7):474–81.
9. Yao H, Song Y, Chen Y, et al. Molecular architecture of the SARS-CoV-2 virus. *Cell* 2020;**183**(3):730–8.
10. Bosch A, Sánchez G, Abbaszadegan M, et al. Analytical methods for virus detection in water and food. *Food Anal Methods* 2011;**4**(1):4–12.
11. Takenaka K, Togashi S, Miyake R, Sakaguchi T, Hide M. Airborne virus detection by a sensing system using a disposable integrated impaction device. *J Breath Res* 2016;**10**(3):036009.
12. Tseng C-C, Li C-S. Collection efficiencies of aerosol samplers for virus-containing aerosols. *J Aerosol Sci* 2005;**36**(5–6):593–607.
13. Kim SW, Ramakrishnan M, Raynor PC, Goyal SM. Effects of humidity and other factors on the generation and sampling of a coronavirus aerosol. *Aerobiologia* 2007;**23**(4):239–48.
14. Ijaz M, Karim Y, Sattar S, Johnson-Lussenburg C. Development of methods to study the survival of airborne viruses. *J Virol Methods* 1987;**18**(2–3):87–106.
15. Van Droogenbroeck C, Van Risseghem M, Braeckman L, Vanrompay D. Evaluation of bioaerosol sampling techniques for the detection of *Chlamydophila psittaci* in contaminated air. *Vet Microbiol* 2009;**135**(1–2):31–7.
16. Li J, Leavey A, Wang Y, et al. Comparing the performance of 3 bioaerosol samplers for influenza virus. *J Aerosol Sci* 2018;**115**:133–45.
17. Jonges M, Van Leuken J, Wouters I, Koch G, Meijer A, Koopmans M. Wind-mediated spread of low-pathogenic avian influenza virus into the environment during outbreaks at commercial poultry farms. *PLoS One* 2015;**10**(5):e0125401.
18. Sawyer MH, Chamberlin CJ, Wu YN, Aintablian N, Wallace MR. Detection of varicella-zoster virus DNA in air samples from hospital rooms. *J Infect Dis (JID)* 1994;**169**(1):91–4.
19. Fabian P, McDevitt J, Houseman E, Milton D. Airborne influenza virus detection with four aerosol samplers using molecular and infectivity assays: considerations for a new infectious virus aerosol sampler. *Indoor Air* 2009;**19**(5):433.
20. Li H-W, Wu C-Y, Tepper F, Lee J-H, Lee CN. Removal and retention of viral aerosols by a novel alumina nanofiber filter. *J Aerosol Sci* 2009;**40**(1):65–71.
21. Zhao Y, Aarnink A, Wang W, Fabri T, Groot Koerkamp PW, de Jong MC. Airborne virus sampling-Efficiencies of samplers and their detection limits for infectious bursal disease virus (IBDV). *Ann Agric Environ Med* 2014;**21**(3).
22. Andersen AA. New sampler for the collection, sizing, and enumeration of viable airborne particles. *J Bacteriol* 1958;**76**(5):471.
23. Booth TF, Kournikakis B, Bastien N, et al. Detection of airborne severe acute respiratory syndrome (SARS) coronavirus and environmental contamination in SARS outbreak units. *J Infect Dis* 2005;**191**(9):1472–7.
24. Gralton J, Tovey E, McLaws M-L, Rawlinson WD. The role of particle size in aerosolised pathogen transmission: a review. *J Infect* 2011;**62**(1):1–13.
25. Xu Z, Wu Y, Shen F, Chen Q, Tan M, Yao M. Bioaerosol science, technology, and engineering: past, present, and future. *Aerosol Sci Technol* 2011;**45**(11):1337–49.
26. Riemenschneider L, Woo MH, Wu CY, et al. Characterization of reaerosolization from impingers in an effort to improve airborne virus sampling. *J Appl Microbiol* 2010;**108**(1):315–24.

27. Hogan Jr C, Kettleson E, Lee MH, Ramaswami B, Angenent L, Biswas P. Sampling methodologies and dosage assessment techniques for submicrometre and ultrafine virus aerosol particles. *J Appl Microbiol* 2005;**99**(6):1422–34.
28. Lednicky J, Pan M, Loeb J, et al. Highly efficient collection of infectious pandemic influenza H1N1 virus (2009) through laminar-flow water based condensation. *Aerosol Sci Technol* 2016;**50**(7):i–iv.
29. Cooper CD, Alley FC. *Air pollution control: a design approach.* Waveland Press; 2010.
30. Kenny L, Thorpe A, Stacey P. A collection of experimental data for aerosol monitoring cyclones. *Aerosol Sci Technol* 2017;**51**(10):1190–200.
31. Cao G, Noti J, Blachere F, Lindsley W, Beezhold D. Development of an improved methodology to detect infectious airborne influenza virus using the NIOSH bioaerosol sampler. *J Environ Monit* 2011;**13**(12):3321–8.
32. Corzo CA, Allerson M, Gramer M, Morrison RB, Torremorell M. Detection of airborne influenza a virus in experimentally infected pigs with maternally derived antibodies. *Transbound Emerg Dis* 2014;**61**(1):28–36.
33. Alonso C, Raynor PC, Davies PR, Torremorell M. Concentration, size distribution, and infectivity of airborne particles carrying swine viruses. *PLoS One* 2015;**10**(8):e0135675.
34. Xu Z, Yao M. Monitoring of bioaerosol inhalation risks in different environments using a six-stage Andersen sampler and the PCR-DGGE method. *Environ Monit Assess* 2013;**185**(5):3993–4003.
35. Kettleson EM, Ramaswami B, Hogan Jr CJ, et al. Airborne virus capture and inactivation by an electrostatic particle collector. *Environ Sci Technol* 2009;**43**(15):5940–6.
36. Zhen H, Han T, Fennell DE, Mainelis G. Release of free DNA by membrane-impaired bacterial aerosols due to aerosolization and air sampling. *Appl Environ Microbiol* 2013;**79**(24):7780–9.
37. Hong S, Bhardwaj J, Han C-H, Jang J. Gentle sampling of submicrometer airborne virus particles using a personal electrostatic particle concentrator. *Environ Sci Technol* 2016;**50**(22):12365–72.
38. Hatagishi E, Okamoto M, Ohmiya S, et al. Establishment and clinical applications of a portable system for capturing influenza viruses released through coughing. *PLoS One* 2014;**9**(8):e103560.
39. Kim S-H, Chang SY, Sung M, et al. Extensive viable Middle East respiratory syndrome (MERS) coronavirus contamination in air and surrounding environment in MERS isolation wards. *Rev Infect Dis* 2016;**63**(3):363–9.
40. Liu Y, Ning Z, Chen Y, et al. Aerodynamic analysis of SARS-CoV-2 in two Wuhan hospitals. *Nature* 2020;**582**(7813):557–60.
41. Guo Z-D, Wang Z-Y, Zhang S-F, et al. Aerosol and surface distribution of severe acute respiratory syndrome coronavirus 2 in hospital wards, Wuhan, China, 2020. *Emerg Infect Dis* 2020;**26**(7):10.3201.
42. Cheng VC-C, Wong S-C, Chan VW-M, et al. Air and environmental sampling for SARS-CoV-2 around hospitalized patients with coronavirus disease 2019 (COVID-19). *Infect Control Hosp Epidemiol* 2020:1–32.
43. Ong SWX, Tan YK, Chia PY, et al. Air, surface environmental, and personal protective equipment contamination by severe acute respiratory syndrome coronavirus 2 (SARS-CoV-2) from a symptomatic patient. *Jama* 2020;**323**(16):1610–2.
44. Burton NC, Grinshpun SA, Reponen T. Physical collection efficiency of filter materials for bacteria and viruses. *Ann Occup Hyg* 2007;**51**(2):143–51.
45. Lindsley WG, Green BJ, Blachere FM, et al. *Sampling and characterization of bioaerosols. NIOSH manual of analytical methods.* 5th ed. Cincinnati (OH): National Institute for Occupational Safety and Health; 2017.

46. Pan M, Lednicky JA, Wu CY. Collection, particle sizing and detection of airborne viruses. *J Appl Microbiol* 2019;**127**(6):1596–611.
47. Chia PY, Coleman KK, Tan YK, et al. *Detection of air and surface contamination by severe acute respiratory syndrome coronavirus 2 (SARS-CoV-2) in hospital rooms of infected patients.* medRxiv; 2020.
48. Faridi S, Niazi S, Sadeghi K, et al. A field indoor air measurement of SARS-CoV-2 in the patient rooms of the largest hospital in Iran. *Sci Total Environ* 2020:138401.
49. Razzini K, Castrica M, Menchetti L, et al. SARS-CoV-2 RNA detection in the air and on surfaces in the COVID-19 ward of a hospital in Milan, Italy. *Sci Total Environ* 2020;**742**:140540.
50. Organization WH. *Surface sampling of coronavirus disease (COVID-19): a practical "how to" protocol for health care and public health professionals.* World Health Organization; 2020.
51. McFee R. *COVID-19: TESTING-laboratory testing-CDC guidelines.* Disease-a-Month; 2020.
52. Mouchtouri VA, Koureas M, Kyritsi M, et al. Environmental contamination of SARS-CoV-2 on surfaces, air-conditioner and ventilation systems. *Int J Hyg Environ Health* 2020;**230**:113599.
53. Peyrony O, Ellouze S, Fontaine J-P, et al. Surfaces and equipment contamination by severe acute respiratory syndrome coronavirus 2 (SARS-CoV-2) in the emergency department at a university hospital. *Int J Hyg Environ Health* 2020;**230**:113600.
54. Lv J, Yang J, Xue J, Zhu P, Liu L, Li S. Detection of SARS-CoV-2 RNA residue on object surfaces in nucleic acid testing laboratory using droplet digital PCR. *Sci Total Environ* 2020;**742**:140370.
55. Wu S, Wang Y, Jin X, Tian J, Liu J, Mao Y. Environmental contamination by SARS-CoV-2 in a designated hospital for coronavirus disease 2019. *Am J Infect Control* 2020;**48**(8):910–4.
56. Ye G, Lin H, Chen L, et al. Environmental contamination of SARS-CoV-2 in healthcare premises. *J Infect* 2020;**81**(2):e1–5.
57. Chia PY, Coleman KK, Tan YK, et al. Detection of air and surface contamination by SARS-CoV-2 in hospital rooms of infected patients. *Nat Commun* 2020;**11**(1):1–7.
58. Santarpia JL, Rivera DN, Herrera VL, et al. Aerosol and surface contamination of SARS-CoV-2 observed in quarantine and isolation care. *Sci Rep* 2020;**10**(1):1–8.
59. Tan L, Ma B, Lai X, et al. Air and surface contamination by SARS-CoV-2 virus in a tertiary hospital in Wuhan, China. *Int J Infect Dis* 2020;**99**:3–7.
60. Foladori P, Cutrupi F, Segata N, et al. SARS-CoV-2 from faeces to wastewater treatment: what do we know? A review. *Sci Total Environ* 2020;**743**:140444.
61. Randazzo W, Truchado P, Cuevas-Ferrando E, Simón P, Allende A, Sánchez G. SARS-CoV-2 RNA in wastewater anticipated COVID-19 occurrence in a low prevalence area. *Water Res* 2020:115942.
62. Zhang D, Ling H, Huang X, et al. Potential spreading risks and disinfection challenges of medical wastewater by the presence of Severe Acute Respiratory Syndrome Coronavirus 2 (SARS-CoV-2) viral RNA in septic tanks of Fangcang Hospital. *Sci Total Environ* 2020;**741**:140445.
63. Wu F, Xiao A, Zhang J, et al. *SARS-CoV-2 titers in wastewater are higher than expected from clinically confirmed cases.* medRxiv; 2020.
64. Kumar M, Patel AK, Shah AV, et al. First proof of the capability of wastewater surveillance for COVID-19 in India through detection of genetic material of SARS-CoV-2. *Sci Total Environ* 2020;**746**:141326.
65. Or IB, Yaniv K, Shagan M, et al. *Regressing SARS-CoV-2 sewage measurements onto COVID-19 burden in the population: a proof-of-concept for quantitative environmental surveillance.* medRxiv; 2020.

66. Wurtzer S, Marechal V, Mouchel J-M, et al. *Evaluation of lockdown impact on SARS-CoV-2 dynamics through viral genome quantification in Paris wastewaters*. MedRxiv; 2020.
67. Green H, Wilder M, Middleton FA, et al. *Quantification of SARS-CoV-2 and cross-assembly phage (crAssphage) from wastewater to monitor coronavirus transmission within communities*. medRxiv; 2020.
68. Ahmed W, Angel N, Edson J, et al. First confirmed detection of SARS-CoV-2 in untreated wastewater in Australia: a proof of concept for the wastewater surveillance of COVID-19 in the community. *Sci Total Environ* 2020:138764.
69. Kocamemi BA, Kurt H, Hacioglu S, Yarali C, Saatci AM, Pakdemirli B. *First data-set on SARS-CoV-2 detection for Istanbul Wastewaters in Turkey*. medRxiv; 2020.
70. Medema G, Heijnen L, Elsinga G, Italiaander R, Brouwer A. Presence of SARS-Coronavirus-2 RNA in sewage and correlation with reported COVID-19 prevalence in the early stage of the epidemic in The Netherlands. *Environ Sci Technol Lett* 2020;**7**(7):511−6.
71. Trottier J, Darques R, Mouheb NA, et al. Post-lockdown detection of SARS-CoV-2 RNA in the wastewater of Montpellier, France. *One Health* 2020;**10**:100157.
72. Rimoldi S, Stefani F, Gigantiello A, et al. *Presence and vitality of SARS-CoV-2 virus in wastewaters and rivers*. medRxiv; 2020. 05. 01.20086009. In:2020.
73. Haramoto E, Malla B, Thakali O, Kitajima M. *First environmental surveillance for the presence of SARS-CoV-2 RNA in wastewater and river water in Japan*. medRxiv; 2020.
74. Cheung KS, Hung IF, Chan PP, et al. Gastrointestinal manifestations of SARS-CoV-2 infection and virus load in fecal samples from the Hong Kong cohort and systematic review and meta-analysis. *Gastroenterology* 2020;**159**(1):81−95.
75. Jeong HW, Kim S-M, Kim H-S, et al. Viable SARS-CoV-2 in various specimens from COVID-19 patients. *Clin Microbiol Infect* 2020;**26**(11):1520−4.
76. Kim Y-I, Kim S-G, Kim S-M, et al. Infection and rapid transmission of SARS-CoV-2 in ferrets. *Cell Host Microbe* 2020;**27**(5):704−9.
77. Olaimat AN, Shahbaz HM, Fatima N, Munir S, Holley RA. Food safety during and after the era of Covid-19 pandemic. *Front Microbiol* 2020;**11**:1854.
78. Rizou M, Galanakis IM, Aldawoud TM, Galanakis CM. Safety of foods, food supply chain and environment within the COVID-19 pandemic. *Trends Food Sci Technol* 2020;**102**:293−9.
79. Wu L, Yin W, Tang K, et al. Highly sensitive enzyme-free immunosorbent assay for porcine circovirus type 2 antibody using Au-Pt/SiO_2 nanocomposites as labels. *Biosens Bioelectron* 2016;**82**:177−84.
80. Ceylan Z, Meral R, Cetinkaya T. Relevance of SARS-CoV-2 in food safety and food hygiene: potential preventive measures, suggestions and nanotechnological approaches. *VirusDisease* 2020:1−7.
81. Pressman P, Naidu AS, Clemens R. COVID-19 and food safety: risk management and future considerations. In: *LWW*; 2020.
82. Yekta R, Vahid-Dastjerdi L, Norouzbeigi S, Mortazavian AM. Food products as potential carriers of SARS-CoV-2. *Food Control* 2020:107754.
83. Fisher D, Reilly A, Zheng AKE, Cook AR, Anderson D. *Seeding of outbreaks of COVID-19 by contaminated fresh and frozen food*. BioRxiv; 2020.
84. Rahmani AR, Leili M, Azarian G, Poormohammadi A. Sampling and detection of corona viruses in air: a mini review. *Sci Total Environ* 2020;**740**:140207.
85. Peng L, Liu J, Xu W, et al. *Novel Coronavirus can be detected in urine, blood, anal swabs and oropharyngeal swabs samples*. medRxiv; 2019. 2020.
86. Wang W, Xu Y, Gao R, et al. Detection of SARS-CoV-2 in different types of clinical specimens. *Jama* 2020;**323**(18):1843−4.
87. Falzone L, Musso N, Gattuso G, et al. Sensitivity assessment of droplet digital PCR for SARS-CoV-2 detection. *Int J Mol Med* 2020;**46**(3):957−64.

88. Seo G, Lee G, Kim MJ, et al. Rapid detection of COVID-19 causative virus (SARS-CoV-2) in human nasopharyngeal swab specimens using field-effect transistor-based biosensor. *ACS Nano* 2020;**14**(4):5135−42.
89. Santarpia JL, Rivera DN, Herrera V, et al. *Transmission potential of SARS-CoV-2 in viral shedding observed at the University of Nebraska Medical Center.* MedRxIV; 2020.
90. Ryu B-H, Cho Y, Cho O-H, Hong SI, Kim S, Lee S. Environmental contamination of SARS-CoV-2 during the COVID-19 outbreak in South Korea. *Am J Infect Control* 2020;**48**(8):875−9.
91. Kenarkoohi A, Noorimotlagh Z, Falahi S, et al. Hospital indoor air quality monitoring for the detection of SARS-CoV-2 (COVID-19) virus. *Sci Total Environ* 2020;**748**:141324.
92. Corman VM, Landt O, Kaiser M, et al. Detection of 2019 novel coronavirus (2019-nCoV) by real-time RT-PCR. *Euro Surveill* 2020;**25**(3):2000045.
93. Shirato K, Nao N, Katano H, et al. Development of genetic diagnostic methods for novel coronavirus 2019 (nCoV-2019) in Japan. *Jpn J Infect Dis* 2020;**73**(4):304−7. JJID. 2020.2061.
94. Kitajima M, Ahmed W, Bibby K, et al. SARS-CoV-2 in wastewater: state of the knowledge and research needs. *Sci Total Environ* 2020:139076.
95. Haramoto E, Kitajima M, Hata A, et al. A review on recent progress in the detection methods and prevalence of human enteric viruses in water. *Water Res* 2018;**135**:168−86.
96. Nao N, Shirato K, Katano H, Matsuyama S, Takeda M. *Detection of second case of 2019-nCoV infection in Japan (corrected version).* 2020.
97. World Health Organization. *Laboratory testing for 2019 novel coronavirus (2019-nCoV) in suspected human cases: interim guidance. (WHO/2019-nCoV/laboratory/2020.2), 14 January 2020.* Geneva: World Health Organization; 2020.
98. Control CfD, Prevention RVB, Division of Viral Diseases. *2019-novel coronavirus (2019-nCoV) real-time rRT-PCR panel primers and probes.* Department of Health & Human Services; 2020. p. 24.
99. La Rosa G, Iaconelli M, Mancini P, et al. First detection of SARS-CoV-2 in untreated wastewaters in Italy. *Sci Total Environ* 2020:139652.
100. Zhao B, Ni C, Gao R, et al. Recapitulation of SARS-CoV-2 infection and cholangiocyte damage with human liver ductal organoids. *Protein Cell* 2020:1−5.
101. Al Huraimel K, Alhosani M, Kunhabdulla S, Stietiya MH. SARS-CoV-2 in the environment: modes of transmission, early detection and potential role of pollution. *Sci Total Environ* 2020:140946.
102. Di Maria F, Beccaloni E, Bonadonna L, et al. Minimization of spreading of SARS-CoV-2 via household waste produced by subjects affected by COVID-19 or in quarantine. *Sci Total Environ* 2020;**743**:140803.
103. Huang WE, Lim B, Hsu CC, et al. RT-LAMP for rapid diagnosis of coronavirus SARS-CoV-2. *Microb Biotechnol* 2020;**13**(4):950−61.
104. Go YY, Kim Y-S, Cheon S, et al. Evaluation and clinical validation of two field−deployable reverse transcription-insulated isothermal PCR assays for the detection of the middle east respiratory syndrome−coronavirus. *J Mol Diagn* 2017;**19**(6):817−27.
105. Hashemzadeh MS, Rasouli R, Zahraei B, et al. Development of dual TaqMan based one-step rRT-PCR assay panel for rapid and accurate diagnostic test of MERS-CoV: a novel human coronavirus, ahead of Hajj pilgrimage. *Iran Red Crescent Med J* 2016;**18**(11).
106. Jalandra R, Yadav AK, Verma D, et al. Strategies and perspectives to develop SARS-CoV-2 detection methods and diagnostics. *Biomed Pharmacother* 2020:110446.
107. Dong L, Zhou J, Niu C, et al. *Highly accurate and sensitive diagnostic detection of SARS-CoV-2 by digital PCR.* medRxiv; 2020.

CHAPTER 3

Mechanism and transmission routes of COVID-19

Lokeshwaran Kirubananthan[1], Ramanaiah Illuri[2], Ramesh Rajendran[3], Prasanna Raja Chandrasekaran[4]
[1]Research and Development, Bhat Biotech India (P) Limited, Bangalore, Karnataka, India; [2]Department of Pharmacology, Vidya Herbs Pvt. Ltd, Bangalore, Karnataka, India; [3]Centre for Bioseparation Technology, Vellore Institute of Technology, Vellore, Tamil Nadu, India; [4]Department of In Vitro Pharmacology, Natural Remedies Private Limited, Bangalore, Karnataka, India

3.1 Origin and transmission of Sars-CoV-2

Acute respiratory disease caused by novel coronavirus, SARS-CoV-2, is a β-coronavirus with a nonsegmented genome consisting of positive-sense, single-stranded RNA enclosed within the viral capsid. Coronaviruses (CoVs) are divided into four genera, such as α, β, γ and δ-CoV; α- and β-CoV cause infection in mammals, while γ- and δ- CoV may cause infection in birds. Six forms of coronaviruses have been shown to cause infections in humans, namely HCoV-229E, SARS-CoV, HCoV-OC43, HCoV-NL63 MERS-CoV, and HCoV-HKU1. Infection with HCoV-229E, HCoV-NL63, HCoV-HKU1, and HCoV-OC43 are usually characterized by moderate respiratory symptoms while SARS-CoV and MERS-CoV are known to cause severe respiratory illness advancing to fatal conditions like multiple organ failure in some of the cases.[1] Genome sequencing data has revealed that SARS-CoV-2 is closely linked to bat-derived SARS-like coronavirus, bat-SL-CoVZC45, and bat-SL-CoVZXC21, with >85% similarity, but with reduced homology of 79% and 50% with SARS-CoV and MERS-CoV, respectively.[2,3] Based on these results and phylogenetic analysis, it is likely that SARS-CoV-2 originated in bats and was probably spread to humans through an (unidentified) intermediate host animal. The genome structure, the encoded structural and nonstructural proteins, and the key host of SARS-CoV-2 are illustrated in Fig. 3.1. Experimental evidences have clearly shown that SARS-CoV-2 infects humans by binding to angiotensin-converting enzyme 2 (ACE2) expressed in the respiratory tract, in a mechanism similar to that of. SARS-CoV.[4]

Environmental Management of COVID-19
ISBN 978-0-323-85780-2
https://doi.org/10.1016/B978-0-323-85780-2.00013-5

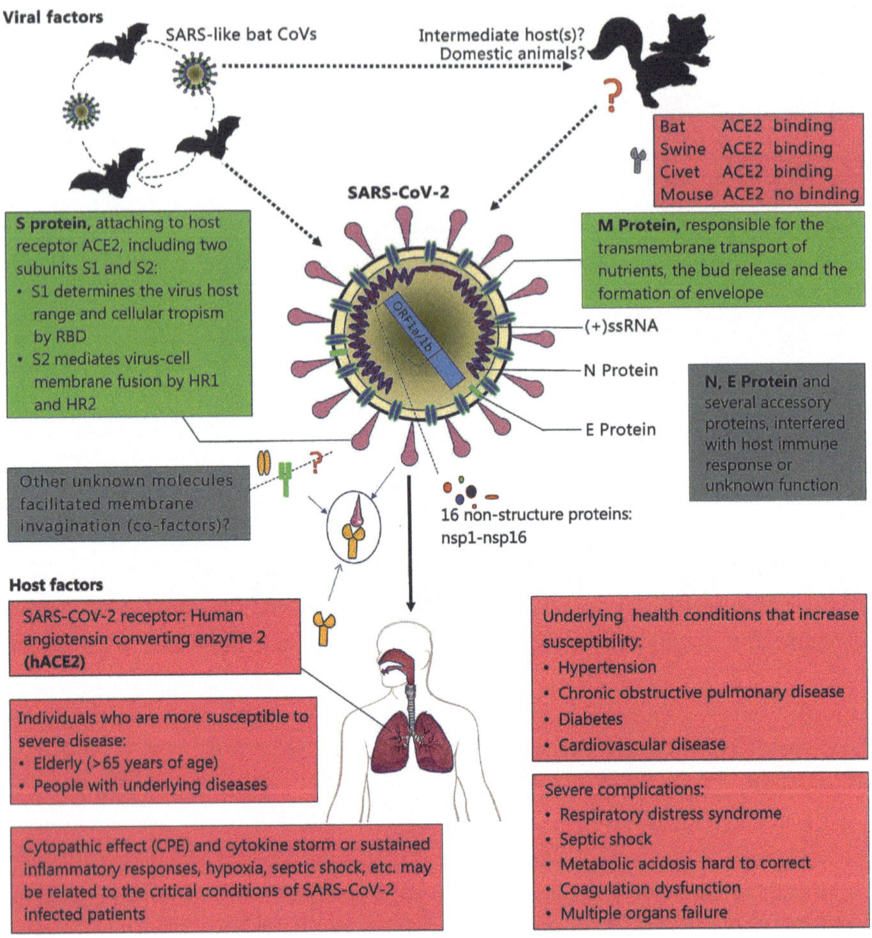

Figure 3.1 Top panel depicts the structural and genomic features of SARS-CoV-2. The genetic segment ORF 1a/b is known to code for various nonstructural proteins (nsp 1–16). The remainder genetic material codes for different structural proteins: spike glycoprotein (S), envelope (E), matrix (M), and nucleocapsid protein (N). Bottom panel represents the different host conditions that can affect the susceptibility and severity of SARS-CoV-2 infection. *(Image reproduced with kind permission from Springer Nature, New York, USA.[5])*

SARS-CoV-2 is found to exhibit higher infectivity than SARS and MERS but has less virulence in terms of morbidity and mortality. Evidences indicate that COVID-19 had a mortality rate of 3.4%, while SARS and MERS had a mortality rate of 9.6% and 35%, respectively. Human-to-human transmission of COVID-19 majorly occurs inside the members of a family and those who are closely associated (friends) with the infected person.[5]

Several studies indicated that symptomatic individuals are the primary cause of COVID-19 transmission. It majorly spreads among humans by respiratory droplets during a cough or sneezes from an infected individual.[6] Transmission of SARS-CoV and MERS-CoV among health care workers was reported to occur predominantly through the nosocomial transmission. Infection among health care staff accounted for 33%−42% of SARS cases, while it was 62%−79% during the MERS-CoV outbreak.[7] Viral transmission is believed to have occurred through direct contact with the host or interaction with the (unidentified) intermediary carrier. Moreover, it is also clear that asymptomatic individuals could also transmit the virus. It has been reported that elderly individuals represent a specific group of patients with an elevated risk of infection with swift clinical deterioration.[8] Additional experiments are required to explain the transmission pathways, the incubation time and the period of COVID-19 infectivity.

Mutations are common in the SARS-CoV-2 genome because the virus replicates and socializes in the anthropological population. They mount up at a rate of about one to two mutations per month in worldwide phylogeny. The new SARS-CoV-2 variant named VUI-202012/01 is the first "Variant Under Investigation" in December 2020 and is specified by a set of 17 changes or mutations. As a result of this ongoing process, many thousands of mutations have already occurred in the SARS-CoV-2 genome since the virus emerged in 2019.[9] Although SARS-CoV-2 shows evidence of some seasonal decline, the persistence of the pandemic may allow the accumulation of immunologically relevant mutations in the population as vaccines develop.[10,11] As mutations continue to develop, new combinations are increasingly observed. Over the last few weeks, the United Kingdom (UK) has faced a rapid increase in COVID-19 cases, especially in South East England, leading to enhanced epidemiological and virological investigations as tracked by COVID-19 genomics—UK consortium (COG-UK). Analysis of viral genome sequence data showed that a large proportion of cases belonged to a new single phylogenetic cluster. The new variant is defined by multiple spike protein mutations (deletion 69−70, deletion 144, N501Y, A570D, D614G, P681H, T716I, S982A, and D1118H) as well as mutations present in other genomic regions. Although it is known and expected that viruses are constantly mutating, leading to the emergence of new variants, preliminary analysis in the UK suggests that this variant is significant—more communicable than previously circulating variants, with an estimated potential to increase the reproductive number by 0.4 or more, with an estimated increased transmutability of up to 70%. This new variant emerged at the time of the year when family and social mixing has traditionally been increased; however, at this point, there is no indication of increased severity of infection associated with the new variant.[12]

3.2 Incidence and risk factors for COVID-19 severity

The COVID-19 pandemic has become a severe burden on the global health care system due to the exponentially mounting infections and death.[13] Clinical and epidemiological data from different populations have provided vital information about the factors determining the COVID-19 susceptibility. The patients diagnosed with COVID-19, significantly older patients, often had a coexisting condition such as cardiovascular disorder, hypertension, and cardiovascular disease.[14] Emerging reports showed that older age (>65 years) and the incidence of comorbidities are significantly associated with the COVID-19 severity.[15] Some of the reported comorbidities are listed in Table 3.1. It is evident that hypertension, diabetes, and cardiovascular disease are comparatively more prevalent and represented a significant proportion of the comorbidities in COVID-19.[16–18] Chronic lung disease, chronic kidney disease, and cancer are some of the other comorbidities reported in SARS-CoV-2 infected patients.[19,20]

3.2.1 Hypertension

Hypertension is a severe condition in which the walls of the arterial blood vessels persistently experience elevated pressure. If left untreated, the results

Table 3.1 Some of the prevalent comorbidities in COVID-19.

Incidence of comorbidities	References
Hypertension (21%), Diabetes (11%), and Cardiovascular diseases (CVD)(7%)	Singh et al.[17]
Hypertension (15.8%), CVD (11.7%), Diabetes (9.4%), Cancer(1.5%), Coronary Obstructive Pulmonary Disorder (COPD) (1.4%), and Chronic Kidney Disease (CKD) (0.8%)	Sanyaolu et al.[19]
Hypertension (30%), Diabetes (19%), Cardiovascular disease (8%), COPD (3%), Cancer (1%), and CKD (1%)	Zhou et al.[20]
Hypertension (15%), Diabetes (7.4%), Coronary heart disease (2.5%), Cerebrovascular disease (1.4%), COPD (1.1%), Cancer (0.9%), and CKD (0.7%)	Guan et al.[16]
Hypertension (31.2%), CVD (14.5%), Diabetes (10.1%), Cancer (7.2%), Cerebrovascular disease (5.1%), COPD (2.9%), CKD (2.9%), and Chronic liver disease (2.9%)	Wang et al.[18]

could be deleterious, leading to ischemic heart disease or stroke.[21] Approximately 18 million global cardiovascular deaths result from hypertension.[22] Epidemiological studies show that hypertension is more common in COVID-19 patients among the other prevailing conditions viz. diabetes and cardiovascular disease. A Chinese COVID-19 study reported the prevalence of three significant comorbidities in the infected patients: hypertension (30), diabetes (19%), and coronary heart disease (8%).[20] Wu et al. reported a higher incidence of hypertension (23%) in COVID-19 patients who suffered acute respiratory distress syndrome (ARDS) compared to other comorbidities in the study such as diabetes and cardiovascular disease (CVD).[23] A large COVID-19 case-control study demonstrated a 6% case fatality rate (CFR) in patients with hypertension as compared to the total case fatality rate (2.3%) of the subjects.[24] In a subsequent Chinese population study, it was reported that 35.8% of COVID-19 patients with hypertension reached the composite endpoints (intensive care unit [ICU], mechanical ventilation, or death) as compared to the 13.7% nonhypertensive group.[16] From the above reports, a causal association between hypertension and COVID-19 is evident, but the underlying pathological mechanism remains ambiguous. Further research in the area is warranted for a better understanding of the pathophysiological role in hypertensive COVID-19 patients.

The SARS-CoV-2 is shown to primarily infect the respiratory tract of humans. The outer transmembrane spike glycoprotein of SARS-CoV-2 interacts with ACE2 expressed in the respiratory tract epithelium, thus facilitating the viral entry.[4] ACE inhibitors (ACEI) and angiotensin receptor blockers (ARB), the contemporary drugs used in the treatment of hypertension, have been shown to elevate ACE2 expression levels in experimental models.[25] This has raised concerns on whether the treatment would increase the susceptibility to SARS-CoV-2 infection.[26,27] Due to the lack of clear evidence of the detrimental effects of ACEI and ARB in COVID-19, the American Heart Association and the European Society of Hypertension has recommended the continuation of these drugs.[28]

Interestingly, there are reports on the protective effects of ARB/ACEI in experimental studies. The potential downstream effects on the usage of ACEI and ARB are depicted in Fig. 3.2. ACEI is shown to reduce the formation of angiotensin 2, thus alleviating the proinflammatory effects. The action of ARB blocks the angiotensin 2 receptors thereby incidentally increasing the levels of angiotensins 1 to 7 and thus promoting the antiinflammatory effects.[25] When treated with ARB, mice models with avian H7N9 influenza virus-induced lung injury showed improvement.[29]

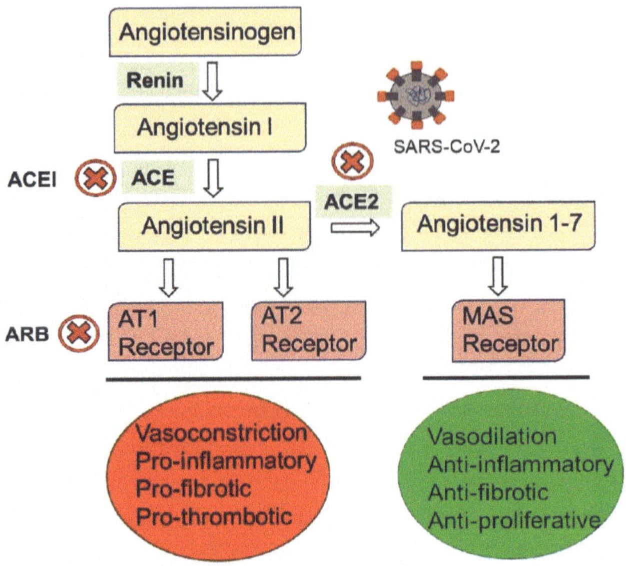

Figure 3.2 Potential effects of ACE inhibitors (ACEI) and angiotensin receptor blockers (ARB) on the renin-angiotensin system. ACE-angiotensin-converting enzyme, ACE-2 angiotensin-converting enzyme 2, AT1/AT2-angiotensin 2 receptors.[25]

Overall, the usage of ACE inhibitors and ARBs could potentially reduce inflammation in the lungs, kidneys, and liver. These drugs can have protective effects on ARDS, myocarditis, and acute kidney injury that are frequent in severe COVID-19 cases.[30,31] Also, a recently concluded study reported that the use of ARB/ACE inhibitors in the hypertensive COVID-19 cohort demonstrated lower mortality compared to the non-ARB/ACE user group.[32] Another case-control study by Liu et al. reported that there was no significant association between the exposure to antihypertensive drugs and COVID-19 severity.[33]

3.2.2 Diabetes

COVID-19 has affected a substantial proportion of diabetic patients apart from patients with hypertension and CVD. With the current data, it is difficult to conclude whether diabetes alone contributes to the disease severity since other factors like age, obesity, and hypertension are also contributing factors. A report from the Centers for Disease Control (CDC) response team, United States[34], showed the prevalence of diabetes in 10.9% of the COVID-19 patients. In a Chinese COVID-19 study, it was reported that 22% of the nonsurvivors were diabetic as compared to the 10% diabetic survivors.[35] In another study, the authors reported that 22.2% of the diabetic COVID-19 patients required ICU care in comparison to the 5.9% nondiabetic COVID-19 controls.[18] The Chinese CDC report on 44672 COVID-19 patients found that the CFR for diabetes was higher (7.3%) than that of hypertension (6%) but lower than CVD (10.5%).[36]

Diabetes is characterized by chronic hyperglycemic conditions arising from impaired insulin secretion or insulin action.[37] An earlier study on SARS has demonstrated diabetes and plasma glucose levels as independent predictors of mortality and morbidity.[38] Several micro- and macrovascular complications observed in diabetic subjects are shown to be significant predictors for the severe outcomes in different viral infections including influenza A virus subtype H1N1 and MERS-CoV.[39,40] As discussed earlier, the binding of ACE2 to spike glycoprotein of SARS-CoV-2 facilitates viral entry into the cells. Drugs that are used to treat diabetes including glucagon-like peptide-1 (GLP-1) agonists, thiazolidinediones, statins, ACEI, and ARB, are known to elevate ACE2 expression.[41–43] Interestingly, a Mendelian randomization study by Rao et al.[44] revealed that elevated ACE2 expression in the lungs was significantly associated with diabetes. The effect of metformin on diabetic COVID-19 patients remains to be studied. However, the use of metformin was shown to alleviate the adverse of diabetes in chronic obstructive pulmonary disease (COPD) patients.[45] This hypothetically suggests a protective role for metformin in diabetic COVID-19 patients since SARS-CoV-2 primarily affects the lungs, causing severe respiratory illness. Furin is a type-1, membrane-bound cellular protease that facilitates the entry of SARS-CoV-2 inside the host cells by cleaving the spike glycoprotein bound to the cellular receptor. Thus, an elevated furin level in diabetic patients could increase the susceptibility to SARS-CoV-2 infection.[46] ACE and ACE2 are two critical regulatory enzymes of the renin-angiotensin system (RAS) (Fig. 3.2). SARS-CoV-2 is known to bind to the ACE2, and this action could potentially result in the accumulation of angiotensin 2, resulting in downstream proinflammatory events. Elevated ACE/ACE2 activity is characteristic of COVID-19 patients with ARDS.[47]

Diabetic patients are known to exhibit impaired adaptive immunity. Abnormalities in T cell response, the impaired phagocytic capacity of macrophages, and delayed antigen presentation lead to chronic inflammation in diabetes.[48] The effect of MERS-CoV infection on experimental mice model with induced diabetes has revealed a significant reduction in CD4 T cell count and elevated levels of interleukin-17a (IL-17a).[49]

Similar observations of reduced CD4 and CD8 T cells (possibly due to lymphopenia) and elevated levels of proinflammatory Th17 cells and cytokines such as IL-6 were reported by Guan and colleagues.[16] Dysregulated T cell response could trigger a "cytokine storm" of proinflammatory cytokines such as tumor necrosis factor- α (TNF-α), IL-6, and interferon-γ (IFN-γ). This hyperinflammation is associated with COVID-19 severity and could lead to multiple organ failure observed in severe cases.[50] Also, diabetic patients are known to exhibit an inherently impaired innate immunity, affecting neutrophil functions such as phagocytosis and chemotaxis.[51] Thus, the synergistic effect of a dysfunctional immune system, elevated ACE2 expression, furin, and the effects of diabetic medications appears to increase the susceptibility and severity in diabetic COVID-19 patients. Diabetes management during the pandemic has become a daunting task for health care professionals (and for the patients) due to the limited information available. Constant glucose monitoring, personalized treatment strategies for diabetic COVID-19 patients, and careful monitoring of drug effects are some of the advocated measures for diabetes management.[52]

3.2.3 Cardiovascular disease

CVDs are considered the foremost contributor to global death than any other disease.[53] The incidence of CVD in COVID-19 is preceded by hypertension and diabetes (Table 3.1). Evidence shows that 7%–19% of COVID-19 patients presented with a history of preexisting cardiovascular complications.[16,20] CVD was found to be the second most prevalent comorbidity, next only to hypertension in some of the reports.[18,19] CDC data from the United States shows that 29% of COVID-19 patients with preexisting CVD required ICU admissions, which were concordant with the Wuhan-based Chinese report of 25% ICU admission.[18,34] Wu and McGoogan[36] reported a 10.5% CFR for CVD in COVID-19 patients compared to 7.3% for diabetes and 6% for hypertension. Shi et al.[54] reported that the risk of in-hospital mortality in COVID-19 patients was associated with the presence of cardiovascular comorbidity. COVID-19 patients are shown to experience multiple cardiac complications. Some of the probable mechanisms leading to cardiac complications and also contributing risk factors are illustrated in Fig. 3.3.

COVID-19 infection is shown to proceed in three phases.[55] The initial phase is marked by the viral entry and proliferation inside the lung cells.

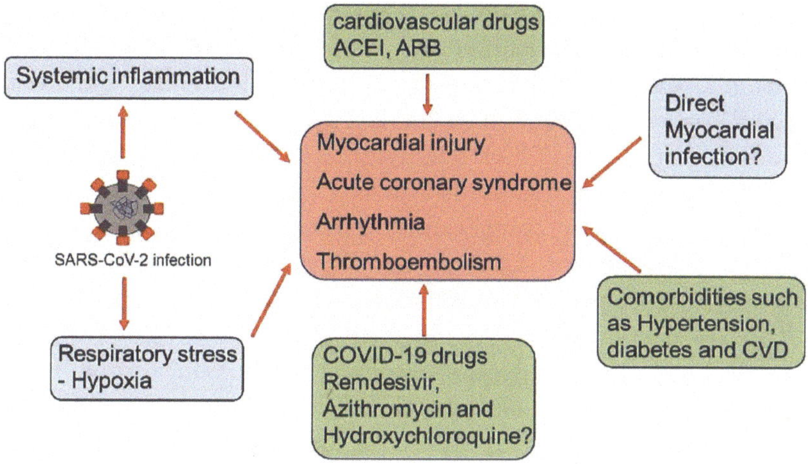

Figure 3.3 Potential mechanisms and risk factors contributing to cardiovascular damage in COVID-19 patients.[59,61,64,65]

Damage during the early phase is evident mainly in the respiratory and gastrointestinal tract with biomarkers such as elevated levels of lactate dehydrogenase (LDH), ferritin, high sensitivity C-reactive protein (hs-CRP), interleukin-6 (IL-6), and mild lymphopenia.[56] The intermediate or pulmonary phase is manifested by shortness of breath, hypoxia due to pulmonary damage, and cardiovascular stress. The tissue damage activates an inflammatory state marked by vasodilation, increased cellular permeability, and leukocyte accumulation.[55] The two significant biomarkers of the intermediate phase are the increased levels of brain natriuretic peptide (BNP) and troponin.[57] The intermediate phase is also evident from an abnormal computed tomography (CT) chest.[58] The inflammatory response of the intermediate stage gets amplified further resulting in a hyperinflammatory state. This stage is marked by a raised level of proinflammatory cytokines and cardiac biomarkers. This critical phase is evident from the fatal "cytokine storm" leading to multiple organ failure, ARDS and sepsis.[59] Xu and colleagues suggest that the heart damage observed in the COVID-19 cases was a result of a hyperactive immune system since no direct evidence of viral infiltration was found in the autopsy.[60]

Some of the patients without preexisting cardiovascular manifestations in COVID-19 are known to experience cardiac damage. Myocardial injury, acute coronary syndrome (ACS), cardiac arrhythmia, and venous thromboembolism are some of the cardiovascular manifestations reported in

SARS-CoV-2 infected individuals.[61] Myocardial injury is the one most prominent cardiovascular manifestation observed in SARS-CoV-2 infected patients. A Chinese study comprising 138 patients found elevated levels of cardiac injury-related biomarkers in COVID-19 patients. The authors observed elevated levels of high sensitivity cardiac troponin I (hs-cTnI) and echocardiographic (ECG) abnormalities in 7.2% of the patients. Also, 22% of patients with cardiac complications required ICU care.[18] Zheng and coworkers reported that 12% of the hospitalized COVID-19 patients without a history of CVD had elevated troponin levels and cardiac damage.[62] Another Chinese study observed that 12% of the COVID-19 patients had experienced myocardial injury and four out of five required ICU care.[63] Zhou and colleagues found that 17% of the COVID-19 subjects in the study had experienced acute cardiac injury with a high death rate of 97% in that particular group of patients.[20] In a similar study, Shi et al. reported that 30.6% of the nonsurvivors had suffered an acute myocardial injury and that the risk of mortality can be independently predicted using biomarkers of cardiac injury.[54]

Myocardial injury is one of the significant cardiac complications experienced by SARS-CoV-2 infected patients. A combined effect of the respiratory stress from hypoxia, cardiac infection, and injury sustained from the systemic inflammatory response is suspected of causing myocardial injury.[64] There is evidence of macrophage infiltration and a low-level of $CD4^{+}$ T cells potentially leading to inflammation-mediated cardiomyocyte damage.[60] Cardiovascular damage could probably result from direct SARS-CoV-2 infection of the heart tissue; however, it is yet to be proved with clinical evidence. ACE2 is also known to be expressed in myocardial pericytes, cardiomyocytes, and endothelial cells.[47] In an earlier study, Oudit and colleagues[65] have demonstrated the presence of the SARS genome in 7 out of 20 infected patients. However, the specific cell type(s) within the cardiac tissues that could be infected by SARS-CoV-2 and/or cell type(s) susceptible to systemic inflammatory attack remains unclear. Also, the degree of damages to the cardiac tissue likely from a direct viral infection or the indirect toxicity from systemic inflammation remains elusive.[61]

ACS is also reported in COVID-19 patients, primarily presenting as ST-elevation myocardial infarction (STEMI).[66] Virus-induced stress or systemic inflammation could lead to plaque rupture triggering myocardial infarction or ischemia.[62,64] A New York-based study[67] reported that in 18 patients with STEMI, variability in presentation, a higher prevalence of non-obstructive disease, and higher mortality was reported. In another study from Italy comprising 28 COVID-19 patients with STEMI, 17 patients presented with evidence of lesions in coronary angiography, demanding revascularization.[68] Activation of endothelial cells and smooth muscle cells, macrophage, expression of tissue factors in the plaque and platelet activation are factors known to trigger ACS.[55] Plaque rupture, coronary spasm, or microthrombi from systemic inflammation are some of the potential mechanisms of ACS in COVID-19.[67]

Cardiomyopathy is the characteristic thickening of heart muscles, making it difficult to pump blood to the parts of the body. Cardiomyopathy can lead to heart failure, blood clots, and cardiac arrest.[69] Zhou et al.[20] reported that heart failure was prominent in nonsurvivors (52%) of the COVID-19 study as compared to the survivors (12%). Heart failure could result from the exacerbation of preexisting conditions such as coronary artery disease or any other undiagnosed cardiac manifestation.[61] Chen et al.[70] reported that 50% of the deceased patients (due to heart failure) had a history of preexisting hypertension or CVD. Inflammatory and respiratory stress can lead to the exacerbation of preexisting cardiac disease or even lead to the onset of new cardiomyopathy.[57] ACS and myocardial injury can induce contractile dysfunction resulting in heart failure. The reduced diastolic function in older patients might trigger heart failure with preserved ejection fraction. Preserved ejection fraction caused by high fever, excessive hydration, and renal dysfunction could lead to acute heart failure.[61,66] Also, acute response to viral infection by the host immune system could also lead to stress-induced cardiomyopathy or systemic inflammation-mediated myocardial dysfunction.[64]

Cardiac arrhythmia is also reported in a group of COVID-19 patients. A Chinese study comprising 138 patients reported that cardiac arrhythmia was observed in 17% of the total subjects and a higher prevalence of 44% in the patients in ICU care.[18] Also, subsequent Wuhan-based COVID-19 study[71] observed that elevated levels of troponin T were associated with ventricular tachycardia/ventricular fibrillation, whereas a similar observation was not noted with normal troponin levels. Arrhythmia can be triggered by hypoxia, fever, electrolyte imbalance, antibiotics, and antiviral drugs.[61]

Abnormalities of the coagulation cascade and elevated thromboembolic events are also reported in COVID-19 patients.[57] Elevated D-dimer is indicative of deep-vein thrombosis, pulmonary embolism, and disturbed intravascular coagulation. Guan et al.[16] reported elevated levels of D-dimer, platelet abnormalities, and mildly extended prothrombin time in SARS-CoV-2 infected patients. The prevalence of hypercoagulation is evident from the increased thromboembolic events. An autopsy report by Wichmann et al.[72] showed deep-vein thrombosis in 7 out of 12 deceased COVID-19 patients while pulmonary embolism was observed in 4 out of 12 patients. The mechanism of these abnormalities remains unclear. However, the inflammation and endothelial damage along with preexisting comorbidities might increase the risk of coagulation abnormalities.[61]

3.2.4 Chronic obstructive pulmonary disease

COPD is a chronic illness of the respiratory system caused by the obstruction of normal airflow through the lungs due to alveolar damage and inflammation.[73] The prevalence of COPD in COVID-19 (1%–3%) is lower than hypertension or diabetes (Table 3.1). However, some of the reports suggest that COPD patients are more susceptible to COVID-19. A variable prevalence of COPD is reported in the hospitalized ICU patients from different regions. Reports from China reported a COPD incidence rate of 0%–10% whereas reports from New York reported an incidence rate of 2.4%–14%.[74] A recent meta-analysis study[75] showed that COPD is significantly associated with COVID-19 severity, with the COPD patients displaying a fivefold increased risk of severe infection (OR: 5.69, 95%; CI: 2.49–13.00). A Chinese nationwide study[76] found that COPD was significantly higher in severe COVID-19 cases (62.5%) compared to nonsevere cases (15.3%). Also, the authors reported that COPD was more prevalent in nonsurvivors (25%) of COVID-19 when compared to the survivors (2.8%).

Cigarette smoking is one of the proven risk factors of COPD. A case-control study by Guan et al.[16] reported a higher COVID-19 related death rate in smokers in comparison to nonsmokers (12.7% versus 4.7%). SARS-CoV-2 is known to enter the respiratory tract through the binding of spike glycoprotein to the ACE2 receptors expressed in the lung epithelial.[77] One of the predisposing factors of COPD for COVID-19 is the elevated levels of ACE2 expression observed in the COPD patients.[78] Tobacco smoking and nicotine exposure are also shown to elevate ACE2 expression in the lungs.[79] The intrinsic immune impairment in COPD could also lead to increased susceptibility to SARS-CoV-2 infection. COPD patients are known to exhibit impaired innate and adaptive immune responses to pathogens.[80] Impaired phagocytic capacity of alveolar macrophages, T cell dysfunction, the release of proinflammatory cytokines, and impaired interferon-β (IFN-β) signaling are some of the potential factors contributing to the severity in COPD patients.[81,82]

Corticosteroids are frequently used in the treatment of COPD. The use of these drugs in SARS and MERS has demonstrated unfavorable effects such as delay in viral clearance, increased mortality, and onset of psychosis.[83,84] Reports on the effect of corticosteroids on COVID-19 are limited. A recent UK-based drug trial found that the use of Dexamethasone reduced the mortality in COVID-19.[85] However, further research on larger cohorts in the different populations is necessary to determine the beneficial/detrimental outcome of Dexamethasone and similar corticosteroid drugs on COVID-19.

3.2.5 Renal diseases

Different reports indicate a lower incidence (0.7%–2.9%) of chronic kidney disease (CKD) in COVD-19.[16,18,23] Although less prevalent (Table 3.1), CKD is shown to be significantly associated with COVID severity. A systematic meta-analysis[86] revealed a strong association of COVID-19 severity in patients with CKD (OR: 3.03, 95%; CI: 1.0–8.4).

SARS-CoV-2 induced acute kidney injury (AKI) is also reported as a subset of COVID-19 patients. A case-control study by Cheng and colleagues[87] reported the biomarkers of renal abnormalities such as proteinuria (43.9%) and hematuria (26.7%) in the COVID-19 subjects. The authors also reported elevated levels of creatinine (14.4%), blood urea nitrogen (13.1%), and a glomerular filtration rate below 60 mL/min/1.73 m^2 in 13.1% in the COVID-19 subjects. In a separate COVID-19 study, it was reported that the prevalence of AKI among nonsurvivors (37.5%) was higher than that of the survivors (15%).[35] The two probable pathological mechanisms contributing to AKI are (1) direct renal damage induced by SARS-CoV-2 infection and/or (2) the systemic inflammation resulting from an activated immune system targeting the kidney. As discussed in the previous sections, organs expressing ACE2 could become the potential targets of SARS-CoV-2. ACE2 is also known to be expressed in the kidney and thus could result in kidney damage from virus-induced cytotropic effects.[88] The detection of SARS-CoV-2 nucleocapsid protein in histology samples and the observation of virus-like particles in the electron microscopy of renal structures indicates the direct vulnerability of the kidney to SARS-CoV-2 infection.[89] However, kidney autopsy analysis from six patients showed acute tubular necrosis due to lymphocyte and macrophage infiltration.[90] The systemic hyperinflammation due to the cytokine storm could potentially target the kidney resulting in AKI.

This mechanism could be linked to the T cell abnormalities, increased proinflammatory cytokines, and lymphopenia that are frequently observed in severe COVID-19 cases.[91]

Recipients of kidney transplants are also shown to be prone to SARS-CoV-2 infection. In a group of 36 COVID-19 positive kidney transplanted recipients, the authors reported a higher mortality rate (at 3 weeks) in kidney transplant recipients (19%–28%) as compared to the mortality rate in the COVID-19 infected general population (1%–5%).[92] Use of Immunosuppressive drugs posttransplantation, comorbidity, older age, and aggravation of systemic inflammatory response to viral infection are the potential contributors to the severity in kidney transplant recipients.[93]

3.2.6 Malignancy

Patients with different types of cancer can be susceptible to COVID-19 due to the immunocompromised nature of the disease resulting from chemotherapy or treatment with immunosuppressive drugs.[94] A large case-control study by Guan and colleagues[16] reported that cancer was prevalent in 0.9% of the COVID-19 patients. Two subsequent COVID-19 studies[18,35] reported a cancer prevalence of 4% and 7.2%, respectively. The effect of cancer on COVID-19 severity was demonstrated using Cox regression analysis by Liang et al.[95] The authors reported that COVID-19 patients with cancer reached the composite endpoints more rapidly (13 days) than those without cancer (43 days) (Hazard ratio: 3·56, 95%; CI: 1·65–7·69). A multicenter COVID-19 study by Dai and colleagues[96] observed a higher frequency of severe events in lung cancer, hematologic cancer, and stage IV metastatic cancer patients, while nonmetastatic cancer patients and patients without cancer showed similar frequency of severity. Also, surgery in cancer patients was shown to increase the severity of COVID-19 compared to those who received radiation therapy. Management strategies in cancer during COVID-19 included a systematic screening of patients for COVID-19 infection, postponing chemotherapy or surgery wherever possible, implementation of robust personal protection mechanisms in and around the patients under treatment, and intensive care for older patients with comorbidities.[76,97]

3.2.7 Management of COVID-19 patients with comorbidities

The treatment of COVID-19 patients has proven to be a challenging task due to the lack of a clinically proven treatment or SARS-CoV-2 vaccine as well as the quick spread of SARS-CoV-2 infection.[98] Given the unprecedented nature of the pandemic, researchers are evaluating the effects of SARS-CoV-2 infection in large populations and are in the process of devising efficient strategies to manage the pandemic. The preexisting comorbidities are shown to be a significant risk factor for COVID-19 morbidity and mortality.[99] The careful evaluation and categorization of comorbidities is an essential step in the management of COVID-19 patients. The clinical evidence pertaining to comorbidities among the COVID-19 patients (especially the elderly) has shown that these patients need to be precautious, and even in the event of hospitalization, require additional care.[100] Also, a different set of guidelines should be established in the hospitals for treating the patients with comorbidities especially those at high risk of severity (hypertension and CVD).[14] The treatment for the preexisting conditions should be continued without disruption until clear evidence of the effects of the particular drug(s) are demonstrated. In other cases, such as cancer and organ transplantation where the use of immunosuppressive drugs is unavoidable, additional care and precaution are mandated.[101,102] Also, care should be taken during surgery and blood transfusion to make sure the donor is uninfected.[103] Special care on personal protection of patients and health care professionals particularly in the outpatient departments (OPD) are mandatory.

3.3 SARS-CoV-2 transmission cycle

The human-to-human COVID-19 transmission occurs through either inhalation of bioaerosol or self-inoculation on the eyes and mouth from the contaminated fomites (surfaces). These two conditions are facilitated in several ways: direct contact, droplets, aerosols, fomite, fecal-oral, bloodborne, and mother-to-child. All the notorious viral diseases (tuberculosis, measles, and chickenpox), including SARS-CoV-2, are shown to spread between individuals primarily through droplets or aerosol and direct contact with infected patients.[104] More than 60% of the air samples obtained from the COVID-19 care centers and the air samples collected a few meters away from the patients produce viral loads (2.5 copies/L). The presence of the highest viral loads (19—48 copies/L) on a patient receiving oxygen nasal cannula during treatment in health care centers provides strong evidence for the airborne spread of COVID-19.[105]

3.3.1 Airborne transmission

Airborne transmission mainly falls into three categories, namely, obligate, preferential, and opportunistic. COVID-19 transmission occurs under preferential or opportunistic pathogenicity through droplets or aerosols. According to the World Health Organization (WHO) and CDC, droplets are 5−10 μm wide and aerosols are <5 μm wide, created by bronchoalveolar lavage aerosolizing procedures. Droplets and aerosols are released while sneezing, coughing, speaking, and during respiration from regular human activities. Larger droplets settle faster than they evaporate, carrying a higher viral load of the infected individual. Physical activities such as talking, coughing, and sneezing produce 3000 to 40,000 droplets lesser than 1 μm in size with 7×10^6 viral copies/ml from lungs rather than the upper respiratory tract. The average time is taken for droplet settlement to the ground ranging from 3 to 17 min with a higher transmission rate.[106] As the droplets are transmitted from the warm and moist environments of the respiratory system to the cooler and drier outside climate, they evaporate and form residual particles from the dry content of the initial droplets. These residual particles are referred to as droplet nucleus or aerosols. Such aerosols remain in the air until a few hours before they settle, bearing even less viral load (bioaerosols).[107] Droplets and aerosols generated from human expiratory activities can travel different distances at different speeds. Aerosols created during sneezing may reach up to 7−8 meters (m) horizontally with an air velocity of 50 m/s, while coughing can reach only 2 m with the same velocity. The exhaling events push the large droplets only up to 1 m horizontally at 1 m/s.[108]

The virulence capacity of aerosols depends on the viral shedding time and environmental factors, the infected patient's viral shedding is at a peak between 10- and 15-days postinfection. Aerosols generated during high viral shedding could contain active virus up to six days at 20°C and 50% relative humidity. In contrast, rapid virus inactivation occurs under 60 min, with only a 5% survival rate at 38°C and 24% relative humidity.[109] Depending on the patient's activity with the susceptible person—for example, through breathing, sneezing, and coughing—aerosols inhaled could cause infection only when it contain sufficient virus quantity. However, the viral dose of viable SARS-CoV-2 required to induce active infection in the host is not clearly understood.[110]

3.3.2 Fomite transmission

The object surfaces made up of plastic, stainless steel, copper, cardboard, aluminum, sterile latex surgical gloves, and sponges act as fomites (contaminated surfaces) when droplets and aerosols expelled from the infected individuals meet the object surfaces. The SARS-CoV-2 virus remains viable in these fomites, ranging from a few hours to days depending on the viral load, the environmental conditions, and the nature of the surface of the object.[111] Like SARS-CoV-1, SARS-CoV-2 remains viable in aerosols for more than 3 h, with the reduction of infectious titer due to time decay. SARS-CoV-2 is more stable in aerosols settled on plastic and stainless steel than in copper and cardboard. It has the maximum viable time of up to 72 h on plastic surfaces and the lowest time on copper of fewer than 4 h. The average half-life of SARS-CoV-2 in aerosols with median projections is about 1.1 h, the most extended half-life of the virus in stainless steel and plastics is 5.6 h in stainless steel and 6.8 h in plastics.[112] The spread of viral infection in fomite mode is more likely due to self-inoculation in the mouth, nose, or eyes following contact with the contaminant. Fomite transmission has the highest possibility in health care facilities, followed by public gathering areas.[111]

3.3.3 Other modes of transmission

SARS-CoV-2 remains in the digestive tract of infected patients ranging from several days to weeks. The excreta of such patients are the primary resource for cross-contamination through fecal-oral transmission, which is facilitated by a bathroom extract ventilation system, turning it into an airborne transmission route. The fecal-oral route of COVID-19 transmission happens through 5-F paths such as fomites, fingers, flies, fluids, and fields.[104] Household transmission contributes to about 23%—33% of secondary infection rate (SIR) through primary patient care and interaction. Precautionary methods, including isolation and contact-free communication with comorbid siblings, may significantly reduce the infection rate.[113] Unpacked foods and food processing during the farm-to-table process and preservation process may contribute to the transmission of the virus.[114,115] Animal forms, such as mink, are potential sources of an animal-to-human transmission; a genome study conducted on residents and employees of the form shown to have 68% infections were caused by the mink form.[116] Zoonotic transmission of the virus occurs through pet animals (dog and cat) of household or farm animals (Pig and rabbit) and their byproducts and handling processes.[117] SARS-CoV-2 is also shown to be present in biological fluids including serum, plasma, and urine up to 17 days after the onset of symptoms. Most considerable plasma viral load of SARS-CoV-2 occurred at four to eight days after the onset of the disease and decreased shortly after viral shedding. These virus-infected specimens might contribute to the possibility of transmission of

SARS-CoV-2 via cross-contamination. However, the odds of this mode of dissemination remain very low due to low viral titer content and half-life.[118] Appropriate and adequate ventilation, particle filtration and disinfection of the air, prevention of air recirculation, and avoiding overcrowding are the engineered controls that play a significant role in controlling COVID-19 transmission. The use of engineering safeguards in public collection facilities, including health care units, markets, industries, educational institutions, libraries, food courts, parks, conference rooms, public utility areas, or public transport, in addition to the use of personal protective equipment (PPE), isolation and quarantine, social distancing, and hand hygiene is vital to lower the chances of transmission between individuals.[105]

3.4 Conclusion

Coronavirus infects people of all ages; it is difficult to establish an early diagnosis as an infected person can remain asymptomatic for a period ranging from a few days to even weeks. The virus transmission predominantly occurs when the infected patient is (unknowingly) exposed to a healthy individual(s). Based on the available scientific evidence on the nature and transmission of SARS-CoV-2 virus infection, the prevention measures and management of SARS-CoV-2 infections differ depending on the person and the environmental conditions in which they reside.

The limited data available on novel SARS-CoV-2 indicates COVID-19 patients with comorbidities accounts for the higher deteriorating outcomes compared with patients without any comorbidity. The adverse impacts of the virus may vary with different disease conditions. Prolonged inflammation and immune cell dysfunction led by SARS-CoV-2 infection is the primary cause of concern in comorbid conditions. A detailed assessment or self-reporting of comorbid conditions during the viral infection can help physicians to establish risk assessment and customize the treatment. This strategic treatment plan for each individual may significantly improve the clinical outcome and success rate.

Personal care is vital in preventing viral infection, as the majority of the infection occurs through airborne transmission. PPE use and frequent washing of hands with soap or alcohol-based sanitizers greatly reduces the infection risk. Social distance and avoidance of public gatherings should be the foremost priority in coping with the pandemic.

References

1. Yin Y, Wunderink RG. MERS, SARS and other coronaviruses as causes of pneumonia. *Respirology* 2018;**23**(2):130–7.
2. Lu R, Zhao X, Li J, et al. Genomic characterisation and epidemiology of 2019 novel coronavirus: implications for virus origins and receptor binding. *Lancet* 2020;**395**(10224):565–74.
3. Ren LL, Wang YM, Wu ZQ, et al. Identification of a novel coronavirus causing severe pneumonia in human: a descriptive study. *Chin Med J* 2020;**133**(9):1015–24.
4. Li F, Li W, Farzan M, Harrison SC. Structure of SARS coronavirus spike receptor-binding domain complexed with receptor. *Science* 2005;**309**(5742):1864–8.
5. Guo YR, Cao QD, Hong ZS, et al. The origin, transmission and clinical therapies on coronavirus disease 2019 (COVID-19) outbreak – an update on the status. *Mil Med Res* 2020;**7**(1):11.
6. Rothan HA, Byrareddy SN. The epidemiology and pathogenesis of coronavirus disease (COVID-19) outbreak. *J Autoimmun* 2020;**109**:102433.
7. Kang CK, Song KH, Choe PG, et al. Clinical and epidemiologic characteristics of spreaders of Middle East respiratory syndrome coronavirus during the 2015 outbreak in Korea. *J Kor Med Sci* 2017;**32**(5):744–9.
8. Perrotta F, Corbi G, Mazzeo G, et al. COVID-19 and the elderly: insights into pathogenesis and clinical decision-making. *Aging Clin Exp Res* 2020;**32**(8):1599–608.
9. Wise J. Covid-19: new coronavirus variant is identified in UK. *BMJ* 2020;**371**:m4857.
10. Korber B, Fischer WM, Gnanakaran S, et al. Tracking changes in SARS-CoV-2 spike: evidence that D614G increases infectivity of the COVID-19 virus. *Cell* 2020;**182**(4):812–27. e819.
11. Sehra ST, Salciccioli JD, Wiebe DJ, Fundin S, Baker JF. Maximum daily temperature, precipitation, ultraviolet light, and rates of transmission of severe acute respiratory syndrome coronavirus 2 in the United States. *Clin Infect Dis* 2020;**71**(9):2482–7. An Official Publication of the Infectious Diseases Society of America.
12. European Centre for Disease Prevention and Control. *Rapid increase of a SARS-CoV-2 variant with multiple spike protein mutations observed in the United Kingdom*. ECDC; 2020.
13. Miller IF, Becker AD, Grenfell BT, Metcalf CJE. Disease and healthcare burden of COVID-19 in the United States. *Nat Med* 2020;**26**(8):1212–7.
14. Wang T, Du Z, Zhu F, et al. Comorbidities and multi-organ injuries in the treatment of COVID-19. *Lancet* 2020;**395**(10228):e52.
15. Oliveros E, Patel H, Kyung S, et al. Hypertension in older adults: assessment, management, and challenges. *Clin Cardiol* 2020;**43**(2):99–107.
16. Guan WJ, Ni ZY, Hu Y, et al. Clinical characteristics of coronavirus disease 2019 in China. *N Engl J Med* 2020;**382**(18):1708–20.
17. Singh AK, Gupta R, Misra A. Comorbidities in COVID-19: outcomes in hypertensive cohort and controversies with renin angiotensin system blockers. *Diab Metaboli Syndr* 2020;**14**(4):283–7.
18. Wang D, Hu B, Hu C, et al. Clinical characteristics of 138 hospitalized patients with 2019 novel coronavirus-infected pneumonia in Wuhan, China. *JAMA* 2020;**323**(11):1061–9.
19. Sanyaolu A, Okorie C, Marinkovic A, et al. Comorbidity and its impact on patients with COVID-19. *SN Compr Clin Med* 2020:1–8.
20. Zhou F, Yu T, Du R, et al. Clinical course and risk factors for mortality of adult inpatients with COVID-19 in Wuhan, China: a retrospective cohort study. *Lancet* 2020;**395**(10229):1054–62.
21. Doyle AE. Hypertension and vascular disease. *Am J Hypertens* 1991;**4**(2 Pt 2):103s–6s.

22. Egan BM, Kjeldsen SE, Grassi G, Esler M, Mancia G. The global burden of hypertension exceeds 1.4 billion people: should a systolic blood pressure target below 130 become the universal standard? *J Hypertens* 2019;**37**(6):1148–53.
23. Wu C, Chen X, Cai Y, et al. Risk factors associated with acute respiratory distress syndrome and death in patients with coronavirus disease 2019 pneumonia in Wuhan, China. *JAMA Intern Med* 2020;**180**(7):934–43.
24. Ma LY, Chen WW, Gao RL, et al. China cardiovascular diseases report 2018: an updated summary. *J Geriatr Cardiol (JGC)* 2020;**17**(1):1–8.
25. Ferrario CM, Jessup J, Chappell MC, et al. Effect of angiotensin-converting enzyme inhibition and angiotensin II receptor blockers on cardiac angiotensin-converting enzyme 2. *Circulation* 2005;**111**(20):2605–10.
26. Fosbøl EL, Butt JH, Østergaard L, et al. Association of angiotensin-converting enzyme inhibitor or angiotensin receptor blocker use with COVID-19 diagnosis and mortality. *J Am Med Assoc* 2020;**324**(2):168–77.
27. Patel AB, Verma A. COVID-19 and angiotensin-converting enzyme inhibitors and angiotensin receptor blockers: what is the evidence? *J Am Med Assoc* 2020;**323**(18):1769–70.
28. Danser AHJ, Epstein M, Batlle D. Renin-angiotensin system blockers and the COVID-19 pandemic. *Hypertension* 2020;**75**(6):1382–5.
29. Yang P, Gu H, Zhao Z, et al. Angiotensin-converting enzyme 2 (ACE2) mediates influenza H7N9 virus-induced acute lung injury. *Sci Rep* 2014;**4**:7027.
30. Kow CS, Hasan SS. Possible protective effect of renin-angiotensin system inhibitors in COVID-19 induced acute kidney injury. *J Am Soc Nephrol* 2020;**31**(8):1917–8.
31. Wang JJ, Edin ML, Zeldin DC, Li C, Wang DW, Chen C. Good or bad: application of RAAS inhibitors in COVID-19 patients with cardiovascular comorbidities. *Pharmacol Ther* 2020;**215**:107628.
32. Zhang P, Zhu L, Cai J, et al. Association of inpatient use of angiotensin-converting enzyme inhibitors and angiotensin II receptor blockers with mortality among patients with hypertension hospitalized with COVID-19. *Circ Res* 2020;**126**(12):1671–81.
33. Liu X, Long C, Xiong Q, et al. Association of angiotensin converting enzyme inhibitors and angiotensin II receptor blockers with risk of COVID-19, inflammation level, severity, and death in patients with COVID-19: a rapid systematic review and meta-analysis. *Clin Cardiol* 2020. https://doi.org/10.1002/clc.23421.
34. Preliminary estimates of the prevalence of selected underlying health conditions among patients with coronavirus disease 2019 – United States, February 12-March 28, 2020. *Morb Mortal Wkly Rep (MMWR)* 2020;**69**(13):382–6.
35. Yang X, Yu Y, Xu J, et al. Clinical course and outcomes of critically ill patients with SARS-CoV-2 pneumonia in Wuhan, China: a single-centered, retrospective, observational study. *Lancet Respir Med* 2020;**8**(5):475–81.
36. Wu Z, McGoogan JM. Characteristics of and important lessons from the coronavirus disease 2019 (COVID-19) outbreak in China: summary of a report of 72 314 cases from the Chinese center for disease control and prevention. *JAMA* 2020;**323**(13):1239–42.
37. Kerner W, Brückel J. Definition, classification and diagnosis of diabetes mellitus. *Exp Clin Endocrinol Diabetes* 2014;**122**(7):384–6. Official Journal, German Society of Endocrinology [and] German Diabetes Association.
38. Yang JK, Feng Y, Yuan MY, et al. Plasma glucose levels and diabetes are independent predictors for mortality and morbidity in patients with SARS. *Diabet Med* 2006;**23**(6):623–8. A Journal of the British Diabetic Association.
39. Banik GR, Alqahtani AS, Booy R, Rashid H. Risk factors for severity and mortality in patients with MERS-CoV: analysis of publicly available data from Saudi Arabia. *Virol Sin* 2016;**31**(1):81–4.

40. Schoen K, Horvat N, Guerreiro NFC, de Castro I, de Giassi KS. Spectrum of clinical and radiographic findings in patients with diagnosis of H1N1 and correlation with clinical severity. *BMC Infect Dis* 2019;**19**(1):964.
41. Romaní-Pérez M, Outeiriño-Iglesias V, Moya CM, et al. Activation of the GLP-1 receptor by liraglutide increases ACE2 expression, reversing right ventricle hypertrophy, and improving the production of SP-A and SP-B in the lungs of type 1 diabetes rats. *Endocrinology* 2015;**156**(10):3559–69.
42. Tikoo K, Patel G, Kumar S, et al. Tissue specific up regulation of ACE2 in rabbit model of atherosclerosis by atorvastatin: role of epigenetic histone modifications. *Biochem Pharmacol* 2015;**93**(3):343–51.
43. Zhang W, Xu YZ, Liu B, et al. Pioglitazone upregulates angiotensin converting enzyme 2 expression in insulin-sensitive tissues in rats with high-fat diet-induced nonalcoholic steatohepatitis. *Sci World J* 2014;**2014**:603409.
44. Rao S, Lau A, So HC. Exploring diseases/traits and blood proteins causally related to expression of ACE2, the putative receptor of SARS-CoV-2: a Mendelian randomization analysis highlights tentative relevance of diabetes-related traits. *Diabetes Care* 2020;**43**(7):1416–26.
45. Ho TW, Huang CT, Tsai YJ, Lien AS, Lai F, Yu CJ. Metformin use mitigates the adverse prognostic effect of diabetes mellitus in chronic obstructive pulmonary disease. *Respir Res* 2019;**20**(1):69.
46. Fernandez C, Rysä J, Almgren P, et al. Plasma levels of the proprotein convertase furin and incidence of diabetes and mortality. *J Intern Med* 2018;**284**(4):377–87.
47. Nicin L, Abplanalp WT, Mellentin H, et al. Cell type-specific expression of the putative SARS-CoV-2 receptor ACE2 in human hearts. *Eur Heart J* 2020;**41**(19):1804–6.
48. Hodgson K, Morris J, Bridson T, Govan B, Rush C, Ketheesan N. Immunological mechanisms contributing to the double burden of diabetes and intracellular bacterial infections. *Immunology* 2015;**144**(2):171–85.
49. Kulcsar KA, Coleman CM, Beck SE, Frieman MB. Comorbid diabetes results in immune dysregulation and enhanced disease severity following MERS-CoV infection. *JCI Insight* 2019;**4**(20).
50. Maddaloni E, Buzzetti R. Covid-19 and diabetes mellitus: unveiling the interaction of two pandemics. *Diabetes Metab Res Rev* 2020:e33213321.
51. Huang J, Xiao Y, Xu A, Zhou Z. Neutrophils in type 1 diabetes. *J Diabetes Investig* 2016;**7**(5):652–63.
52. Hussain A, Bhowmik B, do Vale Moreira NC. COVID-19 and diabetes: knowledge in progress. *Diabetes Res Clin Pract* 2020;**162**:108142.
53. Roth GA, Johnson C, Abajobir A, et al. Global, regional, and national burden of cardiovascular diseases for 10 causes, 1990 to 2015. *J Am Coll Cardiol* 2017;**70**(1):1–25.
54. Shi S, Qin M, Cai Y, et al. Characteristics and clinical significance of myocardial injury in patients with severe coronavirus disease 2019. *Eur Heart J* 2020;**41**(22):2070–9.
55. Akhmerov A, Marbán E. COVID-19 and the heart. *Circ Res* 2020;**126**(10):1443–55.
56. Clerkin KJ, Fried JA, Raikhelkar J, et al. COVID-19 and cardiovascular disease. *Circulation* 2020;**141**(20):1648–55.
57. Dhakal BP, Sweitzer NK, Indik JH, Acharya D, William P. SARS-CoV-2 infection and cardiovascular disease: COVID-19 heart. *Heart Lung Circ* 2020;**29**(7):973–87.
58. Bernheim A, Mei X, Huang M, et al. Chest CT findings in coronavirus disease-19 (COVID-19): relationship to duration of infection. *Radiology* 2020;**295**(3):200463.
59. Tay MZ, Poh CM, Rénia L, MacAry PA, Ng LFP. The trinity of COVID-19: immunity, inflammation and intervention. *Nat Rev Immunol* 2020;**20**(6):363–74.
60. Xu Z, Shi L, Wang Y, et al. Pathological findings of COVID-19 associated with acute respiratory distress syndrome. *Lancet Respir Med* 2020;**8**(4):420–2.

61. Nishiga M, Wang DW, Han Y, Lewis DB, Wu JC. COVID-19 and cardiovascular disease: from basic mechanisms to clinical perspectives. *Nat Rev Cardiol* 2020;**17**(9):543–58.
62. Zheng Y-Y, Ma Y-T, Zhang J-Y, Xie X. COVID-19 and the cardiovascular system. *Nat Rev Cardiol* 2020;**17**(5):259–60.
63. Huang C, Wang Y, Li X, et al. Clinical features of patients infected with 2019 novel coronavirus in Wuhan, China. *Lancet* 2020;**395**(10223):497–506.
64. Bansal M. Cardiovascular disease and COVID-19. *Diabetes Metab Syndr* 2020;**14**(3):247–50.
65. Oudit GY, Kassiri Z, Jiang C, et al. SARS-coronavirus modulation of myocardial ACE2 expression and inflammation in patients with SARS. *Eur J Clin Invest* 2009;**39**(7):618–25.
66. Yang J, Zheng Y, Gou X, et al. Prevalence of comorbidities and its effects in patients infected with SARS-CoV-2: a systematic review and meta-analysis. *Int J Infect Dis (IJID)* 2020;**94**:91–5. Official Publication of the International Society for Infectious Diseases.
67. Bangalore S, Sharma A, Slotwiner A, et al. ST-segment elevation in patients with Covid-19 – a case series. *N Engl J Med* 2020;**382**(25):2478–80.
68. Stefanini GG, Montorfano M, Trabattoni D, et al. ST-elevation myocardial infarction in patients with COVID-19: clinical and angiographic outcomes. *Circulation* 2020;**141**(25):2113–6.
69. Maron BJ, Towbin JA, Thiene G, et al. Contemporary definitions and classification of the cardiomyopathies. *Circulation* 2006;**113**(14):1807–16.
70. Chen T, Wu D, Chen H, et al. Clinical characteristics of 113 deceased patients with coronavirus disease 2019: retrospective study. *BMJ* 2020;**368**:m1091.
71. Guo T, Fan Y, Chen M, et al. Cardiovascular implications of fatal outcomes of patients with coronavirus disease 2019 (COVID-19). *JAMA Cardiol* 2020;**5**(7):811–8.
72. Wichmann D, Sperhake JP, Lütgehetmann M, et al. Autopsy findings and venous thromboembolism in patients with COVID-19: a prospective cohort study. *Ann Int Med* 2020;**173**(4):268–77.
73. Olloquequi J. COVID-19 Susceptibility in chronic obstructive pulmonary disease. *Eur J Clin Invest* 2020;**50**(10):e13382.
74. Leung JM, Niikura M, Yang CWT, Sin DD. COVID-19 and COPD. *Eur Respir J* 2020;**56**(2).
75. Lippi G, Henry BM. Chronic obstructive pulmonary disease is associated with severe coronavirus disease 2019 (COVID-19). *Respir Med* 2020;**167**:105941.
76. Guan WJ, Liang WH, Zhao Y, et al. Comorbidity and its impact on 1590 patients with COVID-19 in China: a nationwide analysis. *Eur Respir J* 2020;**55**(5).
77. Ali A, Vijayan R. Dynamics of the ACE2–SARS-CoV-2/SARS-CoV spike protein interface reveal unique mechanisms. *Sci Rep* 2020;**10**(1):14214.
78. Leung JM, Yang CX, Tam A, et al. ACE-2 expression in the small airway epithelia of smokers and COPD patients: implications for COVID-19. *Eur Respir J* 2020;**55**(5).
79. Cai G, Bossé Y, Xiao F, Kheradmand F, Amos CI. Tobacco smoking increases the lung gene expression of ACE2, the receptor of SARS-CoV-2. *Am J Respir Crit Care Med* 2020;**201**(12):1557–9.
80. Bhat TA, Panzica L, Kalathil SG, Thanavala Y. Immune dysfunction in patients with chronic obstructive pulmonary disease. *Ann Am Thorac Soc* 2015;**12**:S169–75. Suppl 2(Suppl 2).
81. Rovina N, Koutsoukou A, Koulouris NG. Inflammation and immune response in COPD: where do we stand? *Mediat Inflamm* 2013;**2013**:413735.
82. Singanayagam A, Glanville N, Girkin JL, et al. Corticosteroid suppression of antiviral immunity increases bacterial loads and mucus production in COPD exacerbations. *Nat Commun* 2018;**9**(1):2229.

83. Li H, Chen C, Hu F, et al. Impact of corticosteroid therapy on outcomes of persons with SARS-CoV-2, SARS-CoV, or MERS-CoV infection: a systematic review and meta-analysis. *Leukemia* 2020;**34**(6):1503–11.
84. Pelgrim CE, Peterson JD, Gosker HR, et al. Psychological co-morbidities in COPD: targeting systemic inflammation, a benefit for both? *Eur J Pharmacol* 2019;**842**:99–110.
85. Horby P, Lim WS, Emberson JR, et al. Dexamethasone in hospitalized patients with Covid-19 – preliminary report. *N Engl J Med* 2020. https://doi.org/10.1056/NEJMoa2021436.
86. Henry BM, Lippi G. Chronic kidney disease is associated with severe coronavirus disease 2019 (COVID-19) infection. *Int Urol Nephrol* 2020;**52**(6):1193–4.
87. Cheng Y, Luo R, Wang K, et al. Kidney disease is associated with in-hospital death of patients with COVID-19. *Kidney Int* 2020;**97**(5):829–38.
88. Gagliardi I, Patella G, Michael A, Serra R, Provenzano M, Andreucci M. COVID-19 and the kidney: from epidemiology to clinical practice. *J Clin Med* 2020;**9**(8).
89. Su H, Yang M, Wan C, et al. Renal histopathological analysis of 26 postmortem findings of patients with COVID-19 in China. *Kidney Int* 2020;**98**(1):219–27.
90. Diao B, Wang C, Wang R, et al. Human kidney is a target for novel severe acute respiratory syndrome coronavirus 2 (SARS-CoV-2) infection. *medRxiv* 2020. https://doi.org/10.1101/2020.03.04.20031120.
91. Yang L, Liu S, Liu J, et al. COVID-19: immunopathogenesis and Immunotherapeutics. *Signal Transduct Target Ther* 2020;**5**(1):128.
92. Akalin E, Azzi Y, Bartash R, et al. Covid-19 and kidney transplantation. *N Engl J Med* 2020;**382**(25):2475–7.
93. Gandolfini I, Delsante M, Fiaccadori E, et al. COVID-19 in kidney transplant recipients. *Am J Transplant* 2020;**20**(7):1941–3. Official Journal of the American Society of Transplantation and the American Society of Transplant Surgeons.
94. Indini A, Rijavec E, Ghidini M, et al. Coronavirus infection and immune system: an insight of COVID-19 in cancer patients. *Crit Rev Oncol-Hematol* 2020;**153**:103059.
95. Liang W, Guan W, Chen R, et al. Cancer patients in SARS-CoV-2 infection: a nationwide analysis in China. *Lancet Oncol* 2020;**21**(3):335–7.
96. Dai M, Liu D, Liu M, et al. Patients with cancer appear more vulnerable to SARS-CoV-2: a multicenter study during the COVID-19 outbreak. *Canc Discov* 2020;**10**(6):783–91.
97. Gosain R, Abdou Y, Singh A, Rana N, Puzanov I, Ernstoff MS. COVID-19 and cancer: a comprehensive review. *Curr Oncol Rep* 2020;**22**(5):53.
98. Esposito S, Noviello S, Pagliano P. Update on treatment of COVID-19: ongoing studies between promising and disappointing results. *Le Infezioni in Medicina* 2020;**28**(2):198–211.
99. Wang B, Li R, Lu Z, Huang Y. Does comorbidity increase the risk of patients with COVID-19: evidence from meta-analysis. *Aging* 2020;**12**(7):6049–57.
100. Shahid Z, Kalayanamitra R, McClafferty B, et al. COVID-19 and older adults: what we know. *J Am Geriatr Soc* 2020;**68**(5):926–9.
101. Al-Shamsi HO, Alhazzani W, Alhuraiji A, et al. A practical approach to the management of cancer patients during the novel coronavirus disease 2019 (COVID-19) pandemic: an international collaborative group. *Oncol* 2020;**25**(6):e936–45.
102. Gupta R, Misra A. Contentious issues and evolving concepts in the clinical presentation and management of patients with COVID-19 infection with reference to use of therapeutic and other drugs used in Co-morbid diseases (Hypertension, diabetes etc). *Diabetes Metab Syndr* 2020;**14**(3):251–4.
103. Bresadola V, Biddau C, Puggioni A, et al. General surgery and COVID-19: review of practical recommendations in the first pandemic phase. *Surg Today* 2020;**50**(10):1159–67.

104. Arslan M, Xu B, Gamal El-Din M. Transmission of SARS-CoV-2 via fecal-oral and aerosols—borne routes: environmental dynamics and implications for wastewater management in underprivileged societies. *Sci Total Environ* 2020;**743**:140709.
105. Morawska L, Tang JW, Bahnfleth W, et al. How can airborne transmission of COVID-19 indoors be minimised? *Environ Int* 2020;**142**:105832.
106. Tang S, Mao Y, Jones RM, et al. Aerosol transmission of SARS-CoV-2? Evidence, prevention and control. *Environ Int* 2020;**144**:106039.
107. Bourouiba L. Turbulent gas clouds and respiratory pathogen emissions: potential implications for reducing transmission of COVID-19. *JAMA* 2020;**323**(18):1837—8.
108. Jayaweera M, Perera H, Gunawardana B, Manatunge J. Transmission of COVID-19 virus by droplets and aerosols: a critical review on the unresolved dichotomy. *Environ Res* 2020;**188**:109819.
109. Meyerowitz EA, Richterman A, Gandhi RT, Sax PE. Transmission of SARS-CoV-2: a review of viral, host, and environmental factors. *Ann Intern Med* 2020. https://doi.org/10.7326/M20-5008.
110. Dhand R, Li J. Coughs and sneezes: their role in transmission of respiratory viral infections, including SARS-CoV-2. *Am J Respir Crit Care Med* 2020;**202**(5):651—9.
111. Goldman E. Exaggerated risk of transmission of COVID-19 by fomites. *Lancet Infect Dis* 2020;**20**(8):892—3.
112. van Doremalen N, Bushmaker T, Morris DH, et al. Aerosol and surface stability of SARS-CoV-2 as compared with SARS-CoV-1. *N Engl J Med* 2020;**382**(16):1564—7.
113. Lewis NM, Chu VT, Ye D, et al. Household transmission of SARS-CoV-2 in the United States. *Clin Infect Dis* 2020. https://doi.org/10.1093/cid/ciaa1166.
114. Ceylan Z, Meral R, Cetinkaya T. Relevance of SARS-CoV-2 in food safety and food hygiene: potential preventive measures, suggestions and nanotechnological approaches. *VirusDisease* 2020;**31**(2):154—60.
115. Han J, Zhang X, He S, Jia P. Can the coronavirus disease be transmitted from food? A review of evidence, risks, policies and knowledge gaps. *Environ Chem Lett* 2020:1—12.
116. Oude Munnink BB, Sikkema RS, Nieuwenhuijse DF, et al. Transmission of SARS-CoV-2 on mink farms between humans and mink and back to humans. *Science* 2021;**371**(6525):172—7.
117. Yekta R, Vahid-Dastjerdi L, Norouzbeigi S, Mortazavian AM. Food products as potential carriers of SARS-CoV-2. *Food Contr* 2020:107754.
118. Kim JM, Kim HM, Lee EJ, et al. Detection and isolation of SARS-CoV-2 in serum, urine, and stool specimens of COVID-19 patients from the Republic of Korea. *Osong Public Health Res Perspect* 2020;**11**(3):112—7.

SECTION II

Remediation measures in waste and wastewater environments

CHAPTER 4

Presence, detection, and persistence of SARS-CoV-2 in wastewater and the sustainable remedial measures

Bashir Adelodun[1,2], AbdulGafar Olatunji Tiamiyu[3], Fidelis Odedishemi Ajibade[4,5,6], Golden Odey[1], Rahmat Gbemisola Ibrahim[7], Madhumita Goala[8], Hashim Olalekan Bakare[3], Temitope F. Ajibade[4,6,9], Jamiu Adetayo Adeniran[10,11], Kamoru Akanni Adeniran[2], Kyung Sook Choi[1,12]

[1]Department of Agricultural Civil Engineering, Kyungpook National University, Daegu, Korea; [2]Department of Agricultural and Biosystems Engineering, University of Ilorin, Ilorin, Kwara State, Nigeria; [3]Department of Chemical Engineering, University of Ilorin, Ilorin, Kwara State, Nigeria; [4]Department of Civil and Environmental Engineering, Federal University of Technology, Akure, Ondo State, Nigeria; [5]Key Laboratory of Environmental Biotechnology, Research Centre for Eco-Environmental Sciences, Chinese Academy of Sciences, Beijing, PR China; [6]University of Chinese Academy of Sciences, Beijing, PR China; [7]Kwara State Ministry of Health, Ilorin, Kwara State, Nigeria; [8]Nehru College, Pailapool, Affiliated Assam University, Silchar, Cachar, Assam, India; [9]Key Laboratory of Urban Pollutant Conversion, Institute of Urban Environment, Chinese Academy of Sciences, Xiamen, PR China; [10]Environmental Engineering Research Laboratory, Department of Chemical Engineering, University of Ilorin, Ilorin, Kwara State, Nigeria; [11]Atmospheric Chemistry and Modeling Group, Department of Atmospheric and Oceanic Sciences, Peking University, Beijing, China; [12]Institute of Agricultural Science & Technology, Kyungpook National University, Daegu, Korea

4.1 Introduction

The world has experienced another global pandemic, the coronavirus disease 2019 (COVID-19), caused by novel severe acute respiratory syndrome coronavirus 2 (SARS-CoV-2), which was declared a Public Health Emergency of International Concern (PHEIC) by the World Health Organization (WHO) on January 30, 2020.[1] The generally known means of transmission was from direct contact or droplets from respiratory actions. However, the reports on the detection of SARS-CoV-2 in wastewater samples and water bodies may have been present via the gastrointestinal tract of an infected patient and contaminated sewage discharged (Fig. 4.1).[2–4] The possible presence of SARS-CoV-2 in wastewater is a serious issue that needs to be properly attended to and curbed before the widespread of the virus among vulnerable populations;[5] this, in addition to

Environmental Management of COVID-19
ISBN 978-0-323-85780-2
https://doi.org/10.1016/B978-0-323-85780-2.00014-7

a report that the resemblance of SARS-CoV-2 to SARS-coronavirus (SARS-CoV) is about 82%,[6] gives researchers and policymakers an area of focus to expedite actions toward finding lasting solutions.

In 2003, during the outbreak of SARS-CoV, the virus spread rapidly and was detected in feces (surviving for up to four days), water, and sewage systems for days to weeks, and also in the faulty sewage system of an apartment in Hong Kong.[8,9] The complications of SARS-CoV-2 transmission in water and wastewater environment may have become worrisome due to the high possibility of asymptomatic patients and high viral-shedders[10] who may contaminate the environment easily and spill over the transmission to other people, health care personnel, and front-line wastewater treatment plant workers.[11] Other areas of concern are treatment facilities with inefficient treatment processes,[12–14] which may also threaten the public for possible exposure to the virus. For instance, Radazzo et al.[12] confirmed the presence of SARS-CoV-2 RNA in 11% of the secondary-treated water samples investigated. Similarly, Zhang et al.[13] investigated the presence of SARS-CoV-2 in medical wastewater in septic tanks of a hospital in China, where it was revealed that the high load of the virus might be another secondary source of sporadic spread of COVID-19.

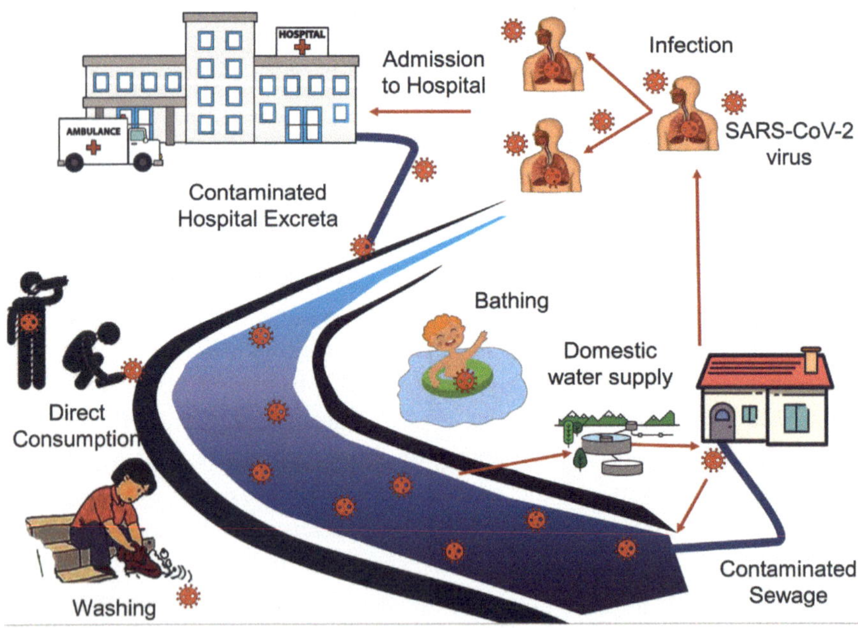

Figure 4.1 The pathways of SARS-CoV-2 in wastewater and water systems. (Adapted from Adelodun et al.[7]).

While the presence of SARS-CoV-2 in wastewater have been studied and confirmed,[15–18] only a few studies have so far investigated its persistence in the wastewater environment,[19] thereby creating an important information gap on the provision of sustainable approaches to mitigate the spread of the virus through this medium. Moreover, the previous knowledge on the virus removal methods, deactivation/inactivation, and treatment[20–22] can be further explored targeting the novel SARS-CoV-2. Therefore, it is important to review recent advances in wastewater treatment and the protection of wastewater plant personnel in the face of the pandemic and proffer effective techniques.

4.2 Occurrence, detection, and persistence of SARS-CoV-2 in wastewater, feces, slurry, or biosolids

4.2.1 Occurrence and detection of coronaviruses in wastewater

Viruses in wastewater have been studied extensively for nonenveloped enteric viruses, such as adenoviruses, polioviruses, enteroviruses, rotaviruses, and noroviruses, primarily due to their fecal-oral transmission routes. With the recent development on the detection of coronaviruses (both SARS-CoV and SARS-CoV-2) in wastewater, there is a need for a replica of the previous efforts on the studies of nonenveloped viruses for the enveloped ones.[23] Due to the report on the isolation of viable SARS-CoV-2 in the feces of infected patients and possible transmission through secondary routes,[10,24,25] the concern and awareness in the persistence of coronaviruses in wastewater has increased, leading to the adoption of the wastewater-based epidemiology (WBE) as a concept for wastewater analysis to serve as a caution and projection for a possible disease epidemic.[17,23,24,26] Fig. 4.2 shows the schematic flow of detection and quantification of SARS-CoV-2 from wastewater sources.

WBE is a new tool for the environmental engineers and researchers to harness the possibility of contaminants (including viruses) surviving in a relatively stable environment and subsequently released in the sewage/wastewater system by sampling and quantifying to get real-time information in early detection of viral outbreaks and preventing such occurrence via a well-defined inactivation and removal techniques.[2,27] The methods often used for the detection after sampling are the variants of polymerase chain reaction (PCR); reverse transcription-polymerase chain reaction (RT-PCR), and reverse transcription real-time polymerase chain

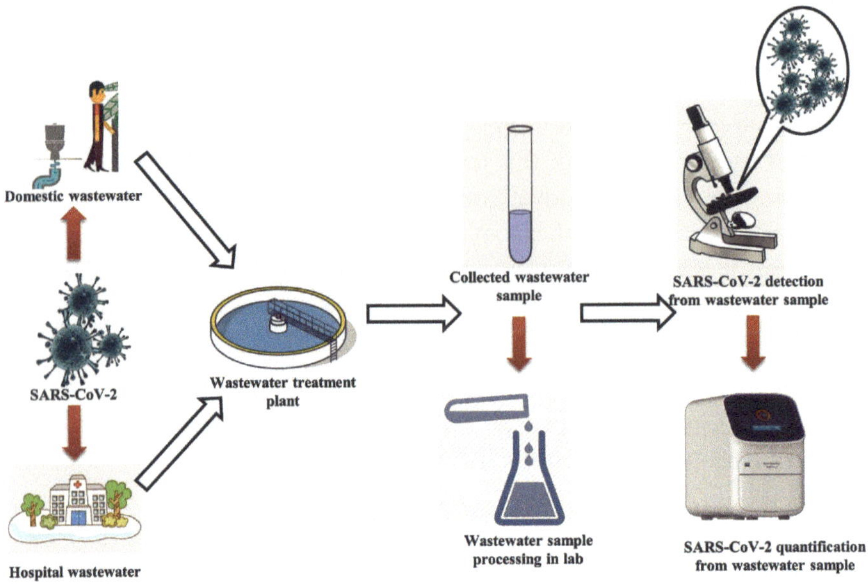

Figure 4.2 Schematic flow of detection and quantification of SARS-CoV-2 from wastewater sources.

reaction (RT-qPCR). Medema et al.[19] suggested surveillance for the sewage system as a sensitive tool to closely monitor for the circulation of the virus in the community. The authors studied the presence of SARS-CoV-2 in sewage samplings in six separate cities and an airport in the Netherlands. The confirmation of the genetic material of SARS-CoV-2 in the sampled sewage six days before the officially reported first COVID-19 cases, and the subsequent detection in the collected sewage samples after the announcement, even when the incidence is still low, signifies the importance of the WBE tool, especially for the monitoring of the municipal wastewater/sewage system including the ones at the airports, ports, and land-border facilities.

Municipal wastewater/sewage systems can be a hotspot for diseases outbreak due to the high dependence and usage by the community and, as such, can be monitored for viral outbreak containment. Therefore, wastewater-based epidemiological studies are important in this regard while improving the initial treatment techniques such as nanofiber filters as wastewater pretreatment techniques and upgrading the existing ones to have an efficient surveillance tool.[28] The limitation of clinical diagnostics in terms of the population coverage makes WBE an important tool for identifying disease hotspots and monitoring a campaign on the extent and duration of the spread for the population in a specific location.[28] Wastewater-based epidemiology can be properly used in areas where patients are asymptomatic, which may not immediately be known via clinical surveillance and possible underdiagnoses. This limitation is of growing concern because of retesting a large number of infected persons to ascertain valid results.[29]

Further, there are instances where the clinical diagnostic methods cannot detect viruses at low levels of prevalence due to a low level of viral loads/concentrations below 100 copies/μL (>40 Ct) among the infected individuals.[12,30,31] Hence, wastewater surveillance could adequately be employed to fill this important gap. The WBE is considered effective during the introduction of a new virus in the community or when there are seasonal variations and the survivability of such viruses is to be determined.[29] WBE is a tool that needs extensive research so that different regions can have an idea of the persistence, occurrence, and survival of different pathogenic viruses in hospital wastewater and rural waterways, as a means of signaling early detection to help in combating future outbreaks.[3]

Sharif et al.[6] investigated the existence of SARS-CoV-2 in 78 wastewater samples from 38 districts across Pakistan using the existing polio environment surveillance sites. The samples of SARS-CoV-2 infected areas and a drainage system at a COVID-19 quarantine center were analyzed using RT-qPCR, with 27% of the samples from 13 districts reporting positive. It was suggested that since the studies were able to replicate the surveillance network of polio sites, the potential for WBE is promising and may be necessary for situations where early detection is vital. At the same time, door-to-door contact tracing may prove difficult and strenuous, especially in densely populated areas.[6]

4.2.2 Methods for SARS-CoV-2 detection in wastewater

The urgency and dynamics of actions in responding to the challenges of containment, assessment, and risk mitigation of SARS-CoV-2 have revealed the shortcomings, which may need to be addressed in earnest, especially when there are emerging indications of the presence of the virus in water and wastewater environments, posing a serious threat to wastewater/sewage treatment workers and the general public.[32] In the preprint by Singer and Wray,[33] a detailed review of recent works showed that there are reports of SARS-CoV and SARS-CoV-2 presence in stool and urine samples with both virus detections carried out using either or both of

RT-PCR and real-time reverse-transcriptase quantitative polymerase chain reaction (RT-qPCR). Other methods of SARS-CoV-2 detection available are categorized based on diagnostic tests such as nucleic acid detection and amplification (examples are PCR, RT-PCR, and qPCR); isothermal amplification technologies (examples are NASBA, LAMP, HAD, RCA, NEAR, SDA, and TMA); immunoassays (examples fluorescent antibody staining, EIA/ELISA); deoxyribonucleic acid (DNA) sequencing (such as sanger sequencers, next-generation sequencers, and DNA microarrays); mass spectrometric methods (MALDI-TOF); direct visualization of viruses (electron microscopy), and microelectronics and microfluidics-based techniques (such as lab-on-a-chip [LOC], point of care [POC] testing, and surface plasmon resonance [SPR] technique).[34] In this section, the RT-PCR and its variants will be discussed considering their widely used quick method for virus detection in water and wastewater, compared to other methods listed due to their disadvantages in the pilot setup stage, scalability for mass production, and cost implications of laboratory setup, reagents, highly trained and qualified personnel, and time.

4.2.2.1 Virus detection in wastewater using RT-PCR/RT-qPCR

The WHO technical brief statement on March 3, 2020 read,

> *[T]he presence of the virus (SARS-CoV-2) in drinking water is a possibility; however, there is no evidence as at the time of the brief, that surrogate human SARS-CoV-2 are present in surface or underground water routes or transmitted via contaminated drinking water; the statement went further to state that the virus has not been detected in the mentioned water supplies and the risk is thereby low.*[35]

The statement has many pointers that need further critical analysis to not misguide the nonscientist reader because "there is no current evidence" may imply that research is still ongoing in this area (water and wastewater). The low-risk claim is as a result of little or no evidence at the moment and may tend to become a high-risk source depending on the action taken over time. Lastly, the similarity between SARS-CoV and SARS-CoV-2 mean guards should not be let down because a secondary source of SARS-CoV was via an inadequate sewage system.[9] Therefore, researchers are vigorously working to understand the possible persistence of SARS-CoV-2 in water and wastewater, and devising some robust mitigation approaches to avoid its widespread.[7]

Recently, there have been different methods of SARS-CoV-2 detection, such as two-phase (PEG-dextran method) separation and aluminum hydroxide adsorption-precipitation method; unfortunately, these methods do not report the percentage recovery.[18,36] The investigation on the presence of SARS-CoV-2 in wastewater was carried out by Sherchan et al.[18] using two concentration methods. The SARS-CoV-2 detection in wastewater was by RT-qPCR with mean recovery efficiency of about 73% using a *Pseudomonas* bacteriophage as a control. Subsequently, the treated wastewater tested negative, thereby indicating the effectiveness of both the detection and treatment techniques. Westhaus et al.[15] carried out the detection of SARS-CoV-2 in wastewater treatment plants (WWTP) using RT-qPCR methods. The specificity and sensitivity were verified by comparing the different viral genes (N-gene, E-gene, M-gene, and RdRP) with three different sewage samples. The performance of RT-qPCR was notably good with the RdRP gene, and other false-positive results were not considered with suggestions for proper RT-qPCR sequencing or design of other suitable methods.[15] The efficiency and sensitivity of RT-qPCR were tested using three different assays, tagged as N1, N2, and N3 for several water samples collected before (October 2019) and after the emergence of 2019-nCoV (within two months).[12] The different inconsistencies were observed in the results, with few samples showing positive results, while others indicated false-positive results. It was concluded that while considering WBE for the emergence of SARS-CoV-2 in water and wastewater, digital RT-qPCR should be considered despite its cost implications.[12]

4.2.3 Persistence of coronaviruses in wastewater

There are several factors that may affect the survival of SARS-CoV-2 in wastewater, and these factors may be intrinsic or extrinsic depending on the environmental conditions. These factors include viral structure, the composition of the wastewater, pH, and temperature.

4.2.3.1 Composition of wastewater

The composition of wastewater is vital to the survival of the viruses in the medium and the level of their reduction.[37] The log reduction level of the viruses in the wastewater depends on the composition of such wastewater. For instance, in a reagent grade water, the 4 $\log_{10}$ (99.99%) reduction of transmissible gastroenteritis virus (TGEV) was achieved in 44 days, while it took 35 days at 25°C for murine hepatitis virus (MHV) to get inactivated under the same inactivation/reduction conditions.[23] Polo et al.[38] opined that the composition of the water, including other microorganisms, can influence the persistence of enveloped viruses in wastewater compared to nonenveloped viruses, thus affecting the virus recovery, inactivity assay results, underestimated infectivity rate, and concentration techniques.

4.2.3.2 pH

Extensive research is needed to investigate varying pH levels to determine the survivability of viruses in wastewater. A recent study that made a comparison between enveloped and nonenveloped viruses found that the stability conditions for MHV was within the pH ranges of 5—7.4 and 3—10 at temperatures of 37 and 4°C, respectively, which are regarded as the most stable conditions that depend on the removal, inactivation, and survival.[23] The stability and persistence of SARS-CoV-2 raised more concerns with the different levels of pH. For example, Tran et al.[39] observed that SARS-CoV-2 was stable at a pH range of 3.0—10 under room temperature, while that of SARS-CoV survived for one day at pH of 8.0, 5 days at pH of 9.0, and only for 3 h at pH of 6.0. Similarly, Chin et al.[40] confirmed that SAR-CoV-2 was stable at pH values of 3—10 at room temperature, indicating the importance of pH as an environmental factor in the survivability of the virus in different environmental conditions, including wastewater.

4.2.3.3 Temperature

Temperature is another important factor that largely determines the survivability of viruses in wastewater. Coronaviruses often get inactivated at a certain temperature (for instance, 20°C) and not at low temperatures (i.e., 4°C and below).[39] It was also reported that SARS-CoV survived in different wastewater samples, such as hospital wastewater, dechlorinated tap water, and domestic sewage for 14 days at 4°C, while the survival was just for two days at 20°C.[41,42] It is safe to say that virus survival will be longer in the colder regions than the tropical regions, which can be ascribed to the protein and nucleic acid denaturation and the upsurge in extracellular enzyme activities.[23]

4.3 Removal of viruses from water and wastewater environment

The consumption of virus-contaminated water can lead to various forms of acute illnesses.[43] According to WHO and UNICEF,[44] about 1.8 billion of the global population consumed contaminated water with pathogens in 2012, leading to gastroenteritis-related diseases.[45,46] The reclamation of water from wastewater for reuse has increased recently due to incessant water scarcity and population surge, especially in developing countries.[47] This has led to an increase in waterborne-related diseases due to poor sanitation and hygiene coupled with a lack of persuasive techniques in treating water and wastewater before usage.[48–50] Therefore, it is important to fashion out some effective treatment methods for the removal of viruses for advanced treatment, which will be made available for reuse for potable water and other agricultural use such as irrigation and food processing.[51]

Having ascertained the presence of SARS-CoV-2 in water and wastewater via several detection methods,[12,16,18] it is imperative to research wastewater treatment methods that are cost-effective and efficient to dispel the fear of the virus spread through secondary transmissions.[45] Technically, virus removal is different from virus inactivation since virus removal may involve some specific steps to concentrate/coagulate the viruses before the disinfection mechanism. However, virus disinfection or inactivation involves applying advanced methods to inactivate the virus before and/or after virus removal. Traditional wastewater treatment methods are primarily aimed at the removal of biodegradable organics and suspended solids,[49] which comprise of physical, biological, and chemical processes such as activated sludge process, filtration (including the use of sand filters), and membranes (nanomaterials, ultrafiltration, microfiltration, membrane bioreactor, ceramic membranes sedimentation, and reverse osmosis).[27,45,50]

Due to the lethargic nature of viruses to humans, there are guidelines sets for handling viruses (especially SARS-CoV-2); these give precautions needed to be adhered to for scientific testing to be carried out. The guidelines for Biosafety and COVID-19 by the Center for Disease Control and Prevention (CDC) for environment specimen testing include "[p]rocedures that concentrate viruses, such as precipitation or membrane filtration can be performed in a BSL-2 laboratory with unidirectional airflow and BSL-3 precautions."[27,52] Due to this, surrogates are used in place of dangerous pathogens and depend on the type of experiment to be carried out. The possible surrogates for SARS-CoV-2 are murine hepatitis virus (MHV), TGEV, feline infectious peritonitis virus (FIPV), bacteriophage φ6, and pepper mild mottle virus (PMMoV).[27]

The use of MHV was considered in four different environments to represent cold, tropical, subtropical, and temperate latitudes using the common RT-qPCR in untreated wastewater, autoclave wastewater, and dechlorinated tap water in order to investigate the survival of both

SARS-CoV-2 and MHV in these environments.[53] It was posited that the two viruses showed similar persistence in the selected environments. The findings indicated that at 37°C, the detection of SARS-CoV-2 RNA was not affected; however, the persistence of the virus was prolonged at 4 and 15°C. The difference in the decay rate between MHV and SARS-CoV-2 was observed and may be due to the gamma-irradiated SARS-CoV-2 used in the study. Conclusively, MHV indicated a viable surrogate for SARS-CoV-2 usage based on the CDC guidelines, with the inability to use viable SARS-CoV-2 considering the required high level of biological safety measures (BSL-3) that must be in place before handling infectious viruses.[53] Mohan et al.[24] also reported the use of bacteriophage φ6 as a surrogate for influenza and coronaviruses, where the inactivation varied based on factors such as temperature, aqueous media composition, and biological activity. It was further explained that both SARS-CoV and SARS-CoV-2 are affected by temperature compared to Poliovirus 1 LSc-2ab (PV-1), while having less survivability in the environment than nonenveloped viruses such as PV-1.

The question will always be on the readiness to test or determine the infection rate, humoral protection, herd immunity, and efficacy of vaccine (during and after clinical trials) using a readily available testing procedure that is faster, economical, and compact. During this pandemic, different challenges range from understanding the virus (infections, survival, replication, and persistence in different environments) to the search for economically viable testing methods and vaccine development. Therefore, researchers have investigated the use of a surrogate virus neutralization test based on antibody-mediated blockage of ACE2-spike protein-protein interaction.[54] In this study, it was revealed that the performance of the surrogate virus neutralization test (sVNT) is significantly high compared to the conventional virus neutralization test (cVNT) and pseudovirus neutralization test (pVNT). This implies that sVNT can be used rapidly in either research or clinical labs without using a live virus and biosafety environment, since it can easily detect total N-specific antibodies (NAbs) in an isotype-independent manner.[54] In addition to the previous position, the performance validation of sVNT was carried out using two cohorts of positive and negative sera from two countries, and the specificity and sensitivity were 100% and 98%, respectively.

Another cohort study was carried out to ascertain the performance level of sVNT with cVNT and pVNT. In this study, the specificity and selectivity was also very much superb, with the rapid screening ability of sVNT ascertained for larger samples. Further, it was posited that sVNT was precise, less complicated, economical, and faster, thereby making it more appropriate for use in a rapid and large number of samples.[55] It was concluded that sVNT at a standardized level will be vital for the selection of convalescent plasma donors for COVID-19 patients treatment, while the assay does not depend on anti-species antibodies. Therefore, it is suitable for use during preclinical testing of SARS-CoV-2 vaccines.[55]

Decision making such as lockdown, rapid testing, treatment, contact tracing, and vaccine developments depend on some conventional procedures, considering the spike in the infection rate of SARS-CoV-2 all around the world. Therefore, in a bid to provide timely information to assist in easy and fast decision making, the use of other surrogate testing procedures was investigated by Sharif et al.[6] The biochemical surrogate point-of-care tests (POCTs) were used to differentiate viral from a bacterial infection in patients with influenza-like illnesses. It was found that FebriDx (the surrogate POCTs) showed some positives for triage early detection, while large population testing and a UK-based studies (two precisely) indicated specificity and selectivity of 93% and 100%; and 86% and 100%, respectively.[6] However, the comprehensive assessment of the POCT field performance was not clearly defined by the authors.

4.4 Virus removal techniques from wastewater

This section is necessary for the SARS-CoV-2 studies and investigations since the knowledge of SARS-CoV-2 transmission routes, survival, and decay in wastewater will give an overview of the effective treatment (removal/inactivation) methods. Therefore, the following paragraph will explain the different removal techniques available and their effectiveness therein. Traditional wastewater treatment techniques are majorly used for organic matter and suspended solids removal.[49,50] However, during the process, it has been reported that some pathogens were also removed, making it considerable for use. Unfortunately, the reports are based on the effective removal of bacteria compared to viruses.[23]

4.4.1 Activated sludge treatment

In the wastewater treatment plant, the activated sludge is one of the most employed biological treatment technology,[56] which comprises multiple unit operations such as sedimentation (primary and secondary settlers), biological decomposition (including aerobic, anoxic, or anaerobic tanks or similar equipment) and followed by disinfection processes such as chlorination, UV irradiation, or ozonation. However, in the treatment plants where the wastewater is to be reused, either for agricultural or public consumptions, sand filters, membrane, and other tertiary treatment techniques may be used.[57] Having understood that the activated sludge treatment (AST) techniques are designed for other purposes such as removing suspended solids, nutrients, and organic substances, it was opined that since viruses are fine particles with colloidal features. It is possible to get absorbed on or within suspended particles in wastewater and further disinfected by other techniques or change of environments.[45,57] The activated sludge process has been widely adopted in most WWTPs. Randazzo et al.[12] employed six WWTPs with activated sludge in all treatment reclamation processes for wastewaters from public use, including irrigation. AST was also used in virus removal from wastewater in a subtropical environment. The removal rate recorded was 3 $\log_{10}$; however, the virus removed an enteric virus.[23]

Further, the use of MHV surrogates for human coronaviruses was investigated with a possible one-fourth of the MHV absorbed to the solids and about 99% at an increased retention time of 0.4−2.9 h at a primary settling stage until saturation is attained.[23] AST virus removal efficacy was put to the test in a study by Arraj et al.[58] where two types of bacteriophages and three types of enteric viruses were removed during sewage treatment. It was found very effective, especially with the enteric viruses, and could be tested for further studies on other types of viruses, especially SARS-CoV-2. This is a proven process that requires further intensive studies at different conditions while considering other factors that affect the persistence of the virus in water and wastewater.

4.4.2 Membrane bioreactor

A membrane bioreactor (MBR) is an example of a biological method of virus removal from wastewater that uses the cellular activity of the microorganisms for organic matter oxidation present in the wastewater. MBR is a combination of a filtration that solely membrane-based and a suspended growth biological reactor for virus removal from wastewater.[45] The major types of filtration used in MBR techniques are microfiltration and ultrafiltration with a size range of 0.1−0.2 μm and 0.005 ≈ 10 μm, respectively. The virus particles Log removal of more than 4 was earlier reported;[59] however, it was opined that the best MBR techniques for coronaviruses removal would be ultrafiltration since the average viral particle diameter and envelop diameter are 120 and 80 nm, respectively.[23]

Nanofiltration was also proposed as part of the processes of MBR techniques. The pore sizes of the membranes are usually less than 10 nm, which is smaller than any virus that may be present in wastewater. Generally, the use of MBR for surrogates may not be extensively applicable for human pathogenic viruses as the removal efficiency may differ, but it is a good pointer as a combination of different techniques will bring about the required efficiency.[27] In wastewater samples obtained for the SARS-CoV-2 detection from hospitals sewage, one of the samples obtained was pretreated using the following sequence of techniques: adjusting tanks → septic tank → adjusting tank → moving bed biofilm reactor (MBBR) → sedimentation → disinfection, while no virus was detected at the MBBR unit.[11]

4.5 Mechanism of inactivation of coronaviruses in water environment using disinfectants

Virus disinfection or inactivation is a general technique for preventing SARS-CoV-2 infection, transmission, and persistence in the environment. There are guidelines for the safe use of these disinfectants in wastewater treatment sources such as laboratories, hospitals, and homes.[60] Wastewater treatment is considered a means of reducing or complete expunging of dissolved and particulate organic matter, suspended solids, nutrients, and heavy metals; and the extent of the treatment is guided by the standards outlined by regulatory bodies such as WHO and local authorities. However, as for the current COVID-19 pandemic situation, there are no additional measures specially designed for the virus by WHO, US Centers for Disease Control and Prevention (CDC), or the Occupational Safety and Health Administration (OSHA).[61] However, it is vital to strictly adhere to the guiding techniques available for wastewater treatment, such as a well-designed and well-functional treatment plant, sufficient to curb the risk posed by fecal pathogens, such as SARS-CoV-2. Meanwhile, there are considerations for tertiary treatments, also known as disinfection procedures, to reduce the pathogen level further. The treatments include physical methods such as ionization by gamma-ray radiation, nonionizing radiation by ultraviolet light, photodynamic oxidation and heat, and chemical methods, which involve chlorine and chlorine dioxide, ozone,

Table 4.1 Comparison of inactivation methods for wastewater treatment.

Inactivation method	Advantage	Disadvantage
Liquid chlorine	Energy consumption is low.	High risk of storage.
UV light	Low costs of investment and operation.	High risks by operators and insufficient level of penetration.
Chlorine dioxide	Highly efficient and minimal operation costs.	Transport and storage risk are high.
Sodium hypochlorite	Less toxic, easy setup and operation and low cost.	Energy consumption is high and highly corrosive and high pollution.
Ozone	Excellent features of deodorizing, decoloring and decomposition of viruses.	Operation cost is high with high generation of harmful disinfection by-products (DBPs).

Source: Adapted from Ghernaout D, Elboughdiri N. Urgent proposals for disinfecting hospital wastewaters during COVID-19 pandemic. *OALib* 2020;**07**(05):1–18. https://doi.org/10.4236/oalib.1106373.

iodine, bromine, and bromine chloride.[61] The comparison of merits and demerits of selected inactivation methods for the virus in wastewater treatment is presented in Table 4.1.

4.5.1 Chlorination pretreatment

In chlorination pretreatment procedures, bleaching powder is mostly used in this regard in which the powder used should be emptied into the septic tanks alongside both the measured residual chlorine and water to ascertain the water quality.[60]

One of the most active disinfectants in the WWTPs is its high reactivity at low concentrations, comparatively low cost, and little residue formation at moderate doses. Chlorine can be used in the gaseous state (which is the most widely used) or hypochlorite. At the gaseous state application, the gas immediately reacts with water to form hypochlorous acid and hydrochloric acid; and the former, which is the primary form of active chlorine for the disinfecting features, subsequently dissociates to produce hypochlorite ion:

$$Cl_2 + H_2O \rightarrow HOCl + HCl$$

$$HOCL \rightleftarrows H^+ + OC^-$$

Source: Lahrich et al.[61]

The application of chlorine in virus disinfection was demonstrated by Wang et al.,[42] where SARS-CoV in low-level wastewater was inactivated using chlorine with 0.5 mg/L as free chlorine residual. However, a standard dosage was advised to reduce the rate of disinfection byproduct (DBP). The use of chlorine was advised in locations with high population density, especially if it is liquid chlorine, which possesses a high risk as regards storage and other concerns raised with its usages such as the use of chlorinator and corrosion-resistance materials (copper pipes and hard PVC pipes).[14]

4.5.2 Chlorine dioxide

Chlorine dioxide can efficiently be used as a disinfectant at the acidic level due to its high oxidization capacity. It has a solubility, which is 500% more than chlorine, and an oxidization capacity of 2.63 times compared to chlorine gas.[14] Chlorine dioxide tends to destroy protein's anabolic pathways and exterminate microorganisms including bacteria, fungi, spores, and viruses. It has the wonderful features of decolorization, oxidation, deodorization, and increasing oxygen content in wastewater; however, the storage risk often outweighs numerous advantages.[14]

4.5.3 Sodium hypochlorite

Sodium hypochlorite is more useful in disinfections of different wastewater categories, such as hospital wastewater and municipal wastewater. The ready availability of the material is due to its numerous advantages such as low level of toxicity, steady and less-tedious mode of operation, ease of control, less-tedious operation, and inadequate preparation and operation costs. However, it has some demerits such as higher pollution in terms of DBPs, high energy, and strong corrosiveness.[14] The use of sodium hypochlorite at the designed dosage may not wholly inactivate the viruses except when a high dosage is applied, bringing about more DBPs. It is unsuitable for disinfecting wastewater with amino acids because it can react to produce cyanogen chlorides and dissolved nitrogen substances in wastewater. This can also produce organochloramines, which are all toxic and more stable.[62]

SARS-CoV-2 inactivation from hospital wastewater was investigated using sodium hypochlorite at the prescribed dosage level by WHO and China CDC.[13] It was found that complete inactivation was not achieved at this set of guidelines. In fact, the viruses were embedded in the suspended solid; hence, a higher dosage was used for complete inactivation, causing more concerns for the DBPs. At 800 g/m^3, the free chlorine declines and paved the way for the SARS-CoV-2 viruses attached to the stools in the organic matter to be released slowly.[13] More research is needed by collecting more wastewater samples to have a range of dosage necessary for complete inactivation and reduce the DBPs to avert the release of more toxicity of solid substances into the environment.

4.5.4 Ozone

Disinfection using ozone has spanned many applications in water and wastewater treatment studies, particularly in developed countries. The ozonation procedures (15–20 mg/L for 10–15 min) come after double-staged sedimentation and purification procedures before discharge.[60] Ozone has a competing application with chlorine because of its highly effective inactivation properties. It has a disinfection efficiency, using the concept of log inactivation levels (or log reduction values) between 0.56 for an LRV of 1–1.32 for an LRV of 1.32, indicating a complete virus inactivation to a 4-log at a CT of 1.0 mg/min. However, the International Ozone Association (IOA) has stated its position on the use of ozone for SARS-CoV-2 inactivation.[27] Ozone has been reported to have an oxidation-potential of 2.07 V and often leads to the formation of secondary oxidants (also called hydroxyl radicals). This makes ozone have a high reactivity and can inactivate viruses in a shorter time of reaction. Despite the corrosive/active nature and costs, ozone has the vulnerability of virus inactivation selection.[24] Extensive research on the use of ozone for specific inactivation of SARS-CoV-2 should be considered, which will help in a future pandemic or emerging viruses. Table 4.2 shows the respective treatment for ozone for different viruses. From the table, only one inactivation was done for SARS-CoV-2 with doses of about 10–20 ppm, the contact time of 10–15 min, and the inactivation rate at 3.5 $\log_{10}$. However, the exposure limit of ozone to humans is 0.05 ppm for 8 h and should be closely monitored during application.

4.5.5 Ultraviolet irradiation

Ultraviolet light (UV) is an electromagnetic wave with a length between 200 and 400 nm. The first UV application dated back to 1910 in the disinfection of drinking water.[14] Ultraviolet light can be divided into four types: ultraviolet A, UV-A (320–400 nm; 315–400 nm), ultraviolet B, UV-B (280–320 nm; 280–315 nm), ultraviolet C, UV-C (200–280 nm), and vacuum ultraviolet (100–200 nm). However, the use of vacuum ultraviolet as a disinfectant is limited due to its absorption by the wastewater.[14,62]

Table 4.2 Viruses and their respective treatment with ozone.

Viruses	Ozone	Time	Medium	Log_{10} reduction
Murine coronavirus	10	10	Air (90%)	>3
SARS-CoV-2	15–20	10–15	Air	3
HSV	10	15	Stainless steel	4
Influenza virus	10	10	Glass	4
Rotavirus	10	10	Plastic	4
Hepatitis A virus	0.3–0.4	0.08	Water	3.9
Norovirus	7	15	Wastewater	5

Source: From Quevedo-León R, Bastías-Montes JM, Espinoza-Tellez T, Ronceros B, Balic I, Muñoz O. Inactivation of Coronaviruses in food industry: the use of inorganic and organic disinfectants, ozone, and UV radiation. *Sci Agropecu* 2020;**11**(2):257–266. https://doi.org/10.17268/SCI.AGROPECU.2020.02.14.

Ghernaout and Elboughdiri[60] posited that the wavelength range value of 200–300 nm is enough to disinfect or destroy the DNA and ribonucleic acid (RNA) structure of bacteria, viruses, and single-celled microorganisms, which can thereby block the synthesis of protein. The inactivation of coronaviruses was reviewed for the food industry using UV radiation by Quevedo-León et al.[62] It was reported that for effective inactivation of the viruses using UV-A and UV-B, the exposure time needs to be extended, while the types of microorganisms affecting the effectiveness of UV-C and the exposure limit to UV-C by a human should be maintained at 2.5 J/m^2 in 10–15 min; otherwise, there is a risk of erythema (the redness of the skin or mucous membranes). Exposure to UV-A is safe at a maximum level of 10 W m^2 for 8 h, at the eye level.

Simmons et al.[63] studied the deactivation of SARS-CoV-2 on surfaces (environmental surfaces and personal protective equipment) using a pulsed-xenon UV light. The authors obtained remarkable success in the disinfection rate for hard surfaces at 1, 2, and 5 min with 3.53 $\log_{10}$, >4.54 $\log_{10}$, and >4.12 $\log_{10}$ viral load reduction, respectively, and a viral load reduction of >4.79 $\log_{10}$ in 5 min for N95 respirators. Therefore, it was concluded that using a pulsed-xenon UV light, SARS-CoV-2 was significantly reduced, and this can be replicated on other mediums where the virus may persist.

4.6 Measures to ensure the protection of personnel and wastewater treatment workers from contacting COVID-19

4.6.1 Viral contamination risk assessment

In the aspect of viral risk assessment for viruses or majorly pathogens, the quantitative microbial risk assessment (QMRA), which has dated back to over 35 years for its usefulness in determining and developing measures for safeguarding public health in line with water, food, remediation, and so forth, is employed.[64] The data obtained from exposure assessment and dose-response assessment are initially used and are obtained from viability assays, including plate counts, plaque assays, or animal infectivity. Meanwhile, with increases in the use of molecular techniques for measuring microorganisms in the environment (food, water, soil, and air), it is now pertinent to investigate how to apply such data to estimate infectious disease risks. However, the limitations to the application of these data need redress.[64]

In a study aimed at estimating the viral risk assessment of SARS-CoV-2 (the current pandemic) using the QMRA, it was found that, while using data from the literature, SARS-CoV and SARS-CoV-2 have different exposure scenarios framework.[65] The dose-response model of a surrogate coronavirus was useful as an early health warning tool for the current and future pandemic.[65] While considering the extreme cases, a three-tiered method showed an estimated risk of infection on workers higher than the derived acceptable infection risk of SARS-CoV-2,[66] indicating the urgent need for hands-on emergency response. Since the pandemic could affect the activities in WWTPs, the population that solely depends on the WWTP facility is at varying levels of risk, depending on the level of their exposure to the contaminated wastewater. It is crucial to use essential tools like QMRA to establish risk management strategies, particularly to WWTP personnel's exposure as the risk level may be higher than anticipated.

A perfect example was the laxity in the 2003 outbreak of SARS-CoV, in which the role of wastewater or sewage systems aided the sporadic spread of the virus as a secondary source, which was never anticipated.[9] However, it occurred due to a faulty sewage system but resulted in full community transmission.[23] Exposure to untreated wastewater from the source poses a higher risk to WWTP workers and inhabitants living close to the sewage system line and is of great danger to the general public. The worst could have been imagined in a situation where there are inadequate WWTPs facilities, especially in developing countries where the majority depends on unclean water to drink, cook, and other household usages.[5,23]

4.6.2 Measures to protect wastewater treatment workers

Using the QMRA, it is now possible to safeguard public health and reduce contamination via proper surveillance, detection methods, analysis, and decision making, while also concentrating on the effective molecular methods for discovery, analysis, and policymaking.[66] Using QMRA involves five major steps that help predict the burden associated with waterborne diseases, risk reduction to the community, and measures to protect water safety (users, climate, aquatic animals, etc.). The steps include hazard identification: identification and quality assessment of the microbial hazard; exposure assessment: the estimation of the time of human exposures to the diseases by laid down routes; dose-response assessment: the classification of the connection between dose and incidence of consequential effect in the population unprotected to the pathogens; risk characterization: the incorporation of information from the identification of hazard (here it is the combination of dose-response and exposure assessment that gives the extent of health effects); and risk management and communication: the decision-making section based on all recent steps in the assessment of risks.[66]

Part of the measures to ensure that health and WWTP workers are properly protected against contracting the virus is to improve on the WBE and QMRA tools so that the viruses in wastewater may have been anticipated. The provision of easy to use testing kit that will be available at the WWTPs is essential. Advanced training of personnel on how to treat sewage/wastewater, either contaminated or not, while adequate provisions of PPEs that are disposable and out of reach, i.e., with proper disposal means, should be ensured. Lastly, the sorting of the waste type via proper tagging of the waste disposal points, which will help in proffering the best sewage treatment and disposal methods, should be implemented.

4.7 Conclusion

The presence of coronaviruses like SARS-CoV and the new SARS-CoV-2 (COVID-19) in wastewater has been ascertained via selected molecular testing methods and techniques, including PCR and its variant methods. However, the current knowledge on the persistence of the SARS-CoV-2 in the wastewater environment is not sufficient to make any concluding remarks. Meanwhile, since the research on the viability and virulence of the SARS-CoV-2 is still evolving, further development of important available tools/methods such as WBE and QMRA targeting the SARS-CoV-2 and related emerging disease outbreaks is inevitable to complement the existing molecular methods. These tools would ensure adequate surveillance and monitoring of the prevalence of the SARS-CoV-2 within the population, while also assessing the possible associated risk with the persistence of the virus in wastewater.

While the development of vaccines to combat the COVID-19 is ongoing, sustainable remedial approaches that can be employed to inactivate the virus in wastewater and water environment is essential to forestall probable spread through this medium. For the WWTPs workers that maybe the line of contact with the wastewater, their safety is paramount, and it is important to train them on waste handling management that involves the waste collection, waste sorting, waste treatment, and proper disposal of any resulting DBPs while they ensure strict compliance on the rules of engagement in order to prevent against viral infections of waterborne diseases.

References

1. *WHO. COVID-19 public health emergency of international concern (PHEIC) global research and innovation forum.* World Health Organization; 2020. Published, www.who.int/publications/m/item/covid-19-public-health-emergency-of-international-concern-(pheic)-global-research-and-innovation-forum. [Accessed 10 October 2020].
2. Foladori P, Cutrupi F, Segata N, et al. SARS-CoV-2 from faeces to wastewater treatment: what do we know? A review. *Sci Total Environ* 2020;**743**:140444. https://doi.org/10.1016/j.scitotenv.2020.140444.
3. Mandal P, Gupta AK, Dubey BK. A review on presence, survival, disinfection/removal methods of coronavirus in wastewater and progress of wastewater-based epidemiology. *J Environ Chem Eng* 2020;**8**(5):104317. https://doi.org/10.1016/j.jece.2020.104317.
4. Guerrero-Latorre L, Ballesteros I, Villacrés-Granda I, Granda MG, Freire-Paspuel B, Ríos-Touma B. SARS-CoV-2 in river water: implications in low sanitation countries. *Sci Total Environ* February 2020;**743**:140832. https://doi.org/10.1016/j.scitotenv.2020.140832.
5. Adelodun B, Ajibade FO, Ighalo JO, et al. Assessment of socioeconomic inequality based on virus-contaminated water usage in developing countries: a review. *Environ Res* 2021;**192**:110309. https://doi.org/10.1016/j.envres.2020.110309 (September 2020).
6. Sharif S, Ikram A, Khurshid A, et al. Detection of SARS-Coronavirus-2 in wastewater, using the existing environmental surveillance network: an epidemiological gateway to an early warning for COVID-19 in communities. *medRxiv* 2020. https://doi.org/10.1101/2020.06.03.20121426.
7. Adelodun B, Ajibade FO, Ibrahim RG, Bakare HO, Choi K-S. Snowballing transmission of COVID-19 (SARS-CoV-2) through wastewater: any sustainable preventive measures to curtail the scourge in low-income countries? *Sci Total Environ* 2020;**742**:140680. https://doi.org/10.1016/j.scitotenv.2020.140680.
8. Hui DSC. Severe acute respiratory syndrome (SARS): epidemiology and clinical features. *Postgrad Med* 2004;**80**(945):373–81. https://doi.org/10.1136/pgmj.2004.020263.
9. McKinney KR, Gong YY, Thomas G. Lewis. Environmental transmission of SARS at Amoy Gardens. *J Environ Health* 2006;**68**(9):26–30. quiz 51.
10. Wu Y, Guo C, Tang L, et al. Prolonged presence of SARS-CoV-2 viral RNA in faecal samples. *Lancet Gastroenterol Hepatol* 2020;**5**(5):434–5. https://doi.org/10.1016/S2468-1253(20)30083-2.
11. Zhang D, Yang Y, Huang X, et al. SARS-CoV-2 spillover into hospital outdoor environments. *medRxiv* 2020;**86**(0). 2020.05.12.20097105.

12. Randazzo W, Truchado P, Cuevas-Ferrando E, Simón P, Allende A, Sánchez G. SARS-CoV-2 RNA in wastewater anticipated COVID-19 occurrence in a low prevalence area. *Water Res* 2020;**181**:115942. https://doi.org/10.1016/j.watres.2020.115942.
13. Zhang D, Ling H, Huang X, et al. Potential spreading risks and disinfection challenges of medical wastewater by the presence of Severe Acute Respiratory Syndrome Coronavirus 2 (SARS-CoV-2) viral RNA in septic tanks of Fangcang Hospital. *Sci Total Environ* 2020;**741**:140445. https://doi.org/10.1016/j.scitotenv.2020.140445.
14. Wang J, Shen J, Ye D, et al. Disinfection technology of hospital wastes and wastewater: suggestions for disinfection strategy during coronavirus Disease 2019 (COVID-19) pandemic in China. *Environ Pollut* 2020;**262**(114665). https://doi.org/10.1016/j.envpol.2020.114665. In press.
15. Westhaus S, Weber FA, Schiwy S, et al. Detection of SARS-CoV-2 in raw and treated wastewater in Germany – suitability for COVID-19 surveillance and potential transmission risks. *Sci Total Environ* 2021;**751**:141750. https://doi.org/10.1016/j.scitotenv.2020.141750.
16. Medema G, Heijnen L, Elsinga G, Italiaander R, Brouwer A. Presence of SARS-coronavirus-2 RNA in sewage and correlation with reported COVID-19 prevalence in the early stage of the epidemic in The Netherlands. *Environ Sci Technol Lett* 2020. https://doi.org/10.1021/acs.estlett.0c00357.
17. Ahmed W, Angel N, Edson J, et al. First confirmed detection of SARS-CoV-2 in untreated wastewater in Australia: a proof of concept for the wastewater surveillance of COVID-19 in the community. *Sci Total Environ* 2020;**728**:138764. https://doi.org/10.1016/j.scitotenv.2020.138764.
18. Sherchan SP, Shahin S, Ward LM, et al. First detection of SARS-CoV-2 RNA in wastewater in North America: a study in Louisiana, USA. *Sci Total Environ* 2020;**743**:140621. https://doi.org/10.1016/j.scitotenv.2020.140621.
19. Bivins A, Greaves J, Fischer R, et al. Persistence of SARS-CoV-2 in water and wastewater. *Environ Sci Technol Lett* 2020. https://doi.org/10.1021/acs.estlett.0c00730.
20. Girones R, Carratalà A, Calgua B, Calvo M, Rodriguez-Manzano J, Emerson S. Chlorine inactivation of hepatitis e virus and human adenovirus 2 in water. *J Water Health* 2014;**12**(3):436–42. https://doi.org/10.2166/wh.2014.027.
21. Shimabuku QL, Arakawa FS, Fernandes Silva M, et al. Water treatment with exceptional virus inactivation using activated carbon modified with silver (Ag) and copper oxide (CuO) nanoparticles. *Environ Technol* 2017;**38**(16):2058–69. https://doi.org/10.1080/09593330.2016.1245361.
22. Van der Laan H, van Halem D, Smeets PWMH, et al. Bacteria and virus removal effectiveness of ceramic pot filters with different silver applications in a long term experiment. *Water Res* 2014;**51**:47–54. https://doi.org/10.1016/j.watres.2013.11.010.
23. Amoah ID, Kumari S, Bux F. Coronaviruses in wastewater processes: source, fate and potential risks. *Environ Int* July 2020;**143**:105962. https://doi.org/10.1016/j.envint.2020.105962.
24. Mohan SV, Hemalatha M, Kopperi H, Ranjith I, Kumar AK. SARS-CoV-2 in environmental perspective: occurrence, persistence, surveillance, inactivation and challenges. *Chem Eng J* 2021;**405**:126893. https://doi.org/10.1016/j.cej.2020.126893 (September 2020).
25. Wang W, Xu Y, Gao R, et al. Detection of SARS-CoV-2 in different types of clinical specimens. *J Am Med Assoc* March 2020:3–4. https://doi.org/10.1001/jama.2020.3786.
26. Sims N, Kasprzyk-Hordern B. Future perspectives of wastewater-based epidemiology: monitoring infectious disease spread and resistance to the community level. *Environ Int* February 2020;**139**:105689. https://doi.org/10.1016/j.envint.2020.105689.

27. Lesimple A, Jasim SY, Johnson DJ, Hilal N. The role of wastewater treatment plants as tools for SARS-CoV-2 early detection and removal. *J Water Process Eng* July 2020;**38**:101544. https://doi.org/10.1016/j.jwpe.2020.101544.
28. Venugopal A, Ganesan H, Sudalaimuthu Raja SS, et al. Novel wastewater surveillance strategy for early detection of coronavirus disease 2019 hotspots. *Curr Opin Environ Sci Heal* 2020;**17**:8–13. https://doi.org/10.1016/j.coesh.2020.05.003.
29. Kitajima M, Ahmed W, Bibby K, et al. SARS-CoV-2 in wastewater: state of the knowledge and research needs. *Sci Total Environ* 2020;**739**:139076. https://doi.org/10.1016/j.scitotenv.2020.139076.
30. Jung Y, Park GS, Moon JH, et al. Comparative analysis of primer-probe sets for RT-qPCR of COVID-19 causative virus (SARS-CoV-2). *ACS Infect Dis* 2020;**6**(9):2513–23. https://doi.org/10.1021/acsinfecdis.0c00464.
31. Vogels CBF, Brito AF, Wyllie AL, et al. Analytical sensitivity and efficiency comparisons of SARS-CoV-2 RT–qPCR primer–probe sets. *Nat Microbiol* 2020;**5**(10):1299–305. https://doi.org/10.1038/s41564-020-0761-6.
32. Barcelo D. An environmental and health perspective for COVID-19 outbreak: meteorology and air quality influence, sewage epidemiology indicator, hospitals disinfection, drug therapies and recommendations. *J Environ Chem Eng* 2020;**8**(4):104006. https://doi.org/10.1016/j.jece.2020.104006.
33. Singer AC, Wray R. Detection and survival of SARS-coronavirus in human stool, urine, wastewater and sludge. *Preprints* 2020:1–29. https://doi.org/10.20944/preprints202006.0216.v2.
34. Samson R, Navale GR, Dharne MS. Biosensors: frontiers in rapid detection of COVID-19. *3 Biotech* 2020;**10**(9):1–9. https://doi.org/10.1007/s13205-020-02369-0.
35. WHO. *Water, sanitation, hygiene and waste management for the COVID-19 virus: technical brief.* World Heal Organ; March 2020. p. 1–9.
36. Rusiñol M, Martínez-Puchol S, Forés E, Itarte M, Girones R, Bofill-Mas S. Concentration methods for the quantification of coronavirus and other potentially pandemic enveloped virus from wastewater. *Curr Opin Environ Sci Health* 2020;**17**:21–8. https://doi.org/10.1016/j.coesh.2020.08.002.
37. Gundy PM, Gerba CP, Pepper IL. Survival of coronaviruses in water and wastewater. *Food Environ Virol* 2009;**1**(1):10–4. https://doi.org/10.1007/s12560-008-9001-6.
38. Polo D, Quintela-Baluja M, Corbishley A, et al. Making waves: wastewater-based epidemiology for COVID-19 – approaches and challenges for surveillance and prediction. *Water Res* 2020;**186**:1–7. https://doi.org/10.1016/j.watres.2020.116404.
39. Tran HN, Le GT, Nguyen DT, et al. SARS-CoV-2 coronavirus in water and wastewater: a critical review about presence and concern. *Environ Res* 2020. https://doi.org/10.1016/j.envres.2020.110265.
40. Chin AWH, Chu JTS, Perera MRA, et al. Stability of SARS-CoV-2 in different environmental conditions. *The Lancet Microbe* 2020;**1**(1):e10. https://doi.org/10.1016/s2666-5247(20)30003-3.
41. Yeo C, Kaushal S, Yeo D. Enteric involvement of coronaviruses: is faecal–oral transmission of SARS-CoV-2 possible? *Lancet Gastroenterol Hepatol* 2020;**5**(4):335–7. https://doi.org/10.1016/S2468-1253(20)30048-0.
42. Wang XW, Li J, Guo T, et al. Concentration and detection of SARS coronavirus in sewage from Xiao Tang Shan Hospital and the 309th hospital of the Chinese People's Liberation Army. *Water Sci Technol* 2005. https://doi.org/10.2166/wst.2005.0266.
43. World Health Organisation (WHO). *Potable reuse: guidance for producing safe drinking-water. Geneva.* 2017. https://www.who.int/water_sanitation_health/publications/potable-reuse-guidelines/en/.
44. WHO and UNICEF. *Progress on sanitation and drinking water: 2015 update and MDG assessment.* 2015.

45. Bhatt A, Arora P, Prajapati SK. Occurrence, fates and potential treatment approaches for removal of viruses from wastewater: a review with emphasis on SARS-CoV-2. *J Environ Chem Eng* 2020;**8**(5):104429. https://doi.org/10.1016/j.jece.2020.104429.
46. Bosch A, Guix S, Sano D, Pintó RM. New tools for the study and direct surveillance of viral pathogens in water. *Curr Opin Biotechnol* 2008;**19**(3):295–301. https://doi.org/10.1016/j.copbio.2008.04.006.
47. Santos SD, Adams EA, Neville G, et al. Urban growth and water access in sub-Saharan Africa: progress, challenges, and emerging research directions. *Sci Total Environ* 2017:607–8. https://doi.org/10.1016/j.scitotenv.2017.06.157. 497–508.
48. Pooi CK, Ng HY. Review of low-cost point-of-use water treatment systems for developing communities. *npj Clean Water* 2018;**1**(1). https://doi.org/10.1038/s41545-018-0011-0.
49. Adelodun B, Ajibade FO, Ogunshina MS, Choi K-S. Dosage and settling time course optimization of *Moringa oleifera* in municipal wastewater treatment using response surface methodology. *Desalin Water Treat* 2019;**167**:45–56. https://doi.org/10.5004/dwt.2019.24616.
50. Adelodun B, Ogunshina MS, Ajibade FO, Abdulkadir TS, Bakare HO, Choi KS. Kinetic and prediction modeling studies of organic pollutants removal from municipal wastewater using *Moringa oleifera* biomass as a coagulant. *Water* 2020;**12**(7):2052. https://doi.org/10.3390/w12072052.
51. Morrison CM, Betancourt WQ, Quintanar DR, Lopez GU, Pepper IL, Gerba CP. Potential indicators of virus transport and removal during soil aquifer treatment of treated wastewater effluent. *Water Res* 2020;**177**:115812. https://doi.org/10.1016/j.watres.2020.115812.
52. CDC. Interim Laboratory. *Biosafety guidelines for handling and processing specimens associated with coronavirus disease 2019 (COVID-19)*. Centers for Disease Control and Prevention; 2020. Published, https://www.cdc.gov/coronavirus/2019-ncov/lab/lab-biosafety-guidelines.html. [Accessed 9 October 2020].
53. Ahmed W, Bertsch PM, Bibby K, et al. Decay of SARS-CoV-2 and surrogate murine hepatitis virus RNA in untreated wastewater to inform application in wastewater-based epidemiology. *Environ Res* August 2020;**191**:110092. https://doi.org/10.1016/j.envres.2020.110092.
54. Tan CW, Chia WN, Qin X, et al. A SARS-CoV-2 surrogate virus neutralization test based on antibody-mediated blockage of ACE2–spike protein–protein interaction. *Nat Biotechnol* 2020;**38**(9):1073–8. https://doi.org/10.1038/s41587-020-0631-z.
55. Bošnjak B, Stein SC, Willenzon S, et al. Novel surrogate virus neutralization test reveals low serum neutralizing anti-SARS-CoV-2-S antibodies levels in mildly affected COVID-19 convalescents. *medRxiv* 2020. https://doi.org/10.1101/2020.07.12.20151407. 2020.07.12.20151407.
56. Gernaey KV, Sin G. *Wastewater treatment models*. Elsevier Inc.; 2013. https://doi.org/10.1016/b978-0-12-409548-9.00676-x.
57. Zhang CM, Xu LM, Xu PC, Wang XC. Elimination of viruses from domestic wastewater: requirements and technologies. *World J Microbiol Biotechnol* 2016;**32**(4):1–9. https://doi.org/10.1007/s11274-016-2018-3.
58. Arraj A, Bohatier J, Laveran H, Traore O. Comparison of bacteriophage and enteric virus removal in pilot scale activated sludge plants. *J Appl Microbiol* 2005;**98**(2):516–24. https://doi.org/10.1111/j.1365-2672.2004.02485.x.
59. Chaudhry RM, Nelson KL, Drewes JE. Mechanisms of pathogenic virus removal in a full-scale membrane bioreactor. *Environ Sci Technol* 2015;**49**(5):2815–22. https://doi.org/10.1021/es505332n.

60. Ghernaout D, Elboughdiri N. Urgent proposals for disinfecting hospital wastewaters during COVID-19 pandemic. *OALib* 2020;**07**(05):1–18. https://doi.org/10.4236/oalib.1106373.
61. Lahrich S, Laghrib F, Farahi A, Bakasse M, Saqrane S, El Mhammedi MA. Review on the contamination of wastewater by COVID-19 virus: impact and treatment. *Sci Total Environ* 2021;**751**:142325. https://doi.org/10.1016/j.scitotenv.2020.142325.
62. Quevedo-León R, Bastías-Montes JM, Espinoza-Tellez T, Ronceros B, Balic I, Muñoz O. Inactivation of Coronaviruses in food industry: the use of inorganic and organic disinfectants, ozone, and UV radiation. *Sci Agropecu* 2020;**11**(2):257–66. https://doi.org/10.17268/SCI.AGROPECU.2020.02.14.
63. Simmons S, Carrion R, Alfson K, et al. Deactivation of SARS-CoV-2 with pulsed xenon ultraviolet: implications for environmental COVID-19 control. *Infect Control Hosp Epidemiol* 2020:1–4. https://doi.org/10.1017/ice.2020.399.
64. Haas CN. Quantitative microbial risk assessment and molecular biology: paths to integration. *Environ Sci Technol* 2020;**54**(14):8539–46. https://doi.org/10.1021/acs.est.0c00664.
65. Zaneti RN, Girardi V, Spilki FR, et al. Quantitative microbial risk assessment of SARS-CoV-2 for workers in wastewater treatment plants. *Sci Total Environ* 2021;**754**:142163. https://doi.org/10.1016/j.scitotenv.2020.142163.
66. Ramírez-Castillo FY, Loera-Muro A, Jacques M, et al. Waterborne pathogens: detection methods and challenges. *Pathogens* 2015;**4**(2):307–34. https://doi.org/10.3390/pathogens4020307.

CHAPTER 5

A comprehensive study of COVID-19 in wastewater: occurrence, surveillance, and viewpoints on its remedy

Moumita Bishai
Department of Botany, Gurudas College, Kolkata, West Bengal, India

5.1 Introduction

Wastewater is a multifarious compound containing an extensive series of physical, chemical, and microbial pollutants released due to human activities. Most of the infectious disease spread through inappropriate management of wastewater causing precarious threats to worldwide public health. The crisis is expected to be exaggerated in the near future due to rapid population increases, change of climate, natural disasters, immigration, globalization, and so forth.[1]

Most infectious diseases are related to microbial agents—principally to viruses. The ongoing episode of novel coronavirus (SARS-CoV-2) has represented a huge worldwide general health threat and caused intense social and financial waves. It was declared as a Public Health Emergency of International Concern (PHEIC), officially, by the World Health Organization (WHO). As of November 29, 2020, more than 61,654,661 infected persons and over 1,444,596 deaths was reported globally.[2] The reaction of the global communal health experts demonstrated an unavoidable delay because of absence of understanding of the COVID-19 infection.[3]

The development of wastewater-based epidemiology (WBE) tools are represented as a harmonizing method for current infectious disease surveillance system.[4]

The present chapter provides a comprehensive overview on various aspects of wastewater, mainly municipal and hospital wastewater. It further strengthens our knowledge on the existence, tenacity, and possibility of the public health threat coupled with the spread of SARS-CoV-2 in aquatic

Environmental Management of COVID-19
ISBN 978-0-323-85780-2
https://doi.org/10.1016/B978-0-323-85780-2.00015-9

waste. Finally, the capability to quickly scrutinize the extent of disease for prevention, intervention, and control, while having several limitations, were also discussed.

5.2 Wastewater: characterization and classification

The amalgamation of different squanders in an aquatic system, which were released from households, organizations, commercial, and industrialized institutions, together with groundwater, surface water, and storm water has been termed "wastewater." They comprised of 99.9% water and 0.1% solids and are organic in nature. In a community, the principal sources of wastewater are the human being and animal wastes. Other important sources include industrial wastes, household wastes, groundwater and stormwater. Corresponding sources producing wastewater have precise properties.[5]

Wastewater has been additionally categorized depending on the sources, i.e., municipal, industrial, sanitary, storm water combined, and health care wastewater as described in Fig. 5.1.

Different types of wastewater contain varied substances, which were accustomed for explicit categorization. The varieties of different substances with their amounts or concentration are source dependent.

5.2.1 Physical characterization of wastewater

The physical properties of aquatic waste body depend on solid content, color, turbidity, taste, and odor, along with temperature.[6] With the exception of gases, the entire mass of pollutants of water belongs to the **solid** content. The characterization is based on their form, their chemical properties, and their quantitative distribution. The amount of suspended solids in wastewater treatment is a notable aquatic characteristic parameter,

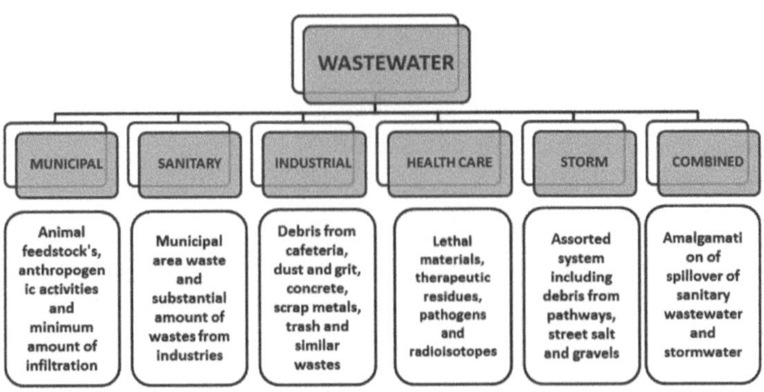

Figure 5.1 Classification of wastewater.

which needs to be utilized for assessing the effluent nature and for screening their performance with regard to numerous processes. **Color** is one additional important physical parameter to use when ruling on the features of water. For the treatment of wastewater, the question is not necessarily the appearance or color; the alternative suggested its use as an indicator to predict the wastewater condition. At the incipient the color of fluid of wastewater is light brownish-gray. The dissolved oxygen (DO) containing wastewater is normally gray in color. With the associated foul odor of wastewater, the color turns black, having no DO and thus depicted as septic in nature. The increase of travel time of the water in the pool and absence of oxygen turns the wastewater from gray to dark gray and ultimately to black. **Turbidity** is yet another measure. It is the degree where light absorption or scattering by suspended material in water takes place. Microbes, vegetable resources, detergents, and soaps, along with various other blending agents, contributed to turbidity. For the management of wastewater, measurements of turbidity are vital when ultraviolet (UV) irradiation is employed during the decontamination processes. **Odors** are generally produced by gases due to putrefaction of biological stuff or by constituents added to the wastewater. Also the **temperature** of wastewater found to be much higher compared to the supply of water. It occurs due to the mixing of warm aquatic wastes from human activities and industrial effluents released; nevertheless, significant amounts temperature fluctuations occurs due to infiltration or storm water flow. During the course of the day, there is variation in the **flow rates** of the wastewater. The phenomenon of variation is known as the diurnal flow variation.[6]

5.2.2 Chemical characterization of wastewater

Generally, the aquatic waste released from different sectors depicted versatile chemical properties. Primarily they were divided into three categories: organic and inorganic matter, along with gases. In the medium strength wastewater, almost three-fourths of the suspended solids along with 40% of solids are filterable and are organic in nature. Proteins, carbohydrates, oil, and grease, along with detergents (surfactants) are included under the organic substances. About 15%–30% of organic matter are **proteinaceous** in nature. Urea with protein are the foremost nitrogenous sources in wastewater. The products released from the breakdown of **carbohydrate** molecules during anaerobic condition are alcohols and organic acids, as well as gases such as hydrogen sulfide and carbon dioxide. Production of huge amounts of organic acid influences the process of treatment by overstraining the protecting capability of the aquatic waste, which causes a drip in pH with the stoppage of biological motion. Categorization of grease includes **oils and fats**, along with waxes and other associated constituents present in aquatic waste. Presence of grease in wastewater is the main reason for obstruction of filters, spigots, and sand beds. **Detergents** are considered to be one of the largest organic molecules. They are soluble in water, and in the wastewater treatment plants they create foams.

Apart from these, there are several other inorganic characters for wastewater, which include determination of pH, alkaline nature, and nitrogenous properties of the aquatic system along with presence of chlorides, phosphates, sulfur, toxic inorganic compounds, and heavy metals, of which quantities vary depending on the type of wastewater.[6]

5.2.3 Biological characterization of wastewater

Organisms that are competent of contaminating or spreading the disease in humans and animals are known as pathogens. The wastewater comprises species of different bacteria, protozoa and parasitic insects (helminths). Viruses were also found, which cause severe infectious diseases.[6] Among the viruses, different types of coronaviruses including SARS-CoV-2 were also found in the wastewater.[7] The existence of these viruses in wastewater is governed by several physicochemical factors such as variations in temperature and pH; the presence of organic matter and oxidants; and plenty of hostile bacteria.[8] The details are discussed in the later part of the chapter.

There are numerous categories of wastewater, but keeping in mind the purview of the topic, the present chapter principally focuses only on different aspects of municipal and health care wastewaters.

5.3 Study of municipal wastewater: characterization and classification

Wastewater created due to "human activities in households" is termed as municipal wastewater. These wastes are discharged from residential areas, workplaces, and factories. They are originated from showers, toilets, and washing machines.[6] They are classified as blackwater and graywater (Fig. 5.2).

1. **Blackwater**: In this case, dual modes of collection of effluents were considered. One is classic and the other one is selective collection. The former consists of entire aquatic collection and excretions discharged from toilets, whereas selective collections are usually from source-separating toilets and are composed of a fecal bin and urine

Figure 5.2 Classification of municipal wastewater.

diverter. Such selective collection allows compartmentalization of yellow water (urine and water from flush) with brown water (feces and toilet paper).

2. **Graywater**: Various emission sources were considered under this category. Effluent consists of kitchen sink and dishwasher discharge, and wash basins in bathrooms, baths, and showers.

Tables 5.1 and 5.2 show the overall physical and chemical characteristics of untreated municipal wastewaters. In many developing countries, municipal wastewater generation has superseded the other types of wastewater generation and has become the major cause of water pollution and eutrophication.

The microbial load of municipal wastewater was found to be versatile in nature, having bacteria, alga, protozoa, and viruses. The pathogenic bacteria include fecal coliform, fecal *Streptococci*, and *Clostridium perfingens*, along with some anaerobic bacteria such as *Clostridium sporogens*, *Bifidobacterium*, and Methanogenic bacteria like *Methanobacterium*. Algal composition includes *Chlorella phormidum*, *Ulothrix*, etc. Different viruses such as Poliovirus, Rotavirus, etc. were found in sewage which acquires access through stools of patients.[6,9] Presence of SARS-CoV-2 virus was also reported by Randazzo et al.[10], Wu et al.[11] and many more.

Table 5.1 Physical characteristics of municipal wastewater.[86]

Parameter	Remarks
Solid	Upended and dissolved organic solids
Color	Dark gray to black
Odor	Foul order due to hydrogen gas decomposition
Temperature	Presence of microbe and difference in the solubility of gases increase the temperature
Turbidity	High turbidity
Flow rate	Leer than pure water because of presence of upended solids

Table 5.2 Chemical characteristics of municipal wastewater.

Parameter	Description of the content	References
pH	6.8–8.3	87,88
Chloride	30–100 mg/L	89
Alkalinity	50–200 mg/L	88
Oil and grease	50–150 mg/L	87
Heavy metals	Ni, Cu, Cr, Cd, Pb, and Hg	87
BOD_5	110–400 mg/L	87,89
COD	250–1000 mg/L	88
Total nitrogen	20–85 mg/L	89
Total phosphorous	4–15 mg/L	87
Total sulfur	20–50 mg/L	89

5.4 Study of health care wastewater: characterization and classification

Hospital wastewater (HWW) usually generates from various types of medical care and its related activities. It comprises a variety of toxic substances, which include pharmaceutical products, radioisotopes, pathogens, solvents, and disinfectants. The characteristics of HWWs are comparable with municipal wastewater to some extent, but individually it also contains some other toxic pollutants.[12]

In the waste released from hospitals, approximately 85% of the waste are equivalent with municipal waste and termed as nonhazardous. The remaining 15% are hazardous.

5.4.1 Hazardous health care waste

These wastes were classified as infectious waste, sharps, pathological waste, pharmaceutical and cytotoxic waste, chemical waste, and waste from radioactive sources, as described in Fig. 5.3.

5.4.2 Nonhazardous health care waste

The wastes other than physical, chemical, biological, or radioactive are known as nonhazardous hospital waste.

Although different physical, chemical, and biological pollutants encompass hospital wastewater, the effluents for both conventional and nonconventional parameters were characterized and summarized in Table 5.3.

The hospital effluents are sources of inorganic/organic loads and were demonstrated by their different concentration ranges of both macro- and micropollutants.[13] Gadolinium (Gd), platinum (Pt), and mercury (Hg) were

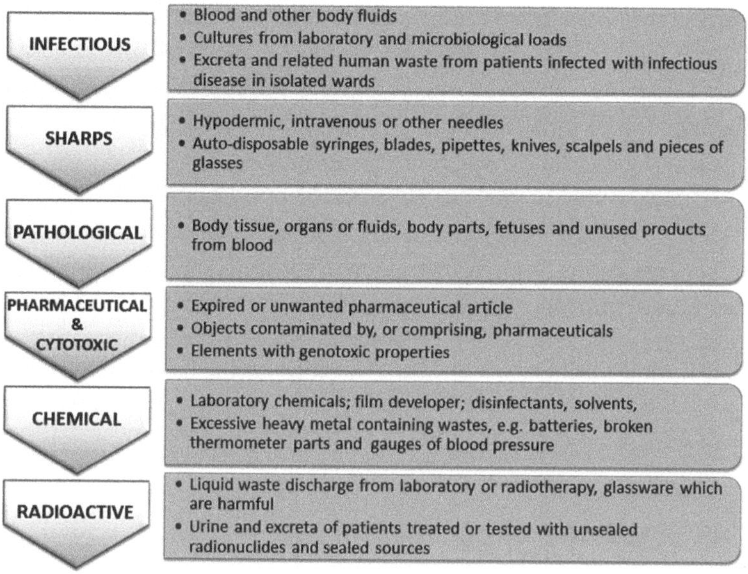

Figure 5.3 Classification of hazardous health care waste.

Table 5.3 Different characteristic properties of hospital wastewater.

Parameter	Concentration	References
pH	6–9	90
Electrical conductivity	300–2700 μS/cm	16
Total suspended solid	116–3260 mg/L	18
Chlorides	80–400 mg/L	90
Fats and oil	50–210 mg/L	91
Total Nitrogen	60–230 mg/L	22
Total phosphate	6–19 mg/L	16
COD	39–7764 mg/L	18
BOD	16–2575 mg/L	90
Heavy toxic metals	Zn, Pb, Ni, Ag, Hg, Gd	91
Total surfactant	4–8 mg/L	16
Microbial load	*E. coli*, enterococcus, adenovirus, rotavirus, norovirus, hepatitis A	22

found to be the main heavy metals in hospital runoffs.[14] Apart from these, Cd, Cu, Fe, Ni, Pb, Zn, and other heavy metals are similar in concentrations as reported in municipal discharge.[15] Various pollutants such as Gd, Hg, Pt, and pharmaceutical products along with acetaminophen, caffeine, ciprofloxacin, and gabapentin are released as hospital waste.[16–21] Along with the other microbes it also shows the presence of the SARS-CoV-2 virus.[22,23]

Hence, it could be predicted that when an infected individual's excreta containing pathogens gets flushed down the toilet or washed down the drain, it travels through the sewage system of a community. Exposure to such pathogens in a wastewater system could potentially cause serious health consequences in the community. Therefore, assessing the potentiality of infection and transmission of the disease through the wastewater system is desired.

5.5 Characteristics of coronavirus

Coronaviruses (CoVs) are a type of enteric virus that contain single-stranded positive RNA genome of the length of 27–32 Kb.[24] They belong to the order *Nidovirales*, family *Coronaviridae*, and subfamily *Coronavirinae*, and are the main cause of the present pandemic.[1] Spikelike projections on their exterior surface seen under the electron microscope are the glycoproteins, imitated as crownlike, hence the name coronavirus (CoV).[25]

Coronavirinae are comprised of four genera, Alpha-, Beta-, Gamma-, and Delta-, among which infection in *Homo sapiens* was instigated by the first two (Alpha- and Beta-) host viruses.[26] Among them, a novel virus, named severe acute respiratory syndrome coronavirus 2 (SARS-CoV-2) belongs to the beta- CoVs category. They cause the disease COVID-19 (CO means Corona, VI is Virus, D is disease, 19 is the year it was first discovered).[1] In December 2019, the virus spread from Wuhan, China, to the rest of the world. They showed high affinity for binding to the human angiotensin-converting enzyme 2 (ACE2) receptors.[1] Principally, SARS-CoV-2, along with ACE2 receptors, penetrate inside the host's system, where the proteins and the viral RNA are synthesized into multiple copies inside the cytoplasm to congregate the new virion gene.[27]

Though the primary modes of the viral transmission are through respiratory droplets and direct or indirect contact, SARS-CoV-2 virus might get incorporated into the wastewater (municipal and hospital) through numerous sources, such as hand washing, sputum and vomiting, fecal waste, etc.[28] Hence, the potentiality of wastewater and its effect on

SARS-CoV-2 was investigated for further understanding the epidemiology of COVID-19.

Existence of the virus in wastewater is highly influenced by subsequent factors, both intrinsic and extrinsic, depending on the wastewater or ecological environment:

- **Structure of the virus**: Enveloped viruses similar to coronavirus have shorter survival periods compared to nonenveloped viruses.[8] This was owing to the activity of detergents and proteolytic enzymes on the peripheral wall of the virus, which is lipid in nature.[29]
- **Composition of the wastewater**: The component of the wastewater likewise plays a vital function in the persistence of viruses. The rapid inactivation of coronaviruses in wastewater is credited to the bacterial proteolytic enzymes and existence of chemicals that are antiviral in nature, along with predation of protozoa and metazoa in the wastewater.[29]
- **Temperature**: Temperature has significant influence on the survival of coronaviruses in wastewater, similar to other microbes. The virus may survive longer in temperate or colder regions compared to tropical regions. With the increasing temperature, there is reduction in the persistence of the virus, which is attributed to the nucleic acid and protein denaturation, along with intensification of extracellular enzyme activity.[30,31]
- **pH**: Significant modifications in the configuration of proteins was achieved by alteration of pH. A pH variation toward its isoelectric point causes precipitation;[32] the steadiness of SARS-CoV-2 in different pH values ranges from 3 to 10 at room temperature.[32]
- **Suspended solid**: The suspended solids act as a shield to protect coronaviruses from inactivation.[8] Studies have shown the effect of suspended solids and organic matter on the survival of coronaviruses where it was reported that coronaviruses can survive longer in primary wastewater than in activated sludge wastewater. Likewise, Zhang et al.[23] reported the presence of organic matter in the patient's feces, which protect the SARS-CoV-2 virus from the disinfection process of medical wastewater.

Thus, the persistence of coronaviruses in wastewater is intensely reliant on the properties of the individual aquatic medium.

5.6 Occurrence and persistence of COVID-19 in wastewater

SARS-CoV-2 strains have also been found to persist inside human beings, such as feces, sputum, and serum, but are unstable in urine, due to the reduction of pH values and the presence of urea. Conversely, saliva showed SARS-CoV-2 positivity and is responsible for the spread of COVID-19 with the help of air. Also, contamination of water occurs when infectious saliva is mistakenly touched by a clinically fit individual. Depending on excessive load of SARS-CoV-2 on saliva as well as in stool samples of COVID-19 patients, contaminated aquatic waste released from the health sector, isolation centers, and municipal areas and from the areas with cases of positive COVID-19[33] finally indicated transmission of SARS-CoV-2 through the fecal route. From feces, the virus gets diluted in toilet water and then in other municipal wastewater, including graywater, to the wider environment. Graywater is anticipated to not be ideal for SARS-CoV-2 transmission, though it contains high viral titers of body fluids.[34] The transport of the viral RNA through a complex sewage system compel them to be exposed to different chemicals and fluctuating temperatures. Diagrammatic route of transmission of the SARS-CoV-2 virus has been presented in Fig. 5.4[35–37]

On the basis of the decaying nature of the pathogens, that is, SARS-CoV-2, the degree of the health threat varies in the aquatic environment. There are possibilities to improve the control measures and

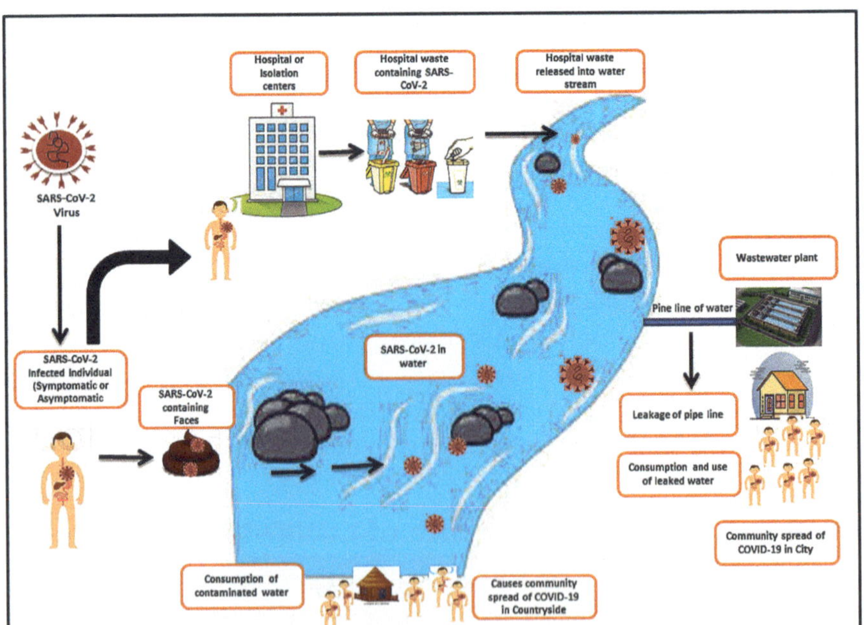

Figure 5.4 Representation of SARS-CoV-2 transmission in wastewater.

wastewater treatment with the analysis of the decomposition of viral RNA. It has been found that the average duration of life of SARS-CoV-2 in the feces was up to 22 days, which was longer than that of SARS-CoV-1. It has also indicated that SARS-CoV-2 can survive longer in the fecal specimens than in respiratory (18 days) and serum (16 days).[38] Van Doremalen et al.[39] confirmed the variation of the SARS-CoV-2 half-life from 0.8 h on copper to 6.8 h on plastic.

Limited data are available on SARS-CoV-2 infection related to its occurrence and its molecular nature in feces and wastewater. Viral shedding in feces has been reported[23,40,41] and summarized in Table 5.4.

After the feces is released into the wastewater, the presence of SARS-CoV-2 in wastewater endorses the fecal transmission route.

Table 5.4 Study on occurrence of SARS-CoV-2 in the fecal sample.

References	Findings
41	In feces, the SARS-CoV-2 positive specimens were 44 out of 153 patients (29%). For the stool specimens, the mean series of threshold value was 31.4, which indicates a viral load of less than 2.6×10^4 copies/mL. In nasal swabs the load was 1.4×10^6 copies/mL.
92	The researchers tested on 10 pediatric SARS-CoV-2 infection cases highlighted that some patients were positive on rectal swabs even after their nasopharyngeal testing had become negative.
93	Fecal samples of nine hospitalized patients with COVID-19 during the course of the disease showed high viral RNA concentrations in initial samples, with a peak during the first week of symptoms. The viral content declined gradually. The viral load in feces varies in the range 10^3–10^7 RNA copies/g feces.
94	The viral RNA in the stool was tested on the day of hospitalization. 15.3% of the patients detected the presence of SARS-CoV-2 RNA. In nine positive patients an average viral load of 104.7 copies/mL was found in the stool.
11	While investigating 74 fecal samples from patients, it was observed that over 50% of patients showed presence of SARS-CoV-2 RNA though the respiratory tract samples became negative
95	The feces of 73 hospitalized patients were examined for SARS-CoV-2 infection and it was evaluated that 53.4% of the patients showed positive results for viral RNA in stool. Additionally, greater than 20% of the patients having SARS-CoV-2 remain positive in feces, even after showing negative results in respiratory samples

Continued

Table 5.4 Study on occurrence of SARS-CoV-2 in the fecal sample.—cont'd

References	Findings
46	The researchers investigated 23 fecal samples, showing 83% positivity for SARS-CoV-2 infection. The average period of virus shedding was 22 days for feces when the titer value for the virus in feces was 5623 copies/mL, but the highest titer at the peak reached 105.8 copies/mL.
40	65 feces samples of hospitalized patients were tested for the presence of SARS-CoV-2, of which 22 (52.4%) were positive. Among them, 6 having GI symptoms that showed esophageal bleeding when subjected to endoscopy. SARS-CoV-2 RNA was detected in esophagus, stomach, duodenum, and rectum specimens for both two severe patients. Whereas, in four nonsevere patients only duodenum showed SARS-CoV-2 positivity.
96	Among the 42 hospitalized patients, 28 showed SARS-CoV-2 infection. Among them, 64.3% of patients remained positive after the pharyngeal swabs turned negative.
97	From 204 patients, 35 confirmed cases of SARS-CoV-2 infected samples were found. The viral loads were less than the respiratory samples (range 550 copies per mL to $1 \times 21 \times 10^5$ copies/mL

One of the most important studies was performed at the Ahmedabad wastewater treatment plant (WWTP) at Gujarat. Kumar et al. in 2020 reported 106 million liters per day wastewater, which receives effluent from the civil hospital treating COVID-19 patients.[42]

Ahmed et al.[43] confirmed the presence of SARS-CoV-2 in untreated wastewater samples collected from pumping stations and wastewater treatment plants in Queensland, Australia.

Alternatively, samples collected from influent and secondary-treated wastewater were investigated for the existence of SARS-CoV-2 virus in Japan.[44] SARS-CoV-2 presence was confirmed in five secondary-treated wastewater samples with a load of 2.4×10^3 copies/L. In contrast, the same was not detected in the influent wastewater samples. The authors explained the result on the basis of the limit of detection (LOD). The LOD value for the secondary-treated wastewater (1.4×10^2–2.5×10^3 copies/L) was unusually lowered compared to influent (4.0×10^3–8.2×10^4 copies/L). This is because the former (200 mL) had a lower filtration volume than the latter (5000 mL).

Likewise, there are many more cases in different parts of the world that corroborate the fact of viral transmission from feces to wastewater[23,45,46] and treated wastewater[10,47] in the recent worldwide outbreak, which was summarized in Table 5.5.

Overall, from the above studies it could be anticipated that the routine wastewater monitoring can help to isolate a noninvasive warning sign to alert the community for new SARS-CoV-2 infections, as a large number of the viral carriers are asymptomatic and are present in a community which might intensify the epidemic abruptly.

Table 5.5 Details of occurrence of SARS-CoV-2 in wastewater environment in different countries.

Name of the place	Findings	References
Australia	Concentrated SARS-CoV-2 RNA copies were classified using RT-qPCR. Two positive cases were detected within a six-day period from the wastewater treatment plant. The observed viral RNA copy numbers are then simulated using Monte Carlo model which estimated a median range of 171–1090 infected persons.	43
China	Three hospitals wastewater samples were used for detection of SARS-CoV-2 viral RNA. It detected 255 copies/L of SARS-CoV-2 viral RNA in adjusting tank of first hospital wastewater. The second hospital wastewater was treated by a series of treatment steps in which SARS-CoV-2 viral RNA was detected in the adjusting tank (633 copies/L), but not detected in MBBR and sedimentation tank. In the third hospital treatment unit, SARS-CoV-2 viral RNA was detected in the range of 557–18,744 copies/L, only in septic tank where 800 mg/L sodium hypochlorite was used. It was not detected with 6700 mg/L of sodium hypochlorite.	23,45
Czech Republic	Concentrated SARS-CoV-2 RNA was detected in 11.6% of samples and more than 27.3% of WWTPs; in many places, SARS-CoV-2 was detected repeatedly.	98

Continued

Table 5.5 Details of occurrence of SARS-CoV-2 in wastewater environment in different countries.—cont'd

Name of the place	Findings	References
France	A time-course quantitative analysis of SARS-CoV-2 by RT-qPCR showed rise of genomic units; six out of eight samples of treated wastewater were positive in crude wastewaters precisely tracked the escalation of human COVID-19 cases observed at the provincial level. The viral genomes were identified before the commencement of the exponential progress of the epidemic. An obvious reduction in the extents of genomic units was observed concurrently with the decrease in the number of new COVID-19 cases.	47
India	All three genes of SARS-CoV-2 were discriminated in the influents with none of the genes were spotted in the effluent. Temporal difference between different samples was of 10× in gene copy loading with corresponding change of 2× in the number active COVID-19 patients in the city. Samples from different local municipal WWTPs and hospital wastewater showed the occurrence of SARS-CoV-2 viral genome at an ambient temperature of 40°C.	3,42
Israel	SARS-CoV-2 RNA was systematically detected in the influent of the primary settler suggesting that, the virus particles have a higher affinity for the sludge. However, in the digested sludge genetic material is not detected. The results further confirm the safety of the sludge after thermal treatment and anaerobic digestion. The combined treatment of thermal hydrolysis and anaerobic digestion also prevented the detection of SARS-CoV-2 in the sludge leaving the plant.	54
Istanbul	Five samples out of seven from wastewater and all samples from manholes were tested positive. SARS-CoV-2 virus titers of manhole were higher than those of inlet WWTPs. Quantitative measurements of SARS-CoV-2 in wastewater were carried out using WBE which showed positive tested viral titer ranges from 1.17×10^4 to 4.02×10^4/L.	99

Table 5.5 Details of occurrence of SARS-CoV-2 in wastewater environment in different countries.—cont'd

Name of the place	Findings	References
Italy	Molecular analysis was undertaken with three nested protocols. SARS-CoV-2 RNA detection was accomplished in volumes of 250 mL of wastewaters collected in areas of high and low epidemic circulation, according to clinical data. Overall, 6 out of 12 samples were positive. One of the positive results was obtained in a Milan wastewater sample collected a few days after the first notified Italian case of autochthonous SARS-CoV-2.	1,55
Japan	SARS-CoV-2 RNA was detected in secondary-treated wastewater samples collected of 2.4 × 103 copies/L by qPCR. SARS-CoV-2 RNA was detected in a secondary-treated wastewater sample when the weekly reported cases in the community were high.	44
Netherlands	RT-PCR performed against three fragments of the nucleocapsid protein gene and one fragment of the envelope protein gene. The fragments were detected in sewage sites, even when the COVID-19 prevalence is low.	56
Spain	By using the real-time RT-PCR three regions of the virus nucleocapsid (N) gene were targeted. Quantification of SARS-CoV-2 RNA was performed. The titers reported for wastewater samples as 5.4 ± 0.2 $\log_{10}$ genomic copies/L on average. Two secondary water samples were positive and all tertiary water samples tested as negative. This environmental surveillance reveals that members of the community were shedding SARS-CoV-2 RNA in their stool even before the first cases were reported by local or national authorities.	10
USA	RT-qPCR technique was operated, which confirmed the presence of SARS-CoV-2 at high titers approximately of 100 viral particles per mL of sewage. Monitoring of wastewater demonstrated that SARS-CoV-2 RNA by RT-qPCR correlated with prevalence of viral infections in the community. Data indicated that symptom onset precedes detection of SARS-CoV-2 RNA in wastewater by five to eight days.	11,100

5.7 Detection of coronavirus in wastewater

Because the virus is novel and has fatal outbreak, detection of SARS-CoV-2 in various environmental matrices is a major bottleneck at present. There is an urgency to identify highly specific and sensitive diagnostic measures for infected areas or people to avoid its further spread. It was observed that the testing rates for COVID-19 were quite low.[48,49] Ever since the feces or urine showed early detection of the virus,[50] such sample testing from a community on a regular basis would minimize the spread of the virus. At present, various other diagnostic and detection methods are needed.

For prompt, regular governing of viral outbreaks, WBE holds greater potential by regularly monitoring the variety and concentration of the virus in wastewater. The approach of WBE is thought to be effective for underdeveloped countries to scrutinize the predominance of the COVID-19 infectious spread in the community. Another advantage of WBE is that with the help of phylogenetic studies it supported the detection of variations in the viral strains.[51] Since many cases of SARS-CoV-2 infection are asymptomatic, the recognition of the inhabitants, which act as vectors to avert dissemination of the virus, is highly needed.

For accurate detection, the samples need to be processed for measuring SARS-CoV-2 appropriately. This involves preparation of the sample, concentration, RNA extraction, and finally genomic analysis. Proper biosafety protocols for processing the wastewater samples that may contain SARS-CoV-2 should be followed as recommended by CDC.[52]

The process of concentration uses microbiological and molecular approaches prior to the detection process.[53] The recovery of virus by concentrating methods is way more challenging. There are various methods for concentration, such as PEG-based separation methods, applied electronegative membranes filtration methods, and applied ultrafiltration methods,[10,43,54–56] among them, concentration using PEG is the most prevalent method used for the COVID-19 WBE.[57]

Followed by concentration is the method of detection or analysis. The most common methods are the reverse transcription-polymerase chain reaction (RT-PCR) and the reverse transcriptase quantitative polymerase chain reaction (RT-qPCR). For example, Ahmed et al.[43] applied RT-qPCR to detect SARS-CoV-2 infected individuals in the catchment basin. Similar findings have been reported elsewhere.[4,23,44,56]

Although these PCR techniques have been applied as a yardstick for detection of SARS-CoV-2, they are unable to distinguish between infectious and inactive fractions.[58] Due to such inadequacies, another complementary method was adopted where direct absolute quantification of virus genome copy numbers in a sample without the necessity of external calibration can be performed. Digital PCR (dPCR) has recently been used for SARS-CoV-2 detection.[59]

Owing to the rapid growth of confirmed cases of COVID-19, there is a serious need to develop a robust tool to address the challenge.

The biosensing technique has emerged with an interdisciplinary approach, which enables point-of-care diagnosis for swift and fast viral detection and prevents epidemics at an initial stage. In this context, Bhalla et al.[60] and Samson et al.[61] has reviewed the opportunities and technical challenges related to biosensors and analytical tools.

In a biosensor-based method, as reported,[62] a paper-based analytical device was constructed for the speedy determination of the pathogens. The small analytical tool has different functional areas with a wax printer that integrates all the processes required for nucleic acid testing into an inexpensive paper material. In a way, a paper-based device has the potential for the diagnosis of COVID-19 in aquatic waste in real-time for tracking the viral carriers in the community.[32,63]

Yet another technology, the CRISPR-based DETECTR, has been recognized to reconfigure SARS-CoV-2 detection within a few days. The assay was performed by simultaneous reverse transcription and isothermal amplification using loop-mediated amplification (RT–LAMP). The extracted RNA was taken in universal transport medium (UTM), followed by Cas12 detection of predefined coronavirus sequences, after which detection of the virus was confirmed by the cleavage of a reporter molecule.[64] The CRISPR-based DETECTR innovation is discovered to be very adaptable to reconfigure within a short time to distinguish SARS-CoV-2. The future improvement of different versatile microfluidic based cartridges and lyophilized reagents could empower purpose of care testing outside the indicative research facility.[64]

For inactivating SARS-CoV-2 viral genome, photocatalytic nanoparticles (NPs) were also considered where titanium dioxide (TiO_2) showed photocatalytic properties under UV light. They remain dormant and are harmless in nature. If effective, the TiO_2 photocatalysis could be used for the deactivation of COVID-19.[65,66]

Electrospun nanofiber membranes were also found to be effective monitoring tools for screening disease causing pathogens. They play vital role in sieving intestinal *Enterococci*, *E. coli*, and coliform microorganisms.[67]

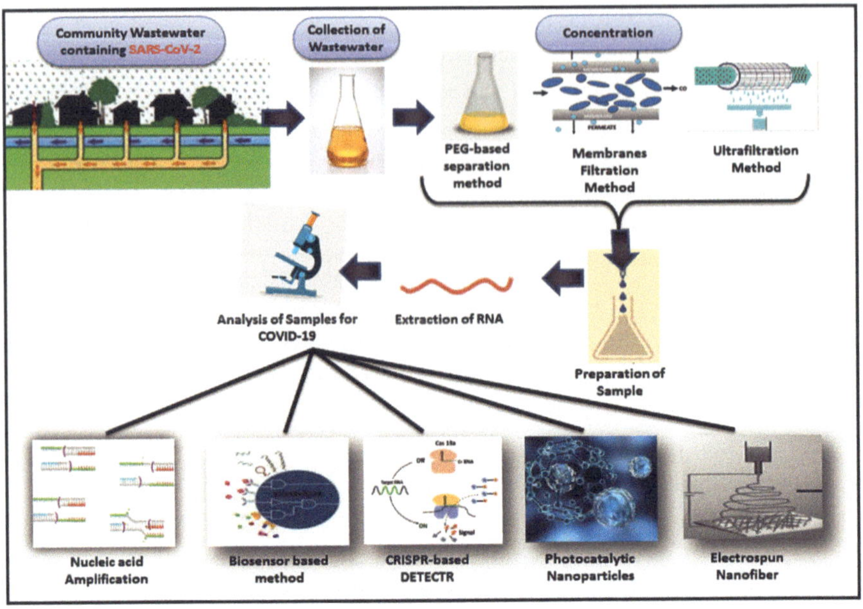

Figure 5.5 Promising methodologies for SARS-CoV-2 detection in wastewater.

These bioengineering-based nanofibers would especially increase the binding sites explicitness toward its objective, in this case COVID-19 viral RNA.

Numerous methods/devices have been applied for SARS-CoV-2 detection in different types of waste aquatic body as summarized in Fig. 5.5, but the standardized protocols and authentication for such methods/devices are yet to be accessible in the public domain.

5.8 Inactivation mechanism of COVID-19 in wastewater

The emergence of SARS-CoV-2 has become a global health concern. from the perspective of virologists, the intestinal tropism of SARS-CoV-2 is extraordinary. As reported by an ongoing study,[68] the intestinal tropism demonstrated the hostile nature of SARS-CoV-2 compared to other viruses. Initially specific treatments were unavailable, until Chan et al.[69] and Yeo et al.[31] deciphered that there is a phylogenetic relationship between the genome of both SARS-CoV-2 and other bat-related SARS coronaviral genome. Also, the nucleotides of the spike protein show 78% similarity with the human SARS-CoV-1.[31,69] With this assumption, scientists all over the world primarily hypothesized that SARS-CoV-2 would be reactive against various environmental aspects or disinfectants. Disinfectants typically act on the viral genome, capsid, or the protein layer, in a way destroying COVID-19.[7]

Sporadic reports show the use of different disinfectants in this field. Initially, a study of hospital wastewater[23] indicated nonappearance of SARS-CoV-2 viral RNA in the influent, whereas the effluent was verified positive for the presence of viral RNA. Afterward, Zhang and his group investigated another study where three clinics of China were considered for the identification of SARS-CoV-2 virus by RT-qPCR.[45] The clinical study reported the presence of 255 copies/L of viral titers in the adjusting tank. The aquatic waste of the second hospital followed a series of treatment steps in the order of adjusting tank—septic tank adjusting tank—moving bed biofilm reactor (MBBR)—sedimentation—disinfection, in which SARS-CoV-2 viral RNA was detected in three units: the adjusting tank (633 copies/L), MBBR (not detected-505 copies/L), and sedimentation tank (not detected-2208 copies/L). In the third clinical treatment unit, it consists of two units (preliminary disinfection tank followed by septic tank), SARS-CoV-2 was detected in the range of 557—18,744 copies/L only in the septic tank when 800 mg/L sodium hypochlorite was used. However, no detection of SAR-CoV-2 was found; the sodium hypochloride concentration range was 6700 mg/L. Several similar studies were reviewed by researchers to further confirm the treatment of SARS-CoV using various disinfection treatments and its infectivity on inanimate surfaces.[41,70—72]

Yet another disinfectant is hypochlorous acid (HOCl), which is a productive mediator against viruses for disrupting their genomic and proteomic mechanism.[73] Block and Rowan depicted the probable use of HOCl against SARS-CoV-2 virus disinfection.[74]

Another disinfectant, ethanol, has been used for disrupting the cell wall of the virus as reported by.[71] They reported that 62%—71% ethanol along with 0.5% hydrogen peroxide, or 0.1% sodium hypochlorite (NaOCl) as surface disinfection rapidly reduces the infectivity of SARS-CoV-2 on different surfaces within 1 min exposure time. According to WHO recommendation, higher concentration of NaOCl or 70% ethyl alcohol are suitable for disinfecting surfaces. Therefore, the ethanol and sodium hypochlorite are highly recommended for killing SARS-CoV-2 and protecting water workers from COVID-19 infection.[75]

The List N of the EPA has registered 431 commercial disinfectants which ascertain COVID-19 inactivation as of June 2020. Most of the products that follow the criteria chosen by EPA for its effectiveness against the removal of the virus possess some important component.[76]

Despite the promising results from treatment using disinfectant as reported above, various disadvantages accompanied them. Hence, there is a need for further research for deactivating SARS-CoV-2 viral infection.

The COVID-19 also demonstrated a high sensitivity toward ozone treatment in indoor spaces. Ozone has the ability for the disruption of protein and the lipid of the spikes of the enveloped viruses, specifically fatty acids, cysteine, methionine, and tryptophan along with linoleic and oleic acid with N-glycopeptides present on the spikes of subunits 1 and 2 proteins.[77] With the gradual enhancement of ozone from 48.83 to 94.67 $\mu g/m^3$, temperature from 13.17 to 19°C and a reduction of RH from 23.33% to 82.67% there is a decrease in the transmission of viral infections.[78]

Ultraviolet C (UV—C) could be a promising alternative for SARS-CoV-2 strain inactivation as it damages the viral genome.[79,80] Recently Inagaki and his coworker in 2020 have illustrated burgeoning deactivation method of COVID-19 using DUV-LED.[80] Another report revealed the inactivation of SARS-CoV-2 strain achieved with monochromatic UV-C (254 nm) irradiation, which is associated with multiple doses of illumination (3.7, 16.9, and 84.4 mJ/cm^2) in contrast to a sequence of different viral titers (0.05, 5, and 1000).[79]

Using the technology, International Advanced Research Center for Powder Metallurgy and New Material (ARCI) India has put forward a positive effort to develop a device. The device contains three disinfection systems that execute various physical, thermal, and chemical processes for sterilization of a selected area and surfaces efficiently at different health care departments as well as public locales for combating the COVID-19 pandemic. The system has a UV-C-based trolley for disinfection, an air heater of honeycomb type, and a fogging system to provide effective inactivation against SARS-CoV-2.[81] Therefore, it could be predicted that in the near future, if used for wastewater treatment, it would be credible to provide benefit to the society in alleviating the viral outburst.

Recently, another group of researchers from US Department of Homeland Security provided the first data on the effect of simulated sunlight on the survival of SARS-CoV-2 suspended in simulated saliva or culture media and suggested that sunlight could directly influence the survival of the virus.[82] The study showed that under simulated sunlight, 90% of infectious virus gets inactivated in every 6.8 min in simulated saliva

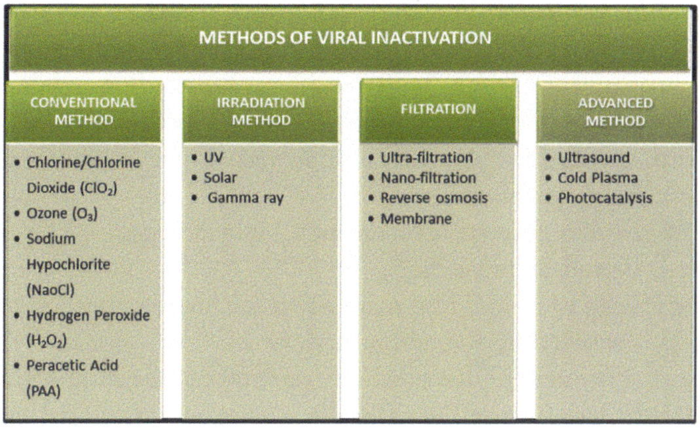

Figure 5.6 Different methods of inactivation of SARS-CoV-2 in wastewater.

and in every 14.3 min in culture media. Further, confirmation is yet to be explored.

Beside this, various advanced oxidation processes (AOPs) including photocatalysis, ultrasonic process, and Fenton processes have also been effectively utilized in treating wastewater.[7] Different deactivation methods used for COVID-19 were represented in Fig. 5.6. However, their use in wastewater treatment needs more momentum to gain robust solutions to meticulously eliminate the SARS-CoV-2 viral infection.

5.9 Remedial approach

In a recent report, the municipal and hospital wastewater estimated the SARS-CoV-2 load ranges from 56.6 million to 11.3 billion per infected individuals in a single day.[4] Such huge viral loads need to be addressed. Hence, apart from eradicating the disease, various prevention-based approaches has been undertaken for self-protection and also for the protection of personnel working in different sectors. WHO in 2020 had provided standard precautionary measures and guidelines to be followed by the workers who were working in the viral-suspected areas[75,83]:

(1) Maintenance of social distancing between two persons of six feet (about two arm's length).

(2) When outdoors, respirator masks (such as FFP2, FFP3, N95, and N99) should be properly used.

(3) Use of soaps for frequent hand washing along with alcohol sanitizer.

(4) Prohibition of touching mouth, eyes, and nose.

Short-term laboratory biosafety guidelines, released by Centers for Disease Control and Prevention (CDC) put forward various instructions regarding management along with specimen dispensation associated with COVID-19.[84] The instructions summarize the facts that

- ➢ Standard precautions should be followed while handling COVID-related clinical specimens.
- ➢ Use of EPA-registered disinfectants, which claims SARS-CoV-2 removal, should be practiced.
- ➢ Proper procedures should be maintained for management of laboratory waste as instructed by the administrator.

Although various effective measures were taken, the stigma in response to COVID-19 has put the life of health care workers and patients along with frontline workers in danger.[35,85] Therefore, the following propaganda was suggested, which might instigate us for future research in this area:

1. The wastewater treatment of the community should be decentralized with the purpose that the release of the waste from the isolation centers, having infected patients residues, will not contaminate the nearby area. Hence, decentralization of a community might reduce the contamination of a neighboring region.
2. The immense populace needs large-scale testing, which is quite strenuous. Consequently timely viral detection of the feces needs to be attained, while scrutinizing the virus genome in the aquatic waste of a community wastewater for early detection of the virus in an individual. This communal testing found to be a better alternative for mitigating the viral explosion.
3. The competent authorities should adequately address the sanitation and improve the water quality required. In a way the aquatic contamination could be surpassed with the avoidance of unpredicted SARS-CoV-2 viral RNA transmission to an infected person. Also, other human enteric viruses and associated viruses which emanate from feces could be diagnosed and prevented.
4. For deactivating and eliminating SARS-CoV-2 viral genome and its related species from the feces present in aquatic waste, an economically cheap point-of-use device should be developed, which eventually will be available to the public.
5. An appropriate legal policy, laws, guidelines, and so forth need to be recommended to certify sufficient agreement for the release of wastewater into the atmosphere with proper treatment. Also, there is a

need for various public awareness campaigns to tackle the stigma derived from the disease properly.

6. The government should implement a strategic plan for wastewater treatment plants at pilot scale, which could be beneficial in the long run. Also, there is a need for reforming the inconsistent health care system and setting proper guidelines for destroying such types of infectious disease outburst in the near future.

The current snowballing spread of COVID-19 can be curtailed with a rising level of community empowerment, with healthy literacy to respond appropriately in the situation.

5.10 Conclusion and future perspectives

The occurrence of the COVID-19 pandemic reminds us that the life in the world is impulsive and challenging. At present, trivial knowledge existed on the potentiality of aquatic waste in the spread of the viral dissemination and its magnitude of infectivity. Studies are under progress on existence and deactivation of COVID-19 in the waste aquatic environment. The background information indicated the firmness of the genome and its analogy with SARS-CoV. Based on the contextual knowledge, the following research directions should be explored:

1. Governments should take responsibility to develop economically cheap wastewater-based epidemiologies for controlling the spread of the pandemic in developed and underdeveloped countries from the global to local level.
2. Though various researchers have shown interesting and positive outcomes addressing the water quality, viral propagation, inactivation mechanisms of SARS-CoV-2 under solar, and other disinfection, evidence regarding their removal still need to be deciphered. Studies on disinfection kinetics with reference to the dosage concentration, activity, etc. should be addressed to gain a comprehensive knowledge regarding the inactivation strategy. Also, environmental impacts must be studied to be considered for disinfectants and other deactivating techniques.
3. Concern associated with the health of wastewater treatment workers and COVID-19 frontline workers need to be attended with priority at regular intervals. Proper guidelines related to hygiene and other preventive measures should be augmented by policymakers and health administration. Also, clear strategies should be guaranteed for the

support of infected health care workers to adequately cope with the disease.

4. Awareness programs for the public related to COVID-19 and its preventive practices need to be adopted to meet goals related to its elimination. For this, educational interventions might be considered.

Ideally, the COVID-19 pandemic has taught us to work together. The integration and synchronization of various actions can contribute to this work on a global scale so that a happy future for all can be achieved.

References

1. La Rosa G, Bonadonna L, Lucentini L, Kenmoe S, Suffredini E. Coronavirus in water environments: occurrence, persistence and concentration methods — a scoping review. *Water Res* 2020a;**179**:115899. https://doi.org/10.1016/j.watres.2020.115899.
2. WHO. *Coronavirus disease (COVID-19) Situation report*. 2020. https://covid19.who.int/. Accessed on 29th November, 2020.
3. Arora S, Nag A, Sethi J, Rajvanshi J, Saxena S, Shrivastava SK, et al. Sewage surveillance for the presence of SARS-CoV-2 genome as a useful wastewater based epidemiology (WBE) tracking tool in India. *Water Sci Tech* 2020;**82**(12):2823—36. https://doi.org/10.2166/wst.2020.540.
4. Hart OE, Halden RU. Computational analysis of SARS-CoV-2/COVID-19 surveillance by wastewater-based epidemiology locally and globally: feasibility, economy, opportunities and challenges. *Sci Total Environ* 2020;**730**:138875.
5. Verlicchi P. *Hospital wastewaters: characteristics, management, treatment and environmental risks*. Cham, Switzerland: Springer International Publishing AG; 2018. p. 243.
6. Spellman FR. *Handbook of water and wastewater treatment plant operations*. 3rd ed. United States: CRC Press, Taylor and Francis Group; 2014. p. 923.
7. Mohan SV, Hemalatha M, Kopperi H, Ranjith I, Kumar AK. SARS-CoV-2 in environmental perspective: occurrence, persistence, surveillance, inactivation and challenges. *Chem Eng J* 2021;**405**:126893. https://doi.org/10.1016/j.cej.2020.126893.
8. Gundy PM, Gerba CP, Pepper IL. Survival of coronaviruses in water and wastewater. *Food Environ Virol* 2008;**1**:10—4.
9. Sincero AP, Sincero GA. *Physical—chemical treatment of water and wastewater*. 1st ed. New York, Washington, DC: IWA Publishing and CRC Press; 2003. p. 856.
10. Randazzo W, Truchado P, Cuevas-Ferrando E, Simón P, Allende A, Sanchez G. SARS-CoV-2 RNA in wastewater anticipated COVID-19 occurrence in a low prevalence area. *Water Res* 2020;**181**:115942. https://doi.org/10.1016/j.watres.2020.115942.
11. Wu Y, Guo C, Tang L, Hong Z, Zhou J, Dong X, et al. Prolonged presence of SARS-CoV-2 viral RNA in faecal samples. *Lancet Gastroenterol Hepatol* 2020;**5**:434—5. https://doi.org/10.1016/S2468-1253(20)30083-2.
12. Sharma P, Mathur N, Singh A, Sogani M. Monitoring hospital wastewaters for their probable genotoxicity and mutagenicity. *Environ Monit Assess* 2015;**187**:41—80.
13. Achak M, Bakri SA, Chhiti BY, M'hamdi FE, Noureddine A, Boumya BW. SARS-CoV-2 in hospital wastewater during outbreak of COVID-19: a review on detection, survival and disinfection technologies. *ACS Nano* 2020;**14**:7704—13.
14. Verlicchi P, Aukidy MA, Galletti A, Petrovic M, Barcelo D. Hospital effluents as a source of emerging pollutants: an overview of micropollutants and sustainable treatment options. *J Hydrol* 2010;**389**:416—28.

15. Kummerer K. Drugs in the environment: emission of drugs, diagnostic aids and disinfectants into wastewater by hospitals in relation to other sources—a review. *Chemosphere* 2001;**45**(6−7):957−69.
16. Amouei A, Asgharnia H, Fallah H, Faraji H, Barari R, Naghipour D. Characteristics of effluent wastewater in hospitals of Babol University of medical sciences, Babol, Iran. *Health Scope* 2015;**4**(2):e23222.
17. Azuma T, Arima N, Tsukada A, Hirami S, Matsuoka R, Moriwake R, et al. Detection of pharmaceuticals and phytochemicals together with their metabolites in hospital effluents in Japan, and their contribution to sewage treatment plant influents. *Sci Total Environ* 2016:189−97. 548−549.
18. Ferrando-Climent L, Rodriguez-Mozaz S, Barcelo D. Incidence of anticancer drugs in an aquatic urban system: from hospital effluents through urban wastewater to natural environment. *Environ Pollut* 2014;**193**:216−23.
19. Kovalova L, Siegrist H, Singer H, Wittmer A, McArdell C. Hospital wastewater treatment by membrane bioreactor: performance and efficiency for organic micropollutant elimination. *Environ Sci Technol* 2012;**46**:1536−45.
20. Lin AY-C, Yu T-H, Lin C-F. Pharmaceutical contamination in residential, industrial, and agricultural waste streams: risk to aqueous environments in Taiwan. *Chemosphere* 2008;**74**:131−41.
21. Oliveira TS, Aukidy MA, Verlicchi P. Occurrence of common pollutant and pharmaceutical in hospital effluent. In: Verlicchi P, editor. *Hospital wastewaters- characteristics, management, treatment and environmental risks*. Springer International Publishing AG; 2017. p. 17−32.
22. Verlicchi P, Aukidy MA, Galletti A, Petrovic M, Barcelo D. Hospital effluent: investigation of the concentrations and distribution of pharmaceuticals and environment risk assessment. *Sci Total Environ* 2012;**430**:109−18.
23. Zhang D, Ling H, Huang X, Li J, Li W, Yi C, et al. Potential spreading risks and disinfection challenges of medical wastewater by the presence of severe acute respiratory syndrome coronavirus 2 (SARS-CoV-2) viral RNA in septic tanks of Fangcang Hospital. *Sci Total Environ* 2020a;**741**:140445. https://doi.org/10.1016/j.scitotenv.2020.140445.
24. Cui J, Li F, Shi ZL. Origin and evolution of pathogenic coronaviruses. *Nat Rev Microbiol* 2019;**17**(3):181−92.
25. Singhal T. A review of coronavirus disease-2019 (COVID-19). *Indian J Pediatr* 2020;**87**:281−6. https://doi.org/10.1007/s12098-020-03263-6.
26. Gorbalenya AE, Baker SC, Baric RS, de Groot RJ, Drosten C, Gulyaeva AA, et al. The species severe acute respiratory syndrome-related coronavirus: classifying 2019-nCoV and naming it SARS-CoV-2. *Nat Microbiol* 2020;**5**:536−44.
27. Munster VJ, Koopmans M, van Doremalen N, van Riel D, de Wit E. A novel coronavirus emerging in China—key questions for impact assessment. *N Engl J Med* 2020;**382**:692−4.
28. Kitajima M, Ahmed W, Bibby K, Carducci A, Gerba CP, Hamilton KA, et al. SARS-CoV-2 in wastewater: state of the knowledge and research needs. *Sci Total Environ* 2020. https://doi.org/10.1016/j.scitotenv.2020.139076.
29. Ye Y, Ellenberg RM, Graham KE, Wigginton KR. Survivability, partitioning and recovery of enveloped viruses in untreated municipal wastewater. *Environ Sci Technol* 2016;**50**(10):5077−85.
30. Amoah ID, Kumari S, Bux F. Coronaviruses in wastewater processes: source, fate and potential risks. *Environ Int Oral Maxillofac Surg* 2020;**143**:105962.
31. Yeo C, Kaushal S, Yeo D. Enteric involvement of coronaviruses: is faecal−oral transmission of SARS-CoV-2 possible? *Gastroenterol Hepatol* 2020;**5**(4):335−7.
32. Chin AWH, Chu JTS, Perera MRA, Hui KPY, Yen H-L, Chan MCW, et al. Stability of SARS-CoV-2 in different environmental conditions. *Lancet Microbe* 2020;**1**(1):e10. https://doi.org/10.1016/S2666-5247(20)30003-3.

33. Xu R, Cui B, Duan X, Zhang P, Zhou X, Yuan Q. Saliva: potential diagnostic value and transmission of 2019-nCoV. *Int J Oral Sci* 2020a;**11**:12.
34. Bogler A, Packman A, Furman A, Gross A, Kushmaro A, Ronen A, et al. Rethinking wastewater risks and monitoring in light of the COVID-19 pandemic. *Nat Sustain* 2020. https://doi.org/10.1038/s41893-020-00605-2.
35. Amirian ES. Potential fecal transmission of SARS-CoV-2: current evidence and implications for public health. *Int J Infect Dis* 2020;**95**:363–70. https://doi.org/10.1016/j.ijid.2020.04.057.
36. Heller L, Mota CR, Greco DB. COVID-19 faecal-oral transmission: are we asking the right questions? *Sci Total Environ* 2020;**729**:138919.
37. Quilliam RS, Weidmann M, Moresco V, Purshouse H, O'Hara Z, Oliver DM. COVID-19: the environmental implications of shedding SARS-CoV-2 in human faeces. *Environ Int* 2020;**140**:105790.
38. Zheng S, Fan J, Yu F, Feng B, Lou B, Zou Q, et al. Viral load dynamics and disease severity in patients infected with SARS-CoV-2 in Zhejiang province, China, January-March 2020: retrospective cohort study. *BMJ* 2020. https://doi.org/10.1136/bmj.m1443.
39. van Doremalen N, Bushmaker T, Morris D, Holbrook M, Gamble A, Williamson B, et al. Aerosol and surface stability of SARS-CoV-2 as compared with SARS-CoV-1. *N Engl J Med* 2020;**382**:1564–7. https://doi.org/10.1056/NEJMc2004973.
40. Lin L, Jiang X, Zhang Z, Huang S, Zhang Z, Fang Z, Gu Z, et al. Gastrointestinal symptoms of 95 cases with SARS-CoV-2 infection. *Gut* 2020;**69**:997–1001. https://doi.org/10.1136/gutjnl-2020-321013.
41. Wang J, Shen J, Ye D, Yan X, Zhang Y, Yang W, Li X, et al. Disinfection technology of hospital wastes and wastewater: suggestions for disinfection strategy during coronavirus disease 2019 (COVID-19) pandemic in China. *Environ Pollut* 2020;**262**:114665. https://doi.org/10.1016/j.envpol.2020.114665.
42. Kumar M, Patel AK, Shah AV, Raval J, Rajpara N, Joshi M, Joshi CG. The first proof of the capability of wastewater surveillance for COVID-19 in India through the detection of the genetic material of SARS-CoV-2. *Sci Total Environ* 2020;**746**:141326.
43. Ahmed W, Angel N, Edson J, Bibby K, Bivins A, O'Brien JW, et al. First confirmed detection of SARS-CoV-2 in untreated wastewater in Australia: a proof of concept for the wastewater surveillance of COVID-19 in the community. *Sci Total Environ* 2020;**728**:138764. https://doi.org/10.1016/j.scitotenv.2020.138764.
44. Haramoto E, Malla B, Thakali O, Kitajima M. First environmental surveillance for the presence of SARS-CoV-2 RNA in wastewater and river water in Japan. *Sci Total Environ* 2020;**737**:140405.
45. Zhang D, Yang Y, Huang X, Jiang J, Li M, Zhang X, et al. *SARS-CoV-2 spillover into hospital outdoor environments*. MedRxiv; 2020b. https://doi.org/10.1101/2020.05.12.20097105.
46. Zhang N, Yuhuan G, Fanping M, Yuhai B, Penghui Y, Fusheng W. *Virus shedding patterns in nasopharyngeal and fecal specimens of COVID-19 patients*. 2020. https://doi.org/10.1101/2020.03.28.20043059.
47. Wurtzer S, Marechal V, Mouchel JM, Moulin L. *Time course quantitative detection of SARS-CoV-2 in Parisian wastewaters correlates with COVID-19 confirmed cases*. medRxiv; 2020. https://doi.org/10.1101/2020.04.12.20062679.2020.04.12.20062679.
48. Kavanagh MM, Erondu NA, Tomori O, Dzau VJ, Okiro EA, Maleche A, et al. Access to life saving medical resources for African countries: COVID-19 testing and response, ethics, and politics. *Lancet* 2020;**395**:1735–8. https://doi.org/10.1016/S0140-6736(20)31093-X.
49. Nkengasong J. Let Africa into the market for COVID-19 diagnostics. *Nature* 2020. https://doi.org/10.1038/d41586-020-01265-0.

50. Orive G, Lertxundi U, Barcelo D. Early SARS-CoV-2 outbreak detection by sewage-based epidemiology. *Sci Total Environ* 2020. https://doi.org/10.1016/j.scitotenv.2020.139298.
51. Venugopal A, Ganesan H, Sudalaimuthu Raja SS, Govindasamy V, Arunachalam M, et al. Novel wastewater surveillance strategy for early detection of coronavirus disease 2019 hotspots. *Curr Opin Environ Sci Health* 2020;**17**:8–13. https://doi.org/10.1016/j.coesh.2020.05.003.
52. CDC. *Wastewater surveillance testing methods*. 2020. https://www.cdc.gov/coronavirus/2019-ncov/cases-updates/wastewater-surveillance/testing-methods.html#:~:text=C%20or%20below.-,RNA%20measurement,also%20possible%20but%20less%20common. Accessed on 29th November, 2020.
53. Farkas K, Hillary LS, Malham SK, McDonald JE, Jones DL. Wastewater and public health: the potential of wastewater surveillance for monitoring COVID-19. *Curr Opin Environ Sci Health* 2020a;**17**:14–20.
54. Bar-Or I, Yaniv K, Shagan M, Ozer E, Erster O, Mendelson E, Mannasse B, et al. Regressing SARS-CoV-2 sewage measurements onto COVID-19 burden in the population: a proof-of concept for quantitative environmental surveillance. *medRxiv* 2020. https://doi.org/10.1101/2020.04.26.20073569.
55. La Rosa G, Iaconelli M, Mancini P, Bonanno Ferraro G, Veneri C, Bonadonna L, Lucentini L, Suffredini E. First detection of SARS-COV-2 in untreated wastewaters in Italy. *Sci Total Environ* 2020b;**736**:139652.
56. Medema G, Heijnen L, Elsinga G, Italiaander R, Brouwer A. Presence of sars-coronavirus 2 RNA in sewage and correlation with reported COVID-19 prevalence in the early stage of the epidemic in The Netherlands. *Environ Sci Technol Lett* 2020;**7**:511–6.
57. Lu D, Huang Z, Luoa J, Zhang X, Sha S. Primary concentration – the critical step in implementing the wastewater based epidemiology for the COVID-19 pandemic: A Mini-review. *Sci Total Environ* 2020;**747**:141245. https://doi.org/10.1016/j.scitotenv.2020.141245.
58. WEF. *Current priority: coronavirus*. 2020. https://www.wef.org/coronavirus. Accessed on 30th August, 2020.
59. Farkas K, Mannion F, Hillary LS, Malham SK, Walker DI. Emerging technologies for the rapid detection of enteric viruses in the aquatic environment. *Curr Opin Environ Sci Health* 2020b;**16**:1–6.
60. Bhalla N, Pan Y, Yang Z, Payam AF. Opportunities and challenges for biosensors and nanoscale analytical tools for pandemics: COVID-19. *ACS Nano* 2020;**14**(7):7783–807. https://doi.org/10.1021/acsnano.0c04421.
61. Samson R, Navale GR, Dharne MS. Biosensors: frontiers in rapid detection of COVID-19. *3 Biotech* 2020;**10**:385. https://doi.org/10.1007/s13205-020-02369-0.
62. Mao K, Zhang H, Yang Z. Can a paper-based device trace COVID-19 sources with wastewater-based epidemiology? *Environ Sci Technol* 2020;**54**:3733–5.
63. Hata A, Honda R. Potential sensitivity of wastewater monitoring for SARS-CoV-2: comparison with norovirus cases. *Environ Sci Technol* 2020;**54**(11):6451–2.
64. Broughton JP, Deng X, Yu G, Fasching CL, Servellita V, Singh J, et al. CRISPR–cas12-based detection of SARS-CoV-2. *Nat Biotechnol* 2020;**38**:870–4. https://doi.org/10.1038/s41587-020-0513-4.
65. Talebian S, Wallace GG, Schroeder A, Stellacci F, Conde J. Nanotechnology-based disinfectants and sensors for SARS-CoV-2. *Nat Nanotechnol* 2020;**15**:618–21. https://doi.org/10.1038/s41565-020-0751-0.
66. Weiss C, Carriere M, Fusco L, Capua I, Regla-Nava JA, Pasquali M, et al. Toward Nanotechnology-enabled approaches against the covid-19 pandemic. *ACS Nano* 2020;**14**:6383–406.

67. Xagoraraki I, O'Brien E. Wastewater-based epidemiology for early detection of viral outbreaks. In: O'Bannon D, editor. *Women in water quality*. Springer Nature Switzerland; 2020. p. 75–97.
68. Xing Y-H, Ni W, Wu Q, Li W-J, Li G-J, Wang W-D, et al. Prolonged viral shedding in feces of pediatric patients with coronavirus disease 2019. *J Microbiol Immunol Infect* 2020;**53**(3):473–80.
69. Chan JF, Kok KH, Zhu Z, Chu H, To KK, Yuan S, Yuen KY. Genomic characterization of the 2019 novel human-pathogenic coronavirus isolated from a patient with a typical pneumonia after visiting Wuhan. *Emerg Microb Infect* 2020;**9**(1):221–36. https://doi.org/10.1080/22221751.2020.1719902.
70. Ciejka J, Wolski K, Nowakowska M, Pyrc K, Szczubiałka K. Biopolymeric nano/microspheres for selective and reversible adsorption of coronaviruses. *Mater Sci Eng C* 2017;**76**:735–42.
71. Kampf G, Todt D, Pfaender S, Steinmann E. Persistence of coronaviruses on inanimate surfaces and their inactivation with biocidal agents. *J Hosp Infect* 2020;**104**(3):246–51. https://doi.org/10.1016/j.jhin.2020.01.022.
72. Mandal P, Gupta AK, Dubey BK. A review on presence, survival, disinfection/removal methods of coronavirus in wastewater and progress of wastewater-based epidemiology. *J Environ Chem Eng* 2020;**8**(5):104317. https://doi.org/10.1016/j.jece.2020.104317.
73. Kataki S, Chatterjee S, Vairale MG, Sharma S, Dwivedi SK. Concerns and strategies for wastewater treatment during COVID-19 pandemic to stop plausible transmission. *Resour Conserv Recycl* 2020;**164**. https://doi.org/10.1016/j.resconrec.2020.105156.
74. Block MS, Rowan BG. Hypochlorous acid—a review. *J Oral Maxillofac Surg* 2020;**78**(9):1461–6.
75. WHO. *Coronavirus disease (COVID-19) advice for the public*. 2020. https://www.who.int/emergencies/diseases/novel-coronavirus-2019/advice-for-public. Accessed on 30th August, 2020.
76. EPA. *List N: disinfectants for use against SARS-CoV*. 2020. https://www.epa.gov/pesticide-registration/list-n-disinfectants-use-against-sars-cov-2. Accessed on 9th, September, 2020.
77. Tizaoui C. Ozone: a potential oxidant for COVID-19 virus (SARS-CoV-2). *Ozone Sci Eng* 2020;**42**(5):378–85. https://doi.org/10.1080/01919512.2020.1795614.
78. Abajo FJG, Hernandez RJ, Kaminer I, Meyerhans A, Rosell-Llompart J, Sanchez-Elsner T. *Back to normal: an old physics route to reduce SARS-CoV-2 transmission in indoor spaces*. 2020.
79. Bianco A, Biasin M, Pareschi G, Cavalieri A, Cavatorta C, Fenizia C, et al. UV-C irradiation is highly effective in inactivating and inhibiting SARS-CoV-2 replication. *MedRxiv* 2020. https://doi.org/10.1101/2020.06.05.20123463.
80. Inagaki H, Saito A, Sugiyama H, Okabayashi T, Fujimoto S. Rapid inactivation of SARS-CoV-2 with deep-UV LED irradiation. *Emerg Microb Infect* 2020;**9**(1):1744–7.
81. Sarada BV, Vijay R, Johnson R, Rao TN, Padmanabham G. Fight against COVIDss-19: ARCI's technologies for disinfection. *Trans Indian Natl Acad Eng* 2020. https://doi.org/10.1007/s41403-020-00153-3.
82. Ratnesar-Shumate S, Williams G, Green B, Krause M, Holland B, Wood S. Simulated sunlight rapidly inactivates SARS-CoV-2 on surfaces. *J Infect Dis (JID)* 2020;**222**(2):214–22.
83. Tomar A, Gupta N. Prediction for the spread of COVID-19 in India and effectiveness of preventive measures. *Sci Total Environ* 2020;**728**:138762. https://doi.org/10.1016/j.scitotenv.2020.138762.
84. CDC. *Interim laboratory biosafety guidelines for handling and processing specimens associated with coronavirus disease 2019 (COVID-19)*. 2020. https://www.cdc.gov/coronavirus/2019-nCoV/lab/lab-biosafety-guidelines.html. Accessed on 1st September, 2020.

85. Adelodun B, Ajibade FO, Ibrahim RG, Bakare HO, Choi KS. Snowballing transmission of COVID-19 (SARS-CoV-2) through wastewater: any sustainable preventive measures to curtail the scourge in low-income countries? *Sci Total Environ* 2020;**742**:140680. https://doi.org/10.1016/j.scitotenv.2020.140680.
86. Henze M, Harremoes P, Jansen JC, Arvin E. *Wastewater treatment: biological and chemical processes*. 3rd ed. Berlin: Springer-Verlag; 2002.
87. Fahad A, Mohamed RMS, Radhi B, Al-Sahari M. Wastewater and its treatment techniques: an ample review. *Indian J Sci Technol* 2019;**12**(25):1–13.
88. Badejo AA, Omole DO, Ndambuki JM. Municipal wastewater management using *Vetiveria zizanioides* planted in vertical flow constructed wetland. *Appl Water Sci* 2018;**8**(110):1–6.
89. Panikkar A, Shrestha P, Hackney P, Riley S. A residential blackwater and municipal waste treatment system — safety issues and risk management, organic recovery and biological treatment. In: *Proceedings of the fourth international conference on biological processing of organics: advances for a sustainable society, Perth, Australia*; 2003. p. 118–24.
90. Daouk S, Chevre N, Vernaz N, Widmer C, Daali Y, Fleury-Souverain S. Dynamics of active pharmaceutical ingredients loads in a swiss university hospital wastewaters and prediction of the related environmental risk for the aquatic ecosystems. *Sci Total Environ* 2016;**537**:244–53.
91. Nour-eddine A, Lahcen B. Estimate of the metallic contamination of the urban effluents by the effluents of the Mohamed v hospital of Meknes. *Eur Sci J* 2014;**10**(3):71–8.
92. Xu Y, Xufang L, Zhu B, Liang H, Fang C, Gong Y, et al. Characteristics of pediatric SARS-CoV-2 infection and potential evidence for persistent fecal viral shedding. *Nat Med* 2020b;**26**(4):502–5.
93. Wolfel R, Corman VM, Guggemos W, Seilmaier M, Zange S, Muller MA, et al. Virological assessment of hospitalized patients with COVID-2019. *Nature* 2020;**581**:465–9.
94. Cheung KS, Hung IFN, Chan PPY, Lung KC, Tso E, Liu R, et al. Gastrointestinal manifestations of SARS-CoV-2 infection and virus load in fecal samples from the Hong Kong cohort and systematic review and meta-analysis. *Gastroenterology* 2020;**159**(1):81–95. https://doi.org/10.1053/j.gastro.2020.03.065.
95. Xiao F, Meiwen T, Xiaobin Z, Ye L, Xiaofeng L, Hong S. Evidence for gastrointestinal infection of SARS-CoV-2. *Gastroenterology* 2020;**158**:1831–3. https://doi.org/10.1053/j.gastro.2020.02.055.
96. Chen Y, Liu Q, Guo D. Emerging coronaviruses: genome structure, replication, and pathogenesis. *J Med Virol* 2020;**92**(4):418–23.
97. Pan L, Mu M, Yang P, Sun Y, Wang R, Yan J, Li P, et al. Clinical characteristics of COVID-19 patients with digestive symptoms in Hubei, China. *Am J Gastroenterol* 2020;**115**(5):766–73. https://doi.org/10.14309/ajg.0000000000000620.
98. Mlejnkova H, Sovova K, Juranova E. Preliminary study of SARS-CoV-2 occurrence in wastewater in the Czech Republic. *Int J Environ Res Publ Health* 2020;**17**(15):5508. https://doi.org/10.3390/ijerph17155508.
99. Kocamemi BA, Kurt H, Hacioglu S, Yarali C, Saatci AM, Pakdemirli B. *First data-set on SARS-CoV-2 detection for Istanbul wastewaters in Turkey. medRxiv*. 2020. https://doi.org/10.1101/2020.05.03.20089417.
100. Nemudryi A, Nemudraia A, Surya K, Wiegand T, Buyukyoruk M, Wilkinson R, et al. Temporal detection and phylogenetic assessment of SARS-CoV-2 in municipal wastewater. *medRxiv* 2020. https://doi.org/10.1101/2020.04.15.20066746.

Further reading

1. Topare NS, Attar J, Manfe MM. Sewage/wastewater treatment technologies: a review. *Sci Rev Chem Commun* 2011;**1**:18–24.

2. Xiao H, Guan D, Chen R, Chen P, Monagin C, Li W, et al. Molecular characterization of echovirus 30 associated outbreak of aseptic meningitis in Guangdong in 2012. *Virol J* 2013;**10**:263.

CHAPTER 6

Route of SARS-CoV-2 in sewerage and wastewater treatment plants: dilution, decay, removal, and environmental transmission

Paola Foladori[1], Francesca Cutrupi[1], Maria Cadonna[2], Serena Manara[3], Federica Maestrini[1]

[1]Department of Civil, Environmental and Mechanical Engineering (DICAM), University of Trento, Trento, Italy; [2]ADEP – Agenzia per la Depurazione, Autonomous Province of Trento, Trento, Italy; [3]Department of Cellular Computational and Integrative Biology (CIBIO), University of Trento, Trento, Italy

6.1 Introduction

Although the most common clinical presentation of COVID-19, the disease caused by SARS-CoV-2,[1] is fever and respiratory tract symptoms, this viral infection may cause other extrarespiratory manifestations, including gastrointestinal problems—such as diarrhea, nausea, vomiting, and abdominal pain—that often represent an early symptom of the disease.[2–4] However, SARS-CoV-2 positivity can be seen also in the stools of people without gastrointestinal symptoms[5] or with asymptomatic or mild disease.[6]

To detect SARS-CoV-2 in fecal samples, the gold standard is the quantitative reverse-transcription polymerase chain reaction (RT-PCR) that is based on the amplification of specific genomic regions that are specific for SARS-CoV-2[7–12] and the simultaneous monitoring of the amplification process through sequence-specific oligonucleotides labeled with fluorescent probes, which allow the quantification of the viral load. The viral load is then expressed as the number of SARS-CoV-2 RNA copies in a known volume and the units are genomic units (GU) per L or mL.

The viral load measured in the stool of COVID-19 patients depends on the day of sampling after infection initiation and can vary largely in the range of 10^3–10^7 GU/g feces.[13,14] The highest concentrations of viral RNA in stool samples were found during the first week of symptoms, while it gradually decreases over time. However, patients may remain RNA-positive

Environmental Management of COVID-19
ISBN 978-0-323-85780-2
https://doi.org/10.1016/B978-0-323-85780-2.00005-6

in stool over three weeks[14] and even for up to 50 days or more,[6] despite the complete resolution of symptoms.[11,12,14] In general, the presence of SARS-CoV-2 in the stool may be due to: (1) swallowing of respiratory secretions, (2) residues of infected cells, or (3) replication of the virus in the intestine.[15,16] The recent demonstration of the active replication of SARS-CoV-2 in human intestinal organoids suggests that the human intestinal tract might be a transmission route of SARS-CoV-2.[17]

Although urine is frequently found negative for SARS-CoV-2[10,18] the literature confirming the presence of SARS-CoV-2 in the urine of patients affected by COVID-19 is continuously increasing.[5,19,20] However, when detected, the SARS-CoV-2 load in urine is low[20] and thus the contribution of urine to the transmission route appears less relevant than that originated from feces.

The detection of SARS-CoV-2 RNA in the feces of infected patients raises the question of potential fecal-oral transmission.[21] Theoretically, the presence of RNA fragments does not imply a priori the presence of infectious virions, but recent studies report the presence of infectious virions isolated from feces,[10,16,22,23] supporting thus the possibility of fecal-oral or fecal-respiratory transmission through aerosolized feces.[16] This latter aspect may occur for example in toilets or in the case of faulty sewage pipelines, as happened during the 2003 SARS-CoV pandemic, when hundreds of residents of a building in Hong Kong were infected due to the aerosolization of contaminated feces.[24]

Jeong et al.[25] demonstrated that viable SARS-CoV-2 can be shed through stool and urine of COVID-19 patients. Using the feces and urine from COVID-19 patients, the authors did not observe viable virus using cell culture isolation (Vero cells), despite some naso/oropharyngeal swabs and saliva specimens appeared positive for virus isolation. Conversely, the presence of infectious virus in fecal and urine specimens was identified after the inoculation in an animal model highly susceptible to SARS-CoV-2. These findings suggested that SARS-CoV-2 may be potentially transmitted via routes such as fecal-oral contact and urine.[25]

When released in infected urine/feces, the path of spread of SARS-CoV-2 RNA begins with the virus being transported from the toilets to the sewer network, where a relevant dilution of feces occurs, and then to the subsequent wastewater treatment plants (WWTPs), where the concentration of SARS-CoV-2 can be significantly reduced before the discharge of the treated effluents into the receiving water bodies.

This chapter focuses on the fate of the viral load of SARS-CoV-2 from feces of positive persons to the sewer network and further in WWTPs and treated effluents. In particular, the progressive drop of the load of SARS-CoV-2 along this route was evaluated on the basis of dilution, decay, removal in WWTPs, and experimental data available in raw wastewater or treated effluents.

6.2 Dilution of SARS-CoV-2 from the feces to the sewerage

When the feces of COVID-19 infected persons are discharged into the sewerage, they undergo a relevant dilution in water and the virus is subjected to several transformations due to the presence of suspended solids, organic substances, micropollutants, and bacteria, in particular, enteric.

The dilution of feces in the sewerage (D) can be estimated by the ratio between the volume of feces per day (V) and the wastewater flow rate (Q), both expressed per capita (*cap*) and per day (*d*), according to Eq. (6.1).

$$D = \frac{V[\text{L cap}^{-1}\text{d}^{-1}]}{Q[\text{L cap}^{-1}\text{d}^{-1}]} \tag{6.1}$$

V can be calculated as the ratio between the wet mass of feces (M) and the density of feces (ρ). M assumes a typical value of 128 g cap^{-1} d^{-1} in high-income countries, but can double in low-income countries where the fiber intake of the population is more abundant.[26] The typical density of feces is in the range from 1.06 to 1.09 g/cm^3 or g/mL.[27]

Q is calculated taking into account the average daily water consumption per capita (w, expressed as liters per person per day) and assuming the sewage discharge as 0.8 times the water supplied (using the coefficient α of 0.8). For example, in the case of households in Europe, w is 144 L cap^{-1} d^{-1},[28] while the value is 50–100 L cap^{-1} d^{-1} or less in developing countries. According to these assumptions, Eq. (6.1) can be rewritten according to Eq. (6.2).

$$D = \frac{\dfrac{M[\text{g cap}^{-1}\text{d}^{-1}]}{\rho[\text{g/mL}]}}{\alpha[-]\cdot w[\text{L cap}^{-1}\text{d}^{-1}]} \tag{6.2}$$

Using the values indicated above, the result of the calculation in Eq. (6.3) indicates a dilution factor of 10^{-3}.

$$D = \frac{\dfrac{128 \text{ g cap}^{-1}\text{d}^{-1}}{1.09\text{g/mL}}}{0.8 \cdot 144\text{L cap}^{-1}\text{d}^{-1} \cdot 1000 \text{ mL/L}} = 0.0010 \tag{6.3}$$

Eq. (6.3) indicates that the SARS-CoV-2 load in the feces of a COVID-19 infected person undergoes an important dilution in the sewer network, and it is reduced by 1000 times passing from feces to wastewater.

6.3 SARS-CoV-2 load of in raw wastewater

The sewer network collects municipal and industrial wastewater and a variable amount of rainwater, the amount of which depends on the presence of a separate or combined wastewater system. The sewerage flow rate may be further increased by the infiltration of parasitic waters (originated by wrong sewer connections or groundwater infiltration).

Therefore the sewerage is a complex network where the SARS-CoV-2 load in the stool of COVID-19 patients that may vary in the range from $5 \cdot 10^3$ to more than 10^7 GU/mL,[14,29,30] is hugely reduced by the following factors:

1. The dilution of feces of COVID-19 positive persons by 10^3 in wastewater, according to the calculation in Section 6.2.
2. Only a fraction of the entire population connected to the sewerage is positive against COVID-19 and has a significant presence of SARS-CoV-2 in the feces. For example, considering 5% of positive cases in a population, a further reduction of the viral load by 20 times occurs in the sewerage.
3. The presence of stormwater and infiltrations of parasitic waters that increases the amount of water in the sewerage and leads to a further reduction of the viral load.

Taking into account all these factors simultaneously, the theoretical estimation shows that the SARS-CoV-2 load in the sewerage can be reduced by four orders of magnitude or more with respect to the load in stool. The result is a concentration of SARS-CoV-2 around 10^{-1}–10^3 GU/mL wastewater. This estimation is obtained from an approximative theoretical calculation, based on simplified assumptions and uncertain data, but permits to understand why the concentrations of SARS-CoV-2 in the sewerage are so low with respect to the values found in the feces.

Table 6.1 presents a comparison among the studies that investigated the viral load in sewerage up to now. The large variation of the results reported in the literature is associated with different dilution in the sewerage, but also

Table 6.1 Comparison among the studies which detected SARS-CoV-2 in raw wastewater entering the WWTPs. When quantified, the viral load is indicated in terms of genomic units per unit of volume (GU/L or GU/mL).

References	Location, date of sampling	Plants and type of sampling	Details on the study	SARS-CoV-2 load in raw wastewater
Ahmed et al.[31]	South East Queensland, Australia. March 27 and April 1, 2020.	• 2 WWTPs and a pumping station. • Automated samplers or grab sampling.	• 22.2% (2/9) of samples tested positive.	Range: 1.9×10^1 -1.2×10^2 GU/L
Ampuero et al.[32]	Santiago, Chile From March to June 2020.	• 2 WWTPs. • 24-h composite samples taken monthly.	• All samples were negative in March and April 2020 due to the few cases in Santiago. • GU increased progressively from May to June and correlated with the increasing number of COVID-19 confirmed cases.	n.a.
Arora et al.[33]	Jaipur, Rajasthan, India. From May 3 to June 14, 2020.	• 6 WWTPs. • 2 hospitals for treatment of COVID-19 patients.	• 2/6 sites showed two or more target genes in raw wastewater. • Areas with positive results showed a continuous increase in the COVID-19 confirmed cases. • Positive results were detected at ambient temperature very high (45°C).	n.a.

Continued

Table 6.1 Comparison among the studies which detected SARS-CoV-2 in raw wastewater entering the WWTPs. When quantified, the viral load is indicated in terms of genomic units per unit of volume (GU/L or GU/mL).—cont'd

References	Location, date of sampling	Plants and type of sampling	Details on the study	SARS-CoV-2 load in raw wastewater
Balboa et al.[34]	Ourense, Spain. From 6 to Apr 21, 2020.	• 24-h composite samples taken twice a week.	• SARS-CoV-2 RNA was systematically detected in the influent wastewater. • The level of SARS-CoV-2 RNA was low in all the samples.	n.a.
Chavarria-Miró et al.[35]	Barcelona, Spain. From April 13 to May 25, 2020.	• 2 large WWTPs. • 24-h composite samples taken weekly. • For 1 WWTP, frozen archival samples from 2018 (Jan–Mar), 2019 (Jan, Mar, Sep–Dec) and 2020 (Jan–Mar) were assayed.	• From 15 Jan to Mar 4, 2020 the presence of SARS-CoV-2 increased, as demonstrated by the analysis of archival samples. • Positive samples were found 41 days (15 Jan) before the 1st COVID-19 positive case (25 Feb). • Despite negative samples in WWTPs around May 18–25, grab samples from urban sewers collected 8–9 AM were positive (as a consequence of different dilution and type of sampling).	12-Mar: Target IP2: 6.4×10^2 GU/L Target IP4: 8.3×10^2 GU/L

Fongaro et al.[36]	Florianopolis, Santa Catalina, Brazil. From October 30, 2019 to March 4, 2020.	• 6 independent sampling taken from a well of the sewerage (well for inspection/cleaning).	• positive samples were detected 97 d before the 1st COVID-19 confirmed case in Santa Catarina (Nov 27, 2020). • the viral load increased by approximately 1 $\log_{10}$ in correspondence of the 1st COVID-19 confirmed case (Mar 4, 2020).	30-Nov: 5.5 $\log_{10}$ GU/L 11-Dec: 5.8 $\log_{10}$ GU/L 20-Feb: 5.6 $\log_{10}$ GU/L 4-Mar: 6.7 $\log_{10}$ GU/L
Gonzalez et al.[37]	Virginia, USA. From March 9 to July 28, 2020.	• 9 WWTPs. • 24 h flow-weighted composite samples taken weekly in 3 WWTPs. • Grab sampling in the other 3 WWTPs.	• 198 samples were analyzed: 98 positive for 3 assays, 22 positive for 2 assays, and 30 positive for only 1 assay. • COVID-19 confirmed cases were 69 on 9 March and they increased to 1,180,000 on 28 July.	Range: 10^2–10^5 GU/L
Green et al.[38]	Syracuse and Onondaga County, NY, USA. May 6 and 13, 2020.	• 24-h composite samples taken from 11 access points (i.e., WWTPs, pump stations or interceptor lines).	• 18/22 samples tested positive. • 13/22 samples were in the quantifiable range.	Average: 42.7 GU/mL Highest: 112 GU/mL
Haramoto et al.[39]	Yamanashi Prefecture, Japan. From March 17 to May 7, 2020.	• Grab sampling of raw and secondary-treated wastewater.	• Despite secondary-treated wastewater was positive, influent samples tested with the same procedure were negative. • The discrepancy may be due to the different volumes used for concentration and the low presence of COVID-19 in the region.	n.a.

Continued

Table 6.1 Comparison among the studies which detected SARS-CoV-2 in raw wastewater entering the WWTPs. When quantified, the viral load is indicated in terms of genomic units per unit of volume (GU/L or GU/mL).—cont'd

References	Location, date of sampling	Plants and type of sampling	Details on the study	SARS-CoV-2 load in raw wastewater
Hata et al.[40]	Ishikawa and Toyama, Japan. From March 5 to Apr 23, 2020.	• 4 WWTPs. • Grab sampling in the morning during the peak flow, weekly or biweekly.	• 7/27 positive for at least one assay. • The detection frequency was 15% (3 positives out of 20 samples) when COVID-19 confirmed cases were below 10 in 100,000 people. • The detection frequency was 57% (4/7) when cases were above 10 in 100,000 people.	Range: 1.2×10^4 -4.4×10^4 GU/L
Hong et al.[41]	Jeddah, Saudi Arabia. From April 15 to Jul 9, 2020.	• Grab sampling at a frequency of 3–5 samples per week. • Hospital wastewater.	• 75.4% (43/57) of samples tested positive for N genes (N1, N2, N3).	N1: 173.7 GU/L N2: 772.1 GU/L N3: 1327.4 GU/L
Kocamemi et al.[42]	Istanbul, Turkey. On April 21 and 25, 2020.	• 7 WWTPs. • 24-h composite samples. • Grab sampling in manholes near pandemic hospitals.	• 5/7 samples from WWTPs tested positive. • All samples from manholes tested positive.	Range in positive WWTPs: 2.9×10^3 -1.8×10^4 GU/L In manholes: 4.5×10^4 GU/L 9.3×10^4 GU/L

Kumar et al.[43]	Gujarat, India. On May 8 and 27, 2020.	• Grab sampling. • A composite sample made from 3 samples taken in each location.	• A 10-fold increase from 8 to 27 May, corresponding to more than double the number of COVID-19 patients (4912 and 10,674 cases on 8 and 27 May, respectively).	Range: 5.6×10 -3.5×10^2 GU/L
La Rosa et al.[44]	Milan and Rome, Italy. From February 3 to April 2, 2020.	• 24-h composite samples.	• 50% (6/12) of samples were positive. • On 24 and 28 February, when the samples were positive in Milan, COVID-19 infections were still limited in Italy.	n.a.
La Rosa et al.[45]	Milan, Turin, Bologna, Italy. From October 9, 2019 to February 28, 2020.	• 5 WWTPs (2 in Milan, 2 in Turin and 1 in Bologna). • 24-h composite samples.	• 88% (23/26) of samples were below the analytical LOQ (5.9×10^3 GU/L).	Range: from <LOD to 5.6×10^4 GU/L
Lodder and de Roda Husman[46]	Amsterdam Airport Schiphol, Netherlands. From February 17, 2020.	• 24-h samples taken once a week.	Samples tested positive 4 days after the 1st (apex) COVID-19 confirmed case (identified in The Netherlands on Feb 27, 2020).	n.a.
Manupati et al.[47]	Hyderabad Metropolitan City, India. From July 7 to August 8, 2020.	• Sampling was performed in the day time, on the days with no report of rainfall during the last 24 h.	• All WWTPs were positive. • In 1 WWTP samples were taken at different times to assess the dynamics of disease spread with time: highly fluctuating data were observed inside a few weeks.	8-Jul: 219.540 GU/L 14-Jul: 30.818 GU/L 29-Jul: 266.360 GU/L

Continued

Table 6.1 Comparison among the studies which detected SARS-CoV-2 in raw wastewater entering the WWTPs. When quantified, the viral load is indicated in terms of genomic units per unit of volume (GU/L or GU/mL).—cont'd

References	Location, date of sampling	Plants and type of sampling	Details on the study	SARS-CoV-2 load in raw wastewater
Medema et al.[48]	The Netherlands. From February 5 to March 16, 2020.	• 24-h flow-dependent composite samples.	• 58% (14/24) of samples were positive. • On 4/5 March (1 week into the epidemic) 4/6 WWTPs were positive, with only 38 and 82 COVID-19 cases confirmed through the health surveillance system.	Range: 2.6×10^3 -2.2×10^6 GU/L.
Or et al.[49]	Tel Aviv and Israel. From March 10 to April 21, 2020.	• Automated samplers at targeted hot-spot areas for 24 h.	• The Ct value measured in rRT-PCR correlated with the number of COVID-19 positive individuals.	n.a.
Randazzo et al.[50]	Murcia, Spain. From March 12 to April 14, 2020.	• 6 WWTPs in the larger municipalities.	• 83% (35/42) of samples were positive for at least one target.	Average values: N1: 5.1 ± 0.3 $\log_{10}$ GU/L N2: 5.5 ± 0.2 $\log_{10}$ GU/L N3: 5.5 ± 0.3 $\log_{10}$ GU/L

Rimoldi et al.[51]	Milan, Monza, Brianza, Italy. On April 14 and 22, 2020.	• 3 WWTPs. • Grab sampling around 1.00 p.m.	• All WWTPs positive on Apr 14, 2020. • Only 1 WWTP positive on 22 Apr, probably following the epidemiological trend in the area.	n.a.
Sherchan et al.[52]	Southern Louisiana, USA. From January to April 2020.	• 2 WWTPs. • 24-h composite and grab samples collected monthly.	• 2/7 samples tested positive.	N1 assay: 7.5×10^3 GU/L N2 assay: 3.1×10^3 -4.3×10^3 GU/L
Trottier et al.[53]	Montpellier, Lattes, France. From May 7 to June 25, 2020.	• 24-h composite samples.	• Viral RNA in wastewater increased from April to 15–25 Jun despite the decrease in the number of COVID-19 patients hospitalized.	n.a.
Westhaus et al.[54]	North Rhine-Westphalia, Germany. April 8, 2020.	• 9 WWTPs. • 24-h flow-weighted composite samples. • Sampling during dry-weather.		Range: 3–20 GU/mL
Wu et al.[55]	Massachusetts, USA. March 18–25, 2020.	• A major WWTP. • 24-h composite samples.	• Samples collected before the 1st known US SARS-CoV-2 case were negative. • Then, all samples taken from 18 to 25 Mar tested positive.	Range: ~10–100 GU/mL

Continued

Table 6.1 Comparison among the studies which detected SARS-CoV-2 in raw wastewater entering the WWTPs. When quantified, the viral load is indicated in terms of genomic units per unit of volume (GU/L or GU/mL).—cont'd

References	Location, date of sampling	Plants and type of sampling	Details on the study	SARS-CoV-2 load in raw wastewater
Wurtzer et al.[56]	Paris, France. From March 5 to Apr 23, 2020.	• Several major WWTPs of the Parisian area.	• All samples tested positive. • The time-course monitoring displayed a 2-log increase.	5-Mar: 5×10^4 GU/L Exponential increase from 5×10^4 GU/L to 3×10^6 GU/L.
Yaqub et al.[57]	Parts of Lahore, Pakistan. From July 13 to 25, 2020.	• Smart lockdown implemented in the area on Jul 9, 2020. • 28 sites (lift stations and sewage lines). • Grab sampling and 24-h sampling.	• 16/28 sample were positive on the 1st day of sampling (13 Jul) with variable load. • Then viral load decreased toward the end of lockdown. • Only a few sites did not follow a clear pattern (due to the complexities in sewage water based surveillance).	13 Jul Positivity: 16/28 cases Range: 2.4 $\log_{10}$−4.55 $\log_{10}$ GU/mL 25 Jul Positivity: 9/28 cases Range: 0.6 $\log_{10}$−3.64 $\log_{10}$ GU/mL

Legend: n.a. = not available.

with the fact that the number of COVID-19 positive persons may vary widely inside the population served by the sewerage network.

On the basis of the most recent literature, summarized in Table 6.1, the concentration of SARS-CoV-2 in wastewater was found to be largely variable, with the maximum values in the order of 4 $\log_{10}$ GU/mL which were found in severe situations or in presence of a lockdown in the urban area monitored. However, the viral loads revealed in wastewater are very low in comparison with those found in feces, which may reach an order of magnitude up to 7 $\log_{10}$ GU/mL.

This in turn highlights the effort needed and the difficulty to detect such low quantities of viral RNA in a complex matrix like wastewater. SARS-CoV-2 RNA is present at concentrations orders of magnitude lower, not only with respect to feces, but also when compared to the standard nasopharyngeal swabs used in COVID-19 diagnostics. This extreme difference in concentrations is the main reason for SARS-CoV-2 RNA detection methods that are extremely well standardized in nasopharyngeal swabs to still lag behind when it comes to detecting viral RNA in wastewater. Before any downstream analysis, it is therefore crucial to concentrate the initial wastewater sample. However, this procedure introduces further potentially problematic steps, including freezing and inactivation of the sample. Since concentration protocols are usually laborious and not scalable to many samples, wastewater is indeed commonly frozen and stored at −20°C (or seldom at −80°C) before processing,[44,58] which greatly impacts RNA integrity and consequently further lowers SARS-CoV-2 concentration.[59] The same is true for the thermal inactivation of the virus,[60,61] a necessary step in many protocols[44] to avoid exposure to aerosols that may generate during concentration in laboratories with limited biosafety measures (e.g., lack of BSL-2 cabinets and/or aerosol-tight centrifuge rotors). However, low viral concentrations and poor RNA integrity are not the only limiting factors for easy SARS-CoV-2 RNA detection from this matrix. Wastewater indeed contains a number of PCR-inhibitory substances like humic and fulvic acids that can prevent the correct amplification of viral RNA and even produce false-negative results.[62] Taken together, this set of limitations explains why the detection of SARS-CoV-2 in sewerage is still an open issue.

6.4 The approach for SARS-CoV-2 detection in sewerage is an open issue

The approach for the detection of SARS-CoV-2 in wastewater includes several steps that start with the sampling in the sewerage or in the WWTPs and the transportation and conservation of the samples. Then, the subsequent analysis can be performed in a Biosafety Level 2 (BSL-2) laboratory, while virus isolation and cultivation require a BSL-3 laboratory.[63] At the moment, some steps in the procedure present critical aspects, not fully investigated yet, when applied to wastewater samples. Despite the quickly evolving situation and the continuous efforts in the research, a standard protocol is not available yet and the quantification of the viral load in wastewater remains an open issue because of the heterogeneity of the approaches proposed in the literature, which may lead to different and not always comparable results.

First, the analysis of SARS-CoV-2 in wastewater requires an appropriate sampling, transportation, and storage of the samples, according to the following steps.

(1) Sampling of wastewater: In order to obtain an accurate estimation of the viral content in the influent wastewater, the collection of composite samples with automated samplers monitors the strong fluctuations of flow rates and contaminants occurring over time. Composite samples[64] can be formed with: (1) aliquots of wastewater with equal volume taken at equal intervals during the day, or (2) 24-h flow proportional samples that are aliquots with volume proportional to the flow rate. Although the second alternative would be preferable,[64] it requires a flow measurement device installed in the inlet channel of a WWTP that is not always available. In some cases, to monitor load peaks during short time intervals, a grab sampling is performed. A comparison on the adoption of the various types of sampling is shown in Table 6.1.

(2) During collection and transportation to the lab, the samples must be maintained at low temperature, i.e., 4°C, with the aim of preserving the number and the viability of the virus.[65] In fact it was widely demonstrated that freeze-thaw cycles affect the integrity of viral RNA, and that storage at −80°C causes a reduction of 1−3 cycle threshold (Ct) in real time RT-PCR of the N1, N2, and N3 gene fragments,[59] and even more in case of storage at −20°C. Conversely, the viral RNA was stable after the storage at 5°C up to 15 days.[59] This led some authors[31,59] to suggest transporting the samples on ice and store them at 4°C until further analysis to preserve the RNA integrity. However, at the moment, there is not full agreement about the temperature and duration of the storage for the conservation of the samples before the analysis. Xiao et al.[16] indicated to maintain the samples at 4°C when the analysis is performed in the subsequent 24 h, otherwise for longer storage (>24 h) the samples are maintained at −70°C or, when not available, at −20°C.

(3) Inactivation of the virus: Some protocols of analysis include a step of inactivation of SARS-CoV-2 performed at 56°C per 30 min[44] to avoid any possible risk for the operator. However, most protocols now avoid the thermal inactivation step to preserve RNA integrity.[60,61]

Then, the whole chain for the quantification of the viral load in wastewater proceeds with molecular biology analysis, as follows.

(4) Viral enrichment is a very important passage because of extremely low viral titers in wastewater samples. Without concentration, detecting the virus in such samples, at the moment, is not achievable. Viral enrichment can be performed through various approaches that include centrifugation steps together with centrifugal ultrafiltration,[31,59] aluminum hydroxide adsorption-precipitation,[50] filtration through 0.22 μm filters combined with PEG precipitation,[66] or ultracentrifugation.[56] For example, La Rosa et al.[44] proposed the use of the standard WHO procedure for Poliovirus surveillance in wastewater (based on the use of polyethylene glycol and dextran) after some modifications in the concentration step.

(5) Viral RNA extraction is the step realized in order to isolate and purify RNA from virus and clean it from pollutants and inhibitors of PCR and is mostly carried out with commercially available RNA extraction kits with custom modifications.[31,44,48,50] In general, extraction consists of four phases: lysis (RNA is released from virus and proteins separated and denatured), binding (RNA is bound on a surface, i.e., silica membrane or magnetic silica), washing (with different special wash buffers RNA is removed of unwanted impurities such as salt, proteins, and other contaminants), and elution (pure RNA is free from the surface bond and ready to use in enzymatic PCR reaction).

(6) Nucleic acid amplification based on rRT-PCR targeting single or multiple regions of the SARS-CoV-2 genome, including the N, S, E, and RdRP genes encoding for the nucleocapsid, spike, envelope, and RNA-dependent RNA polymerase proteins, respectively.[7] The RdRP gene region also covers the ORF1ab gene, specifically targeted by a number of widely used primer sets. The Ct value measured in rRT-PCR is inversely proportional to the virus quantity. For example, a low Ct value means that a low number of amplification cycles are required to obtain a fluorescent signal over the threshold and thus that the SARS-CoV-2 viral load in the sample is higher. In the case of wastewater samples, as a consequence of the relevant dilution of the virus, the Ct value to call a positive sample is often raised with respect to feces or other human specimens, even reaching values around 40.[55]

(7) Finally, the viral load in the wastewater sample can be expressed numerically as GU/mL.

6.5 Decay of SARS-CoV-2 in wastewater due to adverse environmental conditions

A certain reduction of the viral load may occur along the sewerage network, in the wastewater treatment, and in the environment in general, due to adverse conditions caused by pH changes, temperature variations, disinfectants, and various pollutants.[67] In particular, SARS-CoV-2 and the other CoVs belong to the enveloped viruses (characterized by a fragile lipid membrane called envelope) that are more susceptible to environmental factors[68] and more rapidly inactivated than enteric viruses, which are almost all nonenveloped.[69] For example, in wastewater, 90% inactivation of nonenveloped viruses can be observed in a relatively long period ranging from days to months, while it is shorter, from hours to days, in enveloped viruses, as demonstrated by Ye et al.[70] The lower resistance of the enveloped viruses is due to the lysis of the viral envelope—for example, caused by pollutants or detergents that break down the weak fatty layer around the SARS-CoV-2 virion—with the consequent damage of the lipid envelope that is required for the infection of the host cells.[71]

The stability and reduction of SARS-CoV-2 under some different environmental conditions was reported by Chin et al.[72] In particular, with regards to the effect of the sole temperature, SARS-CoV-2 remained highly stable at 4°C (with around a 0.7 log-unit reduction of infectious titer after 14 d), while it was sensitive to heat that inactivated the virus after 5 min at 70°C.[72] For a comparison, the previous SARS-CoV, when seeded into sewage, remained infectious for a period of 14 days at temperature of 4°C and only 2 days at 20°C.[65]

Comparing SARS-CoV-2 with other CoVs, the new coronavirus may have a greater resilience outside the body than expected.[73] In particular, Goh et al.[73] predicted by a model that SARS-CoV-2 could have one of the hardest outer shells among most CoVs. The virus may have the greatest resilience to hostile environmental conditions since the harder shell can protect the virion from damage. This aspect may also affect how long the virus remains infectious in the environment. However, this theoretical study is not exhaustive and further experimental research is needed in this direction.

The difference between SARS-CoV-2 and other CoVs is due to the fact that not all the CoVs have the same persistence or inactivation in water[68]: for example, even if belonging to the same genus, the time required for 90% inactivation of SARS coronavirus may be ninefold longer than another human coronavirus.[70]

It is worth noting that the investigations about the survival of CoVs in the environmental matrices should be carried out using the pathogen of interest, since viral persistence may differ even within different strains of the same viral species. Unfortunately, this is not always possible, and thus the research needs to use "surrogate" (model) viruses with similar properties. However, the choice of a surrogate virus is challenging because the exact relationships between phylogenetically different CoVs is still unclear.[74] The use of multiple surrogates instead of only one permits to cover the multiple aspects of a target virus.[75]

Although surrogates may not be exactly representative of the virus of interest, in absence of experimental data acquired for SARS-CoV-2, surrogates may offer some useful information about the potential behavior of SARS-CoV-2 in the contaminated water, as explained later in the chapter.

CoVs inactivation occurs more rapidly in sewage than in water due to various factors: high presence of organic matter, pollutants, or antagonistic bacteria.[76]

Casanova et al.[74] investigated the stability of two surrogate CoVs (transmissible gastroenteritis virus that is a diarrheal pathogen of swine, and mouse hepatitis virus that is a respiratory and enteric pathogen of laboratory mice) in pasteurized settled sewage at 4°C and 25°C. Experiments were carried out using viral stocks prepared as follows: (1) propagation by infecting confluent layers of host cell cultures in flasks, (2) harvesting of cell lysates, (3) clarification by centrifugation (3000× g, 30 min, 4°C), and (4) storage of supernatants as virus stock at −80°C.[74] In these experiments Casanova et al.[74] observed that the infectivity decreased at rate of 1.5−2.0 $\log_{10}$ per week at temperature of 25°C (99% reduction obtained at 7−9 d), while it decreased more slowly at rate of 0.2−0.3 $\log_{10}$ per week at 4°C.

When using raw sewage instead of pasteurized settled sewage, the rates of viral inactivation could be even faster because the pasteurization process could reduce the role of proteolytic enzymes that contribute to virus inactivation.

Pollution in water has a relevant effect on the virus inactivation. The study of Casanova et al.[74] indicated that the time required for the reduction of infectivity was about half in pasteurized settled sewage compared to clean water. Another study on CoVs observed that 2 d at 23°C permitted to obtain ~2 log reduction in primary effluent, while 7–8 days were needed in tap water,[76] again indicating that the inactivation process may be accelerated in sewage.

The survival of a representative CoV (Human coronavirus 229E) was tested at 23°C in samples of primary effluent (presettled wastewater) and secondary effluent (after activated sludge and secondary settling, but prior to chlorination) taken from a WWTP.[76] The CoV was inactivated in wastewater with 3 log reduction between 2 and 4 d. In comparison with enteroviruses, Gundy et al.[76] indicated that this CoV was less stable in water, more rapidly inactivated, and less transmitted in aqueous environment.

6.6 Reduction of SARS-CoV-2 in wastewater treatment plants

WWTPs are characterized by a sequence of treatments (physical, biological, and disinfection), according to the typical flow-sheet shown in Fig. 6.1.

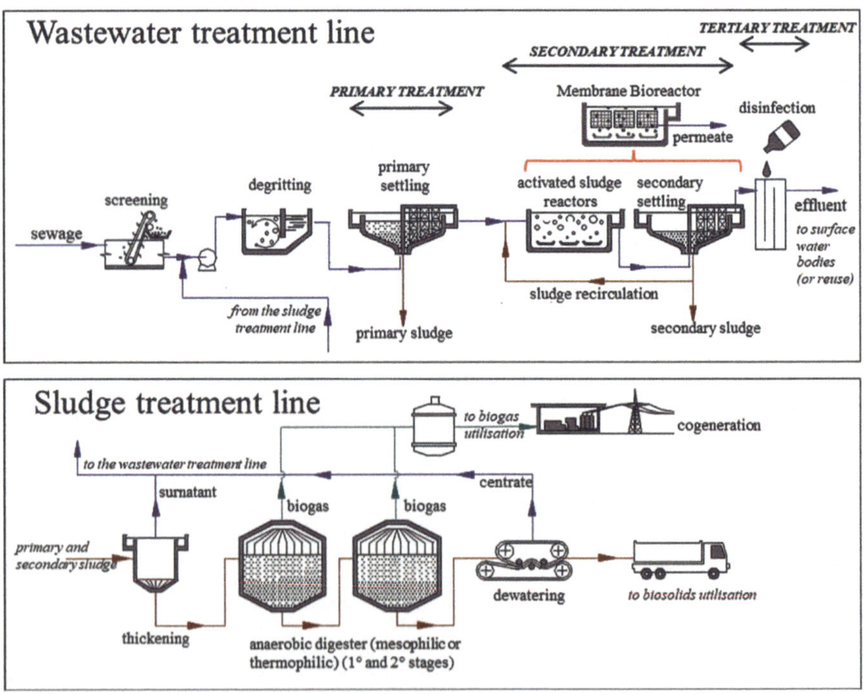

Figure 6.1 Layout of a wastewater treatment plant divided in (1) wastewater treatment line that implements primary, secondary (activated sludge or membrane bioreactor), and tertiary treatments; (2) sludge treatment line.

WWTPs are designed to treat organic matter, solids, nutrients, and bacteria with high removal rates and efficiency, but are expected to be effective also against viruses when disinfection is implemented.

In primary treatment (settling) a reduction of up to 26% of two model enveloped viruses was observed (only 6% of reduction for nonenveloped viruses), according to Ye et al.[68] The viral envelope is hydrophobic and this favors the absorption to solids and thus the separation of the enveloped viruses in the sludge.[76]

Primary treatment and secondary treatment (activated sludge), shown in Fig. 6.1, may contribute only partially to the virus reduction, and removal in these stages is largely variable and viruses can remain viable from hours to days in absence of disinfection. Therefore, to improve the virus inactivation in WWTPs, disinfection must be implemented. In fact, SARS-CoV-2, being an enveloped virus with a fragile outer membrane, is susceptible to disinfection, even more than enteric viruses.[77]

Disinfection can be based on chlorination, ozonation, peracetic acid, or UV light.

With regards to chlorination, free chlorine inactivates SARS-CoV more effectively than chlorine dioxide.[78] The mechanism for the free chlorine inactivation is the reaction with proteins instead of genomes: the free chlorine molecules readily penetrate the lipid membrane, reacting with the proteins in the nucleocapsid.[70]

UV disinfection is based on exposure to UVC light emitted traditionally by UV mercury lamps. The fate of ssRNA viruses, like SARS-CoV-2, after exposure to UV is not completely clear, because the literature on RNA photochemistry is scarce compared to DNA photochemistry. In general, the susceptibility of enveloped viruses to UVC light is comparable to nonenveloped viruses because the inactivation mechanism targets primarily the genome and the lipid membrane cannot protect it from the UV light.[69]

When the secondary treatment is a membrane bioreactor (MBR) (Fig. 6.1), the biological process is coupled with the filtration of solids through a membrane system without the requirement of a secondary settler. Virus removal can be improved significantly[79] due to the small pore size of membranes (0.03–0.40 μm) and the biofouling that is an additional barrier for viruses.[80] The size of virions of SARS-CoV-2 is about 60–140 nm with projections of 9–12 nm, so MBRs installing membranes with a similar absolute pore size are effective in the removal. In particular, ultrafiltration has a cutoff rating of 0.005–0.01 microns and so it is more suitable in virus removal than microfiltration, and this latter requires a subsequent stage of disinfection. No experimental data are available on the removal of SARS-CoV-2 in MBRs, but Lesimple et al.[81] and Tetteh et al.[82] indicate that MBR technology can be an effective technology for the removal of pathogens, including SARS-CoV-2. To have an idea of virus reduction, average removal of some enteric viruses in a full-scale MBR (absolute pore size of 0.1 μm) was 5.5 $\log_{10}$ for human adenovirus, 5.1 $\log_{10}$ per human enterovirus, and 3.9 log10 for norovirus.[83]

A summary of the studies that analyzed SARS-CoV-2 in the treated wastewater effluents or in the sludge separated from full-scale WWTPs is shown in Table 6.2.

In many investigations, the treated effluents were negative for SARS-CoV-2, especially in the plants equipped with tertiary treatments. Conversely, the implementation of a secondary treatment only does not ensure to reach viral loads below the detection limits. All the studies investigating the concentrations of SARS-CoV-2 in sludge, demonstrated the enrichment of the virus in primary, secondary, or thickened sludge.

6.7 Potential fecal-oral transmission associated with sewerage

The presence of SARS-CoV-2 in raw wastewater and treated effluents, measured in GU, is not always associated with viability and infectivity of the virus as indicated by many authors in the literature.[10,11,15,16,21,84,85] It is worth noting that the virus may cause infection only if it is able to retain viability.

Studies have found the presence of infectious virions of SARS-CoV-2 in feces[10,16,22] and in urine.[20] This postulates that SARS-CoV-2 may remain infectious also in wastewater. The infectious potential of untreated and treated wastewater was tested using a cell culture model.[54] Inoculation of differentiated Caco-2 cells for 10 d with wastewater indicated that treated wastewater was noninfectious even though viral RNA fragments were detected. It is important to underline that infectivity of SARS-CoV-2 was not yet observed in WWTPs up to now, either raw or treated samples, despite the presence of viral RNA in the samples.

As far as we know, the scientific literature reports one case of transmission caused by sewerage. Han and He[86] reported that the first known case of a sewage-associated transmission of SARS-CoV-2 was documented in a Chinese website (written in Chinese by Li P. and Bin H.). In particular, this COVID-19 outbreak occurred in a group of a few households, as a

Table 6.2 Comparison among the studies which detected and quantified SARS-CoV-2 in treated effluents or in the sludge separated in the WWTPs.

References	Location, date of sampling	Plants and type of sampling	Details on the study	SARS-CoV-2 load in effluent wastewater and sludge
Ampuero et al.[32]	Santiago, Chile. From March to June 2020.	• 2 WWTPs. • 24-h composite samples.	• SARS-CoV-2 detected in treated wastewater samples.	Effluent from WWTP1: 25 May: 20 GU/mL 15 Jun: 167 GU/mL Effluent from WWTP2: 25 May: 10 GU/mL
Arora et al.[33]	Jaipur, Rajasthan, India. From May 3 to June 14, 2020.	• 6 WWTPs. • Samples collected from different units of the WWTPs.	• The treated wastewater was negative in all the WWTPs, even in those where raw wastewater was positive. • The different wastewater treatments used (Moving Bed Biofilm Reactor and Sequencing Batch Reactor) permitted to reduce the viral load below the detection limit.	Negative effluents.

Continued

Table 6.2 Comparison among the studies which detected and quantified SARS-CoV-2 in treated effluents or in the sludge separated in the WWTPs.—cont'd

References	Location, date of sampling	Plants and type of sampling	Details on the study	SARS-CoV-2 load in effluent wastewater and sludge
Balboa et al.[34]	Ourense, Spain. From April 6 to 21, 2020.	• 24-h composite samples taken twice a week.	• Negative samples in the secondary treatment effluent, confirming that the effluent is safe for reuse and discharge to water bodies. • Viral RNA mainly retained in the sludge from the primary settler. • Rarely detected in the biological sludge. • Its concentration increased in the thickeners due to the long retention time (~24 h) and the high solid content. • Not detected in the digested sludge.	Negative effluents. Positive samples of sludge.
Haramoto et al.[39]	Yamanashi Prefecture, Japan. From March 17 to May 7, 2020.	• WWTP with conventional activated sludge process. • Grab sampling before chlorination.	• SARS-CoV-2 RNA was detected in one (20%) secondary-treated wastewater sample.	Effluent from WWTP: 2.4×10^3 copies/L
Hong et al.[41]	Jeddah, Saudi Arabia. From April 15 to July 9, 2020.	• Grab samples collected in the supernatant of the activated sludge tank located in a hospital.	• 15.4% (8/52) of samples taken from the activated sludge tank (partially treated wastewater) were tested positive for N genes.	Effluent from secondary treatment: N1: 81.1 GU/L N2: 1115.8 GU/L N3: 411.2 GU/L

Kocamemi et al.[42]	Istanbul, Turkey. On May 7, 2020.	• 9 WWTPs. • Grab sampling of primary sludge and waste activated sludge.	• SARS-CoV-2 was more concentrated in primary sludge and waste activated sludge than in the influent.	Range in sludge: 1.17×10^4 -4.02×10^4 GU/L
Kumar et al.[43]	Gujarat, India. On May 8 and 27, 2020.	• Grab sampling of the final effluents after UASB and aeration pond.	• Final effluent samples were negative for 3 genes examined. • The viral load was thus significantly reduced by the UASB treatment and aeration pond.	Negative effluents (CT values > 40)
Peccia et al.[58]	New Haven, Connecticut, USA. From March 19 to May 1, 2020	• Samples taken daily at the outlet of a gravity thickener (solids content 2.6%–5%).	• All five measures traced the rise and fall of SARS-CoV-2 infections during the 10-week period studied.	Range in primary sludge: 1.7×10^3 -4.6×10^5 GU/L
Randazzo et al.[50]	Region of Murcia, Spain. From March 12 to Apr 14, 2020.	• 6 WWTPs. • 18 samples of secondary effluents. • 12 samples of tertiary treated effluents.	• 11% (2/18) of secondary effluents were positive for at least one target. • Tertiary effluent samples were all negative.	Secondary effluent 1 (N2 positive): 5.40 $\log_{10}$ GU/L Secondary effluent 2 (3 targets positive): < quantification limit.
Rimoldi et al.[51]	Milano and Monza e Brianza, Italy. On April 14 and 22, 2020.	• WWTPs equipped with tertiary treatments. • 2 grab samples of treated wastewater.	• Treated wastewater was negative.	Negative effluents

Continued

Table 6.2 Comparison among the studies which detected and quantified SARS-CoV-2 in treated effluents or in the sludge separated in the WWTPs.—cont'd

References	Location, date of sampling	Plants and type of sampling	Details on the study	SARS-CoV-2 load in effluent wastewater and sludge
Sherchan et al.[52]	Southern Louisiana,USA. From January to April 2020.	• 2 WWTPs with conventional activated sludge and chlorine disinfection. • Samples of secondary treated and final effluents after disinfection.	• Secondary-treated wastewater and final effluent samples tested negative. • Thus the virus was removed in WWTPs to undetectable level.	Negative effluents
Westhaus et al.[54]	North Rhine-Westphalia, Germany. Apr 8, 2020.	• 9 WWTPs. • Samples of treated wastewater.	• treated effluents were positive despite the different processes applied in the WWTPs	Range in treated effluents: 2.7–37 GU/mL

consequence of the loss from a private combined sewer pipe that contaminated the surrounding environments causing the diffusion of the infection. The hypothesized route of transmission was confirmed by tracking the overflows from the broken pipe during another heavy rainfall event.[86]

The effects of the COVID-19 pandemic in the field of waste and wastewater services[87] are very complex routes and the knowledge about the potential fecal-oral transmission is only partial.[88] So, due to the presence of infectious SARS-CoV-2 virions in feces and urine and the potential presence in wastewater, the possibility of fecal-oral transmission cannot be excluded.[20]

From the state of the art, at the moment, in high-income countries it is unlikely that the environmental matrices could become a main transmission route for SARS-CoV-2, even if more research in this field is needed and appropriate caution is recommended.[86] Conversely, in low-income countries where pit toilets are the most common system for human excreta disposal, the inadequate sanitation may be a source of contamination of SARS-CoV-2 in soil and groundwater. Similarly, SARS-CoV-2 could be spread in the environment through "open defecation"; in 2017 nearly 950 million people in poor countries did not have basic sanitation and routinely practiced open defecation.[89]

6.7.1 Aerosolization of wastewater

The presence of viral RNA in sewage and the spread of air droplets cannot exclude the possibility of a viral presence in the aerosol. A well-known case occurred in Amoy Gardens in Hong Kong during the previous SARS outbreak, where the aerosols originated from the building pipes and containing the viruses reached a large apartment complex.[24] One study on the viability of SARS-CoV-2 in aerosol[90] indicated that the half-life (that is the time needed to halve the amount of the virus) was approximately 1.1 h.

WHO[77] has developed specific guidance for the workers in WWTPs during the SARS-CoV-2 outbreak. Briefly, and not exhaustively, workers should wear personal protective equipment (PPE; i.e., protective outerwear, gloves, boots, goggles or a face shield, and a face mask or FFP3 respirator mask), which continues to be effective in protecting against pathogens included SARS-CoV-2.[91]

6.7.2 Overflows in municipal combined sewer systems

The events of heavy rainfalls produce large stormwater runoff, which often exceeds the conveyance capacity of the pipes in the combined sewer systems, causing inevitable overflows of untreated sewage (called combined sewer overflows, CSO). There is a large number of municipalities around the world still served by combined sewer systems, often characterized by aged structures and pipes.

CSO is a mixing of contaminated water that includes stormwater runoff and untreated sewage containing feces/urine, pathogenic microorganisms, and other pollutants. Due to the short residence time passed by feces/urine in sewers (approximately in the order of a few hours), the viruses excreted in feces could maintain viability and remain still infective.[68]

When CSO leave the sewerage network, they may spread quickly and widely into public areas or surface waters and may pose potential risks for the public health.[75]

6.7.3 Flooding events in urban areas

A preliminary retrospective analysis was performed by Han and He[86] in some locations characterized by a significant number of COVID-19 infected persons during flooding events in the period May—August 2020. Although any conclusive trends cannot be drawn, Authors discuss the potential risk of SARS-CoV-2 transmission in these flooded areas, where sewage overflows may be an additional factor for the virus spread.

In particular, during flooding events the population may be exposed to overflowed human excreta or human wastes derived from the overload in the combined sewer systems. This could contribute to the sewage-associated transmission of COVID-19 when heavy rainfalls are frequent.[86]

Therefore, in the municipalities served by combined sewer systems in areas subjected to flooding events, especially in summer, the network should be upgraded toward separated systems to reduce the risk of overflows of blackwaters.

6.7.4 Discharge of untreated or treated wastewater into surface water bodies

Although the presence of SARS-CoV-2 in wastewater has been confirmed in various researches, there are only few studies that have investigated on the viral dispersion in the receiving water bodies and the potential associated health risk.

In Rimoldi et al.,[51] three rivers (near Milan, Italy) were surveyed during the peak of pandemic in April 2020. Real-time RT-PCR analysis showed that samples taken from these receiving water bodies were positive for SARS-CoV-2. It is interesting to investigate more in depth the source of the viral contamination. In the study of Rimoldi et al.,[51] caffeine was used as a clear indication of urban-only pollution due to untreated wastewater, because it is highly degradable in wastewater treatment and thus significantly reduced in the treated effluents. When the SARS-CoV-2 presence is associated with a significant detection of caffeine in the rivers, it means that wastewater treatment plants cannot be the origin of the virus presence. Instead the virus could relate to nontreated or inefficiently treated sewage discharged into surface waters as a consequence of illicit discharges, malfunction or leakages in the sewerage systems, noncollected domestic flows, or CSOs.

In some samples SARS-CoV-2 was detected but caffeine was not.[51] These cases suggest that a sporadic release of virus traces in the treated effluents from WWTPs cannot be completely excluded, especially when the WWTP is equipped with secondary treatment only.[51] An important fact is that, even in the cases when the viral RNA was detected, the infectivity and the risk of infection from surface water was not demonstrated.[51]

Similarly, the aim of the study of Haramoto et al.[39] was to detect SARS-CoV-2 RNA in rivers (Yamanashi Prefecture, Japan), in samples collected from March to May. In this study, none of the samples tested positive for SARS-CoV-2 RNA.

The situation in the study of Guerrero-Latorre et al.[92] was quite different. The study concerns urban rivers of Quito (Ecuador) where, as in many other cities worldwide, wastewater is directly discharged into surface water bodies. The samples were taken during a peak of COVID-19 cases and evaluated for water quality parameters, human adenovirus presence and SARS-CoV-2 presence. The water was concentrated, for viral analysis, using skimmed milk flocculation method and the results showed a high presence of SARS-CoV-2 in all the samples. However, the risks for health and ecology, especially in low-income countries should be further assessed.

6.8 Conclusions and future perspective

The recent global outbreak of SARS-CoV-2 has highlighted the limited knowledge on the fate of CoVs in sewerage and WWTPs. There is basis for thinking that SARS-CoV-2 is controlled in WWTPs and that viral loads in treated effluents are very low especially in the presence of a disinfection stage. Conversely, inappropriate discharges of sewerage can be a source of contamination of surface water and groundwater. Fecal-oral transmission of SARS-CoV-2 is unlikely even if caution is recommended in this field. This chapter summarizes the main aspects of the presence of SARS-CoV-2 in wastewater, but alongside certain findings, a number of questions have emerged that need further research. Specific questions include

- viability and infectivity of SARS-CoV-2 in wastewater should be further investigated, since the identification of viral RNA is not evidence of infectious capacity;
- data on SARS-CoV-2 abundance in wastewater (and its variability) are still rare and this limits a shared conclusion on decay and removal in WWTPs;
- the fate of SARS-CoV-2 in biosolids (primary and secondary sludge from WWTPs) and related risks need to be investigated more in depth; and
- a challenge against future epidemic waves is based on the possibility of using the SARS-CoV-2 load in wastewater for epidemiological surveillance, as a tool capable of providing early warning of incipient outbreaks.

References

1. WHO. *Naming the coronavirus disease (COVID-19) and the virus that causes it.* World Health Organization; 2020. Available from: https://www.who.int/emergencies/diseases/novel-coronavirus-2019/technical-guidance/naming-the-coronavirus-disease-(covid-2019)-and-the-virus-that-causes-it.
2. Chen N, Min Z, Xuan D, et al. Epidemiological and clinical characteristics of 99 cases of 2019 novel coronavirus pneumonia in Wuhan, China: a descriptive study. *Lancet* 2020;**395**(10223):507–13.
3. Huang C, Yeming W, Xingwang L, et al. Clinical features of patients infected with 2019 novel coronavirus in Wuhan, China. *Lancet* 2020;**395**(10223):497–506.
4. Lai C-C, Wen-Chien K, Ping-Ing L, Shio-Shin J, Po-Ren H. Extra-respiratory manifestations of COVID-19. *Int J Antimicrob Agents* 2020;**56**(2):106024.
5. Wang L, Xun L, Hui C, et al. Coronavirus disease 19 infection does not result in acute kidney injury: an analysis of 116 hospitalized patients from Wuhan, China. *Am J Nephrol* 2020;**51**(5):343–8.
6. Park S-K, Lee C-W, Park D-I, et al. Detection of SARS-CoV-2 in fecal samples from patients with asymptomatic and mild COVID-19 in Korea. *Clin Gastroenterol Hepatol* 2020;(20):S1542–3565. 30777–1.
7. Corman VM, Landt O, Kaiser M, et al. Detection of 2019 novel coronavirus (2019-nCoV) by Real-Time RT-PCR. *Euro Surveill* 2020;**25**(3):2000045.
8. Ma X, Su L, Zhang Y, Zhang X, Gai Z, Zhang Z. Do children need a longer time to shed SARS-CoV-2 in stool than adults? *J Microbiol Immunol Infect* 2020;**53**(3):373–6.
9. Pan Y, Zhang D, Yang P, Poon LLM, Wang Q. Viral load of SARS-CoV-2 in clinical samples. *Lancet Infect Dis* 2020;**20**(4):411–2.
10. Wang W, Xu Y, Gao R, et al. Detection of SARS-CoV-2 in different types of clinical specimens. *J Am Med Assoc* 2020;**323**(18):1843–4.
11. Wu Y, Guo C, Tang L, et al. Prolonged presence of SARS-CoV-2 viral RNA in faecal samples. *Lancet Gastroenterol Hepatol* 2020;**5**(5):434–5.
12. Xu Y, Li X, Zhu B, et al. Characteristics of pediatric SARS-CoV-2 infection and potential evidence for persistent fecal viral shedding. *Nat Med* 2020;**26**(4):502–5.
13. Foladori P, Cutrupi F, Segata N, et al. SARS-CoV-2 from faeces to wastewater treatment: what do we know? A review. *Sci Total Environ* 2020;**743**:140444.

14. Wolfel R, Corman VM, Guggemos W, et al. Virological assessment of hospitalized cases of coronavirus disease 2019. *Nature* 2020;**581**:465–9.
15. Gu J, Han B, Wang J. COVID-19: gastrointestinal manifestations and potential fecal–oral transmission. *Gastroenterology* 2020;**158**(6):1518–9.
16. Xiao F, Tang M, Zheng X, Liu Y, Li X, Shan H. Evidence for gastrointestinal infection of SARS-CoV-2. *Gastroenterology* 2020;**158**(6):1831–3.
17. Zhou J, Li C, Liu X, et al. Infection of bat and human intestinal organoids by SARS-CoV-2. *Nat Med* 2020;**26**(7):1077–83.
18. Yu F, Yan L, Wang N, et al. Quantitative detection and viral load analysis of SARS-CoV-2 in infected patients. *Clin Infect Dis* 2020;**71**(15):793–8.
19. Kim J-M, Kim HM, Lee EJ, Jo HJ, Yoon Y, Lee N-J, Son J, et al. Detection and isolation of SARS-CoV-2 in serum, urine, and stool specimens of COVID-19 patients from the Republic of Korea. *Osong Public Health Res Perspect* 2020;**11**(3):112–7.
20. Sun J, Zhu A, Li H, et al. Isolation of infectious SARS-CoV-2 from urine of a COVID-19 patient. *Emerg Microb Infect* 2020;**9**(1):991–3.
21. Amirian ES. Potential fecal transmission of SARS-CoV-2: current evidence and implications for public health. *Int J Infect Dis* 2020;**95**:363–70.
22. Zhang Y, Chen C, et al. Isolation of 2019-nCoV from a stool specimen of a laboratory-confirmed case of the coronavirus disease 2019 (COVID-19). *China CDC Weekly* 2020;**2**(8):123–4.
23. Xiao F, Sun J, Xu Y, et al. Infectious SARS-CoV-2 in feces of patient with severe COVID-19. *Emerg Infect Dis* 2020;**26**(8):1920–2.
24. McKinney KR, Gong YY, Lewis TG. Environmental transmission of SARS at Amoy Gardens. *J Environ Health* 2006;**68**(9):26–30.
25. Jeong HW, Kim S-M, Kim H-S, Park S-J, Kim E-H, Choi YK. Viable SARS-CoV-2 in various specimens from COVID-19 patients. *Clin Microbiol Infect* 2020;**26**(11):P1520–4.
26. Rose C, Parker A, Jefferson B, Cartmell E. The characterization of feces and urine: a review of the literature to inform advanced treatment technology. *Crit Rev Environ Sci Technol* 2015;**45**(17):1827–79.
27. Penn R, Ward BJ, Strande L, Maurer M. Review of synthetic human faeces and faecal sludge for sanitation and wastewater research. *Water Res* 2018;**132**:222–40.
28. EEA. Available from: https://www.eea.europa.eu/signals/signals-2018-content-list/articles/water-use-in-europe-2014; 2018.
29. Cheung KS, Hung IFN, Chan PPY, et al. Gastrointestinal manifestations of SARS-CoV-2 infection and virus load in fecal samples from a Hong Kong cohort: systematic review and meta-analysis. *Gastroenterology* 2020;**159**(1):81–95.
30. Zhang N, Gong Y, Meng F, Shi Y, Wang J, Mao P, et al. Comparative study on virus shedding patterns in nasopharyngeal and fecal specimens of COVID-19 patients. *Sci China Life Sci* 2020;**63**.
31. Ahmed W, Angel N, Edson J, et al. First confirmed detection of SARS-CoV-2 in untreated wastewater in Australia: a proof of concept for the wastewater surveillance of COVID-19 in the community. *Sci Total Environ* 2020;**728**:138764.
32. Ampuero M, Valenzuela S, Valiente-Echeverria F, et al. SARS-CoV-2 detection in sewage in Santiago, Chile – preliminary results. *MedRxiv* 2020. https://doi.org/10.1101/2020.07.02.20145177.
33. Arora S, Nag A, Sethi J, et al. Sewage surveillance for the presence of SARS-CoV-2 genome as a useful wastewater based epidemiology (WBE) tracking tool in India. *Water Sci Technol* 2020;**82**(12):2823–36. https://doi.org/10.2166/wst.2020.540.
34. Balboa S, Mauricio-Iglesias M, Rodriguez S, et al. The fate of SARS-COV-2 in WWTPS points out the sludge line as a suitable spot for detection of COVID-19. 2020;**772**:145268. https://doi.org/10.1016/j.scitotenv.2021.145268.

35. Chavarria-Miró G, Anfruns-Estrada E, Guix S, et al. Sentinel surveillance of SARS-CoV-2 in wastewater anticipates the occurrence of COVID-19 cases. *MedRxiv* 2020. https://doi.org/10.1101/2020.06.13.20129627.
36. Fongaro G, Stoco PH, Marques Souza DS, et al. SARS-CoV-2 in human sewage in Santa Catalina, Brazil, November 2019. *MedRxiv* 2020. https://doi.org/10.1101/2020.06.26.20140731.
37. Gonzalez R, Curtis K, Bivins A, Bibby K, Weir MH, Yetka K, Thompson H, Keeling D, Mitchell J, Gonzalez D. COVID-19 surveillance in Southeastern Virginia using wastewater-based epidemiology. *Water Res* 2020;**186**:116296.
38. Green H, Wilder M, Middleton FA, Collins M, Fenty A, Gentile K, et al. Quantification of SARS-CoV-2 and cross-assembly phage (crAssphage) from wastewater to monitor coronavirus transmission within communities. *MedRxiv* 2020. https://doi.org/10.1101/2020.05.21.20109181.
39. Haramoto E, Malla B, Thakali O, Kitajima M. First environmental surveillance for the presence of SARS-CoV-2 RNA in wastewater and river water in Japan. *Sci Total Environ* 2020;**737**:140405.
40. Hata A, Honda R, Hara-Yamamura H, Meuchi Y. Detection of SARS-CoV-2 in wastewater in Japan during a COVID-19 outbreak. *Sci Total Environ* 2021;**758**:143578. https://doi.org/10.1016/j.scitotenv.2020.1435787.
41. Hong P, Rachmadi AT, Mantilla-Calderon D, Alkahtani M, Bashwari YM, Al Qarni H, et al. How early into the outbreak can surveillance of SARS-CoV-2 in wastewater tell us? *MedRxiv* 2020. https://doi.org/10.1101/2020.08.19.20177667.
42. Kocamemi BA, Kurt H, Sait A, Sarac F, Saatci AM, Pakdemirli B. SARS-CoV-2 detection in Istanbul wastewater treatment plant sludges. *MedRxiv* 2020. https://doi.org/10.1101/2020.05.12.20099358.
43. Kumar M, Patel AK, Shah AV, Raval J, Rajpara N, Joshi M, Joshi CG. First proof of the capability of wastewater surveillance for COVID-19 in India through detection of genetic material of SARS-CoV-2. *Sci Total Environ* 2020;**746**:141326.
44. La Rosa G, Iaconelli M, Mancini P, Bonanno Ferraro G, Veneri C, Bonadonna L, Lucentini L, Suffredini E. First detection of SARS-CoV-2 in untreated wastewaters in Italy. *Sci Total Environ* 2020;**736**:139652.
45. La Rosa G, Mancini P, Bonanno Ferraro G, Veneri C, Iaconelli M, Bonadonna L, Lucentini L, Suffredini E. SARS-CoV-2 has been circulating in Northern Italy since December 2019: evidence from environmental monitoring. *Sci Total Environ* 2021;**750**:141711.
46. Lodder W, de Roda Husman AM. SARS-CoV-2 in wastewater: potential health risk, but also data source. *Lancet Gastroenterol Hepatol* 2020;**5**(6):533−4.
47. Manupati H, Kiran U, Kuncha SK, Kopperi H, Gokulan CG, Mishra RK, et al. Comprehensive surveillance of SARS-CoV-2 spread using wastewater-based epidemiology studies. *MedRxiv* 2020. https://doi.org/10.1101/2020.08.18.20177428.
48. Medema G, Heijnen L, Elsinga G, Italiaander R, Brouwer A. Presence of SARS-Coronavirus-2 in sewage and correlation with reported COVID-19 prevalence in the early stage of the epidemic in The Netherlands. *Environ Sci Technol Lett* 2020;**7**(7):511−6.
49. Or IB, Yaniv K, Shagan M, et al. Regressing SARS-CoV-2 sewage measurements onto COVID-19 burden in the population: a proof-of-concept for quantitative environmental surveillance. *MedRxiv* 2020. https://doi.org/10.1101/2020.04.26.20073569.
50. Randazzo W, Truchado P, Cuevas-Ferrando E, Simón P, Allende A, Sánchez G. SARS-CoV-2 RNA in wastewater anticipated COVID-19 occurrence in a low prevalence area. *Water Res* 2020;**181**:115942.
51. Rimoldi SG, Stefani F, Gigantiello A, et al. Presence and infectivity of SARS-CoV-2 virus in wastewaters and rivers. *Sci Total Environ* 2020;**744**:140911.

52. Sherchan SP, Shahin S, Ward LM, et al. First detection of SARS-CoV-2 RNA in wastewater in North America: a study in Louisiana, USA. *Sci Total Environ* 2020;**743**:140621.
53. Trottier J, Darques R, Mouheb NA, et al. Post-lockdown detection of SARS-CoV-2 RNA in the wastewater of Montpellier, France. *One Health* 2020;**10**:100157.
54. Westhaus S, Weber F-A, Schiwy S, et al. Detection of SARS-CoV-2 in raw and treated wastewater in Germany — Suitability for COVID-19 surveillance and potential transmission risks. *Sci Total Environ* 2021;**751**:141750.
55. Wu F, Xiao A, Zhang J, et al. SARS-CoV-2 titers in wastewater are higher than expected from clinically confirmed cases. *mSystems* 2020;**5**(4):e00614—20. https://doi.org/10.1128/mSystems.00614-20.
56. Wurtzer S, Marechal V, Mouchel J-M, et al. Evaluation of lockdown impact on SARS-CoV-2 dynamics through viral genome quantification in Paris wastewaters. *Euro Surveill* 2020;**25**(50):2000776. https://doi.org/10.2807/1560-7917.ES.2020.25.50.2000776.
57. Yaqub T, Nawaz M, Shabbir MZ, et al. A longitudinal survey for genome-based identification of SARS-CoV-2 in sewage water in selected lockdown areas of Lahore City, Pakistan; a potential approach for future smart lockdown strategy. *MedRxiv* 2020. https://doi.org/10.1101/2020.07.31.20165126.
58. Peccia J, Zulli A, Brackney DE, et al. Measurement of SARS-CoV-2 RNA in wastewater tracks community infection dynamics. *Nat Biotechnol* 2020;**38**(10):1164—7.
59. Medema G, Been F, Heijnen L, Petterson S. Implementation of environmental surveillance for SARS-CoV-2 virus to support public health decisions: opportunities and challenges. *Curr Opin Environ Sci Health* 2020;**17**:49—71.
60. Pastorino B, Touret F, Gilles M, de Lamballerie X, Charrel RN. Evaluation of heating and chemical protocols for inactivating SARS-CoV-2. *Viruses* 2020;**12**(6):624.
61. Weidhaas J, Aanderud Z, Roper D, et al. *Correlation of SARS-CoV-2 RNA in wastewater with COVID-19 disease burden in sewersheds*. 2020. Available from: https://www.researchsquare.com/article/rs-40452/v1.
62. Sidstedt M, Jansson L, Nilsson E, et al. Humic substances cause fluorescence inhibition in real-time polymerase chain reaction. *Anal Biochem* 2015;**487**:30—7.
63. CDC. *Centers for disease control and prevention*. 2020. Available from: https://www.cdc.gov/coronavirus/2019-ncov/lab/lab-biosafety-guidelines.html.
64. EPA. *Wastewater sampling. Number: SESDPROC-306-R3*. February 28, 2013.
65. Wang XW, Li J, Guo T, et al. Concentration and detection of SARS coronavirus in sewage from Xiao Tang Shan Hospital and the 309th Hospital of the Chinese People's Liberation Army. *Water Sci Technol* 2020;**52**(8):213—21.
66. Wu F, Xiao A, Zhang J, et al. SARS-CoV-2 titers in wastewater foreshadow dynamics and clinical presentation of new COVID-19 cases. *MedRxiv* 2020. https://doi.org/10.1101/2020.06.15.20117747.
67. Amoah ID, Kumari S, Bux F. Coronaviruses in wastewater processes: source, fate and potential risks. *Environ Int* 2020;**143**:105962.
68. Ye Y, Ellenberg RM, Graham KE, Wigginton KR. Survivability, partitioning, and recovery of enveloped viruses in untreated municipal wastewater. *Environ Sci Technol* 2016;**50**(10):5077—85.
69. Wigginton KR, Boehm AB. Environmental engineers and scientists have important roles to play in stemming outbreaks and pandemics caused by enveloped viruses. *Environ Sci Technol* 2020;**54**(7):3736—9.
70. Ye Y, Chang PH, Hartert J, Wigginton KR. Reactivity of enveloped virus genome, proteins, and lipids with free chlorine and UV254. *Environ Sci Technol* 2018;**52**(14):7698—708.
71. Lin Q, Lim JYC, Xue K, et al. Sanitizing agents for virus inactivation and disinfection. *View* 2020;**1**:e16. https://doi.org/10.1002/viw2.16.

72. Chin A, Chu J, Perera M, Hui K, Yen HL, Chan M. Stability of SARS-CoV-2 in different environmental conditions. *The Lancet, Microbe Correspondence* 2020;**1**(1):E10. https://doi.org/10.1016/S2666-5247(20)30003-3.
73. Goh GK-M, Dunker AK, Foster JA, Uversky VN. Shell disorder analysis predicts greater resilience of the SARS-CoV-2 (COVID-19) outside the body and in body fluids. *Microb Pathog* 2020;**144**:104177.
74. Casanova L, Rutala WA, Weber DJ, Sobsey MD. Survival of surrogate coronaviruses in water. *Water Res* 2009;**43**(7):1893–8.
75. Bibby K, Aquino de Carvalho N, Wigginton K. Research needs for wastewater handling in virus outbreak response. *Environ Sci Technol* 2017;**51**(5):2534–5.
76. Gundy PM, Gerba CP, Pepper IL. Survival of coronaviruses in water and wastewater. *Food Environ Virol* 2009;**1**(1):10.
77. WHO. *Water, sanitation, hygiene and waste management for COVID-19. Technical brief.* March 19 2020. Available from: https://covid19-evidence.paho.org/handle/20.500.12663/843.
78. Wang X-W, Li J-S, Jin M, et al. Study on the resistance of severe acute respiratory syndrome-associated coronavirus. *J Virol Methods* 2005;**126**(1–2):171–7.
79. Hai FI, Riley T, Shawkat S, Magram SF, Yamamoto K. Removal of pathogens by membrane bioreactors: a review of the mechanisms, influencing factors and reduction in chemical disinfectant dosing. *Water* 2014;**6**(12):3603–30.
80. Purnell S, Ebdon J, Buck A, Tupper M, Taylor H. Bacteriophage removal in a full-scale membrane bioreactor (MBR) – implications for wastewater reuse. *Water Res* 2015;**73**:109–17.
81. Lesimple A, Jasim SY, Johnson DJ, Hilal N. The role of wastewater treatment plants as tools for SARS-CoV-2 early detection and removal. *J Water Process Eng* 2020;**38**:101544.
82. Tetteh EK, Amankwa MO, Armah EK, Rathilal S. Fate of COVID-19 occurrences in wastewater systems: emerging detection and treatment technologies—a review. *Water* 2020;**12**:2680.
83. Simmons FJ, Kuo DH-W, Xagoraraki I. Removal of human enteric viruses by a full-scale membrane bioreactor during municipal wastewater processing. *Water Res* 2011;**45**(9):2739–50.
84. Guan W, Ni Z, Hu Y, Liang W, Ou C, He J, Liu L. Clinical characteristics of 2019 novel coronavirus infection in China. *N Engl J Med* 2020;**382**:1708–20.
85. Yuen K-S, Ye Z-W, Fung S-Y, Chan C-P, Jin D-Y. SARS-CoV-2 and COVID-19: the most important research questions. *Cell Biosci* 2020;**10**:40.
86. Han J, He S. Urban flooding events pose risks of virus spread during the novel coronavirus (COVID-19) pandemic. *Sci Total Environ* 2020;**755**:142491.
87. Nghiem LD, Morgan B, Donner E, Short MD. The COVID-19 pandemic: considerations for the waste and wastewater services sector. *Case Stud Chem Environ Eng* 2020;**1**:100006.
88. Heller L, Mota CR, Greco DB. COVID-19 faecal-oral transmission: are we asking the right questions? *Sci Total Environ* 2020;**729**:138919.
89. WHO. *WHO/UNICEF joint monitoring programme for water supply and sanitation.* 2017.
90. van Doremalen N, Bushmaker T, Morris DH, et al. Aerosol and surface stability of SARS-CoV-2 as compared with SARS-CoV-1. *N Engl J Med* 2020;**382**:1564–7.
91. Patel A, Jernigan DB. 2019-nCoV CDC response team. Initial public health response and interim clinical guidance for the 2019 novel coronavirus outbreak. *Wkly Rep* 2020;**69**(5):140–6.
92. Guerrero-Latorre L, Ballesteros I, Villacrés-Granda I, Genoveva Granda M, Freire-Paspuel B, Ríos-Touma B. SARS-CoV-2 in river water: implications in low sanitation countries. *Sci Total Environ* 2020;**743**:140832.

CHAPTER 7

Impediments of coronavirus in healthcare wastewater treatment and ways to ameliorate them

Samuel Jacob, Sneha Mohapatra, Ravi Siddharth, Sukanya Nag, Satish Kumar Santhosh Venkat, Gunasekaran Rajeswari
Department of Biotechnology, School of Bioengineering, College of Engineering and Technology, Faculty of Engineering and Technology, SRM Institute of Science and Technology, Kattankulathur, Tamil Nadu, India

7.1 Introduction

An increase in global population and climate change implications will further create a bottleneck in the health management and sanitation system in the near future. This demands a clear-cut vision, strategies, policies, and regulations to be implemented by the countries to reduce the damage. The recent outbreak of the COVID-19 pandemic has further aggravated the problem, which is speculated to have originated from Wuhan, China, and has spread as a pandemic infectious agent. The SARS-CoV-1 was first identified in 2003 that affected 26 countries and is considered an animal virus passed from animals, such as the civet cat, to humans. There is direct human-to-human transmission of the virus, which majorly takes place in the absence of adequate infection control and lack of hygiene and precautions. Based on what was learned from the 2003 outbreak, the World Health Organization (WHO) recommended appropriate infection control and safety measures that can be applied to curb the recent COVID-19 outbreak.[1] There are different segments in society such as health care and public transport that have to be dealt with frontline importance and where the pandemic may remain a hotspot. Recent research published in *Nature Sustainability* by an international collaboration have assessed the recent studies on coronaviruses in wastewater.[2] Conventional wastewater treatment provides only partial removal of virus particles, which demands meticulous additional treatment strategies for safe disposal or reuse depending on the effectiveness of final disinfection.[3] Wastewater

Environmental Management of COVID-19
ISBN 978-0-323-85780-2
https://doi.org/10.1016/B978-0-323-85780-2.00006-8

containing coronaviruses may be a potential threat, according to a new review of studies that call for additional filtration steps in sewage treatment plants to remove the viruses successfully.[4]

In this context, this chapter focuses on detailed discussions of health care wastewater that reckons to contain infectious virus, mitigation strategies for efficient removal of pathogens, additional precautionary measures for the protection of sanitary, and wastewater treatment operators.

7.2 Municipal wastewater: Definition, classification, and characterization

Municipal wastewater (MWW) can be defined as the water generated as waste either from households or a combination of wastewater from various households, industries, and precipitates as the origin. The content of domestic waste usually constitutes residue from sanitation systems, kitchens, and household chore materials, which are contaminated with bacteria, viruses, synthetic chemicals, etc. These contaminants can be easily degraded with the aid of primary and biological treatments. Other particulates of medicinal and cosmetic origins form micropollutants, which require secondary treatments before discharge. The various types of MWW can be generalized based on the components present in it. The classification of the components based on a myriad of parameters can ease the process of wastewater treatment before discharge.[4]

The composition will vary based on the source and region, but taking into consideration the major components, MWW mostly contains five classified pollutants such as biological oxygen demand (BOD) load (in the form of solids), inorganic nutrients (nitrate and phosphate), toxic xenobiotics, pathogens and inorganic minerals like heavy metals.[4] The typical range of values present in MWW has been represented in Table 7.1.

7.2.1 Characterization of wastewater pertaining to epidemiological index

The characterization of MWW is vital to analyze the constituents present in it to decide on the method of treatment and discharge. It is not a voluntary concept, since following governmental norms, predisposal is mandatory to curb pollution and the spread of diseases. The researchers have analyzed the domestic sewage at the Beishiqiao Wastewater Purification Center, Xian, China, for an extended period and put forth the following characteristic parameters.[6]

Table 7.1 Characteristic range of different pollutant parameters in MWW.[5]

	Strong	Medium	Weak
Total solids	1200	700	350
Dissolved solids	850	500	250
Suspended solids	350	200	100
Nitrogen (as N)	85	40	20
Phosphorous (as P)	20	10	6
Chloride	100	50	30
Alkalinity (as $CaCO_3$)	200	100	50
Grease	150	100	50
BOD	300	200	100

All units are in mg/L.

7.2.1.1 Suspended solids (SS)

In the samples postfiltration through a 0.45 μm membrane, the filter medium showed a negligible amount of SS; hence, most of the raw sewage contained suspended solids above the 0.45 μm as a parameter.[6]

7.2.1.2 Chemical oxygen demand (COD) and biological oxygen demand (BOD)

The microfilter with a pore size of 0.45 μm could retain significant solids that account for the average COD and BOD of 66% and 62%, respectively, in a given wastewater stream. The size of these retainable solids might be their innate nature or attached to other filterable particles.[6]

7.2.1.3 Total phosphorus (TP)

In the analysis of TP in dissolved and suspended solids, it is noticed that suspended matter contained more phosphorus in comparison to the amount in the dissolved matter, with an average concentration of 50.2% and 49.8%, respectively.[6]

7.2.1.4 Total Nitrogen (TN)

It has been observed that a higher amount of TN was present in dissolved fractions rather than suspended solids. The average concentrations in dissolved and suspended fractions were found to be 79.6% and 20.4% respectively.[6]

7.2.1.5 Pathogens

In countries which have prevailing endemics related to helminths diseases such as *Ascaris* and *Trichuris* sp. have shown usage of untreated wastewater in agriculture to irrigate crops or vegetables which are consumed uncleaned or uncooked.[7] A study in West Germany also showed a similar pattern of

Table 7.2 Possible levels of pathogens in wastewater.[8]

Type of pathogen		Possible concentration per liter of domestic wastewater
Viruses	*Enteroviruses*	5000
Bacteria	*Pathogenic E. coli*	—
	Salmonella spp.	7000
	Shigella spp.	7000
	Vibrio cholera	1000
Protozoa	*Entamoeba histolytica*	4500
Helminths	*Ascaria Lumbricoides*	600
	Hookworms	32
	Schistosoma mansoni	1
	Taenia saginata	10
	Truchuris trichina	120

transmission of cholera, leading to an endemic in the region. Pathogenic bacteria, protozoa, viruses, and helminths are found in raw waste and can survive in the water for a significant amount of time. Table 7.2 represents some of the possible common pathogens and their count in raw wastewater. Organisms of pathogenic traits present in wastewater are a significant concern when the wastewater is reused for agricultural irrigation. In epidemiological studies, various adverse health impacts have been attributed to the practice of reuse of wastewater in agricultural fields. The pathogen population could be significantly reduced in the tertiary treatment stage by adoption of ultrafiltration, irradiation with UV, and disinfection by chlorination. Further, the digestion of residual sludge by anaerobic digestion can reduce the pathogenic population due to the heat generated in the digester.[8]

In treatability analysis, it is generally observed that most of the contaminants from the suspended fractions can be easily removed by coagulation or sedimentation procedures. Hence, the effect of primary and enhanced primary methods depends on the fraction of suspended solids present in the wastewater. Biological and advanced treatments are applied for the removal of dissolved contaminants depending on the required parameters of the water quality. BOD and COD parameters change in accordance with the removal of suspended solids, and it can be assumed they are present and coexistent in the wastewater.[7]

7.2.2 Possible disinfection routes of MWW

The SARS pandemic outburst can be linked to the transmission of viruses by water sources and wastewater. With the concept "liter of water, an ocean of information," multiple researchers have studied the presence of

SARS viruses in wastewater and its prolonged presence.[9] The SARS epidemic in 2003 had cases of spread of the virus in a high-rise building in Hong Kong where around 300 people were affected due to malware in the sewage system.[10] Viral particles were also obtained from the enteric tissues of human small intestines and found to be in larger counts than the cultures found in the lung tissue, showing the high probability of viral particles to be present in excretory matters.[11] In the recent case of SARS-CoV-2, the concept of sewage epidemiology, or water-based epidemiology, holds significance to detect the presence of the virus in untreated sewage through fecal matters. The RIVM National Institute for Public Health and the Environment, Netherlands, in their research studies, showed the presence of the virus in the gastrointestinal tract, thus transmitting to the sewage through excreta.[12,13] Similar findings have taken place in Australia, France, and Paris.[14] Considering such research results and analysis, one can conclude that it is of grave importance to treat sewage and wastewater to control the transmission and outbreak of COVID-19.[10]

A varied number of methodologies are applied to disinfect municipal wastewater, which is majorly carried out in the secondary stage of wastewater treatment plants. Though with the increase in numbers of pandemics and epidemics worldwide, a site-specific and conditional approach is required in municipal wastewater treatments. The prevailing technologies and the advent of newer concepts can be broadly divided into four categories as discussed in subsequent sections.[15]

7.2.2.1 UV irradiation

This method applies UV irradiation for the purpose of disinfection. Here, a UV irradiation dense zone is created on a segment of the wastewater by using a UV lamp placed outside the volume of the water. UV radiation has a germicidal wavelength of 253.7 nm, which has the ability to damage and alter the genetic structure or biological macromolecule (DNA/RNA/Protein) of the microbial contaminants thereby rendering them inactive and less virulent.[15]

7.2.2.2 Electrochemical (EC) disinfection

EC disinfector is used for the secondary treatment of wastewater. The method resulted in the highest efficiency of killing 99.9% of coliform bacteria in saline water sources (8% salinity). The efficiency was obtained with a contact time of less than 20 s and less than 0.08 kWh/m^3 power consumption. Even with an increase in power and time, the efficiency of the EC method in freshwater remained minimal. In comparison with direct chlorination methods, it can be concluded that this technique does not use electrochlorination; rather, it produces a powerful germicidal compound.[16]

7.2.2.3 Ozone disinfection

Ozone treatments decolorize and deodorize wastewater through disinfection. Ozone microbubbles (less than 50-μm diameter and high solubility and reactivity) can be sparged into the wastewater due to the balance between surface tension and high tension, and local hotspots are created. These spots generate intense heat as the gas in microbubbles are compressed. There is pyrolytic decomposition leading to the formation of shockwaves and OH radicals in the water. This provides improved and efficient disinfection of wastewater by inactivating the microbial contaminants present in it.[17]

7.2.2.4 Peracetic acid

Peracetic acid acts as a broad-spectrum antimicrobial which shows action on bacteria, fungus, virus, and spores making it a choice for multiple purposes. This method provides ease of application and has no quenching requirements with minimal dependence on pH, short time span, and effectiveness on both primary and secondary effluents. Even with such broad benefits, the major disadvantage of peracetic acid is the formation of acetic acid, which increases the organic load, thereby the microbial growth in the effluent.[18]

7.2.2.5 Ultrasonication

Ultrasound (20 kHz) has been used for the disinfection of domestic wastewater. The primary application of this technology is through the ultrasound probe where the disinfection takes place by the thermal mechanism of heat production (52%), mechanical stress created by the ultrasonic cavitation (36%), and synergistic effects (13%). The disinfection efficiency generally increases with the increase in ultrasound power at a rate of 0.003 log kill/min at 70 W/L to 1.8 log kill/min at 1250 W/L, and this is mainly observed by the increase of thermal inputs.[14]

7.2.2.6 Chlorination

Chlorination is the most widely used method currently in the field of wastewater treatment. It acts as a disinfectant by oxidizing the cellular material of the microbes. The chlorine can be applied in various forms, such as chlorine compounds, gas, hypochlorite solution, etc. The addition of chlorine and chlorine compounds leads to the formation of hypochlorous acid (HOCl) and hypochlorite ions (OCL), which are free available ions that act as the oxidizing agents in water. Though this method has multiple downfalls such as the production of toxic, hazardous materials (THMs) as byproducts, an increase in dissolved solids level, and an increase in organic load reduces the overall effectiveness of this method.[19]

For effective disinfection of nonenveloped viruses, strategies such as thermal treatment of sludge, the addition of lime, drying, and composting were found to be suitable and which can be applied for SARS-CoVs inactivation.[20] However, SARS-CoV-2 is an enveloped virus and can act as a barrier for inactivation; also, during secondary treatment operation, added chlorine/fluorine may get precipitated along with organic matter and nitrogenous compounds, thereby reducing its effectiveness. To overcome such difficulty, separate tertiary treatments that afford combinatorial (physiochemical) inactivation such as membrane filtration, advanced oxidation process, and chlorination shall be provided before release to the environment. To date, there is no specific SARS-CoV-2 disinfection prescribed.[21]

7.3 Healthcare wastewater (HWW)

Water is contaminated severely due to healthcare services, and its discharge is known as healthcare wastewater (HWW). It is majorly composed of solids discarded during treatment of the patients, disinfectants, sewerage, and food waste from healthcare units.[22] Fig. 7.1 represents the sources of healthcare wastewater and the possible contaminants, as indicated in Fig. 7.2. HWW is also contaminated by a photochemical solution like hydroquinone, chemicals from diagnostic labs and pharmacy, absorbable organohalogen compounds in X-ray contrast media, anesthetics, formaldehyde, glutaraldehyde, disinfectants, radioactive wastes, nutrients and nitrates in excess, dental amalgam mercury, and infectious microorganisms from biomedical wastes.[23]

7.3.1 HWW management

HWW has to be handled with utmost care as it is highly contagious and a possible place to spread disease during an epidemic, endemic, and pandemic. Strictly HWW must be limited from being discarded into public sewers. Safe storage, waste minimization, and segregation are to be

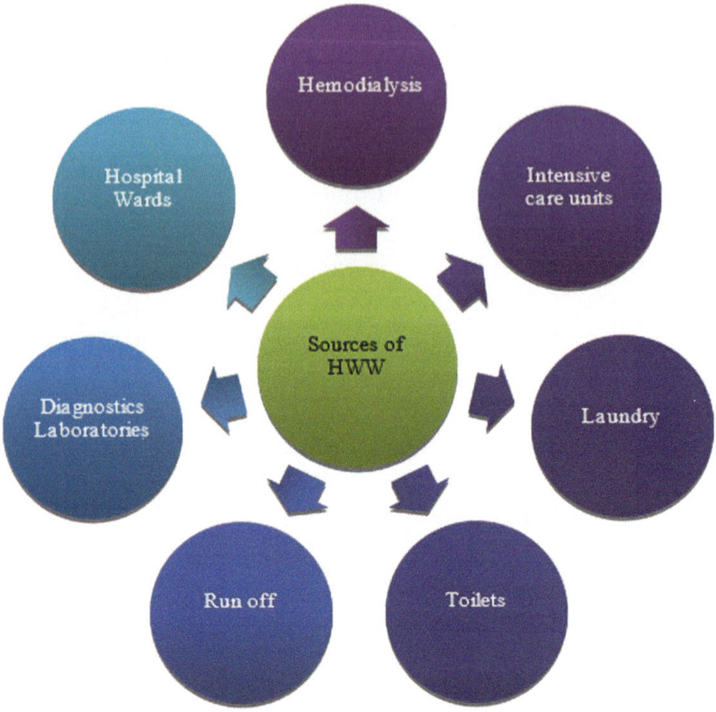

Figure 7.1 Sources of healthcare wastewater (HWW).

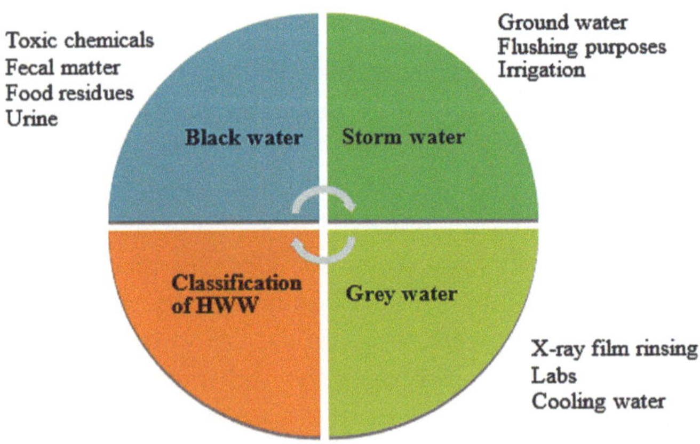

Figure 7.2 Classification of healthcare wastewater (HWW) sources.

performed with equal importance for liquid discards as it is done for solid wastes. Colorants, photographic chemicals, aldehydes, and antibiotics are classified under pharmaceutical and chemical wastes; when directly discarded into sewers, they pose a significant threat.[22] Healthcare sewage systems can be connected to the municipal sewage system and can discharge HWW into it only by meeting demands such as primary, secondary, and tertiary disinfection procedures are met. Approximately 90% of microorganisms are removed with the number of helminth eggs reduced to less than an egg per liter satisfying all regulatory requirements (local). Expired blood bags, including PPE, are disposed of by controlled landfills, high-temperature incinerator treatment, autoclaving with special liquid treatment cycle, prorated pit burying as unopened bags within healthcare establishments, or in a secure location elsewhere.[1]

7.3.2 HWW collection methods

7.3.2.1 Central system

A central system collects wastewater from throughout the healthcare establishment to a central underground spot for disposal and treatment.[1]

7.3.2.2 Decentralized system

Decentralized wastewater collection channels from medical areas carry wastewater to septic tanks, followed by a standard treatment facility for the reduction of the pollutant load. However, this is not a preferred approach (Fig. 7.3).[1]

A sewage system is used for wastewater collection. A stormwater system is installed to collect rainwater on rooftops and pavements of healthcare units, and in turn, use them for flushing, etc. Manholes permit maintenance of 50 or <50 m. Pretreatment is required to reduce contaminants in HWW or manipulate its nature prior to release into the sewer.[1]

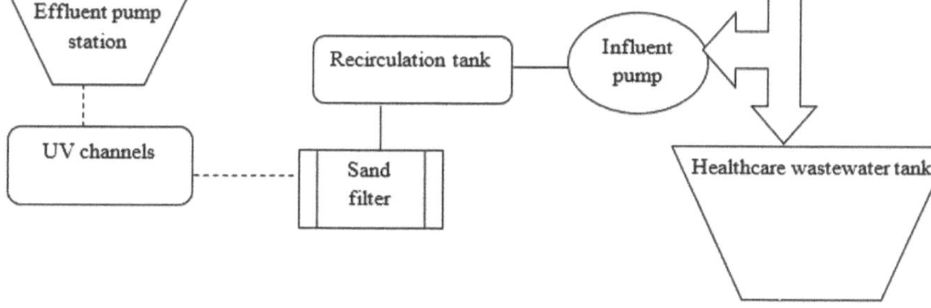

Figure 7.3 Decentralized system of HWW management.

7.3.3 Pretreatment of HWW

Medical labs, feces/vomit during the outbreak, blood, dental departments, and radiotherapy departments require pretreatment, which include acid-base neutralization, filtration, sedimentation, and autoclaving. When there is contamination due to feces in the course of a disease like cholera, the pretreatment can include disinfection with lime milk (calcium hydroxide) administered in a ratio of 1:3. The above mixture is administered for 6 h at least. Then 1:1 urine to lime for at least 2 h is to be incorporated. Blood is pretreated by thermal methods. It is important to note that conventional hypochlorite treatment is not sufficient for blood since blood is a heavy organic load. In the dental department, amalgam separators are used for pretreatment purposes, and safe storage of discarded mercury is done. In the radiotherapy department, collection and storage of radioactive wastewater (e.g., urinary discharge of patients undergoing thyroid treatment) in the die-away basin for decay is done. During this storage time, background concentrations come down, and after meeting the standard storage time, the water is drained into sewers.[1]

7.3.4 Medical wastewater treatment in COVID times

Knowing that this deadly virus can be transmitted via human excreta, disinfecting HWW is very important. The China Ministry of Ecology and Environment (MEE) has issued a couple of standards for healthcare facilities to obey. The use of disinfectants demands more research and guidance. On March 3, 2020 National Health Commission and State Administration of Traditional Chinese Medicine announced, "Diagnosis and treatment protocol for Novel coronavirus pneumonia (Trial Version 7)." The probability of fecal-oral transmission has been added in the diagnosis and treatment plans. SARS-CoV-2 are often found in urine or discharged matter; hence, HWW is to be treated appropriately.[24]

7.3.4.1 Wuhan: A case study

The Water Affairs Bureau of Wuhan's municipality has implemented complete disinfection for sewage and drainage systems. Between January 29 and February 18, 2020, about 196,358,000 Kg of sanitizers were used for disinfection. At all 26 civic run-off treatment units, sodium hypochlorite was approved for disinfection. Frequent evaluation of fecal coliform was made mandatory in all treatment facilities.[25]

7.3.5 Mandatory treatment and disinfection plans

Wastewater from general hospitals also shall be treated the same way as HWW. Distinct festering tanks are mounted for affected zones, and specific guidelines must be followed before the wastewater from these areas is mixed. The hygiene index should be studied by means of fecal coliform group (shall be lower than 100). Requirements for hazardous waste disposal will be followed for sludge disposal, and an account in succession is established for all treatment facilities with proper disinfection.[26]

7.3.5.1 Disinfection procedure

A disinfection procedure removes a majority of or all pathogenic microbes from the HWW. There are many important factors that influence the disinfection process; namely, innate resistance of microorganisms; resistance to chemical germicides, which vary from organism to organism; and the sterilization process, which is also organism-specific. It is important to note that even the population of microorganisms that is most resistant contrast the disinfection time. With prions being an exception, all other bacterial spores are characterized by an inborn resistance toward germicides of chemical origin. The concentration of disinfectants is directly proportional to the efficacy and inversely proportional to the time to kill pathogens. The potency of germicide is essential with respect to the length of disinfection time.[27] Some of the physical and chemical factors influencing disinfection are as follows:

- The activity of disinfectant rises with rising in temperature up to a threshold value of temperature beyond which the disinfectant degrades.[28]
- Disinfectant action is influenced by pH, changing the disinfectant action on all surfaces.[28]
- Water hardness decreases the rate of the killing action of disinfectants as the divalent cations in water result in insoluble precipitates when interacted with disinfectant.[28]

7.4 Occurrence and survival of COVID-19 in wastewater, urine, and biosolids

The ongoing global pandemic caused by SARS-CoV-2 has been a health crisis of global concern. Until recently, the only human coronavirus strain was known to infect the respiratory tract, but now they have been found in the feces of patients too. During the proliferation of an organism like a virus in the host, the infection spread to various organs and bodily fluids. For pathogens transmitted by the fecal-oral route like cholera, the primary site of replication is in the intestinal tract. Infection in other organs leads to the presence of an organism in the blood and then urine after elimination by the kidneys, such as what happens in cases of enteric viruses in urine, and therefore, transported to the community sewage system.[29]

7.4.1 The survival of SARS-CoV-2 in urine and the urogenital system

Most people infected with SARS-CoV-2 exhibit a mild increase in blood urea nitrogen and urine protein along with renal failure and gradual loss of kidney function. Viral nucleic acids are also isolated from the urine samples. Both SARS-CoV-2 and SARS-CoV share the same receptor Angiotensin Converting Enzyme 2 (ACE2). The researchers analyzed the online database and plotted the expression of ACE2 in various organs, including the renal tubular cells, and surmise that SARS-CoV-2 binds to ACE2 to cause kidney dysfunction.[30]

7.4.2 Human pathogens can be monitored in sewage

During an active infection, the pathogens are excreted in a range of body fluids. These pathogens may find their way into sewage systems during day-to-day activities like washing, bathing, and flushing, and thereafter in the collection point where biological, physical, and chemical methods are employed to reduce the population. Since the virus can be detected in fecal and urine examination, a monitoring mechanism that includes determination of the virus concentration, type of metabolites, and biomarkers that enable prior detection of outbreaks can thereby curb the spread. Chinese scientists indicated that coronavirus found in a patient's urine is capable of transmission through the blood and can potentially infect multiple organs.[31]

7.4.2.1 A paper-strip urine test kit was developed to detect COVID-19

The paper-strip urine test is similar to a home pregnancy test kit that could be taken anywhere, be it at home, workplace, or healthcare centers, and can detect the presence of coronavirus protein in a given urine or saliva sample. The lead researcher used electrokinetics to concentrate and isolate the charged particles from the samples. "This protein detection is just like a dam, whereas the fluid flows through the paper strip, the electric field backs up charged molecules including proteins so the test line on the paper is exposed to a higher concentration of them for a longer." The test uses a small power source such as common 9 V batteries.[32]

7.4.3 Novel coronavirus infection and gastrointestinal tract

A recent study affirms the presence of many other members of the virus family (Coronaviridae) in the intestinal tract of other species some of which have been shown to cause gastroenteritis and its related symptoms. The viral nucleic acid fragments can also be found in feces of both symptomatic and asymptomatic carriers.[33] When a patient affected with COVID-19 defecates, a large number of pathogens can persist on the toilet bowl surface. Sequential flushes and sloshing of water can disperse the viruses up to three feet in the air in the form of aerosols and can cause cross-infection among people. Studies of MERS, which shares similarities to SARS-CoV-2, stipulate that this specific virus can persist in the digestive tract of humans, which indicates that the human intestinal tract is susceptible to SARS-CoV-2 and can sustain robust viral replication. One hypothesis suggests that these enveloped viruses can persist in the human gut and protect the germs when the mucus from patients travels down the digestive tract. A Centers for Disease Control and Prevention (CDC) assessment cites one study where they were able to identify the presence of viral nucleic acids in patients' feces. These genetic snippets stipulate that the virus once persisted in the body but has been degraded and can no longer cause illness.[34]

7.4.4 Wastewater-based epidemiology

The US Environmental Protection Agency (EPA) suggests that viruses found in wastewater are an emerging cause of disease outbreaks of unknown origin because of their ability to mutate and undergo rapid transport through aquatic environments. Traditional disease detection is based on a diagnostic analysis of clinical samples or rapid antibody tests. Despite that, these techniques have failed to monitor early signs of a disease outbreak. Wastewater-based analysis is an ideal method to understand the proportion of a particular population found to be affected by the virus present in wastewater as it contains virus particles shed from both symptomatic and asymptomatic carriers and helps to monitor the public health of the entire locality: "This test is like a canary-in-a-coal-mine test, so it gives us a snapshot of what's going on in the entire community."[35]

Dissemination of SARS-CoV-2 through wastewater and sewers is being carefully monitored as there is significant evidence of a presence of viral particles in these environmental samples that can potentially transmit to humans.[36,37] Several reports are available that have detected SARS-CoV-2 RNA in feces, urine, and wastewater; however, the isolation of these strains from these samples seem to be complicated.[38–40] It has been suggested that possible fecal-oral route is through the active replication within the gastrointestinal tract followed by the release of virions in the wastewater streams and sewers.[41] Because the viral particles from the infected patients are usually mixed with organic matter, such as sputum and feces, and are thereby protected, the disinfection mechanism seems to be ineffective.[41]

7.4.5 Gastrointestinal tract may be another site of viral replication

Hubei provincial Renmin Hospital proved that viral nucleic acids are found in fecal samples of infected patients. It was proved in the study that there is a strong convergent evolutionary relationship between the receptor-binding domains of SARS-related coronavirus and COVID-19.[42] Structural analysis indicates that both the receptor-binding domains share similar side-chain properties that are essential for ACE2 binding. The receptor is expressed densely in the luminal surface of the intestine and colon, thus indicating the possible routes for viral transmission. The receptor has other physiological roles, and it also facilitates amino acid transport to the microbial ecology of the digestive tract, where altered gut microbial compositions are exhibited by the ACE2 mutants. Guidance from China's National Health Commission recommended that modulating the gut microbiota by probiotics could be used to maintain the balance of intestinal microecology and could reduce enteritis and ventilator-associated pneumonia.[43]

7.4.6 Viruses and the culture methods

Works have suggested the use of electropositive filter media particles for the recovery of the virus from hospital wastewater.[43] His work illustrates the use of bacteriophage to estimate the concentration recovery of virus from wastewater. Both bacteriophage f2 and SARS-CoV-2 have a single-stranded RNA genome, and the testing procedure for bacteriophage was straightforward and inexpensive.[44] There are various concentration methods carried out for the recovery of viruses from wastewater such as the use of a centrifugal filter or the use of polyethylene glycol concentration methods.[13,36] None of the works stated have estimated the percent yield of virus recovered from wastewater, due to the problems involved with the handling of clinical specimens and the necessity for a laboratory facility of biosafety-level 3. Only one research study has reported the percent yield of SARS-CoV-2 from wastewaters, which was measured to be approximately 1% by using electropositive filter membrane.[43] Since the properties of SARS-CoV-2 are not the same as other waterborne enteric viruses, more study is necessary for the successful recovery of the virus from wastewater. Different test trials may give rise to several contradictory results when the concentration of the virus is low in a given wastewater sample.[43]

7.4.7 Waterborne viral outbreaks related to virus

The range of human deaths per year lies between 2 and 12 million, with waterborne viruses being the cause. Drinking water from wells has caused the majority of waterborne outbreaks. In natural water bodies, recreational water has been known to cause significant outbreaks. In general, an association of viruses with waterborne diseases is overlooked and it is often difficult to identify the agents. In the United States, the cause for more than half of the outbreaks in recreational and drinking water was associated with noroviruses. Since 1980, the CDC also reported outbreaks due to hepatitis A virus. Adenoviruses caused an outbreak in 1982 and a couple more in 1991. All were linked to recreational water and often led to associated diseases such as pharyngitis.

Three outbreaks occurred in 1982, 1990, and 2006, and these were caused by enteroviruses which were etiological agents. Tap water caused an outbreak in Colorado, and rotaviruses were the causative organisms. Hepatitis E outbreaks also pose a severe threat. Swine is a reservoir of hepatitis E and is transmitted to humans.[44]

7.4.8 Transport of virus in the environment

Surface accumulation of virus enables the transfer of virus into the environment. The transport viruses in the environment depend upon factors such as[45]

- pH
- Ionic strength
- Competitors for binding site
- Isoelectric point (IP)
- Hydrophobicity of sorbent

It is noted that clay possesses higher virus surface accumulation capacity, and cation concentration is directly proportional to virus sorption. Organic matter is known to upgrade the transfer of viruses. The positive charge is exhibited by viruses when pH is lesser than IP, and negative charge when pH is more remarkable than IP. IP has been termed to be the dominant controller of virus absorption during transport. Negatively charged surfaces, less than their IP, tend to absorb viruses. Viruses were resistant to water treatment when IP was low.[46]

7.4.9 Survival of virus in the environment

There are many factors influencing the survival of viruses in the environment.[47]

The key factors are as follows:

(1) Soil (granule size and dispersion, content in clay)
(2) Carbon
(3) Chemistry behind solution
(4) Oxides of metals
(5) pH
(6) Ionic strength
(7) Air-water interface
(8) Biological factors

The most important factors are

(1) Temperature
(2) UV rays
(3) Flora of microorganisms

In seawater, polio and adenovirus survive at 15°C for long periods of time. Enteroviruses survive for many weeks in freshwater. Viruses are inactivated by exposure to sunlight or UV.[47]

7.4.10 Microbial tracking tools (viruses)

Microbial source tracking enables us to differentiate between possible sources of contamination due to feces in the environment. Many microorganisms have the potential to be tools for tracking.

Viruses that are potential microbial tracking tools:[48]

(1) Human polyomavirus (HPvV)
(2) Human adenovirus (HAdV)
(3) Human enterovirus (HEV)
(4) Porcine adenovirus (PAdV)

PAdV is highly human-specific and is a reliable tool. They help to detect contamination in river water samples, slaughterhouse slurries, and urban sewage.[49] HAdV and HEV are pollution detectors.[50] To identify sources of water pollution in agriculture, bovine-adenovirus and bovine-enterovirus are useful. Bovine polyomavirus is good when manure is the source of contamination. In freshwater bodies, the contamination of human and animal viruses is relatively low. A concentration procedure is done to filter water in huge amounts while sampling.[51]

7.5 Measures to the protection of personnel and wastewater treatment workers from contracting COVID-19

In response to the growing spread of COVID-19, WHO has released a variety of detailed technical guidance documents on specific topics that include infection prevention and control (IPC). The usual method of virus transmission is through respiratory droplets and the immediate environment of an infected individual (i.e., transmission through contact). Even though the risk of being affected by the COVID-19 infection from the feces of diseased individuals has all the earmarks of being negligible. Proof states that COVID-19 infection prompts bowel diseases and its presence is proven in feces. Until today, only one investigation has refined the COVID-19 infection from a single sample of stool. No case of fecal-oral transmission of the COVID-19 infection has been confirmed.[1] The management aspects have been discussed in subsequent sections.

7.5.1 Safe management of fecal waste and wastewater

At present, no case of SARS-CoV-2 transmissions through sewerage systems has been reported in both treated and untreated wastewater. In addition, there is no evidence that sewage treatment workers have developed SARS-CoV-1 (similar to SARS-CoV-2), which was caused by an acute respiratory disease outbreak in 2003. In properly managed and designed wastewater treatment plants, wastewater in sewerage systems should be handled as part of an integrated public health strategy. Hand hygiene should be done frequently, along with other exterior infection routes such as hair, nose, eyes and mouth should be protected from unwashed hands. Workers have to wear suitable personal protective equipment (PPE), which includes gloves, goggles or a face shield, and a mask.[1]

7.5.2 Personal Protection Equipment (PPE)

All workers likely to have exposure to sewage and biosolids should be supplied with appropriate PPE. The PPE options include gloves, goggles, respirators, face masks that are splash-proof, and liquid-repellent coveralls (Fig. 7.4). In works associated with exposure to sprays and pressured leakage, face shields should be provided. Management and worker representatives must work together to determine which jobs are likely to result from such disclosure, to have sufficient on-site inspections, and to determine the form of PPE that is appropriate in unification with a qualified health and safety profession. A comprehensive program includes respiratory analysis and training.

The goal of PPE is to decrease workers' exposure to hazardous substances when engineering and administrative controls are not feasible or successful in decreasing those risks to acceptable levels. PPE is required when hazards are present in the near vicinity or in direct contact. PPE does have the significant disadvantage that it does not eliminate the hazard on the spot and that workers can be exposed to the hazard if the equipment fails.[52]

PPE acts as a barrier among the wearer/user as well as the work environment. It can impose unnecessary pressures on the user, hinder his capacity to carry out his work, and cause substantial amounts of stress. This would prohibit users from using PPE appropriately, placing them at risk of

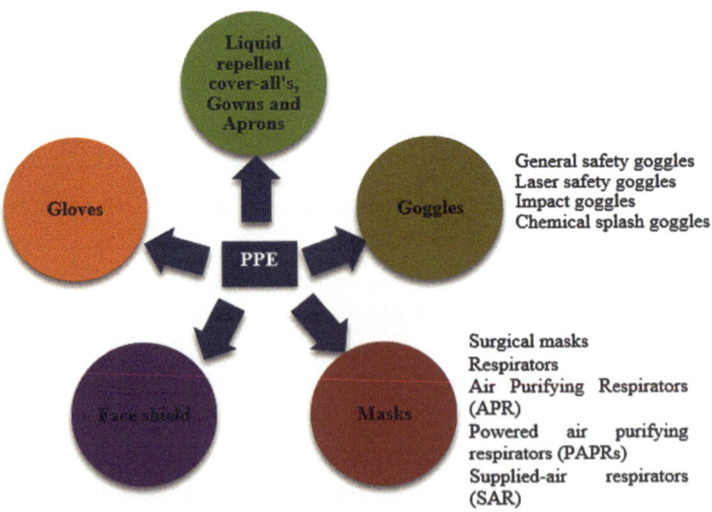

Figure 7.4 Personal protection equipment.[55]

injury, ill-health, or death. Effective ergonomic strategies might support eliminate these obstacles and can also help ensure safe and healthy working environments by using PPE correctly.[53]

Work-related safety and health practices, which practice exposure controls and involvements to mitigate hazards in the workplace that pose a threat to workers' safety and quality of life.[54] The hazard management pyramid offers a policy structure that ranks the kinds of hazard controls in terms of absolute risk mitigation. Elimination and substitution are at the top of the hierarchy, which entirely eliminates and replaces the hazards with a safer alternative. If either of the elimination or substitution steps is not suitable to be enforced, engineering controls and administrative controls would be introduced to create secure systems and guide human behavior. PPE ranks last on the control ladder since the sewage workers, with a protective barrier, are frequently exposed to the hazard. The control pyramid/hierarchy is crucial in recognizing that while PPE has considerable utility, it is not the ideal control for workers' health.[55]

PPE can be categorized into four divisions on the basis of the measure of protection being afforded as given in the following sections.[55]

7.5.2.1 Level A protection

This category is preferred for the maximum degree of skin, eye, respiratory, and mucous membrane protection. A standard level A comprises chemical-resistant gloves, boots, and self-contained respiratory equipment.[55]

7.5.2.2 Level B protection

This category is preferred for the maximum degree of respiratory protection, but a lower level of the skin and an eye shield is required. It is also suggested for initial site entries until the hazards have been recognized and established by inspection. It includes chemical-resistant clothing (coveralls), gloves, boots, and a self-contained breathing apparatus.[55]

7.5.2.3 Level C protection

This category is preferred when the exposure of the skin and eyes is unlikely, with the information on the type of airborne material, level of concentration, and basic requirements of an air-purifying respirator. Air must be tracked regularly. A standard level C comprises chemically resistant clothes, gloves, boots, and an air-purifying respirator.[55]

7.5.2.4 Level D protection

This category is preferred as a standard uniform for sewage workers, where respiratory and skin hazards exist. This category comprises coveralls and safety shoes.[55] The workers who are managing and handling human waste/sewage have to be provided with PPE, instruction on the method of usage of PPE, and facilities for handwashing. Workers must immediately rinse their hands using soap and wash with water as soon as they remove their PPE. For employees treating sewage, the following PPE is recommended.[56]

7.5.2.5 Goggles

Goggles are used to protect eyes from splashes of human waste or sewage. It has transparent glasses, with zero power, well-fitting, and protected on all sides with adjustable holders. Characteristics include:[56]

- Decent seal on the face
- Elastic frame that fits all sizes with little to no pressure
- Mist and abrasion resilient
- Designed with indirect venting so as to prevent or lessen fogging
- Might be reusable (if there are apt arrangements for on-spot) or single-use goggles

There are many types of goggles, such as[56]

(1) **General safety goggles**: Where it protects the eyes from splashes from fluids, like buffers and salts, which are not likely to affect the eye.
(2) **Laser safety goggles**: Where it effectively filters all light entering the eyes and its embedded with protective abilities.
(3) **Impact goggles**: Provide cover from flying debris and have openings on the sides for airflow.
(4) **Chemical splash goggles**: This is the ideal type of goggles for the sewage workers since it protects the eyes from the spray of chemicals or hazardous substances that may enter the eyes and can protect them from flying debris.

Typically, goggles and safety glasses are made using an injection-molding process, which injects a polymer material into a metal mold manufactured to replicate the desired shape of the lenses and frames. Usually, a composite material known as polycarbonate is used for making goggles or glasses. Polycarbonate is a thermoplastic polymer with good optical consistency and twice the impact strength of other comparable plastic materials, making it an easy alternative for protective eyewear. It is known for its impact strength and optical clarity, which also enables the material to be used to create face shields. Prolonged utilization of goggles might maximize worker's discomfort, exhaustion, and skin tissue damage.[56]

Used goggles can be reprocessed and recycled by washing them with water and detergent accompanying sterilization with either 0.1% sodium hypochlorite (accompanied by water rinsing) or wiping with 70% alcohol. Goggles are cleaned on a presterilized surface. Suitable disinfectant with the right time of contact (e.g., 10 min while utilization of 0.1% sodium hypochlorite) should be adhered to prior to goggle reuse. They have to be stored in a clean area after cleaning and disinfection to avoid recontamination.[56] There are companies such as CleanRiver Recycling Solutions, Kimberly-Clark Professional, and many other recycling programs that aid in recycling such PPEs. Further, the global standardization of PPE has been provided in EU PPE Regulation 2016/425 Category III, EN 166, ANSI/ISEA Z87.1.[56]

7.5.3 Masks

Masks are used to protect the workers from contracting respiratory disease-causing microorganisms. There are many types of masks, such as[57]

7.5.3.1 Surgical masks

Used to cover around the mouth and nose and can protect from droplets, splashes, and sprays. Studies confirm that using such a mask alone is not always practical, but it is safer because enduring use and reuse can lead to auto-contamination.[59]

7.5.3.2 Respirators

N95 and variants of FFP2/3 come under respirators. The airborne contaminants are filtered through the fibers that are tangled in the mask. The boundaries around the mouth and nose form a barrier. Such masks should not be used by the general public because they are in limited supply. They should be used only by health care workers. Some of the types of respirators include:[58]

7.5.3.3 Air-purifying respirators (APR)

Through cartridges and filters, air-purifying respirators (APR) scrub pollutants from the ambient air. This form is commercially available in either half-mask or full-face versions with a replaceable one-time use of cartridges and filters or as half-mask face filtering pieces typically intended for one-time use. If APRs are determined by the hazard assessment to provide adequate protection, choosing the correct cartridge and/or filter type should be followed. For safety against airborne hazardous substances and particulates, several cartridges and filter options are to be provided.[58]

7.5.3.4 Powered air-purifying respirators (PAPRs)

Operates using a motor that blows filtered air into the face-piece. This is a safety stage between ordinary APRs and airborne respirators.[58]

7.5.3.5 Supplied-air respirators (SAR)

SAR provides workers with safe breathing air and must be used when workers are in oxygen-deficient atmospheres. There are three basic types available, with breathable air supplied from a cylinder bank, personal (worn) bottle, or a combination of both. Typical SAR industrial use is that workers utilize respirators connected to long lengths of air hose.[58]

Masks consist of several sheets of materials, and it is formed using the process called meltdown, and the fabric that has a range of thickness from 100 to 1000 μm is prepared from microfibers of polypropylene. Fibers from such a meltdown process can be crossed one over another to form a porous three-dimensional (3D) assembly with a permeability of approximately 90%. Such fabrics also have a relatively low pressure drop and can accommodate a considerable amount of particulate matter. Consequently, meltdown fibers are typically charged electrostatically to enhance the particle interaction, causing much greater efficiency in filtration.[58]

Face masks are usually suitable for single-purpose use and are not to be utilized when the inner side gets wet or moist. Usually, the masks are rarely sterilized for reuse. There has been an assessment of the possibility of decontaminating and reusing masks, with early tests showing potential for both steam and UV sterilization. These results, however, are not yet peer-reviewed or published and cannot be applied widely. The repeated steam application resulted in deterioration of filtering efficiency, and the fabric was damaged by alcohol and chlorine-based solutions. While reusing the respirators, the used respirators should be dried for three to four days. The hydrophobic nature of polypropylene present in N95 makes the environment of the mask unsuitable for the survival of the virus, which can usually survive even on thoroughly dried respirators (drying time, 3—4 h). Another method is to disinfect the respirator by overhanging it, in an oven, for 30 min at 70°C. Respirators are degraded by UV light so it should be kept away from UV and direct sunlight.[59]

Some of the global standardizations provided for masks and respirators include (1) surgical masks—EU PPE Regulation 2016/425 Category III, FDA Class 2, EN 14683 Type II, IIR, ASTM F2100 minimum level 1; (2) N95—EU PPE Regulation 2016/425 Category III, FDA Class 2, Minimum "N95" respirator under NIOSH 42 CFR 84, minimum "FFP2" according to EN 149, or demonstrate an equivalent set of standards.[56]

7.5.4 Gloves

Gloves help in preventing direct contact of sewage with sanitation workers. An ideal glove for a sewage worker should be chemical and liquid-resistant and should also have high tensile strength and low reactivity. Various types of gloves are depicted in Fig. 7.5.[60]

Chemical-resistant gloves, which are ideal for sewage workers, are produced from variants of different types of rubber such as neoprene, nitrile, butyl, and fluorocarbon; or variants of plastic such as polyvinyl chloride (PVC), polyvinyl alcohol, and polyethylene. Such components are layered to give rise to the best of their nature. The thicker the gloves are made, the more they are resistant to chemicals, but it comes with a drawback of lack of mobility while using the gloves.[60]

Reprocessing of gloves is possible in nitrile gloves, which were experimented upon. Nitrile inspection gloves from a single new box were used. Gloves were worn, and a coin toss was used to determine whether a 15 s hand wash with nonantibacterial soap followed by drying with a paper towel or applying a hand sanitizer based on alcohol. Handwashing and application of hand sanitizers are made as prescribed by the CDC. Gloves were labeled at wrist level using a marker. Per FDA guidelines, a 1 L water test was used to assess glove integrity. Gloves were connected to a two-inch PVC pipe that was suspended and gradually applied 1 L of normal water followed by an observed period of 2 min to check the leaks and are classified at different levels as follows: 1 (water trickle), 2 (small jet), and 3 (rivaling Niagara falls). A control group of 10 gloves was also tested to assess baseline leak rates, taken directly from the box without a hand hygiene method. The same Fisher test was used to compare the levels of leakage

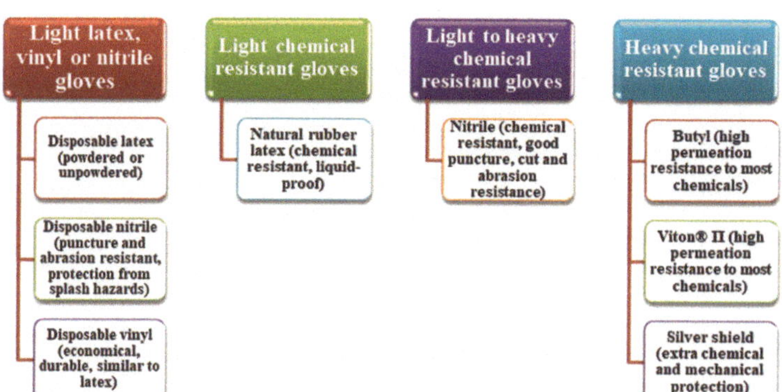

Figure 7.5 Various types of gloves.[60]

between classes. Unfortunately, the recent study found that the procurement of sterile disposable single-use gloves is cheaper than the recycling process.[60] Some of the global standards for gloves include FDA Class 1, EN 455, ASTM D3577.[56]

7.5.5 Face shields

Face shields are used to protect the nose and mouth from hazardous or infectious fluid splashes and human waste or sewage. Face shields of good quality thermoplastics can be made to withstand heat, impact, chemicals, or other hazards.[61] Polycarbonate and polyester (PET) film are two common plastic materials used in the development of surgical face shields. Both materials are transparent and provide high optical clarity. Polycarbonate and PET are both lightweight and can be quickly processed into almost any shape, so manufacturers can design masks and shields to meet the needs of even the most complex applications. Extended use of face shields may increase discomfort, fatigue, and skin tissue damage may occur to the face. Used face shields can be reprocessed and recycled by soap/detergent and washing with water accompanied by disinfection using either 0.1% sodium hypochlorite or 70% alcohol wipes. Face shields may be cleaned immediately after use or placed in designated closed containers for later cleaning and disinfection of the surface prior to shield cleaning. Appropriate disinfectant contact time (e.g., 10 min when using 0.1% sodium hypochlorite) should be adhered to prior to shield reuse.[56] Some of the global standardizations provided for face shields include EU PPE Regulation 2016/425, EN 166, ANSI/ISEA Z87.1.[56]

7.5.6 Liquid-repellent coveralls, gowns, and aprons

Coverall's provide overall protection of the user from any infectious agents and reduce the risk of contracting the disease, especially in workplaces like health care centers, sanitary operation, and wastewater treatment facilities. It is made of fluid-resistant material and includes gowns and aprons. Coverall's are made of fabric, i.e., 100% polyester with PVC or other fluid-resistant coated material. It must be waterproof such that sewage or other debris does not stick to the material. Fabrics guard the user against wounds and cuts from sharp, heavy, or rough materials (if present in sewage). It can be reused if apt provisions are provided for sanitization is in place. But most of the time, the coveralls are disposable and can be for single use only. Reprocessing can be done for cotton gowns and machine washed with warm water of temperature ranges between 60 and 90°C accompanied by laundry detergent. Linen material is stirred with a stick in a huge drum in the presence of soap and hot water. Alternatively. linen is also soaked in 0.05% chlorine for about 30 min. Finally, it is rinsed with water and allowed for drying under sunlight. WHO prefers gowns to aprons for both nonaerosol-generating procedures (non-AGPs) and AGPs during all instances. For AGPs, the guidance suggests clean, long-sleeved, nonfluid resistant gowns and gloves.[62] Some of the global standardizations provided for coveralls include EN ISO 13688, EN 14126-B and partial protection (EN 13034 or EN 14605), EN 343 for water and breathability or equivalent.[56]

7.6 Recycling of PPE

The commercial recycling process involves the following steps:[63]

(1) Sample to be tested from used PPE is directed to recycling plants. Not all PPE can bear the cycle of laundry or dry-cleaning, so care must be taken to ensure the details on the fibers after cleaning. It can lead to shrinkage, but these findings vary from producer to producer.
(2) Collection bins are located on-site where previously "disposable" PPE are housed.
(3) Bins are delivered to the recycling plant at assigned intervals.
(4) Documentation confirms bin receipt and the contents.
(5) Recycling of PPE.
(6) Undesirable PPE is parted from a functional batch.
(7) Before sorting, each PPE is individually scanned for metal traces.
(8) Sorting of PPE is done based on size.
(9) For a new life lease, PPE is returned to the property.
(10) Cleaned PPE is free from contamination or disposed of if it cannot be reprocessed and recovered any more.

7.7 Training for workers

Workers who are handling sewage have to receive drills and training on procedures to prevent contracting diseases due to occupational negligence. The drill will provide knowledge on rudimentary hygienic practices, PPE usage, its disposal, and safe management of sewage. Workers should be qualified to report diseases or symptoms to the appropriate supervisory or health care personnel. This may help in the early diagnosis of health hazards related to the work. Proper training should be given for the sanitation and sewage workers according to the guidelines provided by WHO or other NGOs responsible for international public health. Some of the basic guidelines to be followed by the sanitation and sewage workers include[64]

- Usual handwashing with soap/detergent with water after handling sewage, especially before eating, drinking, and both before and after using the toilet.
- When handling sewage, avoid touching the nose, mouth, and face.
- Contaminated clothing has to be removed before eating.
- Workers are not allowed to chew gum or tobacco during sanitation works.
- Cuts and wounds (if any) have to be covered using clean and dry bandages.
- Water and chemical-resistant gloves are to be used to prevent contact with the sewage.
- Soiled clothing is to be cleaned daily using 0.05% of chlorine solution.

7.8 Vaccination recommendation for workers

Vaccination guidelines for staff exposed to sewage or human waste should be well established in collaboration with the authorities of local health. Tetanus vaccines must be up-to-date, even taking into account the need for vaccines against measles, typhoid, and hepatitis A and B. The potential hazards to employees are likely to vary from one location to another, and a qualified health and safety specialist has to check to develop on-spot health and safety plans for the workforce.[64]

7.9 Conclusion

Characterization and mitigation of any identified SARS-CoV-2 transmission risks through waterborne pathways necessitates addressing the subsequent knowledge gaps. Currently, exploration of molecular approaches for finding SARS-CoV-2 in wastewater is predominantly focused on quantifying the viral RNA; however, quantification of fully functional infectious virions is yet to be resolved. Thus, the emergence of innovative tools perhaps provides adequate scientific evidence in environmental monitoring for policymakers, whereas mitigation of wastewater-associated SARS-CoV-2 transmission could be achieved via optimized disinfection strategies. Based on the existing evidence, PPE usage decreases the disease transmission rates as well as protects the workers. It is noteworthy that the workers understand the function and role of PPE as a part of the disease transmission management program and its appropriate usage to ensure its adequate supply during the epidemic surge.

References

1. World Health Organisation. *Global healthcare waste project-module-23-management of healthcare wastewater.* 2020. https://www.who.int/water_sanitation_health/facilities/waste/module23.pdf?ua=1. [Accessed 11 October 2020].
2. Bogler A, Packman A, Furman A, et al. Rethinking wastewater risks and monitoring in light of the COVID-19 pandemic. *Nat Sustain* 2020;**3**:981–90. https://doi.org/10.1038/s41893-020-00605-2.
3. Qiu Y, Lee BE, Neumann N, et al. Assessment of human virus removal during municipal wastewater treatment in Edmonton, Canada. *J Appl Microbiol* 2015;**119**(6):1729–39. https://doi.org/10.1111/jam.12971.
4. National Research Council. *Use of reclaimed water and sludge in food crop production.* Washington, DC: National Academies Press; 1996.
5. Feachem RG. Sanitation and disease (Health aspects of excreta and wastewater management). *Water Res* 1985;**19**(1):131. https://doi.org/10.1016/0043-1354(85)90337-9.
6. Xiaochang W, Pengkang J, Hongmei Z, Lingba M. Classification of contaminants and treatability evaluation of domestic wastewater. *Front Environ Sci Eng China* 2007;**1**:57–62. https://doi.org/10.1007/s11783-007-0011-7.
7. Shuval H, Adin A, Fattal B, Rawitz E, Yekutiel P. *Integrated resource recovery: wastewater irrigation in developing countries-health effects and technical solutions.* The World Bank; 1986.
8. The United Nations World Water Development Report. *Wastewater the untapped resource.* 2017. https://www.unido.org/sites/default/files/2017-03/UN_World_Water_Development_Report_-_Full_0.pdf. [Accessed 11 May 2020].
9. Gundy PM, Gerba CP, Pepper IL. Survival of coronaviruses in water and wastewater. *Food Environ Virol* 2009;**1**(1):10. https://doi.org/10.1007/s12560-008-9001-6.
10. Peiris JSM, Chu CM, Cheng VCC, et al. Clinical progression and viral load in a community outbreak of coronavirus-associated SARS pneumonia: a prospective study. *Lancet* 2003;**361**(9371):1767–72. https://doi.org/10.1016/s0140-6736(03)13412-5.
11. Leung WK, To KF, Chan PK, et al. Enteric involvement of severe acute respiratory syndrome-associated coronavirus infection. *Gastroenterology* 2003;**125**(4):1011–7. https://doi.org/10.1016/s0016-5085(03)01215-0.
12. RIVM. *Novel coronavirus found in wastewater.* 2019. https://www.rivm.nl/en/news/novelcoronavirus-found-in-wastewater. [Accessed 7 July 2020].
13. Medema G, Heijnen L, Elsinga G, Italiaander R, Brouwer A. Presence of SARS-coronavirus-2 RNA in sewage and correlation with reported COVID-19 prevalence in the early stage of the epidemic in The Netherlands. *Environ Sci Technol Lett* 2020;**7**(7):511–6. https://doi.org/10.1021/acs.estlett.0c00357.
14. Lodder W, de Roda Husman AM. SARS-CoV-2 in wastewater: potential health risk, but also data source. *Lancet Gastroenterol* 2020. https://doi.org/10.1016/s2468-1253(20)30087-x.
15. Madge BA, Jensen JN. Disinfection of wastewater using a 20-kHz ultrasound unit. *Water Environ Res* 2002;**74**(2):159–69. https://doi.org/10.2175/106143002x139875.
16. Li XY, Ding F, Lo PSY, Sin SHP. Electrochemical disinfection of saline wastewater effluent. *J Environ Eng* 2002;**128**(8):697–704. https://doi.org/10.1061/(ASCE)0733-9372(2002)128:8(697).

17. Sumikura M, Hidaka M, Murakami H, Nobutomo Y, Murakami T. Ozone micro-bubble disinfection method for wastewater reuse system. *Water Sci Technol* 2007;**56**(5):53–61. https://doi.org/10.2166/wst.2007.556.
18. Kitis M. Disinfection of wastewater with peracetic acid: a review. *Environ Int* 2004;**30**(1):47–55. https://doi.org/10.1016/S0160-4120(03)00147-8.
19. EPA. *Wastewater technology fact sheet chlorine disinfection.* 1999. https://www3.epa.gov/npdes/pubs/chlo.pdf. [Accessed 6 August 2020].
20. Hurst CJ, Gerba CP. Fate of viruses during wastewater sludge treatment processes. *Crit Rev Environ Sci Technol* 1989;**18**:317–43. https://doi.org/10.1080/10643388909388352.
21. Goddard MRR, Bates J, Butler M. Recovery of indigenous enteroviruses from raw and digested sewage sludges. *Appl Environ Microbiol* 1981;**42**(6):1023–8.
22. World Health Organization. *Water, sanitation, hygiene and waste management for the COVID-19virus.* 2020. https://apps.who.int/iris/bitstream/handle/10665/331305/WHO-2019-NcOV-IPC_WASH-2020.1-eng.pdf. [Accessed 18 December 2020].
23. Joakim Larsson DG, de Pedro C, Paxeus N. Effluent from drug manufactures contains extremely high levels of pharmaceuticals. *J Hazard Mater* 2007;**148**:751–5. https://doi.org/10.1016/j.jhazmat.2007.07.008.
24. Kitajima M, Ahmed W, Bibby K, et al. SARS-CoV-2 in wastewater: state of the knowledge and research needs. *Sci Total Environ* 2020;**739**:139076. https://doi.org/10.1016/j.scitotenv.2020.139076.
25. Han Z, Wang J, Zhang K, Tang Q. The ethics of COVID-19 clinical trials: new considerations in a controversial area. *Integr Med Res* 2020;**9**(3):100425. https://doi.org/10.1016/j.imr.2020.100425.
26. Centres for Disease Control and Prevention. *Guidelines for environmental infection control in health-care facilities.* 2003. https://www.cdc.gov/infectioncontrol/guidelines/environmental/index.html. [Accessed 26 October 2020].
27. Chahal C, van den Akker B, Young F, et al. Pathogen and particle associations in wastewater: significance and implications for treatment and disinfection Processes. *Adv Appl Microbiol* 2016;**97**:63–119. https://doi.org/10.1016/bs.aambs.2016.08.001.
28. Centre for Disease Control and Prevention. *Factors affecting the efficacy of disinfection and sterilization.* 2016. https://www.cdc.gov/infectioncontrol/guidelines/disinfection/efficacy.html. [Accessed 12 November 2020].
29. Pichichero M. Clinical and economic impact of enterovirus illness in private pediatric practice. *Pediatrics* 1998;**102**(5):1126–34. https://doi.org/10.1542/peds.102.5.1126.
30. Huang C, Wang Y, Li X, et al. Clinical features of patients infected with 2019 novel coronavirus in Wuhan, China. *Lancet* 2020;**395**(10223):497–506. https://doi.org/10.1016/S0140-6736(20)30183-5.
31. Li Q, Guan X, Wu P, et al. Early transmission dynamics in Wuhan, China, of novel coronavirus–infected pneumonia. *N Engl J Med* 2020;**382**:1199–207. https://doi.org/10.1056/NEJMoa2001316.
32. Carter LJ, Garner LV, Smoot JW. Assay techniques and test development for COVID-19 diagnosis. *ACS Cent Sci* 2020;**6**(5):591–605. https://doi.org/10.1021/acscentsci.0c00501.
33. Gao Q. Novel coronavirus infection and gastrointestinal tract. *Dig Dis Sci* 2019;**21**(3):125–6. https://doi.org/10.1111/1751-2980.12851. 2020.
34. Ahmed W, Angel N, Edson J, et al. First confirmed detection of SARS-CoV-2 in untreated wastewater in Australia: a proof of concept for the wastewater surveillance of COVID-19 in the community. *Sci Total Environ* 2020;**728**:138764. https://doi.org/10.1016/j.scitotenv.2020.138764.
35. Sinclair R. Pathogen surveillance through monitoring of sewer systems. *Adv Appl Microbiol* 2008;**65**:249–69. https://doi.org/10.1016/s0065-2164(08)00609-6.

36. Wu FQ, Xiao A, Zhang JB, et al. SARS-CoV-2 titers in wastewater are higher than expected from clinically confirmed cases. *medRxiv* 2020;**5**(4):e00614−20. https://doi.org/10.1101/2020.04.05.20051540.
37. Wölfel R, Corman VM, Guggemos W, et al. Virological assessment of hospitalized patients with COVID-2019. *Nature* 2020;**581**:465−9. https://doi.org/10.1038/s41586-020-2196-x.
38. La Rosa G, Iaconelli M, Mancini P, et al. First detection of SARS-CoV-2 in untreated wastewaters in Italy. *Sci Total Environ* 2020;**736**:139652. https://doi.org/10.1016/j.scitotenv.2020.139652.
39. Wurtzer S, Marechal V, Mouchel JM, et al. Evaluation of lockdown impact on SARS-CoV-2 dynamics through viral genome quantification in Paris wastewaters. *medRxiv* 2020. https://doi.org/10.1101/2020.04.12.20062679. Submitted for publication.
40. Wang W, Xu Y, Gao R, et al. Detection of SARS-CoV-2 in different types of clinical specimens. *J Am Med Assoc* 2020;**323**:1843−4. https://doi.org/10.1001/jama.2020.3786.
41. Danchin A, Ng PTW, Turinici G. A new transmission route for the propagation of the SARS-CoV-2 coronavirus. *medRxiv* 2020;**10**(1):1−10. https://doi.org/10.1101/2020.02.14.20022939. 33375381.
42. Lu R, Zhao X, Li J, Niu P, et al. Genomic characterisation and epidemiology of 2019 novel coronavirus: implications for virus origins and receptor binding. *Lancet* 2020;**395**(10224):565−74. https://doi.org/10.1016/S0140-6736(20)30251-8.
43. Wang XW. Concentration and detection of SARS coronavirus in sewage from Xiao Tang Shan hospital and the 309th hospital of the Chinese people's liberation army. *Water Sci Technol* 2005;**52**(8):213−21. https://doi.org/10.2166/wst.2005.0266.
44. Borchardt MA, Spencer SK, Kieke Jr BA, Lambertini E, Loge FJ. Viruses in non disinfected drinking water from municipal wells and community incidence of acute gastrointestinal illness. *Environ Health Perspect* 2012;**120**(9):1272−9. https://doi.org/10.1289/ehp.1104499.
45. Jin Y, M Flury. Fate and transport of viruses in porous media. *Adv Agron* 2002;**77**:39−102. https://doi.org/10.1016/S0065-2113(02)77013-2.
46. Dowd SE, Pillai SD, Wang S, Corapcioglu MY. Delineating the specific influence of virus isoelectric point and size on virus adsorption and transport through sandy soils. *Appl Environ Microbiol* 1998;**64**(2):405−10. https://doi.org/10.1128/AEM.64.2.405-410.1998.
47. Boone S, Gerba C. Significance of fomites in the spread of respiratory and enteric viral disease. *AEM* 2007;**73**:1687−96. https://doi.org/10.1128/AEM.02051-06.
48. Albiñana N, Clemente P, Bofill-Mas S, Hundesa A, Ribas F, Gironés R. Distribution of human polyomaviruses, adenoviruses, and hepatitis E virus in the environment and in a drinking-water treatment plant. *Environ Sci Technol* 2007;**40**:7416−22. https://doi.org/10.1021/es060343i.
49. Hundesa A, Motes C, Albiñana N, Rodriguez-Manzano J, Bofill-Mas S, Suñen E, Gironés R. Development of a qPCR assay for the quantification of porcine adenoviruses as an MST tool for swine fecal contamination in the environment. *J Virol Methods* 2009;**158**:130−5. https://doi.org/10.1016/j.jviromet.2009.02.006.
50. Xagoraraki I, Yin Z, Svambayev Z. Fate of viruses in water systems. *J Environ Eng (New York)* 2014. https://doi.org/10.1061/(ASCE)EE.1943-7870.0000827.
51. Wong K, Xagoraraki I. Evaluating the prevalence and genetic diversity of adenovirus and polyomavirus in bovine waste for microbial source tracking. *Appl Microbiol Biotechnol* 2011;**90**:1521−6. https://doi.org/10.1007/s00253-011-3156-z.
52. Parks DG. OSHA's personal protective equipment standard. *Lab Med* 2004;**27**(2):86−8. https://doi.org/10.1093/labmed/27.2.86.

53. Cook TM. Personal protective equipment during the coronavirus disease (COVID) 2019 pandemic – a narrative review. *Anaesthesia* 2020;**75**(7):920–7. https://doi.org/10.1111/anae.15071.
54. Verbeek JH, Rajamaki B, Ijaz S, et al. Personal protective equipment for preventing highly infectious diseases due to exposure to contaminated body fluids in healthcare staff. *Cochrane Db Syst Rev* 2019;**7**(7):CD011621. https://doi.org/10.1002/14651858.CD011621.pub2.
55. Chemical Hazards Emergency Medical Management. *Personal protective equipment (PPE)*. 2019. https://chemm.nlm.nih.gov/ppe.html. [Accessed 4 September 2020].
56. Ministry of Health and Family Welfare. *Novel coronavirus disease 2019 (COVID-19): guidelines on rational use of personal protective equipment*. 2020. https://www.mohfw.gov.in/pdf/GuidelinesonrationaluseofPersonalProtectiveEquipment.pdf. [Accessed 28 August 2020].
57. Medical News Today. *Different types of face mask to use during the COVID-19 pandemic*. 2020. https://www.medicalnewstoday.com/articles/types-of-face-mask#types. [Accessed 18 August 2020].
58. Mine Safety Appliances. *Respiratory PPE for wet well applications*. 2013. http://s7d9.scene7.com/is/content/minesafetyappliances/1000-73-MC_WetWell. [Accessed 22 September 2020].
59. Narayana Health. *Know about proper usage, disposal and reuse of mask*. 2020. https://www.narayanahealth.org/blog/know-about-proper-usage-disposal-and-reuse-of-mask. [Accessed 10 August 2020].
60. Worms and Germs Blog. *Examination glove re-use: a quasi-scientific evaluation*. 2020. https://www.wormsandgermsblog.com/2020/04/articles/miscellaneous/examination-glove-re-use-a-quasi-scientific-evaluation/. [Accessed 5 October 2020].
61. Roberge RJ. Face shields for infection control: a review. *J Occup Environ Hyg* 2016;**13**(4):235–42. https://doi.org/10.1080/15459624.2015.1095302.
62. Centre for Evidence-Based Medicine. *What is the effectiveness of protective gowns and aprons against COVID-19 in primary care settings?*. 2020. https://www.cebm.net/covid-19/what-is-the-effectiveness-of-protective-gowns-and-aprons-against-covid-19-in-primary-care-settings/. [Accessed 7 November 2020].
63. Industrial Safety and Hygiene News. *Recycling PPE can stretch your dollars*. 2015. https://www.ishn.com/articles/102693-recycling-ppe-can-stretch-your-dollars. [Accessed 6 August 2020].
64. Centers for Disease Control and Prevention. *Guidance for reducing health risks to workers handling human waste or sewage*. 2015. https://www.cdc.gov/healthywater/global/sanitation/workers_handlingwaste.html. [Accessed 20 July 2020].

CHAPTER 8

Handling and treatment strategies of biomedical wastes and biosolids contaminated with SARS-CoV-2 in waste environment

Samuel Jacob, Sajesh Nithianandam, Shrestha Rastogi, Simar Sakhuja, Senthil Nathan Sri Laxma Alankar
Department of Biotechnology, School of Bioengineering, College of Engineering and Technology, Faculty of Engineering and Technology, SRM Institute of Science and Technology, Kattankulathur, Tamil Nadu, India

8.1 Introduction

Biomedical wastes (BW) are the potential source of major hazardous environmen tal pollutants. It includes not only solid wastes such as needles, plastics, and syringes from the hospitals but also comprised of the drugs which are highly proactive, few are radioactive, and some pathogens are present in the excreta of various patients. If not managed or handled properly, all the aforementioned wastes may be carriers of a high amount of pathogenic activity and can cause serious health issues such as respiratory illness, viral, bacterial, fungal, parasitic infections, gastronomical problems, and so on.[1] Studies from Tehran show that there is a direct link between the number of coronavirus infected patients and the biomedical waste produced.[2] In simpler words, the hospital waste increases at increments with coronavirus-infected patients. Reports say that approximately 5.2 million people die every year owing to the mismanagement and improper disposal of biomedical wastes.[3]

In India, 600 tons of BW are generated each day, and the treatment of these wastes is indeed a major obstacle.[4] Having the second-highest number of cases during this global pandemic, India has already crossed the saturation stage of the disposal facilities available.[5] Similarly, in Wuhan, China, the numbers have been estimated to increase from 40 tons per day to 240 tons per day of BW. The figures in the United States has increased drastically from 5 million tons per year to 2.5 million tons per month.[1]

Environmental Management of COVID-19
ISBN 978-0-323-85780-2
https://doi.org/10.1016/B978-0-323-85780-2.00012-3

Considering the previous incidences of pandemics, it has been reported that the coronavirus family, which comprises not only SARS-CoV-2 but also SARS-CoV and MERS-CoV that were spread as epidemics in 2002—03 and 2012—13 respectively, continued to remain mostly on surfaces for nine days.[6] It has been reported that SARS-CoV-2 spreads more rapidly than the previous outbreaks from the same coronavirus family.[7,8] The SARS and MERS outbreak unfolded many flaws present in the healthcare system. These flaws were majorly related to the preparedness for a global health emergency that took a toll on healthcare professionals. Digging deeper into the facts, healthcare workers were the among the majorly infected populations at higher rates due to these superspreader situations. About 18.6% and 21% of healthcare workers were infected by MERS and SARS, respectively.[8] Based on these earlier experiences, during the Nipah virus outbreak in 2018 at Kerala, India, the state and central government took immediate actions that were acknowledged as a story of success.[9] When Zika virus cases were found in Brazil in the year 2015, the World Health Organization (WHO) gave immediate guidelines to suppress the virus by spreading information and also strengthened healthcare facilities and research laboratories. Since the transmission of disease is through the vector Aedes mosquito, which cannot travel for more than 400 m, the viral outbreak in a particular region was controlled.[10] The high rate of human-to-human transmissibility of SARS-CoV-2, and with the place of origin in China, being the hub of global trade, has led to the more rapid and uncontrollable spread of the coronavirus compared to SARS-CoV and MERS-CoV.[8]

The COVID-19 wave has been estimated to disrupt the waste recycling industry.[11] It has been reported that 19% of mismanagement of the recycling industry took place pre-pandemic. This overlook of management could aggravate the high risk of this virus to enter the environment.[12] The use of disposable materials such as masks, personal protective equipment (PPE) kits, used towels, and tissues in every sector of society has increased the amount of solid wastes generated, and this has caused a façade in the environment.[13] In addition, personnel and workers involved in municipal waste management are at risk of exposure to the coronavirus. Are we inviting another problem to protect ourselves from one? The question still lingers. According to recent studies, it has become an emergency to ensure handling of solid waste with the utmost care.[14] This chapter is focused on the characteristics and classification of wastes generated from health care sectors and ways to dispose of such highly contagious, virus-infected biosolids.

8.2 Solid wastes from health care units

The wastes that are generated from health care sectors, which includes various establishments, laboratories, and research facilities, is termed as healthcare wastes. It also includes wastes that come from "scattered or minor" sources. These sources may include healthcare courses in home, which may include syringe, dialysis, insulin shots, and so forth.[15]

8.2.1 Classification

Waste can be divided into two general categories. They are:

- **Non-hazardous waste**: General waste that comes from housekeeping and domestic activities, which take place in the health care sector. It comprises 75%–90% of the total waste according to the World Health Organization.[15]
- **Hazardous waste**: As the name suggests, it poses a risk to the health and environment and can be of various types, which have been discussed in detail in the subsequent sections.[15]

8.2.1.1 Infectious waste[15]

These wastes contain a specific amount of disease-causing microorganisms that may be bacteria, parasites, viruses, or fungi that can cause the spread of disease. For example, waste from the isolation wards, bandages, blood-coated tissues, disposable syringes, excreta of patients, and so on.

8.2.1.2 Pathological wastes[15]

These wastes are also called anatomical waste. These consist of blood, organs, tissues, and animal carcasses.

8.2.1.3 Sharps[15]

These wastes can cause wounds and cuts. For example, sharp blades, scalpel, scissors, and hypodermic needles.

8.2.1.4 Pharmaceutical waste[15]

Pharmaceutical waste that has expired, unused, and infected substances from industry, medicines, and vaccines that are not required and need to be disposed of properly. Items to be discarded in the pharmaceutical include drug vials, bottles, and gloves.

8.2.1.5 Chemical wastes[15]

These wastes contain harmful chemicals which may be toxic, flammable, and corrosive in nature. These are discarded from laboratories as chemical forms, and disinfectants from housekeeping work. These can further be organic or inorganic due to their nature of compound. For example, formaldehyde, solvents, and alkalis and acids.

8.2.1.6 Radioactive wastes[15]

The purpose can be to make medicine compositions containing radioactive elements, called nuclear medicine, in pharmaceutical departments for the treatment of hypothyroidism, carcinoma, and so on. It is also used for the clinical research for radioimmunoassay.[16] The discards may contain significant radioactive material that may pose hazards to human health.

8.2.2 Modes of transmission of SARS-CoV-2

COVID-19 shows resemblance to the pandemic outbreak caused by SARS-CoV in 2002 and MERS-CoV in 2012 because they belong to the same family of viruses, coronavirus.[17] These viruses are highly contagious and mainly target the respiratory tract in humans and cause severe infections.[18] The coronavirus can spread directly from one individual to another and indirectly from contaminated surfaces,[19] as shown in Fig. 8.1.

Transmission through droplets is the most common mode. If an infected person sneezes, coughs, or moist talks, the viruses from the mouth escape in the form of droplets and transmit to the other person.[20] The size of respiratory droplets when released by coughing, sneezing, etc. is larger than 5 μm in diameter, while in aerosols, the droplet size is smaller than 5 μm in diameter.[21] The dilution of the virus in aerosols leads to inactivation if left in the environment for a time period of six days.[22] Patients in quarantine homes produce waste in the form of feces or urine discharge. A study shows that the virus was present in fecal specimens collected from infected patients contradicting the WHO reports that suggested no evidence of the presence

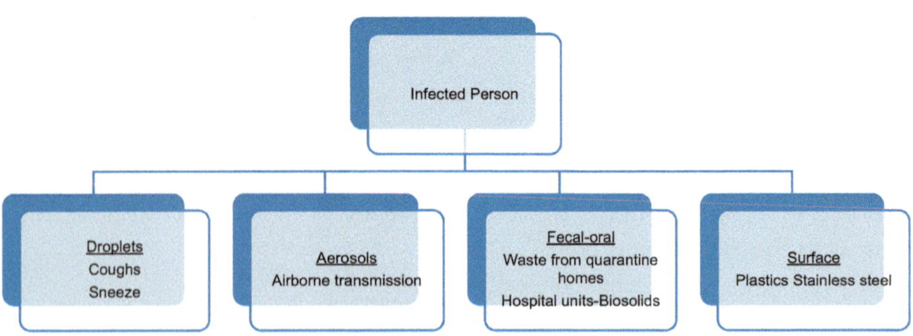

Figure 8.1 Modes of transmission of COVID-19.

and transmission of coronavirus through fecal-oral pathway.[23,24] The coronaviruses have also shown an extended survival rate of up to 17 days in fecal matter and 16.6 days in blood samples.[25] The virus can be transmitted indirectly from the contaminated surface to the healthy person, also known as fomite transmission.[19] It occurs either by the deposition of droplets released by the patient during coughing or sneezing on the surface or by the items such as clothes and bottles used by an infected person that comes in contact with the surface.[21] As stated in a report, there is only 1.3% virus transmission by fomites and 1.3% by aerosols whereas 89.5% transmission through droplets.[26] Hence, the chances of transmission of COVID-19 infection is less through contaminated surfaces as compared to direct contact.

The transmission of SARS-CoV-2 is influenced majorly by three factors: the survival rate of virus on waste surface and biosolids, population density, and socio-economic conditions.[27] The factors of population density and socio-economic conditions are interrelated. Healthcare workers and waste-handlers are in direct contact with SARS-CoV-2. These workers often belong to the lower sections of society, live in close proximity with each other, and also tend to have inferior levels of hygiene practices, which puts them at greater risk of outbreak.

8.3 Occurrence of virus in biomedical waste

The large amount of waste generated from health care centers, laboratories, and isolation homes during the treatment of patients contributes to an increase in transmission. The BW produced from these sources, if not properly marked and segregated, could cause danger to people involved in waste handling and disposal.[1] The BW produced during COVID-19 includes used PPE kits, masks, and sewage from quarantine shelters, gloves, blood collection bags, and laboratory waste. As per the reports from Central Pollution Control Board (CPCB), India there is an increase in BW production in the country by 550 tons per day amid the pandemic.[28]

The tracing of the disposal of BW can help us to keep a check on the spread of infection and reemergence of the virus at different sites. Early detection will thus help to control the outbreak by following preventive measures. As these viruses tend to remain viable on waste surfaces, they may come in contact with workers or other people and lead to infection.[29]

8.4 A comparative study on viruses potentially similar to COVID-19

The zoonotic diseases are dynamic mostly as a result of human activities. The adaptation of species to the environment is due to a genetic mutation. This mutation is constantly leading to the emergence of characters like a crossing of species barrier and transmission of infectious disease from one species to another.[30] The period between the encounter with the virus and onset of symptoms is known as the incubation period. This period varies depending upon the virus encountered. For example, SARS-CoV-1 has an incubation period of 2–8 days,[31] MERS-CoV for 2–15 days,[30] SARS-CoV-2 approximately for 1–19 days,[31] and Ebola virus for 4–10 days.[32] SARS-CoV-1 showed crossed species transmission from bats to palm civets in the live animal trading market in China,[33] whereas MERS-CoV is believed to be transferred cross-species 30 years ago in camels.[34] The similarities included no specific vaccine for controlling the transmission of the diseases; only preventive measures were introduced for both SARS and MERS.[35] The nosocomial transmission of the disease was observed in both the diseases greatly affecting hospital workers and patients.[36] However, the difference in the death rates were comparatively higher than the MERS epidemic of about 30%–40% than that caused by the SARS pandemic, which was less than 10% of deaths.[35] The mortality rate differed greatly since MERS originated as a point source and then followed an explosive epidemiological curve.[37]

Recent studies show that the survival rate and transmission route of SARS-CoV-2 is equivalent to SARS-CoV-1 and differs from MERS-CoV since it was observed that transmission of MERS-CoV was difficult from one person to another and was also confined to Middle Eastern countries.[35,37] The people suffering from chronic respiratory infections, diabetes, high blood pressure, or have low immunity have reported higher mortality risk of 9.6% and 36% when infected with SARS-CoV-1 or MERS-CoV, respectively.[30]

According to the WHO report,[24] no cases were reported for transmission of SARS-CoV-2 through food but they can be present in raw food products, especially in frozen items.[38] The SARS-CoV-2 was observed to survive on imported frozen food products, which include packaged meat, fish, and dairy items at 4°C for about 14–21 days.[39] The studies also suggest that there is a possibility of food contamination by fecal transmission or unclean practices adopted by infected people, which includes improper hand and mouth hygiene;[41] it is advisable to follow personal and food hygiene along with safety precautions. Although there are many incidents of indirect foodborne transmission of SARS-CoV-2, thorough research is required to draw out the direct connection between COVID-19 infection and transmission by food products.[39]

According to a study, about 51% patient's stool tested positive for both SARS-CoV-1 and SARS-CoV-2. In urine samples, 42% and 4% tested SARS-CoV-1 and SARS-CoV-2 tested positive, respectively.[40] The infectious time period of SARS-CoV-1 in human stools is for 2–4 and 17 days in urine.[41] MERS-CoV can survive in organic waste for 2 days at a temperature of 20°C.[25] The human intestinal epithelial layer can stimulate the replication of the virus and act as an optional infective route, which supports the survival of the virus. According to reports,[32] the Ebola virus showed survival on human discharge for a maximum of 8 days at normal temperature. This report also suggested that the waste produced by patients infected from Ebola was directly discharged into sewage after disinfection and the components of organic waste and halogenated groups react to decrease the efficiency of the disinfectants. This increased survival rate of Ebola virus in waste rendering the disinfection process void. As stated earlier, SARS-CoV-1 shown similarity to SARS-CoV-2 hence the survival rate of SARS-CoV in hospital waste can be employed for better understanding of persistence of SARS-CoV-2. SARS-CoV was present in hospital sewage for up to 2 days at 20°C in reactive form and inert isolated form for 8 days. The temperature above 20°C reduces the viability of the virus isolates to a great extent.[38]

The inactivation of human coronavirus (HCoV) was reported by various biocidal products, which includes 71% concentrated ethanol, 0.05% concentrated sodium hypochlorite, and 2% concentrated glutardaldehyde.[6] These chemicals very highly effective with reduction of viral isolates to >3 $\log_{10}$ hence can also be employed for disinfecting SARS-CoV-2. It was reported in the study that the SARS-CoV-2 was weakened at temperature >40°C and could survive only for 24 h.[42]

8.5 Survival of SARS-CoV-2 on different surfaces

SARS-CoV-2 can survive up to 72 h on plastic and steel and is less stable on copper surfaces.[43] The viability of the virus on different surfaces depends on the type of surface, suspending medium, temperature, relative humidity, mode of deposition, and drying time.[44] Studies have demonstrated the

Table 8.1 Survival of SARS-CoV and MERS-CoV on dry surfaces.

Substrate	Test virus	Temperature	Duration
Plastic	SARS-CoV	22–25°C	<5 days
	MERS-CoV	20°C	48 h
Steel	MERS-CoV	20°C	48 h
Metal	SARS-CoV	RT	5 days
Wood	SARS-CoV	RT	4 days
Paper	SARS-CoV	RT	4–5 days
Disposable gowns	SARS-CoV	RT	2 days
Glass	SARS-CoV	RT	4 days

CoV, coronavirus; *MERS*, Middle East respiratory syndrome; *RT*, Room temperature; *SARS*, severe acute respiratory syndrome.

capacity of coronaviruses to survive on different matrices like plastic, copper, and stainless steel as shown in Table 8.1. The survival rate analysis helps to disinfect the surface contaminated by SARS-CoV-2.[6]

The results demonstrated that SARS-CoV, HCoV, and MERS-CoV can effectively and efficiently survive on dry surfaces.[6] However, HCoV has more viability in the cell culture medium as compared to SARS-CoV and MERS-CoV.[38] MERS-CoV can survive at low temperatures (4°C for 72 h); it is also thermolabile making it susceptible to high temperatures.[38]

The research on various viruses comprehended some of the important findings related to SARS-CoV-2. The virus can survive on plastics (2–3 days), hard surface and stainless steel (2–3 days), copper (<4 h), cardboard (<24 h), and air (<3 h). The SARS-CoV-2 is less viable on fabrics and food.[24] However, it is advisable to use PPE kits and coveralls by frontline workers who are in close to infected patients.

The reduced survival of the SARS-CoV-2 in water is due to the flora present in water, which hinders the metabolism of the virus, causing its inactivation.[40] The SARS-CoV have shown an extended survival rate of 96 h when in contact with biosolids.[45] A similar study was carried out for SARS-CoV-2,[48] which suggested the presence of this virus in the primary sludge and thickened sludge (>20%). The sewage sludge is considered as the pool of SARS-CoV-2 as it contains 30%–50% of the total pollutants discharged in sewage system.[46] Therefore, it is important to conduct further studies and analyze the risk for secondary transmission of the virus from contaminated sludge of wastewater treatment plants.

8.6 Safe handling and management of health care waste generated through the care of COVID-19 patients

The safe handling and management of waste produced by the treatment of COVID-19 patients is based on three essential principles. The following principles was used for managing Ebola virus—associated healthcare wastes:[8]

- Safe handling and processing of the waste should be carried out as close to the point of generation as possible. After primary containment personnel should avoid opening containers to control the waste.
- Limit the number of personnel entering the area of care for COVID-19 patients and those handling waste generated before and after primary containment.
- Must use suitable PPE and waste management techniques for off-site inactivation before inactivation or transportation away from hospital.

8.6.1 Preparing a waste management plan as part of COVID-19 patient care

The following waste management plan has to be implemented while handling a COVID-19 patient care unit:[48]

- Comply with the state and local regulations like SPCB (State Pollution Control Board), CBWTF (Common Biomedical Waste Treatment and disposal Facility), and ULB (urban local bodies) on coronavirus-associated waste handling, storage, treatment, and disposal.
- Determine if coronavirus-associated waste should be processed at the hospital for inactivation or moved off-site.
- Identify a waste management team with thorough training on uniform waste handling techniques, including wearing acceptable PPE and secured bag and packaging waste policies, processing waste, and transporting processed waste.
- Make sure the staff in healthcare and environmental services are qualified to wear approved PPE (same for patient care) and follow appropriate procedures.
- The handling and primary waste packaging should take place in the patient room and the area where PPE is removed and performed by primary healthcare workers (i.e., doctors and nurses) wearing PPE as specified in the hospital guidance.

8.7 Collection, transport, and storage of waste

8.7.1 Collection

COVID-19 associated waste should be collected regularly and can be transported in leak/puncture proof containers, labeled with a biohazard sign. The patient's room should be placed with an individual waste bag. The patient's used tissue paper and face masks should be immediately discarded in the waste bag. Gloves and face masks used by the caretaker and the cleaner should be disposed off in the second waste bag immediately next to the patient's room door when the caretaker or cleaner leaves the room.[47] Once the waste bag is three-fourths full the nursing staff and other working personnel must ensure that the waste bags are properly closed and sealed. Fresh bags and containers at all sites where waste is generated should be readily accessible for collecting them regularly. Wastes should not be permitted to accumulate at any point.[47] A regular waste management system must be developed in order to collect the health care wastes as shown in Fig. 8.2.

8.7.2 Storage

Within the healthcare facility, a place of storage for the BW should be allocated. Health care waste that is collected in bags and containers should be stored separately in an area, space, or building in an appropriate size for the quantity of waste generated and depending upon how often the wastes are collected. Storage areas should be clean, safe, and secured against the

Figure 8.2 Health care facility waste management in relation to COVID-19.[15]

weather, pests, and disease-causing vectors. The area where the healthcare wastes are collected and stored should provide easier access to the waste management personnel. It is important that waste disposal vehicles have convenient access to the storage area. The storage should be locked to prevent unauthorized persons from reaching it. Storage bins must have a biohazard sign.[15] Regular inspection and cleaning of the storage area is necessary.

- Storage areas should be clean, secure, and protected from direct contact with other vectors like pests and public.[48]
- The waste that are collected and stored should be handed over to CBWTF personnel.[48]

8.8 Treatment and disposal

8.8.1 Treatment

Best management strategies recommend that the waste should be disinfected by non-incineration processes in particular, by steam-based treatment such as autoclave or microwave.[49] All systems should be checked and reviewed periodically. Organisations like WHO and UNEP (United Nations Environment Program) prefer steam-based or other forms of nonincineration methods for disinfection due to the persistent organic pollutants (POP) produced by incineration. However, incineration is much costlier than steam-based systems and has a higher carbon footprint.[49]

8.8.1.1 Preferred technologies for treating virus infected solid waste

8.8.1.1.1 Autoclave

Autoclave is the most common and efficient wet thermal technology used for treating medical waste, which is based upon pressure and vacuum using high temperature steam. This method is often preferred as it is simple, cheap and has reduced air pollution.[15] The merits and demerits of adoption of this technique have been tabulated in Table 8.2. It is suitable for treating solid healthcare wastes, used PPE, medical instruments that can be reused, glassware, and sharp wastes.[50]

8.8.1.1.2 Sterilization

This method is based upon using a microwave and high temperature steam process with integrated shredding.[51]

Table 8.2 Merits and demerits of autoclave.

Merits	Demerits
Overall handling and operation is very simple It is one of the cost-effective technology It has reduced air pollution	Heat sensitive materials cannot be used No waste volume reduction Remaining waste need to be transported and land-filled

8.8.1.1.2.1 Microwave disinfection In recent days microwave disinfection is considered as an effective supplementary technology for incineration as it has several advantages as mentioned in Table 8.3.[51] In this method the wastes are loaded into a shredder and reduced into small bits, after which the waste is humidified and sent to a chamber containing a series of microwave generators for irradiation process.[15] Merits and demerits of microwave-based disinfection technology have been shown in Table 8.3.

8.8.1.1.2.2 High temperature steam-based technology It is a disinfection technology where pathogenic microorganisms are killed in a high temperature steam (temperature greater than 100°C). For a certain period of time the health care wastes are exposed in a high temperature steam, which contains lot of water vapor. The water vapor emits a latent heat, which results in coagulation and protein denaturation of the microorganisms causing death of the pathogenic microorganisms.[52] This method also has its own merits and demerits which is shown in Table 8.4.

8.8.1.1.3 Shredding

Before carrying out disinfection and the other treatment process it is necessary that the solid BW be taken through a process known as shredding. Shredding increases the contact between the waste and disinfectant by raising the surface area and possibly removing all the enclosed gaps in order to decrease the waste volume. Shredding is a much effective method as it reduces the initial waste volume to a maximum extent (60%–90%).[15]

Table 8.3 Merits and demerits of microwave disinfection technology.

Merits	Demerits
Slow heat loss and rapid action in a reduced space Energy saving technology Low environmental pollution and no residues or toxic wastes after disinfection	Takes longer time when compared to autoclave No waste volume reduction Can sterilize only limited number of materials at a time

Table 8.4 Merits and demerits of high temperature steam-based technology.[50]

Merits	Demerits
Low cost and safe technology	Cannot be used for heat sensitive materials due to melting
Nontoxic and negligible air pollution	May leave instruments wet, causing corrosion
Cycle easy to control and monitor	Requires proper maintenance for controlling high temperature
Rapid cycle action and more efficient	No waste volume reduction

8.8.1.1.4 Incineration

Incineration is one of the widely used technologies for safe management of various wastes that results from healthcare activities.[53] Healthcare waste consists of organic and combustible waste and is reduced to inorganic, incombustible substance in a dry oxidation process using high-temperature range. Initial volume and weight of the waste is significantly reduced by this process. Incineration is a selective process used for treating those types of healthcare waste that cannot be recycled or reused and disposed of directly.[53]

An incinerator usually operates at the temperature range of 900–1200°C at which combustion of organic compounds takes place. The combustion of organic compounds results in the emission of gaseous substances like carbon dioxide, steam, and various other toxic materials in the form of ashes. During combustion, various conditions should be properly controlled or else it may result in the production of harmful carbon monoxide gas. The ash and wastewater generated at the end of the process contains some toxic compounds that are necessary to be treated to avoid harmful effects.[15] The two most widely used incinerators for treating the health care wastes are:

(1) Pyrolytic incinerator and
(2) Rotary kiln incinerator.

8.8.1.1.4.1 Pyrolytic incinerator The pyrolytic incinerator consists of two important chambers: one is a pyrolytic chamber and the other is a post-combustion chamber. Inside the pyrolytic chamber the healthcare waste undergoes a combustion process in the low supply or near absence of oxygen, and they are thermally decomposed through incomplete oxidation, which results in production of gaseous compounds and ash (solid residue) at the end of the process.[15] The operation is carried out using a fuel burner that is present in the pyrolytic chamber. Mixture of gas generated during the process are burned at a high temperature range by the post-combustion chamber fuel burner, which reduces excess smoke and odor. Larger-sized pyrolytic incinerators are equipped in a way that it runs continuously. These may also be able to work entirely automatically, including waste storage and ash removal. In hospitals, limited size pyrolytic incinerators are most widely used since they do not need any special equipment for cleaning. It is also used for disposing various healthcare wastes because they are known for their controlled air incineration.[53]

8.8.1.1.4.2 Rotary kiln incinerator This incinerator works on the rotatory motion basis which consists of a rotating oven and a post-combustion chamber. Rotation helps in mixing the waste, thereby increasing the efficiency of incineration.[55] Waste is fed inside the chamber from the top and the ashes and other solid residues are collected at the bottom end. The gas produced during the process are heated in the post-combustion chamber where the temperature range is very high, resulting in deduction of the harmful volatile organic gaseous compounds.[15] This type of incinerator can operate continuously even with heavy loads. Almost all hazardous substances are destroyed as this incinerator operates at a temperature of 1200°C.[53]

8.8.1.1.4.3 Plasma incineration Plasma incineration is a new waste disposal technology. The aim of this technology is to transfer energy through plasma to rapidly decompose waste into smaller particles or even atoms.[54] Compared with conventional incineration techniques, plasma incineration demonstrates a higher energy efficiency, which demonstrates promising application prospects.[55] The merits and demerits of incineration technology have been tabulated in Table 8.5.

Table 8.5 Merits and demerits of incineration-based technology.

Merits	Demerits
High waste volume reduction	It is an expensive technology in terms installation and maintenance
More efficient as it burns almost 90% of the initial waste generated	Releases harmful gases like nitrogen oxide and acid gases during combustion which causes environmental pollution
Handles different types of wastes by its uniform action	Ash waste can potentially harm both public health and the environment
Prevents transportation of wastes and less dependence on landfills	Releases POP and has a high carbon footprint

8.8.2 Disposal

After various treatments for disinfection the harmful health-care wastes become non-risky and are safe to be disposed at landfills. A landfill site should be properly designed and engineered for disposal of the reduced hazardous waste. Regular monitoring of such sites is mandatory inorder to avoid leaching and other environmental problems like groundwater contamination. General wastes which are not contaminated can be disposed normally as solid waste. Used masks and gloves are to be kept in a paper bag for atleast 72 h before disposing it. Before disposing of the used masks it is better to cut or mutilate it to prevent its reuse.[49] Access to the disposal sites should strictly be allowed only for the authorized persons. Encapsulation is one of the cheap, safe, and easier methods for disposal of health care wastes, where the waste containing metallic drums or containers are added with an immobilizing material, which acts as a medium for immobilizing the waste. The medium is allowed to get dried completely and then, the immobilized containers can be tightly sealed and disposed of into landfills.[15]

8.9 Measures to protection of personnel and waste disposal workers from contracting COVID-19

Poor healthcare waste management easily exposes to waste handlers, patients, and the wider population to illness, harmful effects, and environmental pollution risks. Patients who were treated at home are generating infected waste, possibly discarded as domestic waste, which may pose risks to sanitary staff and the environment, depending on the conditions of transport and disposal.[56] Therefore, these wastes should be regularly tracked and properly handled. Sanitation workers are the most insecure as they come into close contact with various forms of waste, including liquid waste, solid waste, organic waste, and hazardous waste.[48] During this pandemic, waste disposal workers should be careful, and close interaction with the waste during this pandemic makes them more vulnerable to the coronavirus, and it may even infect their family members and spread rapidly. Thus, such waste disposal personnel should be adequately trained, and protective equipment to be regularly provided.

The most affected groups from these health care wastes are as follows:

- nurses and other frontline employees in the hospital,
- waste disposal workers,
- workers bringing waste to a disposal or recycling plant, and

- workers in waste treatment plant (incinerator plant or landfills).

Even common people are also at risk when they get in close contact with those healthcare wastes. The most significant problem that leads to this risk is mainly due to the unavailability of protective equipment, and the government or the respective organization should regularly supply these protective essentials. Globally, most workers employed in waste disposal and treatment plants are poor, so the public also should come forward to supply these protective kits free of cost, which may prevent them from infection (Fig. 8.3).[56]

8.9.1 Actions to reduce the risk of exposure

To reduce the risk of exposure to these sanitary wastes, waste disposal workers should follow the following protective measures.[57]

8.9.1.1 Personal protective equipment (PPE) kit

PPE plays an important role in guaranteeing overall safety and health to the waste disposal workers in treatment or in disposal sites. Ensure that all sanitation workers are equipped with protective kits suited to the nature of their work, along with soaps and hand sanitizers. PPE acts as a barrier between the suspected infectious substance and the healthcare worker; PPE eliminates contact with an infectious agent. Disposal workers must wear PPE. Ineffectiveness of PPE may lead to nosocomial transmission of SARS-CoV.[57]

Components of PPE kits include masks, gloves, goggles, glasses, and gowns. Health institutions and waste disposal services should have practices

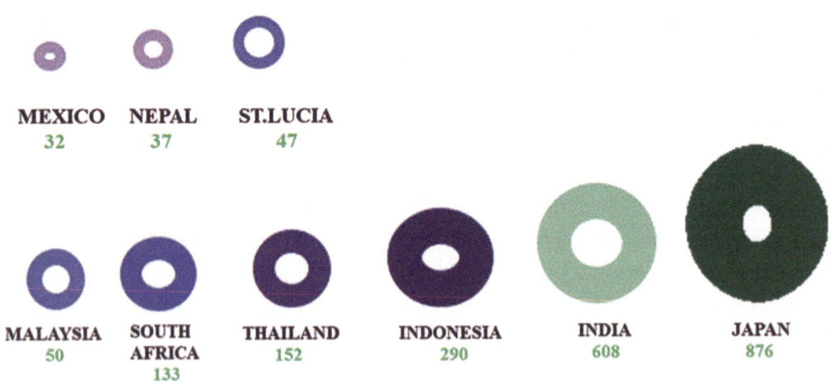

Figure 8.3 Amount of health care waste generation (tons per day) in selected countries.[56]

and regulations that describe the correct order of donning and safely doffing these PPEs. The order for donning the PPE after performing hand hygiene is gown, then mask, goggles, face shield, and gloves; the order for removing the PPE is gloves, face shield, goggles, gown, and then the mask should be removed.[58] PPE kits are not reusable; once they are used they must be disposed, except for goggles (can be reused). Though PPE kits are costlier, service-providing companies should offer workers with high-quality PPE.[58]

Single-use PPE kits should be properly disposed in a plastic bag and should be labeled as hazardous or biomedical waste and segregated from general waste securely. The reusable PPE equipment such as googles, should be washed with detergent and water after use and allowed to dry. Strict adherence to improved hygiene practices, including regular alteration and cleaning of PPE and professional clothing, replacement of professional gloves in case of breakage or other possible incident contamination, routine sanitizing of buildings, cabins and clothes for vehicles. Make sure of strict guidelines on how to put and take off PPE. Avoid incidental contact and contamination.[59]

8.9.1.2 Face masks

Since the COVID-19 pandemic, the wearing of face masks has received numerous guidelines from different public health authorities and government. The WHO and other public health organizations believe that the spread of respiratory viral diseases like COVID-19 can be restricted by the masks.[58] Along with hand washing and social distancing, the face mask has emerged as the most powerful tool in our society against COVID-19. Yet, with billions of people around world using masks every day since COVID-19 began, it is mandatory that governments of different countries and waste disposal organizations provide workers with a continuous supply of face masks and protective apparel to defend themselves against the pandemic. Different kinds of face masks that are feasible to utilize during a pandemic have been represented in Fig. 8.4.[58]

8.9.1.3 Reusable cloth mask

Cloth masks are widely used, and their demand is increasing day by day due to their low costs and reusability nature. Depending on the frequency of use, the WHO recommends that sanitary workers should routinely wash these face cloth masks. After every wear, a reusable cloth mask should be washed.[58] When you take it off, fold it in half to contain the inner secretions and place it in a laundry bin, ideally one with a lid that you can

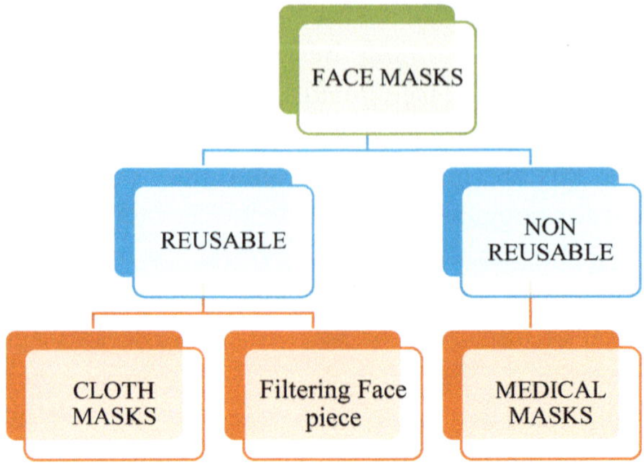

Figure 8.4 Types of masks employed for health care workers.

close if you cannot wash it right away. These cloth masks cannot be used by disposal workers while working, but they can be used when they are away from the disposal site or treatment plant. The collected mass should be safely disposed; recently there has been an increase in mass waste in developing countries (Table 8.6) and governments must devise appropriate strategies to handle it.[62]

8.9.1.4 Gloves

Gloves are the second skin for sanitary workers. The collection of waste from garbage or waste storage yards may contain harmful wastes such as blood-stained injections, cottons, needles, glass bottles, etc. Thus, gloves are essential to protect the sanitary workers from injuries and infections. If the gloves are damaged or contaminated, hand hygiene should be checked and gloves should be changed with new ones. Once it is used it should be disposed in a safe manner.[58] Gloves need not be dried and reused. Strict adherence to improved hygiene practices, including regular alteration and cleaning of gloves and replacement of professional gloves in case of breakage or other possible incidental contamination.[58]

8.9.1.5 Head covers and goggles

Head covers and goggles are also very important because they protect from the entry of infectious droplets and dust by forming a barrier for the medium of the disease, which is necessary for sanitary workers. Eye protection should not be ignored and should be considered part of PPE.[60]

Table 8.6 Estimated daily usage of face mask and medical waste in Asia with confirmed COVID-19 cases.[62]

Country	Population	Total COVID-19 cases	Urban population (%)	Face masks acceptance rate (%)	Number of face mask need of each general population each day	Total daily face mask use (pieces)
India	1,381,085,714	1,643,416	35	80	1	381,179,657
Iran	84,077,062	301,530	75	80	1	50,648,022
Pakistan	221,213,683	278,305	35	80	1	61,762,860
Saudi Arabia	34,855,542	274,219	84	80	1	23,367,155
Bangladesh	164,820,045	234,889	75	80	1	99,155,739
Turkey	84,410,984	229,891	39	80	1	26,066,112
Iraq	40,288,721	121,263	96	80	1	30,973,969
Qatar	2,807,805	110,460	60	80	1	1,341,008

8.9.1.6 Adequate hand washing facilities at the place of work

Appropriate hand washing facilities should be provided at the treatment site, with daily water, soap, and sanitary supplies. Temporary devices such as foot-operated hand-washing systems can be built in the worksite. Bathing facilities close to the place of work may also be required, depending on the nature of the work. Sanitizers must be often used by workers before and after their work. Since sanitizers are alcohol-based they should be continuously used by workers to defend them against germs.[15]

Steps should be followed by both patients and sanitary workers for the disposal of waste:

- Paper tissues, gloves, and face masks among patients must be disposed by putting in the waste bag that was kept in the patient's room.[47]
- Segregate waste at the source, Storage facilities should be clean, secure, and designed to protect from the elements, pathogens, and vectors of disease.[61]
- Gloves and face masks used by the sanitary workers should be immediately put in a second waste bag and should be disposed separately.[47]
- The waste bags should be closed before they are disposed from the patient's room and replaced often; they should never be emptied in another bag.[61]
- Collect the waste at least daily and carry it in air-tight containers, puncture-proof, marked with the biohazard symbol.[61]
- These waste bags can be accumulated and placed in a clean waste yard; the closed waste-containing bags can be directly disposed of in the common garbage yard. No specific collection of waste activity or other disposal method is mandatory.[61]

After disposing of waste bags, strict hand hygiene should be followed by the workers and insist on using water and soap or alcohol-based hand disinfectants.[61]

8.9.1.7 Awareness and training

For waste disposal workers, training and awareness-building are needed in an effort to minimize the transmission of COVID-19. The preparation of workers leads to a more educated workforce, which is the basis for achieving this superior infection prevention performance. Expert staff can also assist patients and other visitors in understanding their role in maintaining good hygiene and becoming more responsible for the waste they generate. Most waste disposal and sanitary workers are from poor economic backgrounds in regions such as the Indian subcontinent and Africa, where they are mostly unaware about these harmful and hazardous wastes.[15] Therefore, they need to educate regarding the handling and transporting of waste securely during this pandemic. The awareness camps for these workers should be conducted on a regular basis by hospitals and treatment service centers. Such camps will help them to gain some knowledge and also prevent them from risk of infection. Education and training are essential parts of the waste management programs for healthcare. When healthcare workers are adequately sensitized to the importance of waste management, they follow professional best practices and help improve and maintain a proper waste management system.[15] Training should be essentially defined and incorporated into the regular functions of the healthcare facility. The trainer who delivers the training and awareness to the workers must be an expert. The trainer should address how to handle these biowastes, cautious of the pandemic, and train them by explaining the facts in their own language so that they can easily understand.[15]

The goals of training and awareness are:

- Increase awareness of health, protection, and environmental concerns related to sanitary waste.[59]
- Ensure that healthcare workers are familiar with best practices and waste care technologies managing them and being able to adapt them to their everyday work.
- Promote responsibility to the control of health care waste among all health care staff.

8.9.2 Offer support for sanitary workers by government

Funds may be obtained at a local level, if necessary, to ensure that workers have proper safety equipment. Government should also support health workers' organizations to ensure that health insurance and regular health check-ups are provided to all health workers. The government should provide food and accommodation to the sanitary workers, if necessary, and they also need to allocate a set member to monitor disposal of waste and to check the health condition of the workers regularly.[15] Governments should also provide a separate disposing yard so that the transmission of the disease can be controlled.[59]

8.10 Conclusion

Recent pandemic situation has altered the ways of waste management and the material recycle process. The sudden hike in BW production during

COVID-19 infection calls for an urgent need for disposal of contaminated waste to ensure sustainable development. SARS-CoV-2 is highly contagious and has the potential to spread through frozen raw food products and fecal-oral routes, and as a result, there is concern among the healthcare employers and waste recyclers. Chemical treatment has proved to be a prominent preventive measure in handling waste surfaces, whereas physical treatment by incineration, microwave disinfection, and autoclave have been utilized for disposal of residual waste products. Incineration is the preferred method by the ignition of waste at high temperatures where SARS-CoV-2 cannot thrive. Further, there is a requirement for public—private partnership and cooperation to safeguard sanitation employers and frontline workers. International organizations like the UNEP, in collaboration with policy makers, have postulated recommendations and guidelines for waste recycling mechanisms to mitigate non-targeted segments of society (sanitary workers) and maintain environment sustainability during pandemic situations.

References

1. Ilyas S, Srivastava RR, Kim H. Disinfection technology and strategies for COVID-19 hospital and bio-medical waste management. *Sci Total Environ* 2020:141652. https://doi.org/10.1016/j.scitotenv.2020.141652.
2. Ramteke S, Sahu BL. Novel coronavirus disease 2019 (COVID-19) pandemic: considerations for the biomedical waste sector in India. *CSCEE* 2020;**1**:100029.
3. Rahman MM, Bodrud-Doza M, Griffiths MD, Mamun MA. Biomedical waste amid COVID-19: perspectives from Bangladesh. *Lancet Glob Health* 2020;**8**(10):e1262. https://doi.org/10.1016/S2214-109X(20)30349-1.
4. Zand AD, Heir AV. Emerging challenges in urban waste management in Tehran, Iran during the COVID-19 pandemic. *Resour Conserv Recycl* 2020;**162**:105051.
5. Kaur S. *Biomedical waste management in the time of coronavirus*. 2020. https://thedailyguardian.com/biomedical-waste-management-in-the-time-of-coronavirus/. [Accessed 12 December 2020].
6. Kampf G, Todt D, Pfaender S, Steinmann E. Persistence of coronaviruses on inanimate surfaces and their inactivation with biocidal agents. *J Hosp Infect* 2020;**104**(3):246—51. https://doi.org/10.1016/j.jhin.2020.01.022.
7. Park M, Thwaites RS, Openshaw PJM. COVID-19: lessons from SARS and MERS. *Eur J Immunol* 2020. https://doi.org/10.1002/eji.202070035.
8. Peeri NC, Shrestha N, Rahman MS, et al. The SARS, MERS and novel coronavirus (COVID-19) epidemics, the newest and biggest global health threats: what lessons have we learned? *Int J Epidemiol* 2020;**49**(3):717—26. https://doi.org/10.1093/ije/dyaa033.
9. World Health Organization. *Nipah virus outbreak in Kerala*. 2018. https://www.who.int/southeastasia/outbreaks-and-emergencies/health-emergency-information-risk-assessment/surveillance-and-risk-assessment/nipah-virus-outbreak-in-kerala. [Accessed 23 December 2020].
10. World Health Organization. *Zika*. 2017. https://www.who.int/news-room/q-a-detail/zika. [Accessed 23 December 2020].

11. Zambrano-Monserrate MA, Ruano MA, Sanchez-Alcalde L. Indirect effects of COVID-19 on the environment. *Sci Total Environ* 2020:138813. https://doi.org/10.1016/j.scitotenv.2020.138813.
12. Vanapalli KR, Sharma HB, Ranjan VP, Samal B, Bhattacharya J, Dubey BK, Goel S. Challenges and strategies for effective plastic waste management during and post COVID-19 pandemic. *Sci Total Environ* 2020;**750**:141514. https://doi.org/10.1016/j.scitotenv.2020.141514.
13. Sarkodie SA, Owusu PA. Impact of COVID-19 pandemic on waste management. *Environ Dev Sustain* 2020:1−10. https://doi.org/10.1007/s10668-020-00956-y.
14. Sharma HB, Vanapalli KR, Cheela VS, Ranjan VP, Jaglan AK, Dubey B, Goel S, Bhattacharya J. Challenges, opportunities, and innovations for effective solid waste management during and post COVID-19 pandemic. *Resour Conserv Recycl* 2020;**1**(162):105052.
15. Chartier Y, Emmanuel J, Pieper U, et al., editors. *Safe management of wastes from health-care activities*. World Health Organization; 2014.
16. Aljabre SH. Hospital generated waste: a plan for its proper management. *J Family Community Med* 2002;**9**(2):61−5.
17. Zhu Z, Lian X, Su X, Wu W, Marraro GA, Zeng Y. From SARS and MERS to COVID-19: a brief summary and comparison of severe acute respiratory infections caused by three highly pathogenic human coronaviruses. *Respir Res* 2020;**21**:224. https://doi.org/10.1186/s12931-020-01479-w.
18. Lai C-C, Shih T-P, Ko W-C, Tang H-J, Hsueh P-R. Severe acute respiratory syndrome coronavirus 2 (SARS-CoV-2) and coronavirus disease-2019 (COVID-19): the epidemic and the challenges. *Int J Antimicrob Agents* 2020;**55**(3):105924. https://doi.org/10.1016/j.ijantimicag.2020.105924.
19. World Health Organization. *Transmission of SARS-CoV-2: implications for infection prevention precautions*. 2020. https://www.who.int/news-room/commentaries/detail/transmission-of-sars-cov-2-implications-for-infection-prevention-precautions. [Accessed 12 December 2020].
20. Liu L, Wei J, Li Y, Ooi A. Evaporation and dispersion of respiratory droplets from coughing. *Indoor Air* 2017;**27**(1):179−90. https://doi.org/10.1111/ina.12297.
21. World Health Organization. *Modes of transmission of virus causing COVID-19: implications for IPC precaution recommendations*. 2020. https://www.who.int/news-room/commentaries/detail/modes-of-transmission-of-virus-causing-covid-19-implications-for-ipc-precaution-recommendations. [Accessed 12 December 2020].
22. Shiu EYC, Leung NHL, Cowling BJ. Controversy around airborne versus droplet transmission of respiratory viruses: implication for infection prevention. *Curr Opin Infect Dis* 2019;**32**(4):372−9. https://doi.org/10.1097/QCO.0000000000000563.
23. Jayaweera M, Perera H, Gunawardana B, Manatunge J. Transmission of COVID-19 virus by droplets and aerosols: a critical review on the unresolved dichotomy. *Environ Res* 2020;**188**:109819. https://doi.org/10.1016/j.envres.2020.109819.
24. World Health Organization. *Coronavirus disease 2019 (COVID-19) situation report-32*. 2020. https://www.who.int/docs/defaultsource/coronaviruse/situation-reports/20200221-sitrep-32-covid-19.pdf?sfvrsn=4802d089_2. [Accessed 12 December 2020].
25. Cevik M, Tate M, Lloyd O, Maraolo AE, Schafers J, Ho A. SARS-CoV-2, SARS-CoV, and MERS-CoV viral load dynamics, duration of viral shedding, and infectiousness: a systematic review and meta-analysis. *Lancet Microbe* 2020. https://doi.org/10.1016/S2666-5247(20)30172-5.
26. Binder RA, Alarja NA, Robie ER, et al. Environmental and aerosolized severe acute respiratory syndrome coronavirus 2 among hospitalized coronavirus disease 2019 patients. *J Infect Dis* 2020;**11**(1):1798−806. https://doi.org/10.1093/infdis/jiaa575.

27. Kulkarni BN, Anantharama V. Repercussions of COVID-19 pandemic on municipal solid waste management: challenges and opportunities. *Sci Total Environ* 2020;**15**(743):140693.
28. Central Pollution Control Board. *Revised guidelines for common bio-medical waste treatment and disposal facilities*. 2016. https://www.cpcb.nic.in/uploads/Projects/Bio-Medical-Waste/BMW-GUIDELINES-COVID.pdf. [Accessed 12 December 2020].
29. Ranjan MR, Tripathi A, Sharma G. Medical waste generation during COVID-19 (SARS-CoV-2) pandemic and its management: an Indian perspective. *Asian J Environ Sci* 2020;**13**(1):10−5. https://doi.org/10.9734/AJEE/2020/v13i130171.
30. Weber DJ, Rutala WA, Fischer WA, Kanamori H, Sickbert-Bennett EE. Emerging infectious diseases: focus on infection control issues for novel coronaviruses (severe acute respiratory syndrome-CoV and middle east respiratory syndrome-CoV), hemorrhagic fever viruses (Lassa and Ebola), and highly pathogenic avian influenza viruses, A(H5N1) and A(H7N9). *Am J Infect Control* 2016;**44**(5 Suppl):e91−100. https://doi.org/10.1016/j.ajic.2015.11.018.
31. Patrikar S, Kotwal A, Bhatti V, et al. Incubation period and reproduction number for novel coronavirus (COVID-19) infections in India. *medRxiv* 2020. https://doi.org/10.1101/2020.06.27.20141424.
32. Centres for Disease Control and Prevention. *Frequently asked questions (FAQs) on interim guidance for managers and workers handling untreated sewage from suspected or confirmed individuals with Ebola in the U.S.* 2014. https://www.cdc.gov/vhf/ebola/prevention/faq-untreated-sewage.html. [Accessed 12 December 2020].
33. Wang L-F, Shi Z, Zhang S, Field H, Daszak P, Eaton BT. Review of bats and SARS. *Emerg Infect Dis* 2006;**12**(12):1834−40. https://doi.org/10.3201/eid1212.060401.
34. Müller MA, Corman VM, Jores J, et al. MERS coronavirus neutralizing antibodies in camels, Eastern Africa, 1983−1997. *Emerg Infect Dis* 2014;**20**(12):2093−5. https://doi.org/10.3201/eid2012.141026.
35. Sampathkumar P. Middle East respiratory syndrome: what clinicians need to know. *Mayo Clin Proc* 2014;**89**(8):1153−8. https://doi.org/10.1016/j.mayocp.2014.06.008.
36. de Wit E, van Doremalen N, Falzarano D, et al. SARS and MERS: recent insights into emerging coronaviruses. *Nat Rev Microbiol* 2016;**14**:523−34. https://doi.org/10.1038/nrmicro.2016.81.
37. Hussain HY. Incidence and mortality rate of Middle East respiratory syndrome - CoronaVirus (MERS-Cov), threatens and opportunities. *J Mycobac Dis* 2014;**4**:4. https://doi.org/10.4172/2161-1068.1000162.
38. Carraturo F, Del Giudice C, Morelli M, Cerullo V, Libralato G, Galdiero E, Guida M. Persistence of SARS-CoV-2 in the environment and COVID-19 transmission risk from environmental matrices and surfaces. *Environ Pollut* 2020;**265**:115010. https://doi.org/10.1016/j.envpol.2020.115010.
39. Han J, Zhang X, He S, Jia P. Can the coronavirus disease be transmitted from food? A review of evidence, risks, policies and knowledge gaps. *Environ Chem Lett* 2020:1−12. https://doi.org/10.1007/s10311-020-01101-x.
40. Singer A, Wray R. Detection and survival of SARS in human stool, urine, wastewater and sludge. *Preprints* 2020. https://doi.org/10.20944/preprints202006.0216.v1.
41. Hung IFN, Cheng VCC, Wu AKL, et al. Viral loads in clinical specimens and SARS manifestations. *Emerg Infect Dis* 2004;**10**(9):1550−7. https://doi.org/10.3201/eid1009.040058.
42. Riddell S, Goldie S, Hill, et al. The effect of temperature on persistence of SARS-CoV-2 on common surfaces. *Virol J* 2020;**17**:145. https://doi.org/10.1186/s12985-020-01418-7.
43. Doremalen N, Bushmaker T, Morris DH, et al. Aerosol and surface stability of SARS-CoV-2 as compared with SARS-CoV-1. *N Engl J Med* 2020;**382**:1564−7. https://doi.org/10.1056/NEJMc2004973.

44. Otter JA, Donskey C, Yezli S, Douthwaite S, Goldenberg SD, Weber DJ. Transmission of SARS and MERS coronaviruses and influenza virus in healthcare settings: the possible role of dry surface contamination. *J Hosp Infect* 2016;**92**(3):235−50. https://doi.org/10.1016/j.jhin.2015.08.027.
45. Wolff MH, Sattar SA, Adegbunrin O, Tetro J. Environmental survival and microbicide inactivation of coronaviruses. *Birkhauser Adv Infect Dis* 2005:201−12. https://doi.org/10.1007/3-7643-7339-3_10.
46. Yang W, Cai C, Dai X. The potential exposure and transmission risk of SARS-CoV-2 through sludge treatment and disposal. *Resour Conserv Recycl* 2020;**162**:105043. https://doi.org/10.1016/j.resconrec.2020.105043.
47. ISWA. *Waste management during the COVID-19 pandemic*. 2020. https://www.iswa.org/fileadmin/galleries/0001_COVID/ISWA_Waste_Management_During_COVID-19.pdf. [Accessed 1 July 2020].
48. Central Pollution Control Board. https://cpcb.nic.in/rules-3/, 2020. [Accessed 7 July 2020].
49. Health Care Without Harm. *Health care waste management: coronavirus update*. 2020. https://noharm-global.org/documents/health-care-waste-management-coronavirus-update. [Accessed 1 June 2020].
50. UNEP. *COVID-19 waste management factsheets*. 2020. https://www.unenvironment.org/resources/factsheet/covid-19-waste-management-factsheets. [Accessed 7 July 2020].
51. Wang J, Shen J, Ye D, et al. Disinfection technology of hospital wastes and wastewater: suggestions for disinfection strategy during coronavirus disease 2019 (COVID-19) pandemic in China. *Environ Pollut* 2020;**262**:114665. https://doi.org/10.1016/j.envpol.2020.114665.
52. Bao ZB, Jin DC, Teng HJ, Li Y. The general process of medical waste high temperature steam sterilization treatment technology. *Adv Mater* 2013;**807−809**:1160−3. doi:10.4028, www.scientific.net/amr.807-809.1160.
53. Ghodrat M, Rashidi M, Samali B. Life cycle assessments of incineration treatment for sharp medical waste. *Energy Technol* 2017:131−43. https://doi.org/10.1007/978-3-319-52192-3_14.
54. Wang J, Shen J, Ye D, Yan X, Zhang Y, Yang W, et al. Disinfection technology of hospital wastes and wastewater: suggestions for disinfection strategy during coronavirus disease 2019 (COVID-19) pandemic in China. *Environ Pollut* 2020:114665. https://doi.org/10.1016/j.envpol.2020.114665.
55. Messerle VE, Mosse AL, Ustimenko AB. Processing of biomedical waste in plasma gasifiers. *Waste Manag*. 2019;**79**:791−9. https://doi.org/10.1016/j.wasman.2018.08.048.
56. UNEP. *Waste management during the covid-19 pandemic from response to recovery*. 2020. https://www.unenvironment.org/ietc/resources/report/waste-management-during-covid-19-pandemic-response-recovery. [Accessed 12 December 2020].
57. Tran K, Cimon K, Severn M, Pessoa-Silva CL, Conly J. Aerosol generating procedures and risk of transmission of acute respiratory infections to healthcare workers: a systematic review. *PLoS One* 2012;**7**(4):e35797. https://doi.org/10.1371/journal.pone.0035797.
58. Ağalar C, Öztürk Engin D. Protective measures for COVID-19 for healthcare providers and laboratory personnel. *Turk J Med Sci* 2020;**50**(SI-1):578−84. https://doi.org/10.3906/sag-2004-132.
59. Malhotra N, Gupta N, Ish S, Ish P. COVID-19 in intensive care. Some necessary steps for health care workers. *Arch Chest Dis* 2020;**90**(1):20201284. https://doi.org/10.4081/monaldi.2020.1284.
60. Seah I, Agrawal R. Can the coronavirus disease 2019 (COVID-19) affect the eyes? A review of coronaviruses and ocular implications in humans and animals. *Ocul. Immunol. Inflamm.* 2020;**28**(3):391−5. https://doi.org/10.1080/09273948.2020.1738501.

61. European Centre for Disease Prevention and Control. *Novel coronavirus disease 2019 (COVID-19) pandemic: increased transmission in the EU/EEA and the UK—sixth update.* 2020. https://www.ecdc.europa.eu/sites/default/files/documents/RRA-sixth-update-Outbreak-of-novel-coronavirus-disease-2019-COVID-19.pdf. [Accessed 12 December 2020].
62. Sangkham S. Face mask and medical waste disposal during the novel COVID-19 pandemic in Asia. *Case Stud. Chem. Environ. Eng.* 2020;**2**:100052. https://doi.org/10.1016/j.cscee.2020.100052.

Further reading

1. CDC. *Procedures for safe handling and management of ebola-associated waste.* 2018. https://www.cdc.gov/vhf/ebola/clinicians/cleaning/handling-waste.html. [Accessed 12 December 2020].
2. Mol MPG, Caldas S. Can the human coronavirus epidemic also spread through solid waste? *Waste Manag Res* 2020;**38**(5):485–6. https://doi.org/10.1177/0734242x20918312.

SECTION III

Environmental and health management aspects

CHAPTER 9

Management of environmental health to prevent an outbreak of COVID-19: a review

Milad Mousazadeh[1,2,†], Zohreh Naghdali[1,2], Neda Rahimian[3], Marjan Hashemi[4], Biswaranjan Paital[5], Zakaria Al-Qodah[6,†], Ahmad Mukhtar[7], Rama Rao Karri[8], Alaa El Din Mahmoud[9,10], Mika Sillanpää[11,12,13], Mohammad Hadi Dehghani[14,15], Mohammad Mahdi Emamjomeh[16]

[1]Student Research Committee, Qazvin University of Medical Sciences, Qazvin, Iran; [2]Department of Environmental Health Engineering, School of Health, Qazvin University of Medical Sciences, Qazvin, Iran; [3]Endocrine Research Center, Institute of Endocrinology and Metabolism, Iran University of Medical Sciences (IUMS), Tehran, Iran; [4]Environmental and Occupational Hazards Control Research Center, Shahid Beheshti University of Medical Sciences, Tehran, Iran; [5]Redox Regulation Laboratory, Odisha University of Agriculture and Technology, College of Basic Science and Humanities, Bhubaneswar, Odisha, India; [6]Al-Balqa Applied University, Faculty of Engineering Technology, Department of Chemical Engineering, Amman, Jordan; [7]Department of Chemical Engineering, Universiti Teknologi PETRONAS (UTP), Seri Iskandar, Perak, Malaysia; [8]Petroleum and Chemical Engineering, Faculty of Engineering, Universiti Teknologi Brunei (UTB), Gadong, Brunei Darussalam; [9]Environmental Sciences Department, Faculty of Science, Alexandria University Alexandria, Alexandria, Egypt; [10]Green Technology Group, Faculty of Science, Alexandria University, Alexandria, Egypt; [11]Chemistry Department, College of Science, King Saud University, Riyadh, Saudi Arabia; [12]Faculty of Science and Technology, School of Applied Physics, University Kebangsaan Malaysia, Bangi, Selangor, Malaysia; [13]Department of Biological and Chemical Engineering, Aarhus University, Nørrebrogade 44, Aarhus C, Denmark; [14]Department of Environmental Health Engineering, School of Public Health, Tehran University of Medical Sciences, Tehran, Iran; [15]Institute for Environmental Research, Center for Solid Waste Research, Tehran University of Medical Sciences, Tehran, Iran; [16]Social Determinants of Health Research Center, Research Institute for Prevention of Non-Communicable Diseases, Qazvin University of Medical Sciences, Qazvin, Iran

9.1 Introduction

Coronaviruses are an influential group of viruses that can infect animals and humans, primarily targeting the human respiratory system.[1,2] As zoonotic diseases or zoonoses, they can spread from animals to people.[3] Two previous zoonotic coronaviruses include the first outbreak started in Guangdong City, China, in 2003, in the form of severe acute respiratory syndrome coronavirus (SARS-CoV), transmitted to humans from civet cats, and Middle East respiratory syndrome coronavirus (MERS-CoV), reported in Jeddah, Saudi

[†] Co-first author, Milad Mousazadeh and Zakaria Al-Qodah contributed equally to this work.

Environmental Management of COVID-19
ISBN 978-0-323-85780-2
https://doi.org/10.1016/B978-0-323-85780-2.00007-X

Arabia, in 2012, transmitted from dromedary camels to humans.[1,3] In January 2020, another new coronavirus was discovered, one which had not been identified in humans until that time.[3,4] In late December 2019, an outbreak of pneumonia with an unknown etiology (cause), was reported in Wuhan City, Hubei Province, China.[5,6] This novel coronavirus provisionally termed 2019 novel coronavirus (2019-nCoV) had been isolated and designated as the causative agent of the outbreak by the Chinese Centers for Disease Control and Prevention (CCDC).[4] The World Health Organization (WHO) named the representing epidemic by 2019-nCoV as coronavirus disease (COVID-19) on February 11, 2020. The virus also has been renamed as severe acute respiratory syndrome coronavirus-2 (SARS-CoV-2) by the Coronavirus Study Group of the International Committee on Taxonomy of Viruses.[4,7,8] SARS-CoV-2 spread rapidly within and outside China, the first case external to China identified on January 13, 2020, in Thailand.[5] By January 2020, a cumulative total of 7818 confirmed cases in China and 18 other countries were reported by WHO; this significant increase in the number of confirmed cases making WHO announce the incident as a Public Health Emergency of International Concern (PHEIC) on January 30, 2020.[9,10] As of March 11, 2020, there were more than 118,000 cases in 114 countries, and 4291 people had lost their lives. WHO, therefore, announced COVID-19 as a pandemic. This virus is still spreading worldwide with mutable prevalence and mortality outbreaks. All basic reproduction numbers (R_0) estimated in different studies are larger than 1, which indicates that COVID-19 is still spreading.[1,8,11–13] As of December 22, 2020, a total number of 77,740,380 cases have been reported, of which 1,709,474 and 21,411,220 have resulted in deaths and active cases, respectively (see Fig. 9.1).[14]

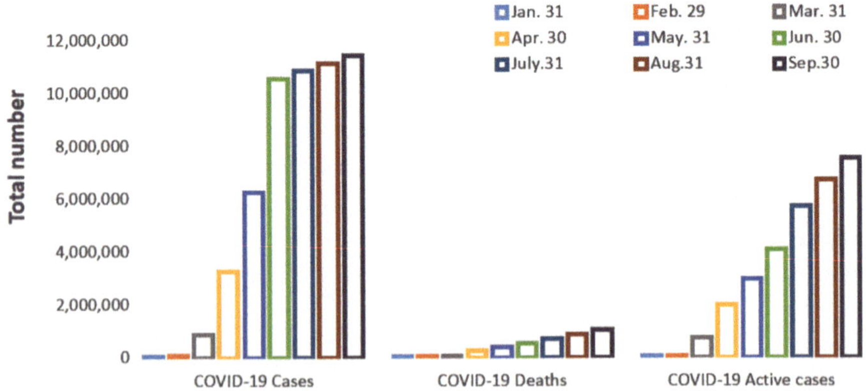

Figure 9.1 The total number of global cases, deaths, and recoveries as of December 22, 2020. *(Modified after Worldometers.[14] https://www.worldometers.info/coronavirus/.)*

The comparative analysis of SARS, COVID-19, and influenza has been tabulated in Table 9.1, which indicates that the fatality ratios of the SARS, COVID-19, and influenza rise with the age of patients. Furthermore, it can be noticed from the listed results in Table 9.1 that the highest fatality ratio, among patients with ages less than 80 years, was caused by SARS, followed by COVID-19 and influenza. The overall fatality ratio for SARS, COVID-19, and influenza was found to be 14%–15%, 1.38%, and 0.0962%, respectively. Epidemiological information has reinforced the evidence on COVID-19 transmission from one individual to another through direct, indirect (infected surfaces and objects), and close contact with infected people by respiratory secretions. There are also some studies investigating airborne transmission, these suggesting the potential transportation of the virus when expelled from infected individuals by aerosol processes in their local environment. However, because not all the necessary data is at hand, the role of airborne transmission remains unclear.[8,15–19] Human-to-human transmission occurs when droplets originating in a COVID-19 case reach the nose, mouth, or eyes of another person through

Table 9.1 Comparison of the mortality rate of coronavirus, influenza, and SARS (*CrI*, credible interval; *SARS*, severe acute respiratory syndrome).[22–25]

	SARS	COVID-19 (95% CrI)	Influenza
Overall	14%–15%	1.38% (1.23–1.53)	0.0962%
Age, years			
0–4	0.0%	0.00260% (0.000312–0.0382)	0.0073%
5–9			0.0028%
10–14		0.0148% (0.00288–0.759)	
15–17	0.5%		
18–19			0.0206%
20–24		0.0600% (0.0317–0.132)	
25–29	1.6%		
30–34		0.146% (0.103–0.255)	
35–39	10.0%		
40–44		0.295% (0.221–0.422)	
45–49	13.0%		
50–54		1.25% (1.03–1.55)	0.0614%
55–59	25.3%		
60–64		3.99% (3.41–4.55)	
65–69	52.5%		0.8315%
70–74		8.61% (7.48–9.99)	
75–79	68.9%		
≥		13.4% (11.2–15.9)	

coughing or exhalation. Furthermore, being that droplets are too heavy to be airborne, they are dropped on surrounding objects and surfaces, and other people may become infected touching these contaminated objects or surfaces.[17,20,21,26]

COVID-19 vaccines production is being developed using sophisticated techniques with high sensitivity and accuracy on a global scale so that all human beings can benefit from it.[27] Even with the development of the COVID-19 vaccines, there will still be challenges and hesitancy in manufacturing, distribution, efficacy, and immunity duration.[28] There is still a long way to go before widespread vaccinations around the world, the primary means of reducing infection is to avoid direct exposure to it, interrupting its chain of transmission.[27,29] This would be possible via a combination of public health measures and infection prevention and control (IPC) in health care settings[30] including hand and respiratory hygiene, rational and appropriate use of personal protective equipment (PPE), environmental cleaning and disinfection, and the safe management of waste resulting from the aforementioned procedures (see Fig. 9.2).[31,32] Following respiratory etiquette, cleaning and washing hands with soap or using alcohol-based hand sanitizers is essential. The appropriate use of PPE, including gloves, medical masks, goggles or a face shield, and gowns and aprons in health care and community settings, is required to reduce the spread of pathogens. In the case of masks and other PPE, appropriate

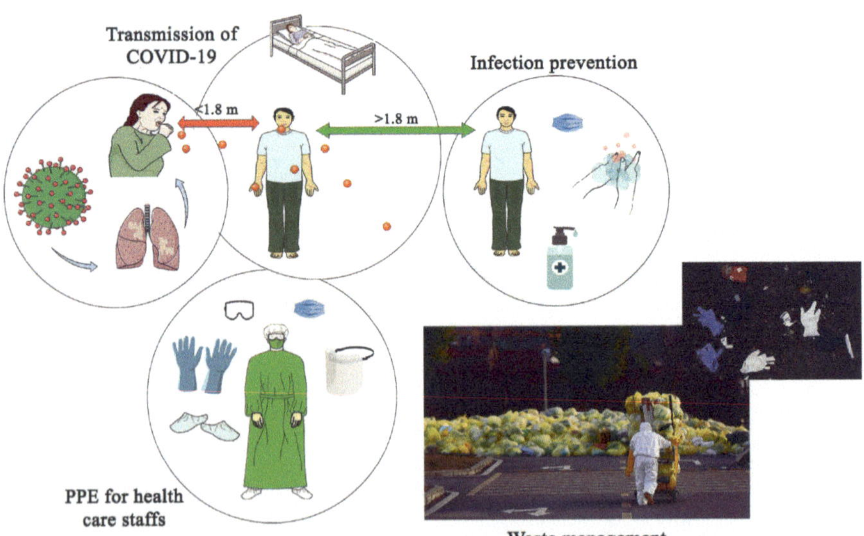

Figure 9.2 Environmental health management against COVID-19.

disposal should be considered to avoid any increase in virus spread and transmission.[32,33] These items are also receptive to typical cleaning and disinfection protocols including sodium hypochlorite, hydrogen peroxide, ethanol, and quaternary ammonium compounds, under the circumstance of using in accordance with the manufacturer's instructions. In addition to the choice of proper disinfectant, contact time, dilution, and expiry date of the prepared solution are also of great importance in the effectiveness of disinfection.[8,34,35]

Different types of medical and hazardous wastes, such as used personal protective equipment and laboratory waste, are being generated during this outbreak, along with a significant volume of noninfected items of the same nature, all of which have to be identified and segregated before decontamination and/or disposal. Sound and safe management of this biomedical and health care waste, in parallel with safe routine procedures, is essential to prevent the dispersal of biological agents and control the risk of infection. Such effective management of this waste includes disinfection procedures, personnel protection, and training.[32,35,36]

The COVID-19 pandemic is spreading daily, and the world is trying to control its increasing impact on both human health and the environment. The implementation of containment protocols and a rapid infection control response are of prime importance to contain and mitigate the risk of virus transmission. SARS-CoV-2 patients could contaminate the surrounding environment significantly, which means that the environment acts as a transmission medium, thus supporting the need for strict adherence to environmental health measures. It is therefore mandatory for health care organizations, public health authorities and also community settings, to have an environmental health management plan to eliminate natural health and environmental hazards due to this virus. In this chapter, environmental health practices have been critically reviewed in order to formulate an integrated framework, assessing the efficiency of current environmental health management processes and disinfection measures designed to contain and prevent further outbreaks.

9.2 Environmental survival and transmission of SARS-CoV-2

9.2.1 Persistence of SARS-CoV-2 in air

Based on one previous study published after the 2003 SARS epidemic, it was concluded that airborne transmission has an important role in the disease spreading.[37] Viruses such as SARS-CoV-2 can be transmitted if an infected person coughs talks and sneezes. The released droplets contain the virus, which infects others when they do not apply social distancing recommendations.[38] Moriyama et al.[39] reported that the viability of SARS-CoV-2 in the droplets increases at low humidity and temperature. Recently, it has been reported that SARS-CoV-2 may persist in aerosols for 180 min.[19] Accordingly, safety considerations should be undertaken in intensive care units (ICUs) and in any infected sector to prevent contamination spreading by aerosol. Another consideration is to inspect carefully and air conditioning systems. It is recommended that air conditioning systems should operate in under negative pressure in polluted areas.[40] Lui et al.[38] revealed that negative pressure ventilation and high air exchange levels within a coronary care unit (CCU) and ICU were effective in reducing airborne SARS-CoV-2. Also, several measures are recommended to reduce airborne contamination risk such as avoiding crowded communities, immediate tracing and diagnosis of asymptomatic bearers for quarantine or treatment, and the use of face masks in crowded places (e.g., public transportation).[38] Van Doremalen et al.[41] evaluated the stability of SARS-CoV-2 and found that the virus could remain in the aerosols up to 3 h. Furthermore, another study by Zhang et al.[19] reported SARS-CoV-2 in the aerosols of a hospital environment. In contrast with some studies, Lednicky et al.[42] detected viable SARS-CoV-2 in the air of a hospital room with COVID-19 patients.

9.2.2 Persistence of SARS-CoV-2 on the surfaces

It is now familiar that SARS-CoV-2 spreads mostly through contaminated surfaces and respiratory droplets. This virus usually persists on some surfaces for several hours to days. For example, Kampf et al.[34] reported that COVID-19 can persist for 9 days at room temperature. Others, such as veterinary coronaviruses can persist up to 28 days. These periods become shorter if the temperature increases to 30°C. Accordingly, the survival period depends on several variables including temperature, surface size, virus tension, and humidity.[34]

Van Doremalen et al.[19] recently studied the survival of the SARS-CoV-2 on different surfaces. They reported that its survival periods were 4, 8, and 72 h on copper, cardboard, plastic, and stainless steel surfaces, respectively. Riddell et al.[43] achieved the half-life of SARS-CoV-2 between 1.7 and 2.7 days at 20°C on surfaces, while dropped to a few hours when it reached 40°C.

Viable virus was also observed for up to 28 days at 20°C from common surfaces such as glass, stainless steel, and both paper and polymer banknotes. Moriarty[44] showed that traces of SARS-CoV-2 remained on these surfaces for even more extended periods. For this reason, WHO recommends that surfaces must be cleaned with water and disinfectant to ensure their disinfection with correct consistent cleaning. It should be noted that sodium hypochlorite is an effective disinfectant.[45]

9.2.3 Persistence of SARS-CoV-2 in water and wastewater

Although at the beginning of the epidemic, there were not any robust pieces of evidence for the presence of SARS-CoV-2 in water or wastewater, it could be speculated that it may exist based on previous research on SARS-CoV-1.[46] According to documented waterborne infection (such as adenoviruses, norovirus, rotaviruses, and hepatitis A), it is likely to become considerably more easily inactivated than the unenveloped human enteric virus. For example, in hospital wastewater and dechlorinated tap water at a temperature of 20°C, a replacement human coronavirus survived only 2 days.[47] A 99.9% die-off was shown in 2 days at 23°C,[48] this rising to 2 weeks at 25°C[49] in the case of human coronaviruses, transmissible gastroenteritis coronavirus, and mouse hepatitis. Temperature, high or low pH, and popular disinfectants, e.g., chlorine, facilitate die-off.

Currently, several researchers have studied the detection of SARS-CoV-2 virus in various water bodies including wastewater treatment plant,[50–52] municipal sewage,[53–55] hospital wastewater,[56] industrial wastewater,[57,58] secondary-treated wastewater,[59–61] and surface water (e.g., river).[59,62,63] To date, SARS-CoV-2 in wastewater samples have been detected using different methods in several cities around the world including Milan and Rome, Italy, in the period from February 24 to April 2, 2020[64]; Louisiana, United States, in the period from January 13 to April 29, 2020[65]; Massachusetts, United States, in the period from March 18 to March 25, 2020[66]; Querétaro, Mexico, in the period from April 23 to July 3, 2020[67]; Quito, Ecuador, on June 5, 2020[62]; Santiago, Chile, in the period from May to June, 2020[68]; Istanbul, Turkey, on May 7, 2020[50]; Tel Aviv, Israel, in the period from March 10 to April 21, 2020[69]; Ahmedabad, India, on May 8 and May 27, 2020[70]; Toyama and Ishikawa, Japan, in the period from March 5 to May 29, 2020[71]; Paris, France, in the period from March 5 to April 23, 2020[72]; Region of Murcia, Spain, in the period from March 12 to April 14, 2020[73]; North Rhine-Westphalia, Germany, in the period from April 8 to April 9, 2020[52]; Milan Metropolitan Area, Italy, in the period from April 14 to April 22, 2020.[63] Recommendations from the WHO and the Environmental Protection Agency (EPA) indicate that the risk of SARS-CoV-2 in drinking water is very low and encourage continued using drinking tap water. However, this concern is still unknown in countries with inadequate water supply system that are not equipped to eradicate viruses, so requires more careful investigation. Because SARS-CoV-2 virus cannot easily overwhelmed the advanced stages of water treatment, including filtration and disinfection.[74]

Wu et al.[66] collected samples from a wastewater treatment plant in Massachusetts in late March and found that the amount of SARS-CoV-2 particles in sewage samples was high. Medema et al.[58] reported that SARS-CoV-2 was found in six sewage samples taken from six different sites in the Netherlands in mid-March. In another study in Australia, Ahmed et al.[75] first confirmed the detection of SARS-CoV-2 in untreated wastewater in Australia. A group of French researchers found SARS-CoV-2 in Paris wastewaters. They found that a significant reduction in the number of genome units coincided with a decrease in the number of new COVID-19 cases, an expected consequence of lockdown. They claimed that the communal spread of COVID-19 could be estimated from sewage surveillance and thus help to predict the new waves of COVID-19 infections that many public health officials have warned about.[72] Based on previous experiences, tracking the pathogen through the sewage system is considered essential. Since Israel and Brazil detected wild poliovirus infection early by monitoring the sewage system, they were able to identify the most likely areas infected. Hence, they rapidly boost vaccination activity and prevent a more severe outbreak.[76,77]

9.3 Transmission of SARS-CoV-2

According to current knowledge, SARS-CoV-2 virus is mainly transmitted between individuals via respiratory droplets and direct contact.[78–82] When the droplets are more than 5–10 μm in diameter, they are known as respiratory droplets, and when they are less than 5 μm in diameter, they are known as droplet sections. Scientifically, the droplet sections can be called the "droplet nuclei."[83] SARS-CoV-2 virus transmission also occurs through direct contact with infected people, in direct contact with materials in the immediate environment, and the use of items on an infected person (e.g., thermometer).[16]

The propagation of airborne compounds is distinct from droplet emissions, related to the presence of bacteria in the droplet nuclei, usually <5 μm in diameter. These may persist for long periods in the environment and be passed on to others across distances of more than 1 m. Airborne delivery may occur with COVID-19 in particular cases, including conditions in which aerosol-generating care or support procedures are conducted. This includes endotracheal intubation, bronchoscopy, open suction, nebulized medication administration, manual ventilation, and patient handling. Reasonable evidence exists that COVID-19 infection can lead to and/or occur in the intestines. At the beginning of the COVID-19 epidemic, there was no assurance of fecal-oral transmission. In April 2020, a study conducted by Heller et al., who raised the fecal-oral hypothesis through the SARS-CoV-2 virus for further research.[84] Zhang et al.[85] extracted data from 14 patients with laboratory-confirmed COVID-19 through January 27, 2020 and observed that out of 62 stool samples, four tests were positive to COVID-19. However, they were not sure of the possibility of fecal-oral transmission, but found that fomite transmission was influential in the rapid spread of COVID-19. A meta-analysis study found that SARS-CoV-2 was detected in 43% of samples from feces or rectal swabs. More than half of the patients (63%) remained positive for 12.5 days even after respiratory PCR was negative. The data from this study suggest that fecal-oral transmission through the mouth is possible and can occur well after negative testing of patients using a respiratory test.[86] Another study by Zhang et al.,[87] which observed 258 stool samples from nine Chinese provinces. They found that a total of 93 of the 258 samples (36%) were positive for SARS-CoV-2 RNA. The presence of the live virus in feces can be an essential infection source, leading to infection and further spread to areas with poor sanitation.

9.3.1 Environmental factors influencing the transmission of SARS-CoV-2

To reduce unsafe conditions and avoid dissemination, environmental scientists, among others, are trying to identify the environmental variables that may affect the propagation of COVID-19. According to Stanford University studies, the actions of viruses and their response to specific conditions, vary greatly. Some viruses may be spread by water, whereas others are spread by air, as stated by Wigginton and Boehm.[88] Although the primary form of dissemination is through direct interaction with infected persons, it is important to consider the exact forms by which COVID-19 can be transmitted to mitigate its spread and avoid more infections. SARS-CoV-2 infections, the origin of which is primarily human transmission, are believed to be distributed by COVID-19. The host releases the virus into the atmosphere where it survives long enough on surfaces, soil, or water to reach another host. During this period, the viruses are enveloped. In addition to recognizing the function of indirect transmission in the current pandemic, scientists add that the length of life of viruses on various surfaces may be affected by specific environmental conditions. Knowledge of the impact of factors such as relative humidity (RH), fomite content, and air temperature will ensure adequate sanitation of surfaces and areas where the virus is less likely to propagate.[88]

RH and temperature are two important factors that can play a role in transmitting virus infection. In general, enveloping viruses have been shown to maintain infection at lower RH ($\leq$50%). At low RH conditions, aerosols evaporate rapidly, droplet size decreases, and droplet virus particles can travel longer distances.[89,90] A recent study showed that due to high humidity, the SARS-CoV-2 virus might be rapidly inactivated due to the interaction of water with the lipid envelope.[91] Furthermore, at low RH, the nasal airway in humans dries out more so that the mucosal layer is less effective in trapping pathogens.[92] Interestingly, Schuit et al.[93] revealed contrary results, which had no significant effect RH of 20%, 37%, 53%, and 70% on the survival of SARS-CoV-2 aerosol.

Other meteorological parameters (e.g., temperature) may also play a role in accelerating the transfer of COVID-19. A study by Bu et al.[94] reported that the transmission of SARS-CoV-2 was faster under hot (13–24°C) and dry climates of RH ~50%, and precipitation <30 mm/month. In the case of aerosol transmission, it is essential to determine the minimum infectious dose of the virus that must be present in severe airborne infection.[90] However, in another study by Gunthe et al.,[95] who analyzed data collected in 80 locations worldwide, they found that no significant relationship was reported between RH and precipitation with the COVID-19 cases. However, a favorable range of temperature and UV index strongly affected the spread and survival of the virus. Studies on the effect of temperature on COVID-19 cases were not in consensus with each other. For instance, the findings of Bannister-Tyrrell et al.[96] and Shi et al.[97] showed a possible reduction in the spread of SARS-CoV-2 with the onset of summer in the Northern Hemisphere. Luo et al. conducted preliminary studies in different countries in the Northern Hemisphere, which did not confirm a decrease in virus transmission with increasing temperature.

9.4 Waste management for COVID-19 and the impact of viral spread

9.4.1 Medical waste management

Wastes that contain infectious materials or potentially infectious materials, generated by health care facilities like hospitals, laboratories, physician's offices, dental practices, medical research facilities, and veterinary clinics are generally called medical wastes and can contain bodily fluids such as blood or other contaminants. The management and supervision of infectious biomedical waste are of paramount importance, usually following a predefined systematic approach. From generation time to final waste disposal, there are several procedures for reducing the risk of infection and improving general waste management. The best procedures for safe health care waste management are those which assign accountability and dispose of waste safely and adequately.[99,100]

9.4.2 Medical waste generated in COVID-19 outbreak

As a result of COVID-19, hospitals, health care facilities, and individuals are continuously producing more COVID-19-related waste than before. These wastes include masks, gloves, gowns, and other protective equipment that could be infected with this virus. Also, there is a significant increase in the use of disposable plastics.[101]

In epidemic outbreaks, the generation of infectious health care waste increases exponentially in a short time due to the drastic transmission of the disease at the first stage of the outbreak. As such, it should be collected and treated meticulously, or it may accelerate the spread of the disease, causing substantial risk for medical staff and patients. In epidemic outbreaks, it is, therefore, necessary to reduce the storage time of medical waste at the production source in order to decrease the risk of the infection spreading.[101] The so-called mean production factor for routine infectious waste was calculated before and during the outbreak of SARS. Over this time, this system saw significant increases in waste from 0.85 to 1.2 kg/patient/day before and during the SARS outbreak, respectively. It was reported that the average daily production of SARS wastes increased to 4.5 kg during the SARS outbreak; this is four times higher than the average daily waste production rate. Nevertheless, after 10 days, health care centers successfully decreased the average daily production of waste to the standard rate by developing and applying effective waste management plans.[102]

Since January 20, 2020, China has disposed of 142,000 tons of medical waste. The country's capacity to process medical waste has now risen from 4000 to 6000 tons per day. To prevent the spread of COVID-19, strict treatment protocols have to be imposed on the disposal of this waste. The use of best management techniques should be undertaken to minimize direct exposure to the waste. If health care waste is not collected and treated periodically in an appropriate process, the possibility of infectious spread increases significantly. For instance, coronaviruses persist on medical waste surfaces for 9 days, so to control the COVID-19 outbreak, medical waste needs collecting and treating periodically, using a safe and effective process to reduce virus spread and its risk to humans.[34,101]

9.4.3 Treatment processes for COVID-19 wastes

The CDC and WHO recommended that waste materials related to COVID-19 and obtained from health care facilities should be managed and performed following the handling procedures used for medical waste. These materials include sharps, cleaning cloths, wipes lab specimens, and single-use microfiber cloths. They advise to continue following the guidance from the CDC, WHO, and the Texas Department of State Health Services (DSHS) on best practice.[103] Approved methods for medical waste treatment are listed in DSHS rules and include incineration, chemical treatments, steam sterilization (autoclave), and shredding. According to these rules, on-site or off-site treatment facilities are first treated by a generator.[104] Medical wastes can be handled and disposed of as average municipal solid wastes after performing specific treatments.[103]

The minimization of all negative impacts on health and the environment, is the first and foremost criterion for the selection of treatment methods, although there is no universal approach for waste treatment. The optimal choice can only be made locally, this inevitably depending on a compromise with the treatment infrastructure available in the vicinity. In the absence of appropriate infrastructure, the responsibility remains with the hospital to treat, or pretreat, its wastes on site, this carrying the additional advantage of avoiding potential complications involved in the transport of hazardous substances. Infectious clinical waste generated from the COVID-19 pandemic does not necessitate incineration, since tested alternative treatment processes can also make the waste safe. The storage time for health care waste in hospitals is 48 h, but for COVID-19 patient waste, it is just 24 h.[105] According to the Biomedical Waste Rules (1998), biomedical waste should be stored for less than 24 h at ordinary room temperature.[106]

One of the measures taken to control medical waste is to treat contagious material by health care centers. Currently, there is no evidence for the need to treat hazardous waste sources owing to the prevalence of coronavirus in health facilities any differently to regular medical waste. For example, medical waste as a hazardous waste may be disposed of in a landfill, including disinfecting pads, personal protection equipment, and discarded foodstuffs. Domestic and medical waste that cannot be collected should be bagged and tagged according to a standard procedure before it is loaded into a garbage collector. These wastes can be deposited in hazardous waste sites which process urban industrial waste.[107] Contrary to popular belief, Dr. Anne Woolridge, Chair of the International Solid Waste Association (ISWA), stated waste is not the main route of transmission of SARS-CoV-2; on the other hand, the pathogen is transmitted by someone who coughs or sneezes to another person or on surfaces where virus particles can be removed and transmitted to the nose and mouth. Therefore, we must observe all health and safety issues.[108]

9.4.4 General principles and guidance for managing infectious waste during the COVID-19 pandemic

Many studies in many countries and international agencies have published some principles and specific guidelines to manage health care and household waste accompanied by the COVID-19 pandemic.[109]

9.4.4.1 Health care waste management

For health care waste management, the following principles should be followed:

1. Wastes resulted from COVID-19 should be managed using suitable protocols for infectious and health care waste management, as found in international and national guidelines and manuals.
2. The urgent use of enhanced treatment and disposal capacities of medical waste in epidemic areas.
3. The gaps in capacity for medical waste treatment and disposal in different locations, in urban versus rural areas, should be identified.
4. Alternatives for interim and emergency treatment and disposal should be identified. Focus for local constraints, COVID-19 infection concerns, and environmental and human health impacts accompanied by improper disposal of wastes.

5. An adequate supply of waste collection bins, bags, and transportation trollies should be ensured. Also, capacity building of the staff dealing with COVID wastes should be improved.

9.4.4.2 Household waste management

For health care waste management, the following principles should be followed:

1. Management services should be continuous.
2. Securing of human resources, equipment, and reducing workers risks in governmental, private companies, and the informal sector is very critical.
3. Apply some changes in waste management operations.
4. Waste generated from households with COVID-19-positive occupants or suspected cases should be collected separately.
5. Use separation and recycling protocols if needed.
6. Awareness among waste management service providers should be improved and strengthened.
7. Keep an adequate supply of waste collection bins, bags, and transportation trollies.

Based on the continuously increasing number of infected people worldwide, COVID-19 has proven to be an unprecedented disaster. It has a destroying effect on human health, social contacts and relations, and the national and international economy. This disaster may spread through medical treatment wastes if managed inappropriately.[110] Accordingly, the following conclusions and recommendations are presented:

1. Improvements in waste collection methods should be developed to minimize human contact. This can be achieved by employing automatic collection systems, including pneumatic machines. Also, biodegradable storage and sealing containers should be used to collect airtight degradable garbage bags.[111]
2. The use of mobile waste incineration stations at the collection sites will minimize the risk of spreading infectious substances found in municipal solid waste (MSW).[112]
3. The treatment of these wastes become sustainable if ashes produced from waste and sludge incineration used as additives to cement used in construction or as adsorbents.[113–115]
4. The currently used sterilization techniques have the drawback of damaging the materials upon the sterilization process. This imposes the need for effective alternatives including the use of supercritical fluids including carbon dioxide ($scCO_2$), which was found to be effective for sensitive materials sterilization.[116,117]

5. It is crucial to understand the bioaerosol generation and transport from open dumps. These open dumps usually become hotspots for hazardous bioaerosol that include viruses, fungi spores, parasite eggs, bacteria, and cysts of protozoa.[118] Generation and releasing of bioaerosols are facilitated by various activities used in waste management starting from loading and unloading to transportation then sorting, onsite, and landfill treatment.
6. It is familiar that pathogens found in the waste can survive for some days in the case of bacteria, to several months in the case of viruses or even years in the case of helminthic eggs.[119] Accordingly, more research studies are required to investigate the viability of viruses in both aerobic and anaerobic conditions, which could be found in landfills, digesters, and composting sites.
7. Finally, the preparedness and response to the pandemic in the waste management sector should consider and assess many issues such as[110]
   a. the levels of disinfectant and chemical concentration in MSW during the COVID-19 pandemic in the last year;
   b. the effect of the disinfectants on the microorganism's activities during MSW degradation, as well as trihalomethane generation;
   c. The characteristics of the volatile organic compounds (VOCs) usually generated during the heat treatment of solid wastes;
   d. recycling some disinfected/sterilized nonbiodegradable wastes such as gloves, masks, and protective equipment to produce some composite construction materials; and
   e. investigating the mechanisms of virus transport and factors affecting its activation in landfill wall material in addition to soil and groundwater.

9.5 Protection and disinfection policies against COVID-19

In the absence of a safe and effective vaccine for COVID-19, there are public health measures that must be taken to prevent the spread of the virus by droplets, close contact, and contaminated surfaces.[120] Common ways the SARS-CoV-2 virus spreads are through transmission from person to person and touching dry surfaces contaminated with nasal, oral, and ocular secretions.[80,121] Considering the rigor of this serious epidemic and the significance of prevention and protection against the outbreak of COVID-19, the following important prevention strategies are reviewed.

9.5.1 Prevention of the spread of COVID-19

Lifestyle, for example, quality of sleep, eating a healthy diet, and exercising, while adhering carefully to personal hygiene, plays the main role in controlling and stopping the spread of COVID-19.[122,123] According to the recommendations issued by the CDC, the main strategies to prevent the spread of the virus include the following:

- Avoiding direct or indirect contact with infected people.
- Using a tissue to prevent the spread of coughing or sneezing drops and throwing contaminated tissue away in contained trash.
- Do not travel to high-risk areas of COVID-19.
- Washing hands with soap and water for at least 20 s.
- Using disinfectants contains at least 60% alcohol if the soap is not available.
- Avoiding contact of the face with contaminated hands.
- Frequent cleaning and disinfecting of contaminated touched surfaces.[124]

9.5.2 Function of face masks in preventing the spread of COVID-19

The largest and smallest sizes of coronavirus-shaped spherical are 0.14 microns and 0.06 microns, respectively, observed using an electron microscope.[125] Physicians and health care professionals commonly use surgical masks to prevent the spread of large droplets and spray on vulnerable patients.[126] The protection of these masks against the SARS-CoV-2 virus, although widely used by the public, is not complete.[127] In health centers, N95 masks filter out at least 95% of microscopic particles (0.3 microns), and should be used by health care workers who are in direct contact with patients.[128] On the other hand, there are concerns that 8 h of continuous or intermittent use of this mask will reduce their performance due to contamination or practical considerations.[129] In contrast, filtering facepiece 2 (FFP2) filters out at least 94% of particles approximately 0.3 microns in size. Both of FFP2 and N95 filters are most effective for protection against microscopic particles.[130,131]

The controversy regarding the use of face masks by healthy people to minimize the risks of respiratory viruses transmission has been addressed in the work of Cheng et al.,[132] where a comparison between mask-wearing countries and nonmask-wearing countries was carried out. The authors concluded that community-wide mask wearing could significantly minimize the spread of COVID-19 among people by minimizing the probability of inhaling infected saliva and respiratory droplets from patients. Since several countries are facing a severe shortage of face masks and timely hand washing can be difficult, this makes it essential to assess the effectiveness of masks. Ma et al.[133] conducted a study to assess the efficiency of face masks and hand wipes. In this study, three different types of masks and wipes were used; the avian influenza virus was used as a coronavirus model. According to this study's findings, 98.36%, 96.62%, and 99.98% of the virus is removed using 1%, 0.05%, and 0.25% of soap, active chlorine solution, and active chlorine (from sodium hypochlorite) solutions, respectively.

Additionally, the authors demonstrated that N95 medical and homemade masks could efficiently block 99.98%, 97.14%, and 95.15% of the virus, respectively. It is noteworthy to mention that the homemade masks consisted of four kitchen paper layers and a layer of cloth. Based on these findings, the authors demonstrated that maintaining mask-wearing and instant hand hygiene (MIH) can effectively slow down, or even inhibit the spread of the virus. As the authors pointed out, this MIH approach has been shown to be useful across seven countries in the fight against COVID 19.

Despite the consensus about using masks by symptomatic patient and health care center staffs, there is a lack of consensus about the general population.[134] Based on the health recommendations in Germany and the UK, supporting evidences about the adequate protection from mask wearing in public is extremely limited. However, wearing masks as a precautionary act during the infected patient care is an agenda for health care workers[135] as WHO recommends too.[136] On the other hand, susceptible individuals ought to avoid attending crowded place and use surgical masks in high-risk conditions.[137] However, According to recent findings by researchers that the virus survives in the air for 3 h,[37] it is necessary to discuss the general public's need to use face masks.

9.5.3 Hand hygiene

In the absence of a vaccine and treatment for COVID-19, people have to socially engineer their behavior to act like a social vaccine. This simply means two things: to not become infected and not infect others. According to the CDC, one of the most effective methods to inhibit the spread of COVID-19 is maintaining hands hygiene using soap[138] as this simple act effectively eliminates viruses. Back in 1975, Ayliffe et al.[139] devised an effective way of hand washing, Graham and Taylor then promoting the Ayliffe technique as the six steps of hand washing.[140] The technique was soon adopted in all UK hospitals and approved by WHO in 2009.[141] The entire procedure is as follows:

> *Wet hands with water and apply enough liquid soap, rub hands palm to palm, right palm over left dorsum with fingers interlaced and vice versa, palm to palm with fingers interlaced, backs of fingers to opposing palm with fingers interlocked, rotational rubbing of right thumb clasped over left palm and vice versa, rotational rubbing backward and forward with clasped fingers of the right hand in left palm and vice versa; hands and wrists were rubbed until the end of the 30 s period, then rinsed under a running tap and dried thoroughly with a single-use towel.*[141]

The use of any soap completely annihilates viruses like SARS-CoV-2, which is encased in what is basically a layer of fats and proteins or a lipid envelope and which will adhere to surfaces like hands easily. If hands are only rinsed with water, the virus will stay in situ as it is quite sticky. Soaps contain amphiphiles, which are similar in composition to the lipids in the envelope of the virus. Amphiphiles are two-sided molecules; one side is attracted to water molecules, and the other side is attracted to fat, which is called hydrophilic and lipophilic, respectively (see Fig. 9.3).[142] When soap molecules are exposed to water and fat, these dual attractions are led to

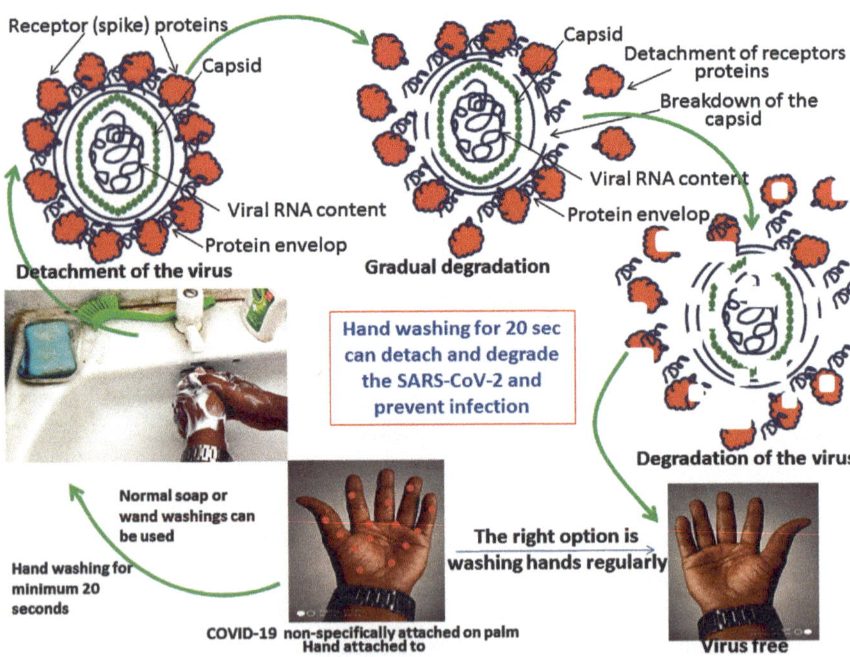

Figure 9.3 Hand washing and virus destruction action.

disperse them in the water. As mentioned earlier, SARS-CoV-2 molecules have a layer of fat holding everything together. Soap literally pulls this fat apart and demolishes the virus. When water is then used to rinse hands, harmless leftover shards of the virus are washed down the drain. However, this takes precisely 20 s for this to happen. If hands are washed with soap for less time, they still contain a virus that causes the person to become ill.[143] It should be noted that it is essential not only to wash hands in public settings but also to make sure that hand washing is done at home. It is also not recommended to use soap bars in public facilities; they are only used in the home and should be replaced with liquid soap. That said, and in contrast to this recommendation, previous published reports[144,145] have shown little risk exists in routine hand washing with bars of soap.

The Food and Drug Administration (FDA) says there is no need to use soap marketed as antibacterial because there is no proof it is any more effective.[146] It is merely essential to wash hands for 20 s using the ultimate virus annihilator: soap. Regular hand washing should, however, be done before and after eating, and after coughing, sneezing, touching one's face, pets, and surfaces. Hands should also be washed after using shared equipment/tools (e.g., computer keyboards), toilets, and shared facilities.

9.5.4 Hand sanitizers

Slowing down the transmission of respiratory viruses through appropriate hand sanitization was possibly the first example of proven medicine,[147] introduced for the first time in 1847 by the Hungarian physician Semmelweis Ignac Fulop. Although isolation, social distancing, and personal protection procedures are not practical for travelers and critical workers (people in direct contact with people and patients), they are classified as effective measures to interrupt the chain of infection. In these circumstances, it is highly recommended to use hand sanitizers such as alcohol-based solutions, to interrupt the infection chain.

Generally, hand sanitizers are classified into two main types: alcohol-based and alcohol-free. Alcohol-free types, less efficient than alcohol-based products, depend on benzalkonium chlorides to remove viruses and other microorganisms, while alcohol-based types rely on alcohol to remove viruses. In general, the concentration of alcohol (mostly in the form of ethyl alcohol), in alcohol-based sanitizers, ranges between 60% and 95% (60% is the minimum required concentration to remove COVID-19 virus). It must be kept in mind that these products do not eradicate viruses through the friction that rubbing hands together and that rinsing provides. That said, they have been found to be effective against many types of bacteria and viruses[148–151] as well as SARS-CoV-2.[152] To break down that fatty layer, there needs to be a high concentration of alcohol, and this can cause skin irritation. Sweaty and dirty hands, however, can dilute and diminish the effectiveness of a sanitizer and may not remove harmful chemicals. For a hand sanitizer to be effective, the hands should be fully covered, rubbing it in for at least 20 s, as is the case when washing with soap and water.

It should be noted that there are some dangers associated with the use of hand sanitizers, especially for children, due to improper use or overuse. The first danger is using hand sanitizer ineffectively on dirty or greasy hands. Infected hands have a layer of grime on which it is difficult to penetrate. The second danger is that when hands are contaminated by chemicals (e.g., pesticides), hand sanitizers cannot effectively deactivate harmful chemicals to prevent them from entering the body, and in this case, should never be used in these situations instead of soap and water. The last risk is that swallowing hand sanitizers can cause alcohol poisoning. Reports of alcohol poisoning is becoming more common in the COVID-19 period.[153–155] In a study, Jing et al.[156] highlighted the side effects of overuse of hand sanitizers on the epidermis, leading to contact dermatitis and worsening of skin lipid barriers. Moreover, Emami et al.[157] concluded that the problematic effects of nonstandard formulas for alcohol-based hand sanitizers are effective in causing skin dryness and irritation. As a result, excessive use of surfactants and soaps can damage the epidermis by negatively affecting lipids, proteins, and creatine compounds.

9.5.5 Disinfecting surfaces

Disinfection of surfaces, especially high contact surfaces (e.g., light switches, phones, doorknobs, and remote controls), is one of the most important measures. Using conventional surface disinfection methods with 62%–71% ethanol, 0.5% hydrogen peroxide, or 0.1% sodium hypochlorite, contaminated surfaces can be quickly disinfected within a minute.[34,99] It is worth noting that the effectiveness of all surface disinfectants depends on their concentration and time of exposure. Hence, it is essential to remember that dirty surfaces should be cleaned with detergent and water to remove dust and germs before disinfection.[158] Of course, it is better to do this in any case so that disinfection is more effective. It is just as important that the disinfectant remains on surfaces to facilitate its effectiveness.

It should be noted that alcohol-based disinfectants (e.g., ethanol) at a concentration of 70%–80% with 1 min of contact time, reduce the infectivity of enveloped viruses, e.g., SARS-CoV-2.[34,159] According to Kampf et al.[34] although antiseptics work well against bacteria, but they do not have much of an effect regarding the reduction in the spread of coronavirus (e.g., SARS-CoV-2). This may be because antiseptics destroy fats in pathogen cells, while SARS-CoV-2 does not contain many fats.

The EPA has recently published a list of disinfectants[160] as per Pradhan et al.,[161] which could be effectively used against coronavirus including hydrogen peroxides, sodium chlorite, ammonia, lactic acid, bleach and soaps.[160] If using bleach, follow the CDC instructions by diluting five tablespoons (1/3rd cup) of bleach per gallon of water or teaspoons of bleach per quart (approximately 1 L) of water. It should be prepared daily, or as needed for consumption, never mixing household bleach with ammonia or any other cleanser. The bleach should remain on surfaces for 10–15 min to work by destroying the protein and the ribonucleic acid (RNA) in the virus.

9.5.6 Disinfectant as a threat to aquatic ecosystems

It is an undeniable fact that substantial amounts of disinfectants have been used to contain the spread of COVID-19. For example, approximately 2 million kg of disinfectants were used by Wuhan City during the pandemic period, the majority of these chlorine-based. Unfortunately, these disinfectants pose a serious threat to the environment, especially to sources of freshwater. Hundreds of free-living animals belonging to 17 different species (e.g., *Turdus merula*) were found to have died owing to abuse of disinfectants in Chongqing, China.[162] These disinfectants get into the infected area's sewage system through runoff and domestic effluents, which in turn, end up in freshwater bodies.[163] Due to the irritating and corrosive effects of disinfectants (e.g., chlorine) on the mucous membranes of the respiratory and gastrointestinal tracts,[164] excessive use of large amounts of these chemicals in urban environments could directly or indirectly pose a significant threat to urban wildlife.[162] Chlorine-based disinfectants impact aquatic life by releasing chlorine that destroys living cells' walls and oxidizes proteins.[163] Chlorine can also react with other chemicals in the aquatic environment, forming harmful compounds such as disinfection byproducts. For instance, chlorine can react with dissolved organic matter, something which widely occurs in freshwater bodies,[165] forming hazardous disinfection byproducts such as trihalomethanes.[163,166] Chlorine also reacts with nitrogen in aquatic environments, forming carcinogenic agents such as chloramine or N-nitrosodimethylamine.[167,168]

As COVID-19 spreads worldwide, the consumption of disinfectants will increase, this having the potential to result in a global environmental disaster. As such, aquatic ecological integrity assessments must be conducted, during and after the pandemic, on a global scale, to control or minimize the environmental impacts of such disinfectants. Hence, disinfectants used during the COVID-19 era to control the virus in urban environments must be selected and used in ways that prevent unnecessary environmental contamination. As a result, the use of disinfectants in areas rich in diversity (e.g., urban parks) should be limited. Instead, human activities in these areas should be suspended, as well as low-risk or nontoxic but effective disinfectants should be used extensively in suburban areas to overcome COVID-19 with minimal environmental side effects.[162]

9.6 Concluding remarks and recommendations

This chapter highlights a multidisciplinary perspective of environmental health management including the environmental transmission and survival of the SARS-CoV-2 virus (e.g., air, surfaces, water, and wastewater), the factors influencing the transmission of virus under different environmental conditions (e.g., temperature and humidity), waste management and its treatment processes, disinfection and protection measures such as the function of masks, hand washing, hand sanitizer, and surface disinfections. From the analysis of the large number of recent papers and reports published in the last year, the following conclusions and recommendations can be summarized:

- There are still many unknowns about the pandemic caused by COVID-19 and the virus itself, including its effects, modes of transmission and genetic mutation to other new forms such as COVID-20, which has recently appeared in Dec 2020 in UK.
- Since the pandemic dramatically changes and covers the entire world, it is very essential to have a continuously updated and integrated information and research results. These results will be of great benefits to all governmental entities that follow up and take care of the pandemic including in the medical sectors.
- It is clear from the COVID-19 pandemic that nothing can be accurately predicted with 100% certainty. For this reason, taking care of the

people's immune systems through the care of personal lifestyles (sleeping, healthy food, and exercise) and health, this could protect people from COVID-19.

- More precautions and preventive measures should be followed to mitigate this pandemic. These preventive procedures, as recommended by WHO and CDC must be applied to fight COVID-19. Facial masks should be worn by workers in the workplace and medical staff when taking care of COVID-19 patients. The essential precautions highlighted in this chapter should be implemented to minimize the transmission of COVID-19.
- Valuable studies have been conducted on tracking SARS-CoV-2 in various water/wastewater around the world, while many countries (e.g., Iran) due to lack of facilities and equipment in this field, could not monitor wastewater-based epidemiology. Because it is one of the low-cost ways to detect and control COVID-19 by preventing the occurrence of subsequent peaks, this issue requires special attention in poor and developing countries.
- Because the present studies are not able to confirm the effect of climate conditions on the spread of COVID-19, more studies are needed to investigate the effect of temperature, humidity, and other factors on it.
- The medical waste materials related to COVID-19 obtained from health care facilities could contain infectious materials. For this reason, their management is of paramount importance since they are classified as dangerous medical wastes. They should be managed and performed following the handling procedures used for medical waste. However, more researches should be investigated on municipal and medical waste management during COVID-19 outbreak and its short- and long-term effects.
- It is clear that disinfectants used to protect people from COVID-19 and to control in urban environments should be of high activity against the virus. These disinfectants must be selected and used in ways that prevent unnecessary environmental contamination. There are very few studies on the effects of overuse of disinfectants against COVID-19 in the short and long term on humans and the environment (e.g., ecosystems of rivers).
- The use of disinfectants in areas rich in diversity such as urban parks should be limited. Instead, human activities in these areas should be suspended, as well as low-risk or nontoxic but effective disinfectants should

be used extensively in suburban areas to overcome COVID-19 with minimal environmental side effects.

- The dissemination of public awareness information related to this epidemic and environment/health is also still needed, including following and keeping up to date with information from WHO and CDC about COVID-19. The key to reducing the spread of COVID-19 is collaboration between health care sectors, authorities, and governments to detect suspected cases everywhere and improve patient care and safety.
- The intensive use of vaccines all over the world gives great hope of stopping the spread of the virus. However, these vaccines should free of charge since most people are not able to buy it.
- The first appearance of a new version of this virus, called COVID-20, derived from the UK and rapidly spread to many countries, puts a heavy burden on researchers to accelerate their intensive studies to find a global vaccine that can attach to all COVID viruses.

Acknowledgments

This research was financially supported by the Research Vice-Chancellor of Qazvin University of Medical Sciences (IR.QUMS.REC.1399.011). Milad Mousazadeh gratitudes the Iran's National Elites Foundation (INEF) for granting the Elite Award, No. 15/11529. The authors would like to thank Dr. Khalid S. Hashim and Dr. Iasonas Hitzazis for assistance in data collection of this chapter.

References

1. Rothan HA, Byrareddy SN. The epidemiology and pathogenesis of coronavirus disease (COVID-19) outbreak. *J Autoimmun* 2020;**109**:102433.
2. Wu D, Wu T, Liu Q, Yang Z. The SARS-CoV-2 outbreak: what we know. *Int J Infect Dis* 2020;**94**:44–8.
3. WHO. Coronavirus. Available from: https://www.who.int/health-topics/coronavirus#tab=tab_1. [Accessed 14 December 2020].
4. Zumla A, Niederman MS. The explosive epidemic outbreak of novel coronavirus disease 2019 (COVID-19) and the persistent threat of respiratory tract infectious diseases to global health security. *Curr Opin Pulm Med* 2020;**26**(3):193–6. https://doi.org/10.1097/MCP.0000000000000676.
5. Boldog P, Tekeli T, Vizi Z, Dénes A, Bartha FA, Röst G. Risk assessment of novel coronavirus COVID-19 outbreaks outside China. *J Clin Med* 2020;**9**(2):571.
6. Mousazadeh M, Paital B, Naghdali Z, Mortezania Z, Hashemi M, Niaragh EK, Aghababaei M, Ghorbankhani M, Lichtfouse E, Sillanpää M, Hashim KS. Positive environmental effects of the coronavirus 2020 episode: a review. *Environ Dev Sustain* 2021:1–23. https://doi.org/10.1007/s10668-021-01240-3.
7. Lai C-C, Liu YH, Wang C-Y, et al. Asymptomatic carrier state, acute respiratory disease, and pneumonia due to severe acute respiratory syndrome coronavirus 2 (SARSCoV-2): facts and myths. *J Microbiol Immunol Infect* 2020;**53**(3):404–12.

8. Lai C-C, Shih T-P, Ko W-C, Tang H-J, Hsueh P-R. Severe acute respiratory syndrome coronavirus 2 (SARS-CoV-2) and corona virus disease-2019 (COVID-19): the epidemic and the challenges. *Int J Antimicrob Agents* 2020;**55**(3):105924.
9. WHO. Available from: https://www.who.int/docs/defaultsource/coronaviruse/situation-reports/20200329-sitrep-69-covid-19.pdf?sfvrsn=8d6620fa_2. [Accessed 14 December 2020].
10. Wang H, Wang Z, Dong Y, et al. Phase-adjusted estimation of the number of coronavirus disease 2019 cases in Wuhan, China. *Cell Discov* 2020;**6**(1):1−8.
11. Hamid S, Mir MY, Rohela GK. Noval coronavirus disease (COVID-19): a pandemic (epidemiology, pathogenesis and potential therapeutics). *New Microbes New Infect* 2020;**35**:100679.
12. Li Q, Guan X, Wu P, et al. Early transmission dynamics in Wuhan, China, of novel coronavirus−infected pneumonia. *N Engl J Med* 2020;**382**:1199−207.
13. Zhai P, Ding Y, Wu X, Long J, Zhong Y, Li Y. The epidemiology, diagnosis and treatment of COVID-19. *Int J Antimicrob Agents* 2020;**55**(5):105955.
14. Worldometers. Available from: https://www.worldometers.info/coronavirus/. [Accessed 14 December 2020].
15. Morawska L, Cao J. Airborne transmission of SARS-CoV-2: the world should face the reality. *Environ Int* 2020;**139**:105730.
16. Ong SWX, Tan YK, Chia PY, et al. Air, surface environmental, and personal protective equipment contamination by severe acute respiratory syndrome coronavirus 2 (SARS-CoV-2) from a symptomatic patient. *JAMA* 2020;**323**(16):1610−2.
17. WHO. Available from: https://www.who.int/docs/default-source/coronaviruse/situation-reports/20200326-sitrep-66-covid-19.pdf?sfvrsn=9e5b8b48_2. [Accessed 14 December 2020].
18. Santarpia JL, Rivera DN, Herrera V, et al. Transmission potential of SARS-CoV-2 in viral shedding observed at the University of Nebraska Medical Center. *medRxiv* 2020. https://doi.org/10.1101/2020.03.23.20039446.
19. Zhang R, Li Y, Zhang AL, Wang Y, Molina MJ. Identifying airborne transmission as the dominant route for the spread of COVID-19. *Proc Natl Acad Sci Unit States Am* 2020;**117**(26):14857−63.
20. Chang L, Yan Y, Wang L. Coronavirus disease 2019: coronaviruses and blood safety. *Transfus Med Rev* 2020;**34**(2):75−80.
21. Ryalino C. Covid-19: what we know so far. *Bali J Anesthesiol* 2020;**4**(1):1.
22. Ruan S. Likelihood of survival of coronavirus disease 2019. *Lancet Infect Dis* 2020;**20**(6):630−1.
23. WHO. Update 49 -SARS case fatality ratio, incubation period. Available from: https://www.who.int/csr/sars/archive/2003_05_07a/en/. [Accessed 14 December 2020].
24. Chan-Yeung M, Xu R. SARS: epidemiology. *Respirology* 2003;**8**(1):9−14.
25. Verity R, Okell LC, Dorigatti I, et al. Estimates of the severity of coronavirus disease 2019: a model-based analysis. *Lancet Infect Dis* 2020;**20**(6):669−77.
26. Wu Y-C, Chen C-S, Chan Y-J. The outbreak of COVID-19: an overview. *J Chin Med Assoc* 2020;**83**(3):217−20.
27. CDC. Interim Clinical Guidance for Management of Patients with Confirmed Coronavirus Disease (COVID-19). Available from: https://www.cdc.gov/coronavirus/2019-ncov/hcp/clinical-guidance-management-patients.html. [Accessed 14 December 2020].
28. Abila DB, Dei-Tumi SD, Humura F, Aja GN. We need to start thinking about promoting the demand, uptake, and equitable distribution of COVID-19 vaccines NOW! *Public Health Pract* 2020;**1**:100063.

29. WHO. Coronavirus disease 2019 (COVID-19) Situation Report – 42. Available from: https://www.who.int/docs/default-source/coronaviruse/situation-reports/20200302-sitrep-42-covid-19.pdf?sfvrsn=224c1add_2. [Accessed 14 December 2020].
30. WHO. Coronavirus disease 2019 (COVID-19) Situation Report – 52. Available from: https://www.who.int/docs/default-source/coronaviruse/situation-reports/20200312-sitrep-52-covid-19.pdf?sfvrsn=e2bfc9c0_4. [Accessed 14 December 2020].
31. WHO. Rational use of personal protective equipment for coronavirus disease 2019 (COVID-19). Available from: https://apps.who.int/iris/bitstream/handle/10665/331215/WHO-2019-nCov-IPCPPE_use-2020.1-eng.pdf?sequence=1&isAllowed=y. [Accessed 14 December 2020].
32. WHO. Infection prevention and control during health care when COVID-19 is suspected. Available from: https://apps.who.int/iris/rest/bitstreams/1272420/retrieve. [Accessed 14 December 2020].
33. WHO. *Advice on the use of masks in the context of COVID-19: interim guidance*. April 6, 2020. https://apps.who.int/iris/handle/10665/331693. [Accessed 14 December 2020].
34. Kampf G, Todt D, Pfaender S, Steinmann E. Persistence of coronaviruses on inanimate surfaces and its inactivation with biocidal agents. *J Hosp Infect* 2020;**104**(3):246–51.
35. WHO. *Laboratory biosafety guidance related to coronavirus disease 2019 (COVID-19): interim guidance*. February 12, 2020. https://apps.who.int/iris/handle/10665/331138. [Accessed 14 December 2020].
36. UNEP. Waste management an essential public service in the fight to beat COVID-19. Available from: https://www.unenvironment.org/news-and-stories/press-release/waste-management-essential-public-service-fight-beat-covid-19. [Accessed 14 December 2020].
37. Yu IT-S, Qiu H, Tse LA, Wong TW. Severe acute respiratory syndrome beyond Amoy Gardens: completing the incomplete legacy. *Clin Infect Dis* 2014;**58**(5):683–6.
38. Liu Y, Ning Z, Chen Y, et al. Aerodynamic characteristics and RNA concentration of SARS-CoV-2 aerosol in Wuhan hospitals during COVID-19 outbreak. *bioRxiv* 2020. https://doi.org/10.1101/2020.03.08.982637.
39. Moriyama M, Hugentobler WJ, Iwasaki A. Seasonality of respiratory viral infections. *Annu Rev Virol* 2020;**7**:83–101.
40. Khalil EE. Energy Efficient Hospitals Air Conditioning Systems. *Open Journal of Energy Efficiency* 2012;**1**(1):1–7. https://doi.org/10.4236/ojee.2012.11001.
41. Van Doremalen N, Bushmaker T, Morris DH, et al. Aerosol and surface stability of SARS-CoV-2 as compared with SARS-CoV-1. *N Engl J Med* 2020;**382**(16):1564–7.
42. Lednicky JA, Lauzard M, Fan ZH, et al. Viable SARS-CoV-2 in the air of a hospital room with COVID-19 patients. *Int J Infect Dis* 2020;**100**:476–82.
43. Riddell S, Goldie S, Hill A, Eagles D, Drew TW. The effect of temperature on persistence of SARS-CoV-2 on common surfaces. *Virol J* 2020;**17**(1):1–7.
44. Moriarty LF. Public health responses to COVID-19 outbreaks on cruise ships—worldwide, February–March 2020. *Morb Mortal Wkly Rep* 2020;**69**(12):347–52.
45. Asgharzadeh M, Mazloumi A, Kafil HS, Ghazanchaei A. Mannose-binding lectin gene and promoter polymorphism in visceral leishmaniasis caused by *Leishmania infantum*. *Pakistan J Biol Sci* 2007;**10**(11):1850–4.
46. Tran HN, Le GT, Nguyen DT, et al. SARS-CoV-2 coronavirus in water and wastewater: a critical review about presence and concern. *Environ Res* 2020;**193**:110265. https://doi.org/10.1016/j.envres.2020.110265.
47. Wang X-W, Li J-S, Jin M, et al. Study on the resistance of severe acute respiratory syndrome-associated coronavirus. *J Virol Methods* 2005;**126**(1–2):171–7.
48. Gundy PM, Gerba CP, Pepper IL. Survival of coronaviruses in water and wastewater. *Food Environ Virol* 2009;**1**(1):10.

49. Casanova L, Rutala WA, Weber DJ, Sobsey MD. Survival of surrogate coronaviruses in water. *Water Res* 2009;**43**(7):1893–8.
50. Kocamemi BA, Kurt H, Sait A, Sarac F, Saatci AM, Pakdemirli B. SARS-CoV-2 detection in Istanbul wastewater treatment plant sludges. *medRxiv* 2020. https://doi.org/10.1101/2020.05.12.20099358.
51. Mlejnkova H, Sovova K, Vasickova P, Ocenaskova V, Jasikova L, Juranova E. Preliminary study of SARS-CoV-2 occurrence in wastewater in the Czech Republic. *Int J Environ Res Publ Health* 2020;**17**(15):5508.
52. Westhaus S, Weber F-A, Schiwy S, et al. Detection of SARS-CoV-2 in raw and treated wastewater in Germany—suitability for COVID-19 surveillance and potential transmission risks. *Sci Total Environ* 2020;**751**:141750.
53. Nemudryi A, Nemudraia A, Wiegand T, et al. Temporal detection and phylogenetic assessment of SARS-CoV-2 in municipal wastewater. *Cell Rep Med* 2020;**1**(6):100098.
54. Peccia J, Zulli A, Brackney DE, et al. Measurement of SARS-CoV-2 RNA in wastewater tracks community infection dynamics. *Nat Biotechnol* 2020;**38**(10):1164–7.
55. Prado T, Fumian TM, Mannarino CF, Maranhão AG, Siqueira MM, Miagostovich MP. Preliminary results of SARS-CoV-2 detection in sewerage system in Niterói municipality, Rio de Janeiro, Brazil. *Mem Inst Oswaldo Cruz* 2020;**115**:e200196. https://doi.org/10.1590/0074-02760200196.
56. Zhang D, Ling H, Huang X, et al. Potential spreading risks and disinfection challenges of medical wastewater by the presence of severe acute respiratory syndrome Coronavirus 2 (SARS-CoV-2) viral RNA in septic tanks of Fangcang Hospital. *Sci Total Environ* 2020;**741**:140445. https://doi.org/10.1016/j.scitotenv.2020.140445.
57. Ahmed W, Bertsch PM, Angel N, et al. Detection of SARS-CoV-2 RNA in commercial passenger aircraft and cruise ship wastewater: a surveillance tool for assessing the presence of COVID-19 infected travellers. *J Trav Med* 2020;**27**(5). taaa116.
58. Medema G, Heijnen L, Elsinga G, Italiaander R, Brouwer A. Presence of SARS-Coronavirus-2 in sewage. *Environ. Sci. Technol. Lett.* 2020;**7**(7):511–6. https://doi.org/10.1021/acs.estlett.0c00357.
59. Haramoto E, Malla B, Thakali O, Kitajima M. First environmental surveillance for the presence of SARS-CoV-2 RNA in wastewater and river water in Japan. *Sci Total Environ* 2020;**737**:140405. https://doi.org/10.1016/j.scitotenv.2020.140405.
60. Peccia J, Zulli A, Brackney DE, et al. SARS-CoV-2 RNA concentrations in primary municipal sewage sludge as a leading indicator of COVID-19 outbreak dynamics. *medRxiv* 2020. https://doi.org/10.1101/2020.05.19.20105999.
61. Randazzo W, Truchado P, Ferrando EC, Simon P, Allende A, Sanchez G. SARS-CoV-2 RNA titers in wastewater anticipated COVID-19 occurrence in a low prevalence area. *Water Res.* 2020;**181**(115942). https://doi.org/10.1016/2Fj.watres.2020.115942. https://www.ncbi.nlm.nih.gov/pubmed/32425251.
62. Guerrero-Latorre L, Ballesteros I, Villacrés-Granda I, Granda MG, Freire-Paspuel B, Ríos-Touma B. SARS-CoV-2 in river water: implications in low sanitation countries. *Sci Total Environ* 2020;**743**:140832. https://doi.org/10.1016/j.scitotenv.2020.140832.
63. Rimoldi S, Stefani F, Gigantiello A, et al. Presence and infectivity of SARS-CoV-2 virus in wastewaters and rivers. *Sci Total Environ* 2020:744. 140911. https://doi.org/10.1016/2Fj.scitotenv.2020.140911.
64. La Rosa G, Iaconelli M, Mancini P, et al. First detection of SARS-CoV-2 in untreated wastewaters in Italy. *Sci Total Environ* 2020;**744**:140911. https://doi.org/10.1016/j.scitotenv.2020.139652.
65. Sherchan SP, Shahin S, Ward LM, et al. First detection of SARS-CoV-2 RNA in wastewater in North America: a study in Louisiana, USA. *Sci Total Environ* 2020;**743**:140621. https://doi.org/10.1016/j.scitotenv.2020.140621.

66. Wu F, Zhang J, Xiao A, et al. SARS-CoV-2 titers in wastewater are higher than expected from clinically confirmed cases. *mSystems* 2020. https://doi.org/10.1128/mSystems.00614-20.
67. Carrillo-Reyes J, Barragán-Trinidad M, Buitrón G. Surveillance of SARS-CoV-2 in sewage and wastewater treatment plants in Mexico. *J Water Process Eng* 2020:101815. https://doi.org/10.1016/j.jwpe.2020.101815.
68. Ampuero M, Valenzuela S, Valiente-Echeverria F, et al. SARS-CoV-2 detection in sewage in Santiago, Chile-preliminary results. *medRxiv* 2020. https://doi.org/10.1101/2020.07.02.20145177.
69. Or IB, Yaniv K, Shagan M, et al. Regressing SARS-CoV-2 sewage measurements onto COVID-19 burden in the population: a proof-of-concept for quantitative environmental surveillance. *medRxiv* 2020. https://doi.org/10.1101/2020.04.26.20073569.
70. Kumar M, Patel AK, Shah AV, et al. First proof of the capability of wastewater surveillance for COVID-19 in India through detection of genetic material of SARS-CoV-2. *Sci Total Environ* 2020;**746**:141326. https://doi.org/10.1016/j.scitotenv.2020.141326.
71. Hata A, Hara-Yamamura H, Meuchi Y, Imai S, Honda R. Detection of SARS-CoV-2 in wastewater in Japan during a COVID-19 outbreak. *Sci Total Environ* 2020:143578. https://doi.org/10.1016/j.scitotenv.2020.143578.
72. Wurtzer S, Marechal V, Mouchel J-M, et al. Evaluation of lockdown effect on SARS-CoV-2 dynamics through viral genome quantification in waste water, Greater Paris, France, 5 March to 23 April 2020. *Euro Surveill* 2020;**25**(50):2000776. https://doi.org/10.2807/1560-7917.ES.2020.25.50.2000776.
73. Randazzo W, Truchado P, Cuevas-Ferrando E, Simón P, Allende A, Sánchez G. SARS-CoV-2 RNA in wastewater anticipated COVID-19 occurrence in a low prevalence area. *Water Res* 2020;**181**:115942. https://doi.org/10.1016/j.watres.2020.115942.
74. Langone M, Petta L, Cellamare C, et al. SARS-CoV-2 in water services: presence and impacts. *Environ Pollut* 2020;**268**:115806. https://doi.org/10.1016/j.envpol.2020.115806.
75. Ahmed W, Angel N, Edson J, et al. First confirmed detection of SARS-CoV-2 in untreated wastewater in Australia: a proof of concept for the wastewater surveillance of COVID-19 in the community. *Sci Total Environ* 2020;**728**:138764.
76. WHO. Poliovirus detected from environmental samples in Israel - update. Available from: https://www.who.int/csr/don/2013_07_15/en/. [Accessed 14 December 2020].
77. WHO. *Detection of poliovirus in sewage, Brazil.* Geneva: WHO; 2014. https://www.who.int/csr/don/2014_6_23polio/en/. [Accessed 14 December 2020].
78. Bai Y, Yao L, Wei T, et al. Presumed asymptomatic carrier transmission of COVID-19. *JAMA* 2020;**323**(14):1406–7.
79. Burke RM. Active monitoring of persons exposed to patients with confirmed COVID-19—United States, January–February 2020. *MMWR Morb Mortal Wkly Rep* 2020;**69**(9):245–6.
80. Chan JF-W, Yuan S, Kok K-H, et al. A familial cluster of pneumonia associated with the 2019 novel coronavirus indicating person-to-person transmission: a study of a family cluster. *Lancet* 2020;**395**(10223):514–23.
81. Huang C, Wang Y, Li X, et al. Clinical features of patients infected with 2019 novel coronavirus in Wuhan, China. *Lancet* 2020;**395**(10223):497–506.
82. Liu J, Liao X, Qian S, et al. Community transmission of severe acute respiratory syndrome coronavirus 2, Shenzhen, China, 2020. *Emerg Infect Dis* 2020;**26**(6):1320–3.

83. WHO. Modes of transmission of virus causing COVID-19: implications for IPC precaution recommendations. Available from: https://www.who.int/news-room/commentaries/detail/modes-of-transmission-of-virus-causing-covid-19-implications-for-ipc-precaution-recommendations. [Accessed 14 December 2020].
84. Heller L, Mota CR, Greco DB. COVID-19 faecal-oral transmission: are we asking the right questions? *Sci Total Environ* 2020:138919. https://doi.org/10.1016/j.scitotenv.2020.138919.
85. Zhang J, Wang S, Xue Y. Fecal specimen diagnosis 2019 novel coronavirus—infected pneumonia. *J Med Virol* 2020;**92**(6):680—2.
86. van Doorn AS, Meijer B, Frampton CM, Barclay ML, de Boer NK. Systematic review with meta-analysis: SARS-CoV-2 stool testing and the potential for faecal-oral transmission. *Aliment Pharmacol Ther* 2020;**52**(8):1276—88.
87. Zhang Y, Chen C, Song Y, et al. Excretion of SARS-CoV-2 through faecal specimens. *Emerg Microb Infect* 2020;**9**(1):2501—8.
88. Wigginton KR, Boehm AB. Environmental engineers and scientists have important roles to play in stemming outbreaks and pandemics caused by enveloped viruses. *Environ Sci Technol* 2020;**54**(7):3736—9.
89. EPA. Exposure pathways to high-consequence pathogens in the wastewater collection and treatment systems. Available from: https://cfpub.epa.gov/si/si_public_record_Report.cfm?dirEntryId=341856&Lab=NHSRC. [Accessed 14 December 2020].
90. Mohapatra S, Menon NG, Mohapatra G, et al. The novel SARS-CoV-2 pandemic: possible environmental transmission, detection, persistence and fate during wastewater and water treatment. *Sci Total Environ* 2020;**765**:142746. https://doi.org/10.1016/j.scitotenv.2020.142746.
91. Kumar M, Kuroda K, Dhangar K. The most eagerly awaited summer of the Anthropocene: a perspective of SARS-CoV-2 decay and seasonal change. *Groundwater Sustain Dev* 2020;**11**:100400. https://doi.org/10.1016/j.gsd.2020.100400.
92. Sun Z, Thilakavathy K, Kumar SS, He G, Liu SV. Potential factors influencing repeated SARS outbreaks in China. *Int J Environ Res Publ Health* 2020;**17**(5):1633.
93. Schuit M, Ratnesar-Shumate S, Yolitz J, et al. Airborne SARS-CoV-2 is rapidly inactivated by simulated sunlight. *J Infect Dis* 2020;**222**(4):564—71.
94. Bu J, Peng D-D, Xiao H, et al. Analysis of meteorological conditions and prediction of epidemic trend of 2019-nCoV infection in 2020. *medRxiv* 2020. https://doi.org/10.1101/2020.02.13.20022715.
95. Gunthe SS, Swain B, Patra SS, Amte A. On the global trends and spread of the COVID-19 outbreak: preliminary assessment of the potential relation between location-specific temperature and UV index. *J Public Health* 2020:1—10.
96. Bannister-Tyrrell M, Meyer A, Faverjon C, Cameron A. Preliminary evidence that higher temperatures are associated with lower incidence of COVID-19, for cases reported globally up to 29th February 2020. *medRxiv* 2020. https://doi.org/10.1101/2020.03.18.20036731.
97. Shi P, Dong Y, Yan H, et al. The impact of temperature and absolute humidity on the coronavirus disease 2019 (COVID-19) outbreak-evidence from China. *medRxiv* 2020. https://doi.org/10.1101/2020.03.22.20038919.
98. Luo W, Majumder MS, Liu D, et al. The role of absolute humidity on transmission rates of the COVID-19 outbreak. *medRxiv* 2020. https://doi.org/10.1101/2020.02.12.20022467.
99. WHO. *Water, sanitation, hygiene, and waste management for the COVID-19 virus: interim guidance*. April 23, 2020. https://apps.who.int/iris/handle/10665/331846. [Accessed 14 December 2020].

100. UNEP. Healthcare waste: what to do with it? Available from: https://www.unenvironment.org/news-and-stories/story/healthcare-waste-what-do-it. [Accessed 14 December 2020].
101. Yu H, Sun X, Solvang WD, Zhao X. Reverse logistics network design for effective management of medical waste in epidemic outbreaks: insights from the coronavirus disease 2019 (COVID-19) outbreak in Wuhan (China). *Int J Environ Res Publ Health* 2020;**17**(5):1770.
102. Chiang CF, Sung FC, Chang FH, Tsai CT. Hospital waste generation during an outbreak of severe acute respiratory syndrome in Taiwan. *Infect Contr Hosp Epidemiol* 2006;**27**(5):519–22.
103. CDPH. Novel Coronavirus disease 2019 (COVID-19) medical waste management - interim guidelines. Available from: https://www.cdph.ca.gov/Programs/CID/DCDC/Pages/COVID-19/MedicalWasteManagementInterimGuidelines.aspx. [Accessed 14 December 2020].
104. TCEQ. COVID-19: waste disposal guidance. Available from: https://www.tceq.texas.gov/response/covid-19/waste-disposal-guidance. [Accessed 14 December 2020].
105. Xinhuanet. Properly dispose of medical waste and guard the last line of defense for epidemic prevention and control. Available from: http://www.xinhuanet.com/2020-02/18/c_1125589479.htm. [Accessed 14 December 2020].
106. Singh A, Singh A, Neelesh Kumar M. Health-care waste management. *Int J Sci Res Rev* 2019;**16**(6):1060.
107. CDPHE. Impacts of COVID-19 on solid waste management. Available from: https://www.colorado.gov/pacific/cdphe/impacts-covid-19-solid-waste-management. [Accessed 14 December 2020].
108. ISWA. Covid-19: a letter concerning healthcare waste. Available from: https://www.iswa.org/home/news/news-detail/article/covid-19-and-healthcare-waste/109/. [Accessed 14 December 2020].
109. Keith A. *Waste management during the COVID-19 pandemic from response to ecovery.* United Nations Environment Programme; 2020. https://reliefweb.int/sites/reliefweb.int/files/resources/WMC-19.pdf. [Accessed 14 December 2020].
110. Vaverková MD, Paleologos EK, Dominijanni A, et al. Municipal solid waste management under COVID-19: challenges and recommendations. *Environ Geotech* 2020:1–16. https://doi.org/10.1680/jenge.20.00082.
111. Adamcová D, Zloch J, Brtnický M, Vaverková MD. Biodegradation/disintegration of selected range of polymers: impact on the compost quality. *J Polym Environ* 2019;**27**(4):892–9.
112. Wajs J, Bochniak R, Golabek A. Proposal of a mobile medical waste incinerator with application of automatic waste feeder and heat recovery system as a novelty in Poland. *Sustainability* 2019;**11**(18):4980.
113. Al-Qodah Z. Adsorption of dyes using shale oil ash. *Water Res* 2000;**34**(17):4295–303.
114. Al-Qodah Z, Lafi W. Adsorption of reactive dyes using shale oil ash in fixed beds. *J Water Supply Res Technol - Aqua* 2003;**52**(3):189–98.
115. Rutkowska G, Fronczyk J, Wichowski P. Research on the possibility of using fly ashes from combustion of municipal sewage sludge on properties of ordinary concretes. *Rocznik Ochrona Srodowiska* 2018;**20**:1113–28.
116. Knez Hrnčič M, Cör D, Knez Ž. *Supercritical Fluids as a Tool for Green Energy and Chemicals. Advanced Applications of Supercritical Fluids in Energy Systems.* IGI Global; 2017. p. 554–87.
117. Ribeiro N, Soares GC, Santos-Rosales V, et al. A new era for sterilization based on supercritical CO_2 technology. *J Biomed Mater Res B Appl Biomater* 2020;**108**(2):399–428.

118. Thomas KV, Bijlsma L, Castiglioni S, et al. Comparing illicit drug use in 19 European cities through sewage analysis. *Sci Total Environ* 2012;**432**:432–9.
119. Breza-Boruta B. The assessment of airborne bacterial and fungal contamination emitted by a municipal landfill site in Northern Poland. *Atmos Pollut Res* 2016;**7**(6): 1043–52.
120. Whitworth J. COVID-19: a fast evolving pandemic. *Trans R Soc Trop Med Hyg* 2020;**114**(4):241–8.
121. Otter J, Donskey C, Yezli S, Douthwaite S, Goldenberg S, Weber D. Transmission of SARS and MERS coronaviruses and influenza virus in healthcare settings: the possible role of dry surface contamination. *J Hosp Infect* 2016;**92**(3):235–50.
122. Besedovsky L, Lange T, Haack M. The sleep-immune crosstalk in health and disease. *Physiol Rev* 2019;**99**(3):1325–80.
123. Warren-Gash C, Fragaszy E, Hayward AC. Hand hygiene to reduce community transmission of influenza and acute respiratory tract infection: a systematic review. *Influenza Other Respir Viruses* 2013;**7**(5):738–49.
124. Bialvaei AZ, Kafil HS, Asgharzadeh M. Role of treatment cost on transmission of multidrug-resistant tuberculosis into Iran. *Clin Infect Dis* 2015;**61**(6):1029–30.
125. Zhu N, Zhang D, Wang W, et al. A novel coronavirus from patients with pneumonia in China, 2019. *N Engl J Med* 2020;**382**(8):727–33.
126. Sandaradura I, Goeman E, Pontivivo G, et al. A close shave? Performance of P2/N95 respirators in health care workers with facial hair: results of the BEARDS (adequate respiratory defences) study. *J Hosp Infect* 2020;**104**(4):529–33.
127. Xiao Y, Torok ME. Taking the right measures to control COVID-19. *Lancet Infect Dis* 2020;**20**(5):523–4.
128. Sadreddini S, Safaralizadeh R, Baradaran B, et al. Chitosan nanoparticles as a dual drug/siRNA delivery system for treatment of colorectal cancer. *Immunol Lett* 2017;**181**:79–86.
129. CDC. Recommended guidance for extended use and limited reuse of N95 filtering facepiece respirators in healthcare settings. Available from: https://www.cdc.gov/niosh/topics/hcwcontrols/recommendedguidanceextuse.html. [Accessed 14 December 2020].
130. Loeb M, Dafoe N, Mahony J, et al. Surgical mask vs N95 respirator for preventing influenza among health care workers: a randomized trial. *JAMA* 2009;**302**(17): 1865–71.
131. Radonovich LJ, Simberkoff MS, Bessesen MT, et al. N95 respirators vs medical masks for preventing influenza among health care personnel: a randomized clinical trial. *JAMA* 2019;**322**(9):824–33.
132. Cheng VC, Wong S-C, Chuang VW, et al. The role of community-wide wearing of face mask for control of coronavirus disease 2019 (COVID-19) epidemic due to SARS-CoV-2. *J Infect* 2020;**81**(1):107–14.
133. Ma QX, Shan H, Zhang HL, Li GM, Yang RM, Chen JM. Potential utilities of mask-wearing and instant hand hygiene for fighting SARS-CoV-2. *J Med Virol* 2020;**92**(9):1567–71.
134. Feng S, Shen C, Xia N, Song W, Fan M, Cowling BJ. Rational use of face masks in the COVID-19 pandemic. *Lancet Respir Med* 2020;**8**(5):434–6.
135. Gholizadeh P, Aghazadeh M, Asgharzadeh M, Kafil H. Suppressing the CRISPR/Cas adaptive immune system in bacterial infections. *Eur J Clin Microbiol Infect Dis* 2017;**36**(11):2043–51.
136. WHO. Coronavirus disease (COVID-19) advice for the public: when and how to use masks. Available from: https://www.who.int/emergencies/diseases/novel-coronavirus-2019/advice-for-public/when-and-how-to-use-masks. [Accessed 14 December 2020].

137. Adams JG, Walls RM. Supporting the health care workforce during the COVID-19 global epidemic. *JAMA* 2020;**323**(15):1439–40.
138. Mathur P. Hand hygiene: back to the basics of infection control. *Indian J Med Res* 2011;**134**(5):611.
139. Ayliffe G, Babb J, Bridges K, et al. Comparison of two methods for assessing the removal of total organisms and pathogens from the skin. *Epidemiol Infect* 1975;**75**(2):259–74.
140. Wilson J, Loveday H. Obituary: professor Graham Ayliffe 1926–2017. *J Infect Prevent* 2017;**18**(4):168.
141. WHO. WHO Guidelines on Hand Hygiene in Health Care: a Summary Available from: https://www.who.int/gpsc/5may/tools/who_guidelines-handhygiene_summary.pdf. [Accessed 14 December 2020].
142. Pisarek S, Maximova K, Gryko D. Strategies toward the synthesis of amphiphilic porphyrins. *Tetrahedron* 2014;**38**(70):6685–715.
143. CDC. Healthcare facilities: managing operations during the COVID-19 pandemic. Available from: https://www.cdc.gov/coronavirus/2019-ncov/hcp/guidance-hcf.html. [Accessed 14 December 2020].
144. Heinze JE, Yackovich F. Washing with contaminated bar soap is unlikely to transfer bacteria. *Epidemiol Infect* 1988;**101**(1):135–42.
145. Todd EC, Michaels BS, Holah J, Smith D, Greig JD, Bartleson CA. Outbreaks where food workers have been implicated in the spread of foodborne disease. Part 10. Alcohol-based antiseptics for hand disinfection and a comparison of their effectiveness with soaps. *J Food Protect* 2010;**73**(11):2128–40.
146. U.S.FDA. Available from: https://www.fda.gov/consumers/if-soap-and-water-are-not-available-hand-sanitizers-may-be-good-alternative. [Accessed 14 December 2020].
147. Best M, Neuhauser D. Ignaz Semmelweis and the birth of infection control. *BMJ Qual Saf* 2004;**13**(3):233–4.
148. Malherbe H, Nugier A, Clément J, Lamboy B. Evidence-based and promising interventions for preventing infectious diseases among young people as a result of poor hand hygiene in schools: a literature review. *Sante Publique* 2013;**1**(1):57–63.
149. Ramasethu J. Prevention and treatment of neonatal nosocomial infections. *Maternal Health Neonatol Perinatol* 2017;**3**(1):5.
150. Stebbins S, Cummings DA, Stark JH, et al. Reduction in the incidence of influenza A but not influenza B associated with use of hand sanitizer and cough hygiene in schools: a randomized controlled trial. *Pediatr Infect Dis J* 2011;**30**(11):921.
151. Tamimi AH, Maxwell S, Edmonds S, Gerba CP. Impact of the use of an alcohol-based hand sanitizer in the home on reduction in probability of infection by respiratory and enteric viruses. *Epidemiol Infect* 2015;**143**(15):3335–41.
152. Kratzel A, Todt D, V'kovski P, et al. Inactivation of severe acute respiratory syndrome coronavirus 2 by WHO-recommended hand rub formulations and alcohols. *Emerg Infect Dis* 2020;**26**(7):1592–5.
153. Bonner L. CDC report calls attention to hand sanitizer risk in children. *Pharm Today* 2017;**23**(5):34.
154. Hadaway A. Handwashing: clean hands save lives. *J Consum Health Internet* 2020;**24**(1):43–9.
155. CDC. Handwashing: clean hands save lives. Available from: https://www.cdc.gov/handwashing/index.html. [Accessed 14 December 2020].
156. Jing JLJ, Pei Yi T, Bose RJ, McCarthy JR, Tharmalingam N, Madheswaran T. Hand sanitizers: a review on formulation aspects, adverse effects, and regulations. *Int J Environ Res Publ Health* 2020;**17**(9):3326.

157. Emami A, Javanmardi F, Keshavarzi A, Pirbonyeh N. Hidden threat lurking behind the alcohol sanitizers in CoVID-19 outbreak. *Dermatol Ther* 2020:e13627.
158. Khan MH, Yadav H. Sanitization during and after COVID-19 pandemic: a short review. *Trans Indian Natl Acad Eng* 2020:1–11.
159. Siddharta A, Pfaender S, Vielle NJ, et al. Virucidal activity of world health organization–recommended formulations against enveloped viruses, including Zika, Ebola, and emerging coronaviruses. *J Infect Dis* 2017;**215**(6):902–6.
160. U.S.EPA. List N: disinfectants for Coronavirus (COVID-19). Available from: https://www.epa.gov/pesticide-registration/list-n-disinfectants-use-against-sars-cov-2. [Accessed 14 December 2020].
161. Pradhan D, Biswasroy P, Ghosh G, Rath G. A review of current interventions for COVID-19 prevention. *Arch Med Res* 2020;**51**(5):363–74.
162. Nabi G, Wang Y, Hao Y, Khan S, Wu Y, Li D. Massive use of disinfectants against COVID-19 poses potential risks to urban wildlife. *Environ Res* 2020;**188**:109916. https://doi.org/10.1016/2Fj.envres.2020.109916.
163. Sedlak DL, von Gunten U. The chlorine dilemma. *Science* 2011;**331**(6013):42–3.
164. Dumas O, Varraso R, Boggs KM, et al. Association of occupational exposure to disinfectants with incidence of chronic obstructive pulmonary disease among US female nurses. *JAMA Network Open* 2019;**2**(10):e1913563.
165. Baker LA. *Environmental chemistry of lakes and reservoirs*. American Chemical Society; 1994.
166. Liu J, Zhang X. Comparative toxicity of new halophenolic DBPs in chlorinated saline wastewater effluents against a marine alga: halophenolic DBPs are generally more toxic than haloaliphatic ones. *Water Res* 2014;**65**:64–72.
167. Bei E, Shu Y, Li S, et al. Occurrence of nitrosamines and their precursors in drinking water systems around mainland China. *Water Res* 2016;**98**:168–75.
168. Krytopoulos S. IARC monographs on the evaluation of the carcinogenic risk of chemicals to humans, vol. 17: some N-nitroso compounds (1978). *Br J Cancer* 1979;**40**(4):666.

CHAPTER 10

A review of deciphering the successes and learning from the failures in preventive and health policies to stop the COVID-19 pandemic

Milad Mousazadeh[1,2], Zohreh Naghdali[1,2], Zahra Mohammadi Goldar[1], Marjan Hashemi[3], Alaa El Din Mahmoud[4,5], Zakaria Al-Qodah[6], Miguel A. Sandoval[7,8], Khalid S. Hashim[9], Mika Sillanpää[10,11,12], Chaudhery Mustansar Hussain[13], Mohammad Mahdi Emamjomeh[14]

[1]Student Research Committee, Qazvin University of Medical Sciences, Qazvin, Iran; [2]Department of Environmental Health Engineering, School of Health, Qazvin University of Medical Sciences, Qazvin, Iran; [3]Environmental and Occupational Hazards Control Research Center, Shahid Beheshti University of Medical Sciences, Tehran, Iran; [4]Environmental Sciences Department, Faculty of Science, Alexandria University, Alexandria, Egypt; [5]Green Technology Group, Faculty of Science, Alexandria University, Alexandria, Egypt; [6]Al-Balqa Applied University, Faculty of Engineering Technology, Department of Chemical Engineering, Amman, Jordan; [7]Universidad de Santiago de Chile USACH, Facultad de Química y Biología, Departamento de Química de los Materiales, Laboratorio de Electroquímica Medio Ambiental, LEQMA, Santiago, Chile; [8]Universidad de Guanajuato, División de Ciencias Naturales y Exactas, Departamento de Ingeniería Química, Guanajuato, Mexico; [9]Department of Civil Engineering, Liverpool John Moores University, Liverpool, United Kingdom; [10]Chemistry Department, College of Science, King Saud University, Riyadh, Saudi Arabia; [11]Faculty of Science and Technology, School of Applied Physics, University Kebangsaan Malaysia, Bangi, Selangor, Malaysia; [12]Department of Biological and Chemical Engineering, Aarhus University, Nørrebrogade 44, Aarhus C, Denmark; [13]Department of Chemistry and Environmental Science, New Jersey Institute of Technology, Newark, NJ, United States; [14]Social Determinants of Health Research Center, Research Institute for Prevention of Non-Communicable Diseases, Qazvin University of Medical Sciences, Qazvin, Iran

10.1 Introduction

Coronavirus disease 2019 (COVID-19) originated in the city of Wuhan, China, in December 2019. It is an infectious disease caused by a new type of coronavirus named severe acute respiratory syndrome coronavirus 2 (SARS-CoV-2), which can be potentially fatal. The virus has rapidly spread all over the world.[1] On March 12, 2020, the World Health Organization

Environmental Management of COVID-19
ISBN 978-0-323-85780-2
https://doi.org/10.1016/B978-0-323-85780-2.00003-2

(WHO) declared the infection a pandemic. The COVID-19 pandemic seems to be a major global health threat due to the increasing number of cases and associated mortality. As of August 11, 2020, COVID-19 had affected 213 countries and territories around the world, infecting 20,394,035 human beings and, unfortunately, 741,807 deaths confirmed worldwide.[2,3] The rate of virus spread and people-to-people transmission is high, and as a newly emerging virus still much remains to be understood about its transmission.[4,5]

Although there are reports of using different medicines and efforts to develop a successful vaccine, at this moment there is continuously research novel approaches to developing proven treatment for COVID-19. The vaccines for COVID-19 have been developed using novel methods intended to increase the volume and speed of production and bring about strong immune responses.[5–7] The disease spread is ongoing, with a rising death toll with various trajectories in countries with a different population (see Fig. 10.1). Since there is neither a confirmed vaccine nor effective pharmaceutical treatment, the best way to reduce the transmission rate of the disease seems to be nonpharmaceutical interventions (NPIs) as preventive measures.[4,6,8,9]

Different countries are at different stages of the pandemic, so governments and health departments have chosen different strategies to contain the spread of the virus. For example, social distancing and movement restrictions, commonly known as lockdown in China, Italy, India, and the UK[1]; massive testing to identify new and asymptomatic cases in South Korea[1,11]; and mass masking policies as a preventive measure even long before WHO recommendations in many Asian countries such as China, South Korea, and Japan.[11–13] The main idea behind all these implementations was to reduce the transmission rate to flatten the epidemic curve and distribute medical treatment demand over time. This last point allows health systems to deal with a larger number of patients at the same time. Fig. 10.2 shows the impact of five public health interventions implemented individually and in combination to suppress infection transmission and health care demand based on epidemiological modeling in the UK.[3,4,14] Data show that countries, which implemented such containment and preventive measures, experienced lower infection rates and fewer deaths.[3,8,15] Since wide testing would not be affordable and feasible in many low and middle-income countries, and poor countries such as Brazil and Ghana cannot tolerate lockdown consequences, herd immunity might be chosen as an alternative solution in this pandemic. However, the duration of such immunity is not known and might offer serious risks and

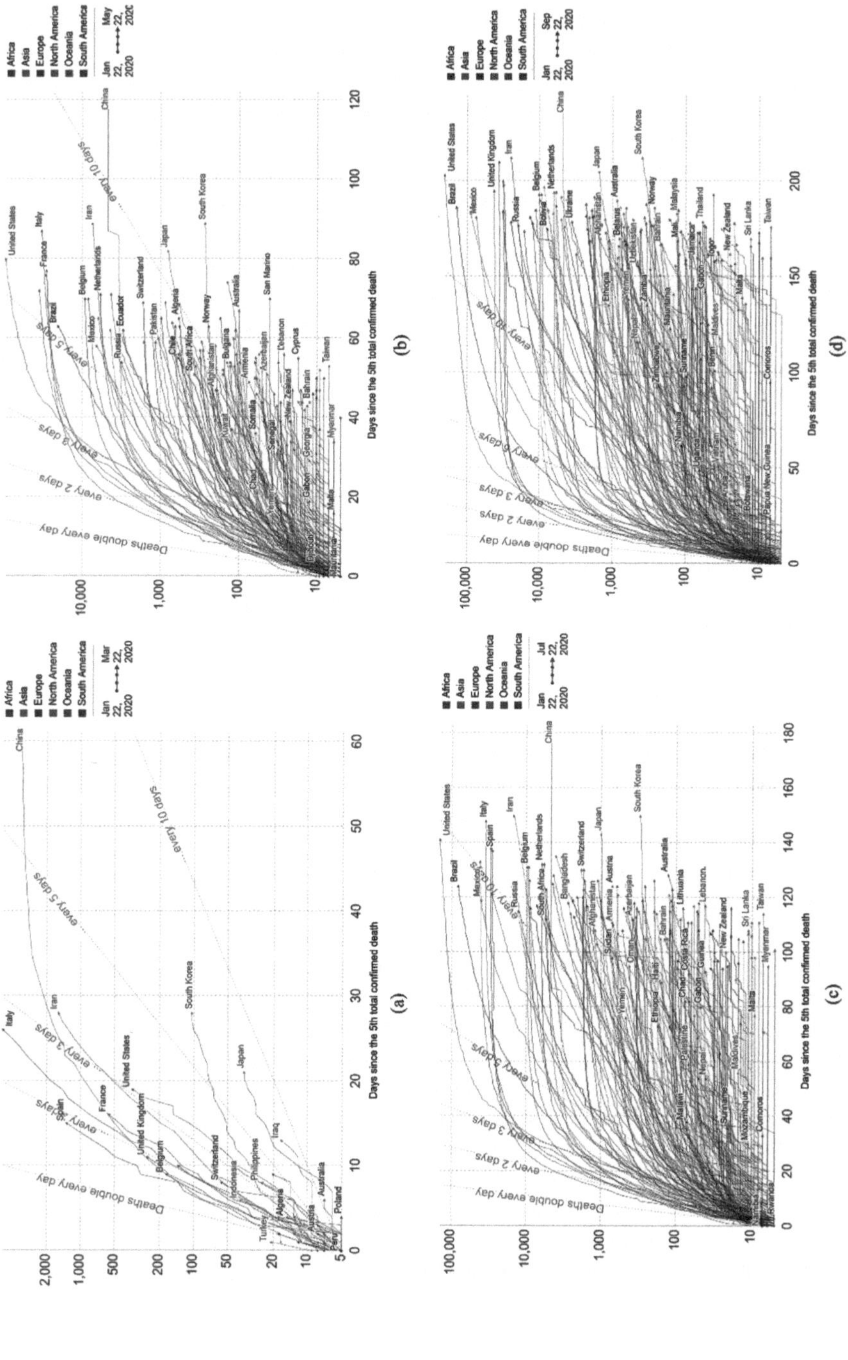

Figure 10.1 Total confirmed COVID-19 deaths in two-month periods from January 22 to September 22, 2020. (A) From January 22 to March 22; (B) from March 22 to May 22; (C) from May 22 to July 22; and (D) from July 22 to September 22. *(Modified after* Our World in Data.[10])

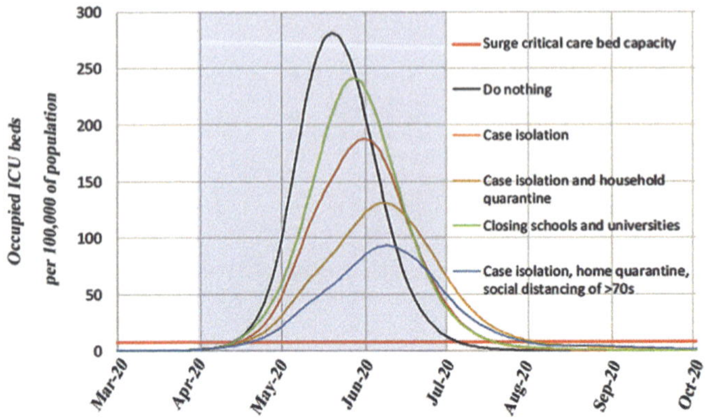

Figure 10.2 Intensive care unit (ICU) bed requirements in Britain with and without mitigation strategies. Modified after Ferguson et al. [4]

raising mortality.[16] Therefore, it is of utmost importance to evaluate the accurate time and duration of social distancing policy to be effective; further, the use of multiple interventions is required to diminish the transmission rate. In addition, the effectiveness of different strategies critically depends on society's response and acceptance, which varies in different countries.[4,14] COVID-19 has had a profound impact on different socioeconomic aspects and posed the fundamental question of what the most effective and sustainable policies are to contain its spread.

The Swiss cheese model of COVID-19 prevention called "Swiss Cheese Respiratory Pandemic Defense" sketched by the Australian virologist, Ian MacKay, is very effective in analyzing the reasons for countries' failure and success in controlling COVID-19. According to this model (as shown in Fig. 10.3), COVID-19 prevention is based on several principles as layers including social distancing, mask wearing, regular hand washing, rapid and accurate testing to diagnose people infected with COVID-19, ventilation, quarantine, proper information, government funding, and ultimately vaccination. These layers are complementary to each other and must exist together to help control and prevent COVID-19 because none of these methods of prevention is absolute and certain.[17,18] Based on this important model, the strategic role of herd immunity, quarantine, social distancing, and masking was challenged by various governments.

This chapter reviews the experiences of different countries, including successes and failures, in preventive and health policies against the COVID-19 pandemic to compare their effectiveness, feasibility, and affordability.

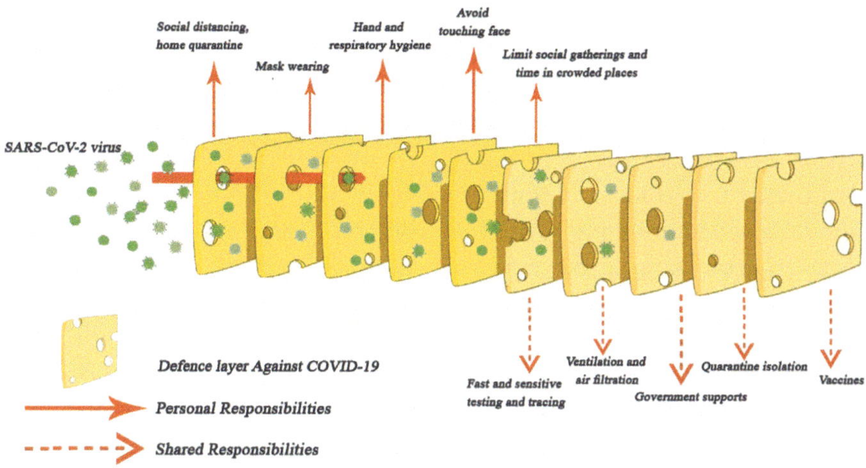

Figure 10.3 Multiple layers of Swiss Cheese Respiratory Pandemic Defense developed by Ian MacKay. *(Modified after Roberts.[17])*

10.2 Herd immunity approach

The story of herd immunity seems to go back to the work of American veterinarians who observed contagious abortion in cattle and sheep. As a result, ranchers slaughtered and sold affected cows, but George Potter, an American veterinarian, realized in 1916 that this was the wrong solution and proposed the theory of herd immunity. He had the idea that by keeping the cattle immune, raising the calves, and preventing new cows from entering, the herd would be protected. In the same way, further work was done until the researchers achieved the concept of herd immunity. In 1919, Topley created an epidemic in the mice population and found that the growing prevalence of immune individuals would put an end to an epidemic if there was no continued influx of susceptible mice. In 1923, Topley and Wilson defined the phenomenon as "herd immunity."[19] In 1924, Dudley established herd immunity to humans. Dudley prefaced his report in 1934 entitled "Active Immunization against Diphtheria" with photos of "the human herd" and "bacterial herds." Finally, in 1935, Topley stated a broader concept of herd immunity.[19] Topley explained that herd immunity, in addition to involving safety distribution, also determines the social factors that affect herd exposure.[19]

The concept of "herd immunity" increased after 1990 when health officials sought to achieve acceptable levels of universal vaccine coverage. As the world tackles with the COVID-19 pandemic, some countries are urging people to follow the wrong herd immunity policy to get sick or killed in pursuit of herd immunity. Herd immunity is a key concept for pandemic control in that only part of the population should be immune to an infectious agent to prevent the spread of the disease from person to person. As a result, not only those who are immune but also the whole community are protected against the disease. According to Marc Lipsitch,[20] the fundamental principle of herd immunity is how many safety people are needed in the population and how many people are immune to the infection. Usually, a percentage of people should be susceptible to the disease, which is called the herd immunity threshold. The number of susceptible individuals in a community should be low; otherwise, if the portion of vulnerable persons in a population is too high, herd immunity cannot be achieved. Therefore, the pathogen spreads and its prevalence increases.[20] In conclusion, herd immunity occurs when immune individuals are above the threshold; consequently, the susceptible individuals receive indirect protection against infection. In this regard, the herd immunity threshold relies on a factor, namely the basic reproduction number (R_0) or the effective reproduction number (R_e or R_t).[21] R_0 relates to the average number of secondary infections caused by only one infected person entering a highly susceptible population. Herd immunity threshold (P_{crit}) is mathematically expressed as $1 - 1/R_0$. It can be found that for instance, if $R_0 = 4$, the herd immunity threshold is 0.75.[22] This means that more than two-thirds of the population need to be immune.

According to a published study by the Italian Institute for International Political Studies (ISPI) (Fig. 10.4), as of May 4, 2020, 12 countries concentrated the highest mortality rates, while the prevalence of COVID-19 remained quite low and certainly far from the required share for herd immunity (between 70% and 80%). The results showed that Belgium, which tops the list, had only 84% (with 6.2% of R_t) of its population needed to be immune, while in Sweden—without mandatory lockdown measures—was even 60% (with 2.5% of R_t) of the population. In the major EU countries including Italy, UK, France, and Spain, 70%–80% (with 3%–4% of R_t) of population needed to be immune. Related to Germany and Portugal with R_t of less than 1% was utterly different and the whole population needed to be immune. According to ISPI estimates, acceptable emissions in the 12 countries affected by the virus can never exceed 7%, and in almost all cases remain below the 5% threshold.[23]

The three developed countries, Britain and the Netherlands, and especially Sweden, have dealt with the pandemic COVID-19 with a soft

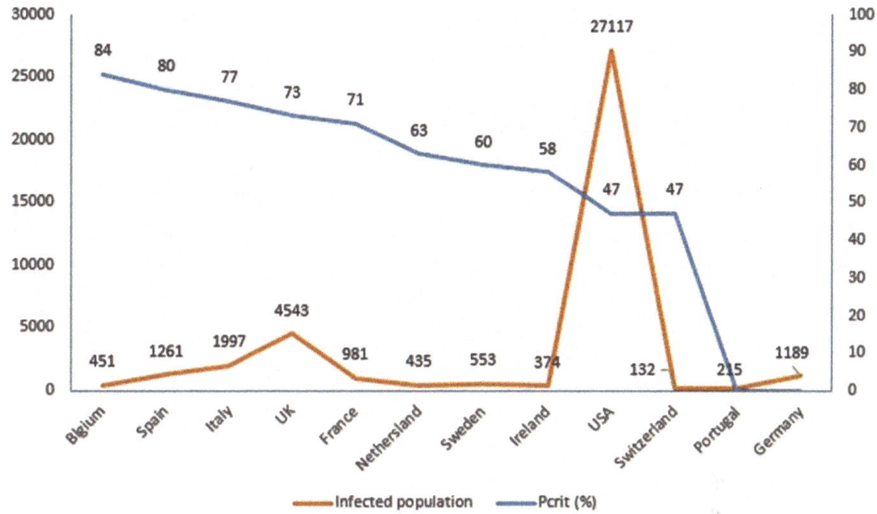

Figure 10.4 As of May 4, 2020, the countries with the highest P_{crit}. Belgium (84%), Spain (80%), Italy (77%), UK (73%), France (71%), the Netherlands (63%), Sweden (60%), Ireland (58%), USA (47%), Switzerland (47%), Portugal (0%), and Germany (0%).[23]

approach and had no lockdown. In contrast, these countries had voluntary compliance and the use of masks. Though the COVID-19 strategy was risky, the government of Sweden believed that with a herd immunity approach it could curb the current pandemic. Britain and the Netherlands, after observing the rapidly increasing trend of deaths, the delayed countrywide lockdown was adopted.[24] However, Sweden was faithful to its method of implementation and continued it. In late March 2020, Sweden abandoned the herd immunity policy and decided to take active interventions to control the infection. Most educational centers were closed to learners, travel limitations were enforced, remote working was urged, and prohibitions on parties of more than 50 were endorsed. Contrary to herd immunity, the prevalence in Stockholm, Sweden in April 2020 was reported to be less than 8%.[25] As of May 2020, the rate of positive antibody testing in Sweden was estimated about 15%, which did not reach public safety at all, while the death rate was more than 556 deaths per million. This was 4.5 times more deaths than its Nordic neighbors (e.g., Denmark).[26,27] Iran, as the first country of COVID-19 confirmed in the Middle East, also used a herd immunity strategy due to the poor economic conditions resulting from US sanctions and internal mismanagement. The Institute for Health Metrics and Evaluation (IHME) predicts that if a herd immunity strategy is pursued in Iran, around 110,000 deaths in the country will be estimated by February 1, 2021.[28] Currently (as of December 13, 2020), the death toll is 51,949.[2]

Hunter and Pankhania[29] still believe that other European countries should not pursue Sweden's pattern because risk remains a major factor. Pankhania stated, "Because the new virus is circulating and we do not know what it does, we cannot test the herd immunity."[29] Herd immunity is only possible when people can be safely vaccinated, thus ensuring that a large number of people are protected.[30] However, the researchers said that this strategy failed in Britain and other European countries but could be successful in poor countries such as India due to the disproportionate population of young people would face less risk of hospitalization and death. They also acknowledged that by spreading the virus compared to European countries such as Italy, it could limit the coronavirus death toll, as 93.5% of the Indian population is under 65 years old. Herd (population) immunity strategy in less developed countries, due to the impossibility of social distancing in crowded environments, the lack of test kits to detect contagions, and the human suffering that occurs in lockdowns, indicates that in these regions a different route is needed.[31]

In a recent correspondence published in The Lancet, more than 80 researchers warned that herd immunity is a dangerous and erroneous method that is not supported by scientific evidence.[32] Studies conducted in June and July 2020 showed that herd immunity is a questioned methodology. Antibody studies performed in Spain and Switzerland showed a low sero-prevalence of less than 10%.[33] Critics have unanimously argued in light of recent findings that any proposed method of herd immunity through natural infection is both immoral and unfeasible. However, advocates of herd immunity have remained. Some researchers believe that antibodies are not necessary because of durable T-cell immunity by the SARS-CoV-2 virus.[34] Others think that if the most vulnerable people in a communal are infected first, they will be able to achieve herd immunity after encountering only 20% of the population. While only a few months remain until possible vaccination as the main hope for controlling the pandemic, there are currently no ideal options other than social distancing, hand hygiene, and the use of masks, as well as quarantine. However, it is well known that these containment measures can harm both the public's physical and mental health and long-term economic disruption.[35] Bhopal[16] examined herd immunity from another perspective and encouraged the use of population immunity.

Bhopal believes that 40%–50% of the population immunity is enough to eradicate the COVID-19 pandemic. Bhopal explicitly thinks that confirmed immunity is crucial, particularly in young people or children.

Vaccines cannot be the only hope as they may be effective only for a short period, particularly if new strains of the virus develop. However, this is just a change of concept and does not eliminate the essence of the problem.[16] In conclusion, Adhanom, the WHO director, finally acknowledged on October 22, 2020 that the population (herd) immunity approach that some countries have taken to combat the COVID-19 epidemic is completely immoral. Adhanom also stated that population immunity is achieved by protecting people from the virus, not by exposure.[36]

Contrary to the herd immunity perspective, Kenyon[37] achieved a herd immunity strategy called "flattening the curve." After analyzing 65 countries, Kenyon proposed the current solution to control the spread of infection through the community. The flattening-the-curve strategy is a kind of gradual affliction of people that can cause reduction of the burden in hospitals and help the economic situation and, subsequently, prevent further complications and mortality. This plan requires the observance of principles such as mask wearing, physical distancing, hand washing, and avoiding mass gatherings.[37]

As a result, by the time a vaccine is available worldwide, many people will die from COVID-19 before they can reach immunity. Overall, instead of having faith in herd immunity, governments should rely on extensive testing, tracking, and treatment of the disease, along with strict rules of social distancing, mask wearing, and quarantine of patients—an experience that has already had a successful impact in several countries.

10.3 Lockdown and quarantine

As COVID-19 can reach and infect any age group at any place and time, strict restrictions should be applied, and one of the forefront of these restrictions is lockdown.[38] Many countries applied and adopted lockdown policies because of the lack of vaccination.[39] He et al.[40] defined the lockdown city when all the following preventive measures were enforced: (I) ban of reopening of unnecessary commercial activities at the community level; (II) prohibition of excessive gatherings by city dwellers; (III) restrictions on the movement of personal vehicles and public transport.

WHO recommended strict lockdown to diminish the reproduction rate of infection.[41] Therefore, "Complete Lockdown" has been adopted by many governments across the world as a measurement to prevent COVID-19 spreading.[42] Lockdown was implemented in Europe, and the well-known example is Italy, being the first country executing a nationwide lockdown.[43] The whole Italian territory was under lockdown restrictions in February 20, 2020.[44] Table 10.1 illustrates the measurement of the quarantine levels by the spent hours per person in the public/day. The Italian government also considered different average household sizes, including large average households of six persons, medium average households of three persons, small average households of two persons, and single-person households. Their model predicted 43 new infections within two weeks for a large household (six persons) and zero secondary cases for a single-person household over the same period.

New Zealand reacted swiftly to the COVID-19 epidemic, which included a severe lockdown that effectively disrupted COVID-19 transmission by the community. However, this success has had significant economic and social effects (e.g., psychological distress).[45] Spain is one of the most affected countries, which announced the national lockdown on March 14, 2020. According to the results of an ecological study, the lockdown period in Spain showed that it took approximately 18.33 days to reduce the mortality rates. Even with the national lockdown, some communities such as Catalonia and Aragon experienced an increase in the daily mortality rate of more than 20%, reaching a maximum value of 34% in the daily death rate in Catalonia. Higher rates in these communities were probably associated with higher population density and by the higher mobility before lockdown.[46] In April 2020, almost all schools, universities,

Table 10.1 The effectiveness of quarantine based on the degree of adherence to quarantine.

Quarantine level	Definition	Hours/day
Complete noncompliance	Individuals do their every-day outdoor activities (i.e., working, shopping, etc.) without restrictions.	10
Medium quarantine	Restricts every-day out-of-household activities to 50% of normal.	5
Complete quarantine	No out-of-household activities at all.	0

industries, and activities were prohibited in Milan. This lockdown was also extended to Germany, Spain, and France to prevent the rapid spread of COVID-19.[47] Some countries had to use excessive force and teargas against the violators of lockdown law as happened in Kenya.[48] Recently, Systems Dynamics models were used to simulate and evaluate different quarantine periods. This model was applied for a population of 100,000 inhabitants to simulate three different scenarios as illustrated in Table 10.2.[49] It was found that the three types of quarantine were effective in lessening infection numbers compared to no quarantine. Nevertheless, scenario II and scenario III could be the best approaches because these scenarios will not negatively affect the capacity of available hospitals.

The stress on the health and administration sections is a major problem for accommodating a large number of patients in a short period with the symptoms of COVID-19.[50] Following social distancing and lockdown restrictions as the major factors in reducing the spread of COVID-19,[51] Ambikapathy and Krishnamurthy[52] assessed the lockdowns for 4, 14, 21, 42, and 60 days as intervention strategies in India. They observed that reductions in the transmission of COVID-19 were detected in 21 days and more significant reductions for 42 days. Extending further to 16 days might not result in the positive effect on transmission. The success of lockdown can be measured on certain parameters. Table 10.3 illustrates these major parameters according to Krishnan et al.[53]

Singapore recorded its first case on January 23, 2020, almost at the same time as the US. Outside of China, Singapore had the strongest contact with Wuhan, China, so it could have been the most threatened by COVID-19, however, the outbreak of COVID-19 in Singapore has been one of the slowest in the world. In the early days of the COVID-19 outbreak,

Table 10.2 Different quarantine scenarios implemented by the Italian government.

Scenario	Initial day	Final day	The initial day of the second lockdown	Final second lockdown
I. Long lockdown	25	85	—	—
II. Two shorts lockdown and one smart	25	55	85	115
III. One medium lockdown and one smart	25	65	66	106

Table 10.3 Parameters of lockdown measurements according to Krishnan et al.[53]

Parameters	Explanation
1. Flattening the curve	• Means that fewer people need treatment at any time. • Guarantees that the infected cases who need to be hospitalized remain below the capacity of the health care system.
2. Reducing the growth rate of new cases	• Helps decision-makers evaluate the effectiveness of containment by making sure it reaches its peak or the number of new items in a downtrend. • Indicate a negative growth rate.
3. Containing the spread of COVID-19	• Prohibiting spread to newer and rural areas during lockdown.
4. Improving health care capacity	• The lockdown period assists to ease the burden on the health care system and provides an opportunity for the government to use the time to build the additional capacity needed to reopen.

Singapore conducted simulations that highlighted the combined efforts of quarantine, social distance from school, and work stoppages to slow down COVID-19. However, Singapore did not immediately implement a national lockdown, and continued to recommend a social distance, which was easily enforceable. Almost four months after the first case was recorded, Singapore was experiencing a second wave of SARS-CoV-2 infections, leading to an official nationwide lockdown that took place on April 28, 2020. The key to Singapore's success in controlling the COVID-19 epidemic, in addition to timely action, can be attributed to its dynamic and strong economy.[54] The economic situation and management in Singapore are not comparable to poor and low-income countries like African countries. In a study, Kuguyo et al.[54] found that lockdown in African countries during the COVID-19 period should not be an option. Poverty has been a major barrier to being affected by the national lockdown in South Africa, a situation that is reflected in most other African economies. The lockdown period slows down the transmission of society and leads to the identification of infection clusters. Unfortunately, this containment policy was not possible for Africa's ailing economy.[54]

As the outbreak of COVID-19 started in Wuhan city, China, the Chinese government locked down one-third of its cities. The government did not enforce a complete lockdown but implemented tailored measures such as the closure of schools and factory shutdowns.[42] A recent study found that Wanzhou, China succeeded in effectively controlling the spread of COVID-19 to disrupt the transmission chain by including social distancing, extensive contact tracing, mask wearing, extensive testing, and strict quarantine. If these containment measures had not been implemented, the number of infections per day could have reached a peak of about 39,000 by January 30, 2020, leading to about 560,000 infected people in the city.[55] During the lockdown in Scotland, they adapted the strategy of "Test and Protect." This approach included: (1) testing people, (2) tracing contacts, and (3) quarantining infected people or those exposed to the virus.[35] Another successful story took place in Germany where a protocol of systematic contact tracing was developed. They could keep the mortality rate low and control COVID-19 spreading.[56] A latest research compared the spreading of COVID-19 between the complete lockdown countries and bordering countries applying social distancing procedures only.[57] It confirmed the absence of any effects of COVID-19.

Sweden (in Europe) and Taiwan (in Asia) are examples of countries that managed the outbreak of COVID-19 without enforcing a lockdown. However, two completely different stories emerged, one successful and the other almost defeated by negligence. For instance, in Sweden, the government relied on the awareness of the citizens. Hence, the citizens are asked to keep social distancing in public places and stay at home if they were sick. As mentioned in the previous section, the Swedish government followed a deadly soft COVID-19 strategy. Sweden's strategy of not implementing quarantine has not been as effective. Anders Tegnell, a chief epidemiologist for the Swedish public health authority who architected the herd immunity strategy, admitted that many people died by urging his country not to pursue a severe lockdown policy.[58] On the other hand, Taiwan, or Chinese Taipei, an island attached to China, achieved a significant victory over COVID-19. Taiwan, unlike most of the countries that undertook strict lockdown, was able to prevent severe lockdown throughout the country with basic measures. Nevertheless, the keys to Taiwan's great success can be traced back to: (1) quick preparation, timely action; (2) border control, travel restrictions, and finding suspicious items: enforcing strict rules in the first few days for travelers, especially the Chinese; (3) timely allocation of resources, which led to the cessation of exports of masks and medical equipment, so increased production of masks in the country; and (4) communication and policy-making measures, including timely and accurate information and prevention of rumors. The key to Taiwan's success is the government's "rapid preparedness and timely action," which is the result of the country's previous experience in fighting the 2002–2004 SARS outbreak. There is also a centralized command center for crisis management called the National Health Command Center (NHCC) that provides data and information seamlessly to officials and experts. On the other hand, the very special feature of the Taiwanese people is to follow the government regulations as the main factor that complemented the decisions and timely actions of the NHCC. Surprisingly, Taiwan has not had any deaths from COVID-19 in the last 200 days, and has recorded a total of 733 cases and 7 deaths so far (as of December 13, 2020). As a result, Taiwan has shown how to learn well from past experience and quickly overcome the crisis. Taiwan became the most successful country in the confrontation of COVID-19 by taking timely, fast, and principled measures and employing skilled and experienced teams of officials.[59,60]

It is noteworthy to highlight that lockdown cannot be imposed forever. The factories cannot be closed and the vehicular movement cannot be restricted for a long time.[61] As the number of infected COVID-19 cases starts to decrease, governments decided to finish the lockdown. Therefore, governments and policy-makers should plan sustainable strategies to continue regular lifestyle after the COVID-19 period. The criteria that governments should follow to lift lockdown include: (1) controlling the transmission of infection; (2) accessing the health system to identify, test, isolate, and ultimately treat the infection; (3) minimizing the risk of vulnerable areas such as nursing homes; and (4) protective measures for essential places such as workplaces and educational institutions, and control of the potential risk of newly imported infections.[62]

Extended periods of lockdown are not recommended because it will increase economic and social issues.[63] However, it was noticeable that lockdown had a positive effect on the environment.[64] During the lockdown across the world, people started to observe a clear blue sky and seas. The effect of lockdown restrictions has been studied and evaluated on the quality of water and air. Fig. 10.5 summarizes the positive impact of lockdown due to COVID-19 on the environment. The first study to demonstrate this positive impact on surface water quality in India was performed by Yunus et al.[65] They found that suspended particulate matter (SPM) in the Vembanad Lake in India decreased by 15.9%, which reflected the enhancement of surface water quality. Mandal and Pal[66] detected an improvement in the water quality of the Ganga River during

Figure 10.5 Lockdown effects owing to COVID-19 on water and air quality.

the lockdown by 40%–50%. Total dissolved solids (TDS) concentration was reduced from 2457 to 987 mg/L and biological oxygen demand (BOD) was <2 mg/L. Depellegrin et al.[67] evaluated the effect of lockdown by studying the human-marine environment interactions and design innovative approaches that can enrich ecosystem-based management of marine resources.

In 2016, WHO reported that air pollution is responsible for around 8% of total death in the world.[68] Air Quality Index (AQI) is usually used to express the magnitude of air pollution of a region according to $PM_{2.5}$, PM_{10}, CO, SO_2, NO_2, and O_3. It categorizes the air quality into six categories from Good (0–50) to Severe (more than 401).[61,69] A recent article estimated the air pollution concentrations through the lockdown in 34 countries using satellite data and a network of more than 10,000 air quality stations.[70] It was found that NO_2 and $PM_{2.5}$ concentrations were reduced by 60% and 31%, respectively. He et al.[40] estimated the effect of lockdown on AQI across China's cities which decreased 19.84 points. It was found that $PM_{2.5}$ was lowered by 14.07 $\mu g/m^3$ in the locked-down cities. The same behavior was detected in other cities without lockdown due to preventive measures. Wang et al.[71] recorded the average of AQI was decreased from 89.6 points to 71.6 points with 20% reductions after the breakout of COVID-19 with imposing some restriction measures. Dantas et al.[72] found that CO emission levels reduced during the partial lockdown with 30.3%–48.5% in Rio de Janeiro, Brazil. Berman and Ebisu[73] indicated a 25.5% reduction of NO_2 compared to historical data and also an 11.3% reduction in fine particulate matter ($PM_{2.5}$). In Italy, the lockdown led to a significant drop in SO_2 levels only in the city of Milan.[74] In India, the most polluted megacities were based on the environmental performance index in Delhi. Since 2011, $PM_{2.5}$ and PM_{10} concentrations are very high and they are beyond the tolerable limits followed by NO_2, SO_2, and CO.[69]

With the onset of the second wave of COVID-19 in Europe, possibly driven by Belgium and the Czech Republic, as of October 30, 2020, European countries have been forced to reimplement quarantine laws to combat COVID-19.[75] The following are some cases: Ireland became the first European Union country to impose a quarantine that allowed people to travel only 5 km from their place of residence. Restrictions imposed in Ireland are among the most severe restrictions in Europe.[76] France announced a general quarantine until December 1, 2020, which might be extended until January. Schools and universities continue to operate under these restrictions, but family members cannot accommodate more than one person. Quarantine was reintroduced in Germany with less intensity than in the spring, as well as the current situation in France.[77] Belgium was the last country to join the convoy of European countries that imposed a national quarantine for the second time. During the new quarantine, which was almost identical to that in France with a duration of 45 days, citizens were not allowed to invite more than one family member to their home and were allowed to leave the house only in groups of up to four people.[78]

In South America, Brazil had a sad story. President Jair Bolsonaro continued to pursue wrong policies, ignoring the danger of an epidemic, and false sense of immunity. At the beginning of COVID-19, Bolsonaro likened the virus to "little flu" and insisted that Brazilians were "immune" to it and had nothing to worry about. In April 2020, Bolsonaro fired the then Brazilian Minister of Health for supporting policies of physical distancing and social quarantine. Bolsonaro saw the damage to quarantine policies and social distancing as worse than COVID-19 and was concerned about the country's economy. The deteriorating situation in Brazil reached a point where in mid-June 2020 the number of COVID-19 cases passed 1 million.[79]

Iran was one of the failed countries at the beginning of the epidemic. Iranian officials identified the city of Qom as the outbreak of the disease, which was announced on February 19, 2020, but Iranian President Rouhani prevented it from being quarantined, calling it a relic of the pre—World War I era. Rouhani considered the recent sanctions imposed by the Trump administration as the main reason for not implementing the quarantine, which creates more fundamental problems for the Iranian economy by creating a quarantine. Rouhani stated that his government did not intend to impose a lock on the eve of the Iranian New Year holiday, March 20, 2020, and there was no such thing at all. Although Rouhani did not believe in quarantine, he attended all meetings in accordance with health protocols and often by video conference.[80] After the virus spread to other cities, the government kept roads to the north of the country open despite requests from local authorities to restrict traffic. The armed forces took further action after many Iranians refused to follow official instructions to prevent the spread of the virus.[81] Finally, as of March 13, 2020, with the deteriorating situation and the increase in the COVID-19 confirmed to more than 11,300 cases, the first quarantine was officially announced by the Iranian government.[82] The quarantine did not last long and eased off after a while. Unplanned policies have led to an increase in the number of patients in recent months and the loss of health care workers. The IHME had predicted that daily mortality would increase in the coming months, peaking at around 825 daily in mid-December, forcing Iran to reapply quarantine.[28] Unsurprisingly, in the face of a new rise in the number of COVID-19, Iranian authorities have ordered hundreds of cities across the country to impose a variety of severe restrictions, again from November 21, 2020. In the month before the quarantine, about 400 deaths were reported daily.[83]

As a result, the decision to use the lockdown must be carefully considered in all respects to indicate a reduction of new COVID-19 cases considering its negative impact of the sudden removal of the lockdown without precautions. The findings presented here emphasize the importance of early and critical decision-making to control the epidemic by considering the negative side effects of lockdown.

10.4 Physical or social distancing

The respiratory viruses are usually transmitted via mucosalivary droplets produced during speech, sneezing, and exhalation.[84] In addition, the transmission of respiratory disease occurs via two routes, namely, droplet route, for large droplets, that assumes they fall on external surfaces and contaminate them, and aerosol or airborne route, for small droplets of less than five microns, that assumes their inhalation. For this reason, COVID-19 is assumed to spread by both direct contact and droplet transfer while the airborne transmission is still under debate.[85] Accordingly, keeping a suitable physical distance between different persons is assumed to effectively reduce COVID-19 transfer by both airborne and droplet transmission. This physical distance, or what is called "social distancing rules," is being now applied in many countries to decrease or prevent the risk of COVID-19 spread.[86–88]

COVID-19 outbreaks in environment are most common when people are outdoors, traveling, in restaurants and catered events, schools and institutional settings, as well as health care facilities settings. Several socio-environmental measures like social distancing and stay-at-home rules has already been taken to manage this crisis and have shown positive results.[89] It means to keep physical distance of at least about two arm's length among individuals which is established as a safe space between individuals. Exact measure of social distancing designate of staying at least 6 feet away from each other, in every aspect of public and professional domains. With respect to environmental health, social distancing offers a very strong and powerful preventive measure against transmission of COVID-19 virus in near vicinity, which ultimately lowers down the possibility of its transmission.[90] The main reason to enforce social distancing is to prevent the contact with infected people and contaminated surfaces. A recent study by researchers at the University of Texas at Austin of COVID-19 outbreaks in 58 Chinese cities found that places that took longer to implement social distancing spent more time with the rapidly expanding virus. The results of the analysis showed that every day that a city is delayed in social distancing measures after the onset of the first case, the duration of the outbreak is increased by 2.4 days.[91]

The droplet size concept could be useful to differentiate the large droplets and small airborne fine particles. It is known that respiratory system infections are caused by exhaled air containing a wide range of droplet sizes. This interval of droplets sizes varies from visible (millimeters) to invisible ones (microns). However, social distancing rules refer to the risk of droplet spread to isolated comparatively large droplet releases only. Accordingly, if COVID-19 is only transferred by large droplets, this would indicate that the risk will be reduced by applying shorter physical distancing. It has been reported that infected peoples with COVID-19 caused high rates of secondary infection to people of close contact with them. This infection was likely to occur within 1–2 m.[92–95] According to the previous assumptions, the WHO has suggested the policy of 1-m social distance.[85] Nonetheless, other countries have applied their own policies. For example, Spain and Canada are now implementing a 2 m distance rule. Similarly, the UK recommended the 2 m rule but it is still under review.[96] The social distance should be appropriate and flexible since too short distance increases the risks, whereas too long distance could be disruptive to society.

10.4.1 Origins of two-meter rule

Studies concerning the transmission of droplets while speaking or when sneezing or coughing have started since in the 19th century by using air samples cultured on agar or glass plates.[84,97] Flugge assumed that approximately 2 m distance is safe based on the fact that samples over this distance usually contain droplets with pathogens.[98] Turner et al.[99] presented some visual representations of these emissions. Subsequently, Hamburger and Robertson[100] reported that 65% of the total participants (48 people) produced large droplets, and only <10% of these droplets traveled about 1.7 m. In contrast, in 10% of participants, the "hemolytic streptococci" bacteria were collected 2.9 m away. These findings reveal the scientific basis of the rule of 1–2 m distance.[101]

However, different distances were used by different studies. A meta-regression conducted by Chu et al.[102] who found that as the distance increases the strength of association becomes larger. It should be noted that 172 studies were analyzed in this systematic review and 44 studies of them are considered as comparative studies with about 25,697 patients, with COVID-19, SARS, and MERS. The main results indicate that a large reduction in infection was associated with social distancing of at least 1 m. On the other hand, it was clear that distances of 2 m might be more effective.

The main conclusions that can be withdrawn from these studies are:

- The 2 m social distancing rule assumes that large droplets falling on surfaces are the dominant routes of transmission of COVID-19.
- Smaller droplets (airborne) contaminated with COVID-19 could spread up to 8 m. These small droplets are concentrated in the exhaled air resulted from infected individuals.
- The risk of COVID-19 transmission increases as the physical distance between people decreases, so for indoor settings, even 2 m might be too close.[103]

Consequently, social distancing should be used in conjunction with other approaches to reduce transmission. These strategies include adapting ventilation to indoor spaces, air hygiene, regular surface cleaning, effective hand washing, and using face masks in addition to fast isolation of affected individuals. It is now familiar that COVID-19 transmission occurs primarily by people having symptoms as well as those who have not yet developed symptoms. This transmission occurs when infected and noninfected people are very close together for long periods. However, since persons who never develop symptoms can transfer the virus to others, more studies are needed in this area.

10.4.2 COVID-19 transmission prevention

The main aim of the "Strategic Preparedness and Response Plan" is to control COVID-19 by preventing or reducing its transmission and the associated illness and consequently death.[104] WHO recommends a comprehensive set of measures to prevent COVID-19 transmission. These measures include:

- Identify of all suspect, test, and isolate all cases of infected people as quickly as possible using appropriate facilities.
- Identify and quarantine all infected individuals in close contact. Then testing the people who develop the symptoms and isolating them in case of infection and needing more care.
- Use face masks in public places where physical distancing is not possible.
- Health workers caring for COVID-19 patients should use all precaution including the use of a medical mask to prevent their infection.
- The frequent use of hand hygiene, keeping suitable physical distancing, and respiratory etiquette should be always practiced. In addition, people should avoid close-contact surroundings and crowded enclosed spaces with poor aeration. Finally, proper ventilation in all enclosed environments and the use of appropriate cleaning and disinfection for the environment should be always applied.[105]

The main conclusions of the above discussion support the application of physical distancing of 1 m or more. In addition, the optimum use of face masks, respirators, and eye protection in public and health care settings should be applied. Moreover, accurate and randomized trials are needed to better inform the evidence for these interventions.

10.5 Mask wearing

Historically, using face masks by people in public places has been a common action for controlling the pandemic of respiratory diseases. For instance, the 1918 H1N1 influenza pandemic,[106,107] 2002, and 2003 SARS epidemic in Asia, and the COVID-19 pandemic in Taiwan.[108] Wearing a mask by a susceptible person is led to infection prevention. Furthermore, if an infected person wears a mask, the face mask prevents the transmission of infection to susceptible people.[109,110] To prevent COVID-19 transmission, the WHO has recommended wearing a mask, according to published guidelines on June 5, 2020. In this recommendation, the governments should encourage people to mask wearing in public places where social distancing is not achieved, as a plenary approach to stop COVID-19 transmission. Many Asian countries have used a face mask much before the guideline setting, while most European countries have reserved face masks for COVID-19 patients, and health care staff.[13]

Using surgical masks assist in decreasing the risk of COVID-19 transmission. In general, face mask wearing has less effect on the control of seasonal respiratory infections. Although, in such a severe pandemic in which there is no definite antiviral drug, face masks can significantly prevent pandemic transmission.[111] Wearing a face mask and observing social distancing can be more impressive in control of COVID-19. Face mask wearing in combination with social distancing with an efficiency of around 50% and observing by just 30% of the population can lead to a decreasing of COVID-19 load considerably. According to a mathematical modeling, the physical distancing processes should carry on to the elimination of COVID-19 in late 2021. Otherwise, the second devastating wave will occur.[112] In a study, three factors were investigated in masks, including rate of aerosol reduction, public coverage, and availability to evaluate the influence of wearing ordinary medical masks in community through combination of mathematical modeling and existing scientific data to struggle over the COVID-19 pandemic. It shows that mask wearing in addition to social distancing has a positive effect to reduce the epidemic.[113]

Due to the beginning of the COVID-19 pandemic and increasing the information on transmission, everyone recognizes the importance of mask wearing in the prevention of transmission and considers it a strong cover against the spread of SARS-CoV-2. Some masks work better than others to help slow the spread of the virus that causes COVID-19. This last depends on the fabrics used and, most important, how the mask is made (such as the type of fabric, the number of layers of fabric, and how well the mask fits). Is there a strong reason why using multiple layers of a face mask is more effective in preventing transmission than a simple layer face mask? To clarify this, more research is needed in the upcoming. However, face shields also may be considered an alternative to mask. Wearing a mask with social distancing and frequent hand washing are effective in prevention of transmission, not alone. The efficiency of several commonly used face masks was studied by laser light and simulation of cough. This study showed that two-layer homemade face masks prevent the respiratory drops.

Therefore, they are recommended for the public. However, the medical staff should wear surgical or N95 masks. Well-fitting is an important issue in using face masks to leach the respiratory droplets form around the face. Multilayered masks with waterproof outer layer have the best efficiency in minimizing virus spread. However, the most urgent and important act in COVID-19 pandemic control is wearing a mask beside hand washing and social distancing.[114,115] A modeling study showed that even face masks with 20% efficiency are useful. Wearing a mask in public has a significant effect on limiting transmission of the pandemic. The universal approval of face mask wearing besides the physical distancing has community-wide benefits.[116] It should be noted that people think mask wearing protects them, but the main objective is to protect others from respiratory droplets because of asymptomatic transmission of the coronavirus.[117] Airborne transmission in the spread of COVID-19 is reduced significantly using mask wearing. For instance, in Italy and New York City 78,000 fewer in a month and 66,000 fewer infections within three weeks, respectively.[118]

Three things determine in the effect of mask wearing: the duplication of the virus in a society, the efficiency of masks in blocking, and the number of wearers. If masks with an efficiency of 60% wear by 80% of people, the reproduction rate of the virus gets to less than 1. In Hong Kong, 80% of people wear a mask in public places, in Israel, Singapore, and the Czech Republic wearing masks in public places has been legally regulated.[119] Face mask wearing rate by Hong Kong Special Administrative Region (HKSAR) population was 96.6%. Hence, the COVID-19 prevalence in HKSAR (129 per million person) was lower compared to some other countries from December 31, 2019 to April 8, 2020.[11] Fig. 10.6 shows shares of Singapore's citizens who have worn masks in public places from February to September 2020 during the COVID-19 pandemic.[120]

The centers for Disease Control and Prevention (CDC) issued new guidance to wearing face masks where close contact is unavoidable between people in public places. Italy, Germany, France, South Korea, Spain, and China have enacted laws to enforce face mask rules in public areas.[121,122] China, because it is the world's largest manufacturer of personal protective equipment (PPE), made it relatively easy for them to make masks. It was also very good for the Chinese to readily adopt mask wearing because of their experience in the previous epidemic.[123] By contrast, the United States refused to wear masks even in June and July, when it was on the rise. Even in late September 2020, President Trump ridiculed Joe Biden's mask wearing as a weakness.[123] However, between April 8 and May 15, 2020,

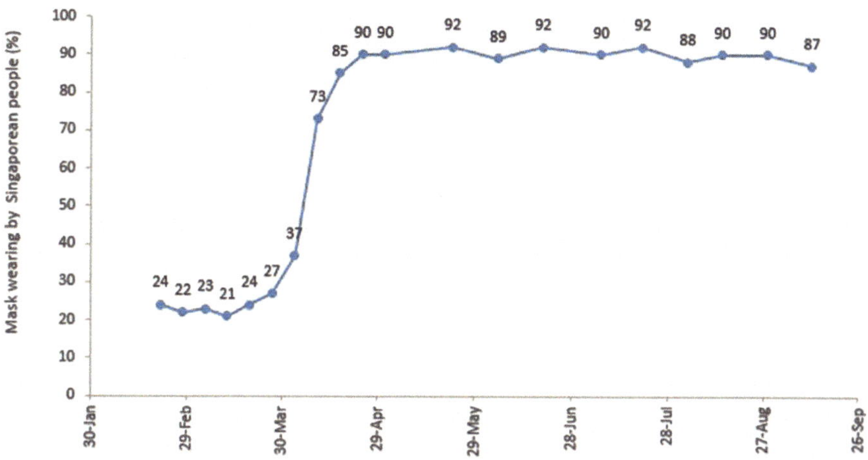

Figure 10.6 From February to September 2020, the participation of Singaporean who used face masks in public places during the COVID-19 pandemic. *(Modified after Statista.[120])*

mask wearing had been mandated in public places where possible social distancing may not be observed in 15 states in the US. The effect of mandating mask wearing in public place showed a significant decline in the COVID-19 growth rate.[124] In Hong Kong, only four deaths have been reported within the COVID-19 pandemic period. Notwithstanding high density, mass transportation, and closeness to Wuhan city. Taiwan reported only six deaths as well due to mandating the population to wear a mask in public places. The country could control the pandemic, and schools have been opened since late February 2020.[119] An observational cohort study by Salvatore et al.[125] examined the possibility of intrauterine mother-to-child transmission as well as breastfeeding during the COVID-19 epidemic at three Presbyterian Hospitals in New York City, United States. In this study, mothers used surgical masks and practiced proper hand hygiene. One hundred sixteen mothers were tested positive for SARS-CoV-2 and 120 neonates were examined. All neonates were tested in the first 24 h and it was observed that none were infected. Eighty-two (68%) neonates were followed for five to seven days of life, of which 68 (83%) neonates were in the same room with breastfeeding mothers, and the test was repeated for 79 (96%) neonates, all of whom were negative. Seventy-two (88%) neonates were tested after 14 days of life, all of which were negative. None of the neonates had COVID-19 symptoms. This study demonstrated the potential role of mask wearing for mothers when near their neonate.[125]

According to the Ipsos survey of 28,000 participants in early April 2020 found that 81% of respondents used a face mask because of COVID-19 epidemic. Data revealed that Vietnam had the highest number of people wearing masks due to the COVID-19 (91%). The survey indicated that people in Asian countries, including Vietnam, China, Japan, and India, performed better than most European countries. Western authorities have been slower in advising to cover their faces to prevent the spread of the coronavirus. Facial masks became mandatory in some parts of Italy in early April, which also reduced the prevalence of the disease. However, in several German states, this practice was only mandatory in congested settings. According to the results, only 20% of Germans had used this kind of mask by April 12th' 2020. The use of the mask was mandatory in seven US states, and after the CDC recommendation in early April 2020, it was encouraged in many countries, and as a result, 50% of Americans have at least tried it. France also recommended the use of masks throughout the country, and 34% have tried them out.[126] The countries of the Eastern Mediterranean Region (EMR) had different knowledge and risk perception of COVID-19, so that the use of masks varied from 6% to 83%.[127] As of November 3, 2020, Iran (628,780), Iraq (478,701), Saudi Arabia (348,037), and Morocco (222,544) have the highest total number of confirmed COVID-19 cases in the EMR. In these countries, the use of masks has been around or even less than 50%. These countries used the wrong policy to create a culture of mask wearing in society. Not wearing enough masks in the whole population was an important reason for the substantial increase in the number of COVID-19 cases in such countries.[128] According to the IHME, if wearing mask in Iran increases to 95%, rendering to the pattern that existed in Singapore and several countries in Latin America, the number of predicted deaths will be close to 68,000 and can be about 13,900 people rescued. That is about a 29% drop in the number of deaths predicted from now until February 1, 2021.[28]

Although Japan is one of the first countries to be infected with COVID-19, it can be considered another strong model in reducing the number of people with COVID-19 infection. Despite Japan's large elderly population and premature infection, infection and death rates are among the lowest in the world. The main reason for this is Japan's strong mask culture, which reduces the COVID-19 to controllable levels.[129] Japan currently (as of September 22, 2020) has about 80,000 cases; in contrast, about 90% of them have recovered, indicating a high level of prevention and mask wearing by them nationwide.[2] Also recently, the Advanced

Telecommunications Research Institute International in Kyoto, Japan took a creative step to deal with the possible third wave of COVID-19 in the country, and by launching a robot called "Robovie" in stores which detected people who do not use a face mask and warn them of other COVID-19 guidelines.[130] In Czechia, which made it mandatory to wear a mask while in public, as of May 24, 2020, there were only 8932 confirmed COVID-19 cases and 315 deaths, making it one of the least infected countries in Europe. Czechia government, like most countries, faced a shortage of masks early in the epidemic, but people began to make masks with unprecedented collective effort, not just for hospitals but also for everyone. In some areas, people created "mask trees" where they would put available additional masks that were up for grabs for others. The order to use the mask—jointly with neighboring Slovakia—quickly became a symbol of the Czechia struggle against the epidemic.[131] As of September 22, 2020, the infection rate in Czechia has reached 55,000 cases and 555 deaths,[2] which shows the significant impact of mask use in this country. Table 10.4 shows countries with mask-wearing requirements in public places.

Table 10.4 Countries, which have required or recommended masks in public places.

Country	Mask requirement/Date
USA	Part of country
Germany	Full country/April 27, 2020
UK	Full country/June 15, 2020
France	Full country/May 11, 2020
Italy	Full country/May 4, 2020
Brazil	Part of country
Russia	Part of country
Mexico	Full country/May 20, 2020
Spain	Full country/May 2, 2020
China	Universal mask usage
Japan	Universal mask usage
India	Universal mask usage
South Korea	Universal mask usage
Switzerland	Recommends masks
Sweden	—
Norway	Recommends masks
Hong Kong	Universal mask usage
Finland	Recommends masks
New Zealand	Recommends masks
Netherland	Public transport/June 1, 2020
Denmark	Public transport

In Mongolia, one of the low-income countries with the outbreak of COVID-19, people have reacted widely with preventive measures by increasing health and safety behaviors, including the use of face masks. The use of face masks has been widely used and seen since February 2020 in most places such as streets, workplaces, and especially in markets and health centers. This led to the emergence of home fabric face masks as the demand for face masks increased. As of mid-February, the Mongolian government had required most organizations and businesses to wear masks. Mongolia, with a low number of infected cases—after 7 months after the pandemic, only 313 infected cases—showed that having strong prevention systems such as wearing a mask as a culture can effectively respond to an epidemic in a low-income or middle-income country.[132] In Inner Mongolia, drones equipped with echoing loudspeakers warned citizens who were not wearing a mask and told them not to talk or walk around without mask wearing.[123] A recent study by Mitze et al.[133] examined the effect of masks on the prevention of COVID-19 in Germany and found that after 20 days of becoming mandatory face masks, the number of new infections have reduced about 45% (between 15% and 75%). In addition, given that the mask has negligible economic costs compared to other public health measures, the mask is a cost-effective means of combating COVID-19. The importance of mask wearing is evident from the recent reports, which concluded that on the first day of COVID-19 infection, the probability of detecting the virus was zero. Even on the fifth day of infection, with symptoms and with a high load of the virus, no more than one-third of cases were diagnosed. As a result, mask wearing is a key parameter even with a negative test.[134]

In conclusion, wearing a mask can be an acceptable NPI to decrease efficiency of transmission for the viral respiratory diseases transmitted through aerosol and droplets, although masks should not be an alternative for social distancing and hand washing. Therefore, countries should introduce a compulsive policy and express the advantages of mask wearing.

10.6 Conclusion

Within a month or two of the onsets of COVID-19, it was often supposed that like other viruses (e.g., MERS and SARS), COVID-19 would have fewer deaths. The herd immunity approach often adopted by the Nordic countries (e.g., Sweden) failed because they did not achieve the expected immunity (above 60%) at all. On May 4, 2020, Belgium topped the list, exposing only 6.2% of its population to COVID-19, which is not expected to exceed 7%. Similarly, in July it was observed that the prevalence in Spain and Switzerland is less than 10%. However, other researchers believe that applying this strategy to low-income countries (e.g., India) might be responsible due to its high youth population.

Enforced lockdown was one of the measures taken by most countries during this epidemic period. In February 2020, Italy became the first country in Europe to adopt a nationwide lockdown policy. They adopted quarantines based on the population of the household, which was effective in reducing the prevalence of the disease. Gradually, different scenarios including (1) long lockdown, (2) two short lockdowns and one smart, and (3) one medium lockdown and one smart were implemented in different countries. Social distancing of 1 m is another precautionary measure proposed by WHO guidelines, while governments have each adopted their own policies in this regard. Canada and Spain, for example, required a social distance of 2 m, which could be even higher in closed areas due to the faster transmission of infection. Currently, reports of deaths indicate that some Asian countries such as Japan, South Korea, Singapore, China, and Hong Kong are smarter than most European and American countries in taking measures such as testing and tracking individuals, lockdown hot spot areas, mask wearing, and social distancing. In most epidemics, mask wearing as a protector dramatically reduces the transmission of infection. Mask wearing will be about 50% effective if accompanied by social distancing. Hong Kong is one of the most successful countries in this field and almost 80% of people use masks in public places, so it is considered as a culture among the people. Taiwan with its ideal management has been a very successful example of how society responds quickly to crises and a protector worthy of the interests of its citizens. The low number of deaths and infected cases in this country compared to other countries shows their performance well.

At the same time, vaccines with high immunization efficiency are developed making COVID-19 vaccines available on a global scale requires the use of complex production techniques, scrupulous quality control and trustworthy distribution channels that ensure the vaccines are effective and accessible to all human kind. Based on experiences of the countries involved with COVID-19, this chapter suggests that current pandemic prevention policies are including isolating individuals, maintaining a social distancing of about 2 m in crowded places, and wearing masks to prevent transmission of infection is the best solution needed by all countries, which should be institutionalized as a culture among them. As of the writing of the chapter (mid-December 2020), there are about 72.1 million cases and about 1.61 million deaths, bringing the epidemic level of the disease to the level of syndemic. Therefore, what scientists have learned so far from COVID-19 indicates that the story behind the disease is not as simple as other epidemics. This shows to protect the health of our communities, an even more subtle approach is needed. In a crisis, governments often make difficult decisions under hesitation and time constrictions. These decisions must be culturally appropriate and sensitive to the people, and the utmost important task that governments must have is to be honest with their people in providing information.

Acknowledgments

The support of the Research Vice-Chancellor of Qazvin University of Medical Sciences (IR.QUMS.REC.1399.405) is wholeheartedly appreciated. Milad Mousazadeh gratitudes the Iran's National Elites Foundation (INEF) for granting the Elite Award, No. 15/11529. The authors would like to thank Dr. Mohammad Saeid Rezaee-Zavareh for assistance in data collection of this chapter.

References

1. Aldila D, Khoshnaw S-H, Safitri E, et al. A mathematical study on the spread of COVID-19 considering social distancing and rapid assessment: the case of Jakarta, Indonesia. *Chaos Solit Fractals* 2020;**139**:110042.
2. Worldometers. *Coronavirus*. Available from: https://www.worldometers.info/coronavirus/. [Accessed 14 December 2020].
3. Yezli S, Khan A. COVID-19 social distancing in the Kingdom of Saudi Arabia: bold measures in the face of political, economic, social and religious challenges. *Trav Med Infect Dis* 2020;**37**:101692.
4. Ferguson N, Laydon D, Nedjati Gilani G, et al. Report 9: Impact of non-pharmaceutical interventions (NPIs) to reduce COVID19 mortality and health-care demand. https://www.imperial.ac.uk/media/imperial-college/medicine/sph/ide/gida-fellowships/Imperial-College-COVID19-NPI-modelling-16-03-2020.pdf. [Accessed 14 December 2020].
5. Shaw R, Kim Y-k, Hua J. Governance, technology and citizen behavior in pandemic: lessons from COVID-19 in East Asia. *Prog Disaster Sci* 2020;**6**:100090.
6. Allegrante JP, Auld ME, Natarajan S. Preventing COVID-19 and its Sequela:"There is no magic bullet... It's just behaviors". *Am J Prev Med* 2020;**59**:288–92.
7. Sardar T, Nadim SS, Rana S, Chattopadhyay J. Assessment of lockdown effect in some states and overall India: a predictive mathematical study on COVID-19 outbreak. *Chaos Solit Fractals* 2020;**139**:110078.
8. Chang K-C, Strong C, Pakpour AH, Griffiths MD, Lin C-Y. Factors related to preventive COVID-19 infection behaviors among people with mental illness. *J Formos Med Assoc* 2020;**119**:1772–80.

9. Raoofi A, Takian A, Sari AA, Olyaeemanesh A, Haghighi H, Aarabi M. COVID-19 pandemic and comparative health policy learning in Iran. *Arch Iran Med* 2020;**23**(4):220–34.
10. Kennedy D, Seale A, Bausch D, Ritchie H, Roser M. How experts use data to identify emerging COVID-19 success stories. *Our World in Data.* Available from: https://ourworldindata.org/identify-covid-exemplars. [Accessed 14 December 2020].
11. Cheng VC, Wong S-C, Chuang VW, et al. The role of community-wide wearing of face mask for control of coronavirus disease 2019 (COVID-19) epidemic due to SARS-CoV-2. *J Infect* 2020;**81**(1):107–14.
12. Feng S, Shen C, Xia N, Song W, Fan M, Cowling BJ. Rational use of face masks in the COVID-19 pandemic. *Lancet Respir Med* 2020;**8**(5):434–6.
13. Fodjo JNS, Pengpid S, de Moura Villela EF, et al. Mass masking as a way to contain COVID-19 and exit lockdown in low-and middle-income countries. *J Infect* 2020;**81**(3):e1–5.
14. Morato MM, Bastos SB, Cajueiro DO, Normey-Rico JE. *An optimal predictive control strategy for COVID-19 (SARS-CoV-2) social distancing policies in Brazil.* 2020. arXiv preprint arXiv:200510797.
15. Okell LC, Verity R, Watson OJ, et al. Have deaths from COVID-19 in Europe plateaued due to herd immunity? *Lancet (London, England)* 2020;**395**(10241):e110–1.
16. Bhopal RS. COVID-19 zugzwang: potential public health moves towards population (herd) immunity. *Public Health Pract* 2020;**1**:100031.
17. Roberts S. *The Swiss Cheese Model of Pandemic Defense.* Available from: https://www.nytimes.com/2020/12/05/health/coronavirus-swiss-cheese-infection-mackay.html. [Accessed 14 December 2020].
18. Wiles S, Morris T. *Covid-19 and the Swiss cheese system.* Available from: https://www.auckland.ac.nz/en/news/2020/10/22/covid-19-and-the-swiss-cheese-system.html. [Accessed 14 December 2020].
19. Jones D, Stefan H. A history of herd immunity. *Lancet* 2020;**396**(10254):810–1.
20. Randolph HE, Barreiro LB. Herd immunity: understanding COVID-19. *Immunity* 2020;**52**(5):737–41.
21. Kwok KO, Lai F, Wei WI, Wong SYS, Tang JW. Herd immunity—estimating the level required to halt the COVID-19 epidemics in affected countries. *J Infect* 2020;**80**(6):e32–3.
22. Delamater PL, Street EJ, Leslie TF, Yang YT, Jacobsen KH. Complexity of the basic reproduction number (R0). *Emerg Infect Dis* 2019;**25**(1):1.
23. ISPI. *Covid-19: How Many Got Infected around the world?* Available from: https://www.ispionline.it/en/pubblicazione/covid-19-how-many-got-infected-25972. [Accessed 14 December 2020].
24. Elizabeth G. *Whose coronavirus strategy worked best? Scientists hunt most effective policies.* Available from: https://www.nature.com/articles/d41586-020-01248-1#ref-CR1. [Accessed 14 December 2020].
25. Omer SB, Yildirim I, Forman HP. Herd immunity and implications for SARS-CoV-2 control. *Jama* 2020;**324**(20):2095–6.
26. Korhonen J, Granberg B. Sweden backcasting, now?—strategic planning for covid-19 mitigation in a liberal democracy. *Sustainability* 2020;**12**(10):4138.
27. Orlowski EJ, Goldsmith DJ. Four months into the COVID-19 pandemic, Sweden's prized herd immunity is nowhere in sight. *J R Soc Med* 2020;**113**(8):292–8.
28. IHME. *Iran Model updates for October 29, 2020.* Available from: http://www.healthdata.org/sites/default/files/files/Projects/COVID/briefing_Iran_20201029.pdf. [Accessed 14 December 2020].

29. Hunter P, Pankhania B. *Expert comments about herd immunity*. Available from: https://www.sciencemediacentre.org/expert-comments-about-herd-immunity/?cli_action=1607185121.104. [Accessed 14 December 2020].
30. Limam A. *Does Sweden's COVID-19 experience support the herd immunity theory?* Available from: https://newseu.cgtn.com/news/2020-09-14/Does-Sweden-s-COVID-19-experience-support-the-herd-immunity-theory–TL82pOjfpe/index.html. [Accessed 14 December 2020].
31. Altstedter A. *Infect Everyone: How Herd Immunity Could Work for Poor Countries*. Available from: https://www.bloomberg.com/news/articles/2020-04-21/a-herd-immunity-strategy-could-actually-work-in-youthful-india. [Accessed 14 December 2020].
32. Alwan NA, Burgess RA, Ashworth S, et al. Scientific consensus on the COVID-19 pandemic: we need to act now. *Lancet* 2020;**396**(10260):e71–2.
33. Eckerle I, Meyer B. SARS-CoV-2 seroprevalence in COVID-19 hotspots. *Lancet* 2020;**396**(10250):514–5.
34. Hellerstein M. What are the roles of antibodies versus a durable, high quality T-cell response in protective immunity against SARS-CoV2? *Vaccine X* 2020;**6**:100076.
35. Sridhar D, Chen A. Why Scotland's slow and steady approach to covid-19 is working. *BMJ* 2020;**370**:m2669.
36. WHO. *WHO Director-General's opening remarks at the media briefing on COVID-19 -October 12, 2020*. Available from: https://www.who.int/director-general/speeches/detail/who-director-general-s-opening-remarks-at-the-media-briefing-on-covid-19—12-october-2020. [Accessed 14 December 2020].
37. Kenyon C. Flattening-the-curve associated with reduced COVID-19 case fatality rates-an ecological analysis of 65 countries. *J Infect* 2020;**81**(1):e98–9.
38. Atalan A. Is the lockdown important to prevent the COVID-19 pandemic? Effects on psychology, environment and economy-perspective. *Ann Med Surg* 2020;**56**:38–42.
39. Praticò AD. COVID-19 pandemic for Pediatric Health Care: disadvantages and opportunities. *Pediatr Res* 2020:1–2. https://doi.org/10.1038/s41390-020-0955-x.
40. He G, Pan Y, Tanaka T. The short-term impacts of COVID-19 lockdown on urban air pollution in China. *Nat Sustain* 2020;**3**:1005–11.
41. Bedford J, Enria D, Giesecke J, et al. COVID-19: towards controlling of a pandemic. *Lancet* 2020;**395**(10229):1015–8.
42. Caduff C. What went wrong: corona and the world after the full stop. *Med Anthropol Q* 2020;**34**(4):467–87. https://doi.org/10.1111/maq.12599.
43. Gualano MR, Lo Moro G, Voglino G, Bert F, Siliquini R. Effects of COVID-19 lockdown on mental health and sleep disturbances in Italy. *Int J Environ Res Publ Health* 2020;**17**(13):4779.
44. Sjödin H, Wilder-Smith A, Osman S, Farooq Z, Rocklöv J. Only strict quarantine measures can curb the coronavirus disease (COVID-19) outbreak in Italy, 2020. *Euro Surveill* 2020;**25**(13):2000280.
45. Every-Palmer S, Jenkins M, Gendall P, et al. Psychological distress, anxiety, family violence, suicidality, and wellbeing in New Zealand during the COVID-19 lockdown: a cross-sectional study. *PloS One* 2020;**15**(11):e0241658.
46. Siqueira CAS, Freitas YNL, Cancela MC, Carvalho M, Oliveras-Fabregas A, de Souza DLB. The effect of lockdown on the outcomes of COVID-19 in Spain: an ecological study. *PloS One* 2020;**15**(7):e0236779.
47. Yuan J, Li M, Lv G, Lu ZK. Monitoring transmissibility and mortality of COVID-19 in Europe. *Int J Infect Dis* 2020;**95**:311–5.
48. Namwaya O. Kenya police abuses could undermine coronavirus fight. *Human Rights Watch Website* 31 March 2020. https://www.hrw.org/news/2020/03/31/kenya-police-abuses-could-undermine-coronavirus-fight. [Accessed 14 December 2020].

49. Ibarra-Vega D. Lockdown, one, two, none, or smart. Modeling containing covid-19 infection. A conceptual model. *Sci Total Environ* 2020;**730**:138917.
50. Tomar A, Gupta N. Prediction for the spread of COVID-19 in India and effectiveness of preventive measures. *Sci Total Environ* 2020;**728**:138762.
51. Babbar S. Battle with COVID-19 under partial to zero lockdowns in India. *medRxiv* 2020. https://doi.org/10.1101/2020.07.03.20145664.
52. Ambikapathy B, Krishnamurthy K. Mathematical modelling to assess the impact of lockdown on COVID-19 transmission in India: model development and validation. *JMIR Public Health Surveill* 2020;**6**(2):e19368.
53. Krishnan S, Deo S, Manurkar S. *50 Days of lockdown: measuring India's success in arresting COVID-19*. 2020.
54. Kuguyo O, Kengne AP, Dandara C. Singapore COVID-19 pandemic response as a successful model framework for low-resource health care settings in Africa? *OMICS A J Integr Biol* 2020;**24**(8):470–8.
55. Shi Q, Hu Y, Peng B, et al. Effective control of SARS-CoV-2 transmission in Wanzhou, China. *Nat Med* 2020:1–8.
56. Mohr M, Datan-Grajewski A. *So stark ist die Krankenhaus–Auslastung mit Corona-Patienten*. MDR website: https://www.mdr.de/sachsenanhalt/corona-daten-update-so-stark-ist-die-krankenhaus-auslastung-coronaviruspatienten-covid-neunzehn-100.html. [Accessed 14 December 2020].
57. Meunier TA. Full lockdown policies in Western Europe countries have no evident impacts on the COVID-19 epidemic. *medRxiv* 2020. https://doi.org/10.1101/2020.04.24.20078717.
58. Bloomberg. *How experts use data to identify emerging COVID-19 success stories*. Available from: https://www.bloomberg.com/news/articles/2020-06-03/man-behind-sweden-s-virus-strategy-says-he-got-some-things-wrong. [Accessed 14 December 2020].
59. Summers J, Cheng HJ, Lin HH, et al. Potential lessons from the Taiwan and New Zealand health responses to the COVID-19 pandemic. *The Lancet Regional Health-Western Pacific* 2020;**4**(100044). https://doi.org/10.1016/j.lanwpc.2020.100044.
60. Wang CJ, Ng CY, Brook RH. Response to COVID-19 in Taiwan: big data analytics, new technology, and proactive testing. *Jama* 2020;**323**(14):1341–2.
61. Lokhandwala S, Gautam P. Indirect impact of COVID-19 on environment: a brief study in Indian context. *Environ Res* 2020;**188**:109807.
62. Shammi M, Bodrud-Doza M, Islam ARMT, Rahman MM. Strategic assessment of COVID-19 pandemic in Bangladesh: comparative lockdown scenario analysis, public perception, and management for sustainability. *Environ Dev Sustain* 2020:1–44.
63. Peto J, Alwan NA, Godfrey KM, et al. Universal weekly testing as the UK COVID-19 lockdown exit strategy. *Lancet* 2020;**395**(10234):1420–1.
64. Mousazadeh M, Paital B, Naghdali Z, Mortezania Z, Hashemi M, Niaragh EK, et al. Positive environmental effects of the coronavirus 2020 episode: a review. *Environ Dev Sustain* 2021. https://doi.org/10.1007/s10668-021-01240-3.
65. Yunus AP, Masago Y, Hijioka Y. COVID-19 and surface water quality: improved lake water quality during the lockdown. *Sci Total Environ* 2020;**731**:139012.
66. Mandal I, Pal S. COVID-19 pandemic persuaded lockdown effects on environment over stone quarrying and crushing areas. *Sci Total Environ* 2020;**732**:139281.
67. Depellegrin D, Bastianini M, Fadini A, Menegon S. The effects of COVID-19 induced lockdown measures on maritime settings of a coastal region. *Sci Total Environ* 2020;**740**:140123.
68. WHO. *Air pollution*. Available from: https://www.who.int/health-topics/air-pollution#tab=tab_1. [Accessed 14 December 2020].

69. Mahato S, Pal S, Ghosh KG. Effect of lockdown amid COVID-19 pandemic on air quality of the megacity Delhi, India. *Sci Total Environ* 2020;**730**:139086.
70. Venter ZS, Aunan K, Chowdhury S, Lelieveld J. COVID-19 lockdowns cause global air pollution declines. *Proc Natl Acad Sci U S A* 2020;**117**(32):18984–90.
71. Wang Y, Yuan Y, Wang Q, Liu C, Zhi Q, Cao J. Changes in air quality related to the control of coronavirus in China: implications for traffic and industrial emissions. *Sci Total Environ* 2020;**731**:139133.
72. Dantas G, Siciliano B, França BB, da Silva CM, Arbilla G. The impact of COVID-19 partial lockdown on the air quality of the city of Rio de Janeiro, Brazil. *Sci Total Environ* 2020;**729**:139085.
73. Berman JD, Ebisu K. Changes in U.S. air pollution during the COVID-19 pandemic. *Sci Total Environ* 2020;**739**:139864.
74. Collivignarelli MC, Abbà A, Bertanza G, Pedrazzani R, Ricciardi P, Carnevale Miino M. Lockdown for CoViD-2019 in Milan: what are the effects on air quality? *Sci Total Environ* 2020;**732**:139280.
75. Time. *Europe's Second Wave of COVID-19 is Being Driven by Two Countries. Here's Why*. Available from: https://time.com/5902172/europe-coronavirus-second-wave-belgium-czech-republic/. [Accessed 14 December 2020].
76. Euronews. *Corona in Ireland; Traffic beyond 5 km from the residence was prohibited*. Available from: https://per.euronews.com/2020/10/23/ireland-enters-new-lockdown-as-new-wave-of-covid-19-hit. [Accessed 14 December 2020].
77. Stevis-Gridneff M. *France and Germany Lock Down as Second Coronavirus Wave Grows*. Available from: https://www.nytimes.com/2020/10/28/world/europe/france-germany-coronavirus-covid.html. [Accessed 14 December 2020].
78. DeutscheWelle. *Coronavirus: belgium hit hard by second pandemic wave*. Available from: https://www.dw.com/en/belgium-coronavirus-measures-lockdown/a-55332941. [Accessed 14 December 2020].
79. Charner F. *Brazil tops 1 million Covid-19 cases. It may pass the US next, becoming the worst-hit country on the planet*. Available from: https://edition.cnn.com/2020/06/19/americas/brazil-one-million-coronavirus-jair-bolsonaro-cases-intl/index.html. [Accessed 14 December 2020].
80. Rasmussen Sune E., Aresu E. *Iran's coronavirus strategy favored economy Over public health, leaving both exposed*. Available from: https://www.wsj.com/articles/irans-coronavirus-strategy-favored-economy-over-public-health-leaving-both-exposed-11584272132. [Accessed 14 December 2020].
81. BBC. *Coronavirus: Iran has no plans to quarantine cities, Rouhani says*. Available from: https://www.bbc.com/news/world-middle-east-51651454. [Accessed 14 December 2020].
82. GradaWorld. *Iran: Natiowide lockdown implemented as over 11,300 COVID-19 cases confirmed March 13/update 12*. Available from: https://www.garda.com/crisis24/news-alerts/322811/iran-natiowide-lockdown-implemented-as-over-11300-covid-19-cases-confirmed-march-13-update-12. [Accessed 14 December 2020].
83. Tahereh H-J. *Iran tries another lockdown as coronavirus cases soar*. Available from: https://www.atlanticcouncil.org/blogs/iransource/iran-tries-another-lockdown-as-coronavirus-cases-soar/. [Accessed 14 December 2020].
84. Parienta D, Morawska L, Johnson GR, et al. Theoretical analysis of the motion and evaporation of exhaled respiratory droplets of mixed composition. *J Aerosol Sci* 2011;**42**(1):1–10.
85. WHO. *Coronavirus disease (COVID-19) advice for the public*. Available from: https://www.who.int/emergencies/diseases/novel-coronavirus-2019/advice-for-public. [Accessed 14 December 2020].

86. Alagoz O, Sethi AK, Patterson BW, Churpek M, Safdar N. Impact of timing of and adherence to social distancing measures on COVID-19 burden in the US: a simulation modeling approach. *medRxiv* 2020. https://doi.org/10.1101/2020.06.07.20124859. https://www.ncbi.nlm.nih.gov/pmc/articles/pmc7302402/.
87. Du Z, Xu X, Wang L, et al. Effects of proactive social distancing on COVID-19 outbreaks in 58 cities, China. *Emerg Infect Dis* 2020;**26**(9):2267.
88. Shiu EY, Leung NH, Cowling BJ. Controversy around airborne versus droplet transmission of respiratory viruses: implication for infection prevention. *Curr Opin Infect Dis* 2019;**32**(4):372–9.
89. Saadat S, Rawtani D, Hussain CM. Environmental perspective of COVID-19. *Sci Total Environ* 2020;**728**:138870.
90. Paital B, Das K, Parida SK. Inter nation social lockdown versus medical care against COVID-19, a mild environmental insight with special reference to India. *Sci Total Environ* 2020:138914.
91. Kent J. *COVID-19 data shows how social distancing impacts virus spread.* https://healthitanalytics.com/news/covid-19-data-shows-how-social-distancing-impacts-virus-spread. [Accessed 14 December 2020].
92. Bi Q, Wu Y, Mei S, et al. Epidemiology and transmission of COVID-19 in 391 cases and 1286 of their close contacts in Shenzhen, China: a retrospective cohort study. *Lancet Infect Dis* 2020;**20**(8):911–9.
93. Chaw L, Koh WC, Jamaludin SA, Naing L, Alikhan MF, Wong J. Analysis of SARS-CoV-2 transmission in different settings, Brunei. *Emerging Infectious Diseases* 2020;**26**(11). https://doi.org/10.3201/eid2611.202263.
94. Li W, Zhang B, Lu J, et al. The characteristics of household transmission of COVID-19. *Clin Infect Dis* 2020;**71**(8):1943–6.
95. Van Doremalen N, Bushmaker T, Morris DH, et al. Aerosol and surface stability of SARS-CoV-2 as compared with SARS-CoV-1. *N Engl J Med* 2020;**382**(16):1564–7.
96. Bourouiba L, Dehandschoewercker E, Bush JW. Violent expiratory events: on coughing and sneezing. *J Fluid Mech* 2014;**745**:537–63.
97. Jones NR, Qureshi ZU, Temple RJ, Larwood JP, Greenhalgh T, Bourouiba L. Two metres or one: what is the evidence for physical distancing in covid-19? *BMJ* 2020;**370**:m3223.
98. Eames I, Tang J, Li Y, Wilson P. Airborne transmission of disease in hospitals. *J R Soc Interface* 2009;**6**(6).
99. Turner CE, Jennison MW, Edgerton HE. Public health applications of high-speed photography. *Am J Public Health Nation's Health* 1941;**31**(4):319–24.
100. Hamburger Jr M, Robertson O. Expulsion of group A hemolytic streptococci in droplets and droplet nuclei by sneezing, coughing and talking. *Am J Med* 1948;**4**(5):690–701.
101. Bourouiba L. Turbulent gas clouds and respiratory pathogen emissions: potential implications for reducing transmission of COVID-19. *Jama* 2020;**323**(18):1837–8.
102. Chu DK, Akl EA, Duda S, et al. Physical distancing, face masks, and eye protection to prevent person-to-person transmission of SARS-CoV-2 and COVID-19: a systematic review and meta-analysis. *Lancet* 2020;**395**(10242):1973–87.
103. Qureshi Z, Jones N, Temple R, Larwood J, Greenhalgh T, Bourouiba L. What is the evidence to support the 2-metre social distancing rule to reduce COVID-19 transmission? *Cent Evid Based Med* 2020:200622.
104. Papineni RS, Rosenthal FS. The size distribution of droplets in the exhaled breath of healthy human subjects. *J Aerosol Med* 1997;**10**(2):105–16.
105. WHO. *Transmission of SARS-CoV-2: implications for infection prevention precautions.* https://www.who.int/news-room/commentaries/detail/transmission-of-sars-cov-2-implications-for-infection-prevention-precautions. [Accessed 14 December 2020].

106. Luckingham B. To mask or not to mask: a note on the 1918 Spanish influenza epidemic in Tucson. *J Ariz Hist* 1984:191–204.
107. Tognotti E. Influenza pandemics: a historical retrospect. *J Infect Dev Ctries* 2009;**3**(05):331–4.
108. Wu J, Xu F, Zhou W, et al. Risk factors for SARS among persons without known contact with SARS patients, Beijing, China. *Emerg Infect Dis* 2004;**10**(2):210.
109. Aiello AE, Murray GF, Perez V, et al. Mask use, hand hygiene, and seasonal influenza-like illness among young adults: a randomized intervention trial. *J Infect Dis* 2010;**201**(4):491–8.
110. Aiello AE, Perez V, Coulborn RM, Davis BM, Uddin M, Monto AS. Facemasks, hand hygiene, and influenza among young adults: a randomized intervention trial. *PloS One* 2012;**7**(1):e29744.
111. WHO. *Coronavirus disease (COVID-19) advice for the public: when and how to use masks.* https://www.who.int/emergencies/diseases/novel-coronavirus-2019/advice-for-public/when-and-how-to-use-masks. [Accessed 14 December 2020].
112. Ngonghala CN, Iboi E, Eikenberry S, et al. Mathematical assessment of the impact of non-pharmaceutical interventions on curtailing the 2019 novel Coronavirus. *Math Biosci* 2020;**325**(108364). https://doi.org/10.1016/j.mbs.2020.108364.
113. Li T, Liu Y, Li M, Qian X, Dai SY. Mask or no mask for COVID-19: a public health and market study. *PloS One* 2020;**15**(8):e0237691.
114. McCabe C. *Face masks really do matter. The scientific evidence is growing.* Available from: https://www.wsj.com/articles/face-masks-really-do-matter-the-scientific-evidence-is-growing-11595083298. [Accessed 14 December 2020].
115. Van Dam A, Romm T. *As Arizona struggles, neighboring New Mexico found a more cautious path to sustained growth.* Available from: https://www.washingtonpost.com/business/2020/07/21/arizona-struggles-neighboring-new-mexico-found-more-cautious-path-sustained-growth/. [Accessed 14 December 2020].
116. Eikenberry SE, Mancuso M, Iboi E, et al. To mask or not to mask: modeling the potential for face mask use by the general public to curtail the COVID-19 pandemic. *Infect Dis Model* 2020;**5**:293–308.
117. Cheng KK, Lam TH, Leung CC. Wearing face masks in the community during the COVID-19 pandemic: altruism and solidarity. *Lancet* 2020:1–2. https://doi.org/10.1016/S0140-6736(20)30918-1.
118. Galasso V, Pons V, Profeta P, Becher M, Brouard S, Foucault M. Gender differences in COVID-19 attitudes and behavior: panel evidence from eight countries. *Proc Natl Acad Sci U S A* 2020;**117**(44):27285–91.
119. Tufekci Z, Howard J, Greenhalgh T. *The real reason to wear a mask.* Available from: https://www.theatlantic.com/health/archive/2020/04/dont-wear-mask-yourself/610336/. [Accessed 14 December 2020].
120. Statista. *Share of Singaporean population who wore face masks in public places during COVID-19 outbreak from February to September 2020.* Available from: https://www.statista.com/statistics/1110983/singapore-wearing-masks-during-covid-19-outbreak/. [Accessed 14 December 2020].
121. Bostock B. *France has made wearing face masks compulsory in public, while maintaining a controversial ban on burqas and niqabs.* Available from: https://www.businessinsider.com/france-face-masks-compulsory-burqas-niqabs-banned-criticism-muslims-2020-5. [Accessed 14 December 2020].
122. Duncan C. *Coronavirus: Spain makes face masks compulsory on public transport as country begins to ease strict lockdown.* Available from: https://www.independent.co.uk/news/world/europe/coronavirus-face-masks-spain-public-transport-lockdown-pedro-sanchez-a9496031.html. [Accessed 14 December 2020].
123. Burki T. China's successful control of COVID-19. *Lancet Infect Dis* 2020;**20**(11):1240–1.

124. Lyu W, Wehby GL. Community Use of Face Masks and COVID-19: evidence from a Natural Experiment of State Mandates in the US: study examines impact on COVID-19 growth rates associated with state government mandates requiring face mask use in public. *Health Aff* 2020;**39**(8):1419–25.
125. Salvatore CM, Han J-Y, Acker KP, et al. Neonatal management and outcomes during the COVID-19 pandemic: an observation cohort study. *Lancet Child Adolesc Health* 2020;**4**(10):721–7.
126. Buchholz K. *Asians still most likely to wear face masks due to COVID-19*. Available from: https://www.statista.com/chart/21452/share-of-people-wearing-face-masks-per-country-covid-19/. [Accessed 14 December 2020].
127. Al-Mandhari AS, Brennan RJ, Abubakar A, Hajjeh R. Tackling COVID-19 in the Eastern Mediterranean region. *Lancet* 2020;**396**(10265):1786–8.
128. Statista. *Total number of confirmed coronavirus (COVID-19) cases in the Eastern Mediterranean region as of November 3, 2020, by country*. Available from: https://www.statista.com/statistics/1105588/eastern-mediterranean-number-of-coronavirus-cases-by-country/. [Accessed 14 December 2020].
129. DW. *Coronavirus: how Japan keeps COVID-19 under control*. Available from: https://www.dw.com/en/coronavirus-how-japan-keeps-covid-19-under-control/a-52907069. [Accessed 14 December 2020].
130. Theguardian. *Japan shop deploys robot to check people are wearing face masks*. Available from: https://www.theguardian.com/world/2020/nov/16/japan-shop-deploys-robot-to-check-people-are-wearing-face-masks. [Accessed 14 December 2020].
131. The guardian. *Czech Republic lifts face mask rule as coronavirus restrictions ease*. Available from: https://www.theguardian.com/world/2020/may/25/czech-republic-face-mask-coronavirus-restrictions-pubs-restaurants-hotels. [Accessed 14 December 2020].
132. Erkhembayar R, Dickinson E, Badarch D, et al. Early policy actions and emergency response to the COVID-19 pandemic in Mongolia: experiences and challenges. *Lancet Glob Health* 2020;**8**(9):E1234–41.
133. Mitze T, Kosfeld R, Rode J, Wälde K. Face masks considerably reduce covid-19 cases in Germany. *Proc Natl Acad Sci USA* 2020;**117**(51):32293–301. https://doi.org/10.1073/pnas.2015954117. 33273115.
134. Kucirka LM, Lauer SA, Laeyendecker O, Boon D, Lessler J. Variation in false-negative rate of reverse transcriptase polymerase chain reaction–based SARS-CoV-2 tests by time since exposure. *Ann Intern Med* 2020;**173**(4):262–7.

CHAPTER 11

Food safety, hygiene, and awareness during combating of COVID-19

Tanima Chowdhury[1], **Shirsendu Nandi**[2]
[1]Quality Department, Regional Head Office, ALPLA India Pvt. Ltd., Hyderabad, Telangana, India;
[2]Operations Management and Quantitative Techniques, FORE School of Management, New Delhi, India

11.1 Introduction

Viruses are gaining particular interest in their behavior and are entirely different from bacteria. It has been noticed that most foodborne viruses are harmful in nature and cause a detrimental effect on human health. Viruses are microorganisms that are generally very small and tiny, having a diameter ranging from 0.02 to 0.4 μm, whereas bacteria usually range between 0.5 and 5 μm. Transmission of viruses to humans may happen in different ways, but most of the foodborne viruses are infected via the gastrointestinal tract and are excreted through feces or vomitus.[1] Though the consumption of contaminated food has been clearly implicated as the source of viral infections, the proportion of infection that can be caused due to the consumption of the contaminated food is still an unknown fact. The most common food-induced viruses are noroviruses (NoV) and hepatitis A virus (HAV), which is a threat to human health worldwide. Some other viruses like rotaviruses, enteroviruses, and astroviruses are also considered critical foodborne viruses; however, to a lesser extent,[2,3] vomiting and diarrhea are the most common symptoms of viral gastroenteritis, which may be severe to some extent and also causes death in a few cases.[3]

The name coronavirus came from the Latin word "corona," which refers to "crown" or "halo" due to the morphological characterization of the virus. Very recently, a novel strain of coronavirus was identified. The current COVID-19 pandemic primarily originated in Wuhan, China.[4–8] The initial infection was originated from a Wuhan seafood market where various types of seafood, as well as live animals, were sold. The seafood market was closed on January 1, 2020 when the initial two to three cases were identified from the same location. The infection started spreading very rapidly, and on January 21, 2020, the transmission of this virus from human to human was

Food Safety Management during COVID-19
ISBN 978-0-323-85780-2
https://doi.org/10.1016/B978-0-323-85780-2.00002-0

first reported by Chinese Health Authorities. As a result, case counts had increased rapidly and have spread all over the world in several countries.[9]

Though it is still unclear if the market is the original source of this virus, the spread of the virus was noticed to increase more rapidly due to the vast crowd and multiple touchpoints in the market area.[10] Coronavirus (CoV) is thought to be zoonotic, which means it may pass from animals to humans, which generally happens if any infected animal is killed and the same is consumed.[11]

CoV has been identified in various animals such as bats, cats, rats, and chickens, which has been reported to cause respiratory tract and gastrointestinal diseases.[12–17] A new strain, namely severe acute respiratory syndrome coronavirus 2 (SARS-CoV-2), has also been identified, which is a virus having positive-stranded RNA covered by a lipid bilayer.[18] Pneumonia and fever are the common symptoms of coronavirus infections, the severity of which depends from person to person. Fever, cough, chest discomfort, and dyspnea are some of the clinical symptoms of these kinds of viruses.[17,19] It has been reported that most of the transmission occurs by the spread of large droplets, which generates when a patient coughs or sneezes.[20,21]

The virus can also spread from the surfaces that are touched by some infected person, including food or food packaging.[22] It may be that the food or the packaging has the virus on it, and when someone touches those infected packs and then touch their own body parts like mouth, nose, or eyes, the person may get infected with the virus. The chance of this virus getting transmitted through food is found to be less common than the transmission from one person to another.[23] Even if contaminated food is consumed, there is no evidence that swallowing the contaminated food leads to any infection.[24] The virus needs to be transmitted to the respiratory system—into the nose, sinuses, or lungs to act on the same and spread the infection. To follow good food safety practices, it is always important to wash hands with hand wash for at least 20 s or at least sanitize hands with 70% alcohol after handling any kind of packages or before preparing or eating food.[25] This may reduce the risk of illness from common foodborne pathogens. It is also believed that properly cooked food may kill the virus.[26]

To manage food safety risks and also the chances of food contamination, the food industry should strictly follow the Food Safety Management Systems (FSMS) based on the Hazard Analysis and Critical Control Point (HACCP) principles.[27,28] According to the FSMS, the prerequisite programs are mandatory requirements, which include good hygiene practices, cleaning, sanitation and sanitization, area zoning (high risk and low risk), supplier evaluation and control, storage, logistics and supply chain, housekeeping, personnel hygiene, and also medical checkup to ensures fitness.[29] These are all part of Good Manufacturing Practices (GMP), which helps to control the hygiene at each stage of the food supply chain to prevent any type of food contamination. The employers should also have FSMS and HACCP team who will identify all the food safety—related risks and take necessary precautions to eliminate the same from the system.[30,31] After the outbreak of this pandemic, there are additional requirements for the food industries which should be maintained effectively to prevent the spread of this virus and also to improve the hygiene and sanitization practices. Food industries should also implement policies as per ISO 22,000[32] to protect food workers from getting affected by COVID-19 and also restrict the transmission of the virus by any means.[33]

There is a lot of interest to strengthen or boost immunity in the wake of the COVID 19 pandemic, which can be the first line of defense to fight against the virus. A well-balanced diet and being active physically and mentally can contribute to improve anyone's immunity and lead a healthy life. Many natural ingredients are known to boost the immune system. Some herbs and spices have been well known since ancient times for their medicinal properties, particularly in Asia. According to the World Health Organization,[34] herbal medicines are used as the primary health care by around 80% of the world's population across Europe and South Asia. India is the hub for many herbs and spices that are used for making this traditional medicine. The herbs have antiinflammatory properties and also help in building natural immunity to overcome the adverse effects of COVID-19. Most of these herbs, such as basil, fenugreek, ginger, garlic, and turmeric, are safe to consume and do not have any side effects like allopathic medicines, e.g., antibiotics.[26]

The main objective of this review is to provide an overview of the most common foodborne viruses and their effects on human health. The possibilities of transmitting COVID-19 through food and food packaging will be discussed along with the precautionary measures to avoid the risk of transmission. This should be adopted by the food handlers or food business owners to mitigate the risk of transmitting COVID-19 in the foodservice and retail sector. The new requirements of the FSMS will be discussed, which needs to be implemented in all food-related industries and followed religiously. Finally, dietary supplements that provide essential sources of vitamins, minerals, zinc, etc. will be focused on to boost the immune system, which can help to fight the virus.

11.2 Viruses and their transmission

11.2.1 Foodborne viruses

Foodborne viruses are derived from the gastrointestinal tract of a human.[37] These viruses are found in food and water as a result of poor hygiene practices, sewage systems, or contamination by food handlers. There are numerous viruses that are found in the human gut, among which only a few are reported as the most common foodborne pathogens, for example, NoV and HAV (Table 11.1).[35–37] Rotaviruses (group B and C) and hepatitis E virus outbreaks also occur frequently due to food and water contamination. Among all these viruses, NoV causes infection in people irrespective of their ages and is considered to be the most common mode of foodborne infection worldwide. Studies report that NoV infection occurs highest in children but also occurs in adults.[35,38] HAV is also a waterborne virus, which is known to occur in several countries.[39] This virus is seen to be more active in developing countries rather than in undeveloped countries where it is less common. The infection very rarely occurs among young children in their early childhood and remains asymptomatic.[4] However, most of the adults also remain susceptible to infection by HAV.[2,40]

11.2.2 Coronavirus disease (COVID-19) transmission via food

It is very unlikely that humans can get infected with COVID-19 through food or its packaging.[23] It causes respiratory illness, which is primarily transmitted through contact of one person to another and also by the spread of respiratory droplets that are generated through coughs/sneezing droplets produced by the infected person and reaches the nose, mouth, or eyes of the other person.[41]

To date, there is no report or evidence available, which confirms that COVID-19 is transmitted via food or its associated packaging.[23] This virus

Table 11.1 List of foodborne viruses and their sources.

Name of virus/ Illness	Source of contamination	Transmission via food products
Hepatovirus A/Hepatitis A	Water, human stool	Meat, raw beef, shellfish, fruits, and vegetables
Orthohepevirus/ Hepatitis E	Pork liver	Pork
Norovirus/ Gastroenteritis	Water, food, human stool	Meat, berry fruit, oyster, and shellfish

cannot grow in food, and they need an animal or human host to grow and multiply. The source of the COVID-19 is understood to be animals, but the exact source is still unknown to all. However, studies have shown that low temperature has a significant influence on the persistence of SARS-CoV-2 and other coronaviruses. It has been reported that the virus remains actively stable in frozen and refrigerated food.[42] The virus is known to be transmitted through droplets of infected persons, which may be present in the air and is a result of sneezing or touching an infected surface, e.g., doorknobs or shaking hands with the infected person and then touching their own face. Recent studies reported that the virus remains viable for up to 72 h on plastic material and stainless steel, up to 4 hours on copper, and up to 24 h on cardboard.[43,44]

In some cases, there may be cases who are asymptomatic and do not show any signs or symptoms of the disease at all or at a very mild level. However, asymptomatic infected people also are capable of spreading the virus.[45] This proves that the practice of personal hygiene and appropriate use of PPE is compulsory for all who work in the food industry, irrespective of their outward health condition. To maintain a virus-free working environment, food business owners should introduce very high-level security and staff management, which should include temperature monitoring, sickness reporting, sanitization, and so forth.[46–49]

11.3 Nanotechnology-based active and intelligent packaging on combating COVID-19

Although we are now aware that transmission of COVID-19 through food is not reported by any researchers, there is a chance that the virus can survive on the surface of the package, which in turn can pass from one person to the other during the entire supply chain. The development of polymers or biopolymers with antiviral properties can lead to an area of interest for scientists to minimize the risk of spreading COVID-19 through the food packaging material.[50] The spread of viruses like HAV and human norovirus (HuNoV) have already been shown to be controlled through various polymers and biopolymers, which contain the antiviral agent.[37] Warnes et al.[51] reported that polymers developed with copper had released the copper ions which helped in inactivating the HuCoV-229E virus on copper surfaces. van Doremalen et al.[52] recently observed that SARS-CoV-2 is most stable and sustained on copper and plastic surfaces and the viability of the virus could be detected up to 72 h. Some other studies

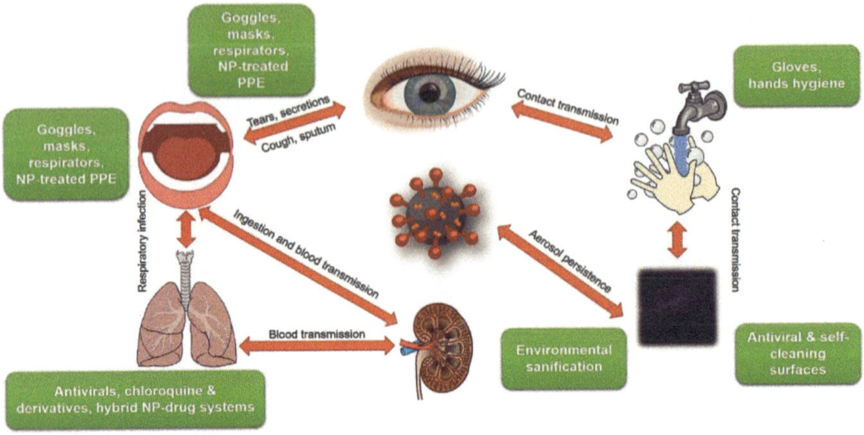

Figure 11.1 Transmission pathways of SARS-CoV-2. *NP*, nanoparticle; *PPE*, personal protective equipment. *(This figure is reproduced with permission from Sportelli.[32])*

showed that the use of nanomaterial coatings or films containing copper, silver, and zinc nanoparticles has the capability to fight against SARS-CoV-2 which prevented the contamination of food packaging material and reduces the transmission of the virus (Fig. 11.1).[53]

11.4 Food safety management system during COVID-19

During the COVID-19 situation, the FSMS should be reviewed by all the food business owners or food packaging-related industries where the implementation of different policies should be documented and the effectiveness to be monitored on a regular basis.[29] Defined or specific food safety HACCP should be in place for managing the emergency situations like COVID-19 outbreak. Following initiatives to be taken by all industries to maintain the Food Safety and Quality Management system.[32,54]

11.4.1 Internal audits

An internal audit program should be conducted to review the changes that have been introduced due to the response of the COVID-19 pandemic. Personal hygiene policy and facility housekeeping policies should be of greater emphasis. A remote site audit can be performed to maintain social distancing norms. There may be a need for emergency supplier approval for raw materials/packaging materials because of the shortage of material production at the supplier end due to fewer available workers or plant shutdowns. In such cases, before approving the alternate supplier, a remote/virtual audit should be conducted and observations to be recorded to overcome the shortage of supplies from the approved suppliers and to ensure continuity of production. However, this may include increased risk, which needs to be managed by the food safety and HACCP team.[32,55]

For food-related raw materials, the risk of allergens, the shelf life of the product, potential fraud or substitution, and product specification should be given more importance during the audit. For changing the packaging material suppliers, it is essential to understand if there are any differences in specification between the new and former product. The review should be focused more on the migration of chemicals into food (where there is direct food contact), product shelf life, product stability during storage, and transport.[56]

11.4.2 Site security and food defense

11.4.2.1 Temporary workers

It is very likely that during the COVID-19 crisis, all manufacturing sites will have a shortage of workers, and they need to make temporary arrangements for workers or increase the working hours due to production demand and slower line speeds. An alternate arrangement of temporary workers may increase the potential risk of the site. There may be violation of site policies, for instance, secret filming on phones or passing confidential data or chances of sabotage-related issues.[57] Site visitor's access to the site needs to be reviewed and may be restricted during the period of COVID-19. Acceptance of visitors to the site should be defined, which should cover general visitors, service providers including pest control, service engineers, laundry services, courier agency, vehicle drivers and emergency services. Health declaration to be taken from all types of visitors before giving access to the site.[29]

11.4.2.2 Entry-exit policy

The entry-exit policy should be reviewed and necessary changes to be implemented following personal hygiene, and also to ensure protection from any kind of allergen or microbial cross-contamination. Sickness reporting methodologies should be part of the daily training and monitoring. The movement of people from one area to the other should be restricted and should not be compromised.[56]

11.4.2.3 Maintenance

Planned preventive maintenance programs should be reviewed and implemented to run the plant without any unacceptable risk of breakdown. The shutdown of machines and equipment should be planned well in advance to meet the production requirement without any downtime or loss in production.[58]

11.4.3 Staff facilities

Changing areas and locker rooms should be designed in such a way that there is effective social distancing maintained at peak times, especially during the start and end of the shift. Additional space for changing or storage should be provided so that there is no person-to-person contact and also to confirm that the safety aspect is not compromised. Personal protective equipment (PPE) should be issued to all workers, and if required, additional PPEs also may be provided during this pandemic.[59] Mostly disposable PPEs are advisable; however, if any PPE is expected to reissue, there should be a provision that those are cleaned appropriately after use, it is numbered for easy identification and traceability, and last but not least, clean and hygienic storage areas should be available to store the PPE between shifts. Hand washing and sanitizing activity is always on high priority in food factories. Management should ensure hands are washed periodically, followed by sanitization.[60] Sanitizers should be provided in abundant quantity and placed near each and every work station. Frequency of handwashing and sanitization should be defined for proper effectiveness; wherever possible, automated soap/sanitizer dispensers should be used to avoid direct contact.[27]

11.4.4 Housekeeping and hygiene practices

Cleaning frequency should be increased with particular attention to direct touchpoints like doors, handles, etc. The effectiveness of cleaning equipment should be monitored during the start and end of every shift to avoid any type of cross-contamination.[61] Specific cleaning or fumigation should be done if any employee is detected with COVID-19 at the workplace to ensure the safety of other employees. The pest management program should also be in place and a record to be maintained.[30,37,62-63]

11.4.5 Vulnerability assessment/fraud prevention

It has been noticed that whenever there is a shortage of supply in materials, there is a rise in price for the raw materials, which increases the risk of fraudsters entering the market with substituted products at a very reasonable rate and also false claims. It is very important to verify all the details if the management decides to buy raw materials from alternative sources. All the technical terms and conditions to be evaluated and potential risk should be assessed to prevent any fraud.[64]

The site should closely monitor and evaluate the suppliers and also review the risk rating and contingency plan to increase testing or inspection procedures to identify the appropriate risk of the materials.[55]

11.4.6 Personnel

As already mentioned, during the COVID-19 crisis, it is very likely that due to sickness and self-isolation during this pandemic, there will be a shortage of workers for which additional employees are appointed for meeting the requirements. Safety and security should be ensured by the owners as well as monitoring or surveillance system should be strong to ensure that there is no fraud or sabotage during any phase of the entire supply chain.[65]

11.4.7 Training

It is particularly important that training is provided for all new and temporary employees taken on, which should additionally include all practices which should include personal hygiene procedures, entry-exit practices, social distancing, allergen management, sickness reporting, and so forth, which are specially implemented for COVID-19.[66] Sickness reporting is the crucial component that is included as one of the new procedures which will reduce the risk of spreading COVID-19. This is mainly for the safety of all employees (permanent and temporary workers), which should be monitored and a record maintained.[67]

11.4.8 Medical screening

Medical screening should be done for the COVID-19 affected employee after his/her return to work. COVID negative test report to be provided to ensure there is no more infection, which should be communicated to all other employees for their safety.[67] Management should ensure all employees get proper treatment and medical facilities to combat and overcome the situation.[66]

11.5 Precautionary measures to control COVID-19 in the food sector

The spread of the COVID-19 virus in the foodservice sector can be controlled by following the below precautionary measures. This will also ensure food safety requirements and will be beneficial for all individuals. The preventive measures, hygiene practices, and safety precautions should be adopted by the food handlers or food business owners to mitigate the risk of transmission of COVID-19.[24,68,102]

11.5.1 Hand hygiene

Human hands are the primary source of spreading this virus.[69] Foodservice operation activities directly involve touching the food or through the food packaging material (utensils, trays, packets/pouches, etc.). Nonfood surfaces like door handles, doorbells, handrails, etc., also come in contact during the supply chain. Hence maintaining hygiene practices by the food handlers is very important for their daily operation.[37]

As per the Centers for Disease Control and Prevention (CDC), washing hands frequently with soap for a minimum of 20 s and sanitization by 70% alcohol is the most essential and critical step to mitigate the risk of spreading the coronavirus during the foodservice operation.[39] In addition to the above precautions, food handlers should also maintain hygiene in the kitchen to wash and sanitize hands frequently, especially after a bio break.[70] Appropriate PPE like gowns, gloves, masks, hair nets, etc., to be provided to all employees who need to be changed as and when necessary to prevent the spreading of any germs.[59] Hands should also be washed and gloves should be changed in case of handling raw meat or poultry products after coughing/sneezing or blowing the nose. Once the disposable gloves, gowns, hair nets, shoe covers are used, those should be scrapped appropriately and the record maintained.[71]

11.5.2 Sanitization process

Sanitizers are the agents that are used to minimize the microbial load from the environment or contact surface to an acceptable level, which has been given by the different organizations and as per food safety regulation.[70] There are different sanitizers and disinfectants which should be used by the food handlers for different applications at various concentrations as per the severity of the infection that can be spread by the coronavirus.[72–74]

Surfaces that directly comes in contact with the food should be sanitized frequently to prevent and reduce the transmission of coronavirus. Before and after each use of the food-contact surfaces, food handlers should follow the proper cleaning and sanitization practices once every 2 h, especially during this pandemic.[75] It has been reported that the use of sodium hypochlorite at 50–100 ppm concentration is the most effective and cheaper sanitizer, which is mostly used in various industries for disinfecting food contact surfaces.[25,76] Surfaces like doors, chairs, tables, etc. which are nonfood, should also be sanitized at regular intervals with an appropriate concentration of sanitizer.[61]

In the retail sector, fresh produce is sold as cut fruits and vegetables in a packaged container. For this, the fruits and vegetables are manually cut, washed, and cleaned before packing the same in containers.[77] In such cases, there is a considerable chance of contamination as there are multiple touchpoints. Hence, food handlers should ensure that the fruits and vegetables are cut, maintaining all hygiene practices like using clean hands, sanitized cutting boards, knives, and sanitization of the containers/packages before packing to eliminate the risk of spreading of COVID-19 through surfaces and equipment.[78]

11.5.3 Social distancing

Maintaining social distance is an essential and crucial guideline that will help to slow down the risk of COVID-19 spread from one person to another. CDC and other food safety authorities have clearly recommended that a space of at least 2 m should be maintained in the workplace or in public.[79,80] Food handlers need to strictly follow the same and limit the presence of workers in close contact during the working hours at the workplace, especially where food is prepared in the kitchen or packed in a particular area which will curtail the hazard of spreading COVID-19.[81] The employees should avoid gathering at one place during a shift change, tea breaks, and training, and should always maintain a safe distance.[37]

11.5.4 Temperature monitoring

All food sectors should follow temperature monitoring by trained/skilled personnel at the workplace by using thermometer guns or infrared thermometers while each person enters the premises and also at an interval of 4 h in between their job.[82] If fever is found to be 100.4°F or higher, the employee should be sent back and not allowed to perform the job until he/she is found to be fit.[83]

Employers/the owners should revise the sick leave policies on account of COVID-19 and extend support to their staff to ensure that they get sufficient time to take care of and recover from the virus infection. Full recovery from COVID-19 is mostly found to be around 14–21 days depending on the patient's immunity, physical health, and, most notably, the severity of the infection.[80,84] All those employees along with their family members who are detected as COVID-19 positive should stay at home in quarantine until they recover fully and test negative.[85,86] Symptoms such as coughing, shortness of breath, loss of smell, and fever are mostly observed as symptoms of COVID-19;[87] if anyone seems to have any of the above symptoms should stay at home until the time they are free from the usual symptoms. If any food worker is tested COVID-19 positive, the employer should inform all the stakeholders, quarantine the infected person as well as associated workers who came in contact with him in the last two to three days. The company should also provide necessary support to the affected person's family in whatever way possible.[88]

11.5.5 Customer safety

Due to the COVID-19 outbreak, people have restricted themselves from going out to restaurants. To attract customers and also to provide no-contact service, many restaurants are now offering no-contact ordering of food and also home delivery to reduce the risk of transmission. Nowadays, contactless deliveries are encouraged by most people than dine-in. All food business operators should maintain and abide by all rules and regulations following food safety authorities to continue their business.[89]

It has been directed that hand sanitizers should be available in sufficient quantities at entrances and also in the waiting lounge of restaurants. Dining room tables and chairs should be sanitized after each use. High touchpoints like doors, restrooms, and computers should be frequently cleaned and sanitized.[61] Contactless orders should be taken through weblink or mobile apps. Social distancing should be maintained between each table so that the virus spread is less even during peak hours. Handling of cash/cards should be avoided and contactless payment gateways to be introduced without which there may be a chance of spreading the new coronavirus.[90]

11.6 Bioactive compounds and their role in the prevention of coronavirus

World Health Organization (WHO) has declared COVID-19 as a pandemic worldwide. It is clear that people all over the world are at high risk due to the community transmission of this virus; however, individuals can take few essential measures that may help them to fight against this pandemic.[37]

The effect or risk of getting affected with COVID-19 is much higher in people having some preexisting diseases or comorbidities like diabetes, cancer, hypertension, cardiac problems, and respiratory illness.[91–93]

The risk level is also high for elderly people as immunity, in general, reduces as we grow older. In the younger generation, if someone is a nonsmoker, COVID 19 may result in a minor infection due to the immunity which helps to combat the virus. Below are some of the measures which we can undertake to improve our immunity. Galanakis et al.[93] has explained in detail the role of food ingredients and bioactive compounds like bioactive peptides, polysaccharides, bioactive lipids, and natural polyphenols, which help in improving the immune system and prevention from COVID 19.[94]

11.6.1 Diet and supplements to boost immunity

The food we eat plays a crucial role in our overall health and immunity. During this pandemic, a low carbohydrate diet and high protein diet is advised by many physicians to keep the balance (Fig. 11.2).[95] Consumption

Figure 11.2 Foods as immune boosters. *(Image has been reproduced from an open access webpage published by Albin,[45] hence no permission required.)*

of fruits and vegetables will boost up immunity, which is usually rich in beta carotene, ascorbic acid, and other vitamins. Also, foods cooked with mushrooms, tomatoes, and other green vegetables are very good options to build flexibility to fight against any kind of bacterial or viral infections. It is also recommended to consume food that includes whole grains, legumes, nuts, and animal source foods.[95]

Supplements rich in omega 3 and 6 fatty acids and also known as some natural immunity boosters like ginger, gooseberries (amla), and turmeric, which can be a part of our daily diet. Several herbs like garlic, black cumin, and different kinds of nuts are known to help in boosting immunity and are excellent sources of protein and vitamin E. Yogurt and fermented foods should also be included in the regular diet. Apart from the above-mentioned diets, few common supplements can be used, which are known to boost immunity and also to fight against the COVID-19 virus.[34]

Vitamin C is considered to be a very crucial supplement in boosting immunity. It is well known to prevent the common cold and also acts as an antioxidant that defends against any kind of damage occurred due to oxidative stress. For severe infections like sepsis and acute respiratory distress syndrome (ARDS), a high dose of intravenous vitamin C has shown significant improvement in managing the symptoms in patients.[96]

In our present lifestyle and also due to the current pandemic situation, everyone is mostly staying at home and not having any exposure to sunlight, which in turn is resulting in a deficiency of vitamin D. Additional supplements of vitamin D are seen to control respiratory tract—related illness. Most people nowadays are deficient in vitamin D, and doctors suggest taking a supplement of vitamin D, which will help to improve the immune system. Food sources of vitamin D are fatty fishes, fish liver oils, cheese, and egg yolks. Fortified foods with vitamin D can also be chosen as an alternative source of vitamin D.[96]

Zinc is essential for white blood corpuscles (WBC), which helps to fight against viral infections. Consumption of zinc supplements improves immunity, which makes people less susceptible to cold, cough, and other viral infections.[97]

The berries and flowers of elderberry contain many antioxidants and vitamins that are very important in boosting our immune system. It is known to ease cold and flu, fight against influenza, lessen stress, inflammation, and also to protect from heart-related diseases. Elderberries contain very important nutrients and minerals like phosphorus, potassium, iron, copper, vitamins, proteins, and dietary fiber. It is also known to have antibacterial and antifungal properties.[98]

Functional food plants like garlic, tea, ginger, turmeric, pomegranate, black pepper, and honey possess immunomodulatory and antiviral properties, which are known to improve immunity in various ways, keeping away common cold and flu.[99]

Garlic has been known as the best for its antiviral properties and herbal remedies from ancient days. Several sulfur-containing compounds, like sulfoxide, proteins, and polyphenols, are present in garlic, which strengthens the immune system. Best results are observed when crushed garlic is consumed with honey.[100,101]

11.7 Conclusion

The COVID-19 disease resulted in a huge outbreak of respiratory illness, which originated from China. The food operators or food business owners should follow all the general hygiene rules to mitigate the risk of transmission of COVID-19 in the retail sector. Precautionary measures like hand sanitization, social distancing, personal hygiene, and contactless approaches should be maintained so that transmission does not happen from one person to another. All food business owners should strictly abide by the FSMS. Dietary supplements such as vitamin C, zinc, probiotics, and other compounds are known to boost immunity, which will reduce the risk of getting affected by COVID-19 or any other virus easily. More research needs to be conducted on novel approaches like the introduction of nanofibers, nanomaterials, and bioactive compounds over the conventional practices of overcoming such pandemic situations and ensuring food safety.

Acknowledgments

Continuous support and infrastructural facility provided to the first author from ALPLA India Pvt. Ltd., Regional Head Quarters, Hyderabad, is acknowledged with thanks.
The infrastructural support provided to the second author by FORE School of Management, New Delhi, in completing this Book Chapter is gratefully acknowledged.

References

1. CDC. *How COVID-19 spreads? Last reviewed.* June 1, 2020.
2. Koopmans M, Duizer E. Foodborne viruses: an emerging problem. *Int J Food Microbiol* 2004;**90**(1):23–41.
3. Bosch A, Gkogka E, Le Guyader FS, Loisy-Hamon F, Lee A, van Lieshout L. Foodborne viruses: detection, risk assessment, and control options in food processing. *Int J Food Microbiol* 2018;**28**:110–28.

4. Li Q, Guan X, Wu P, Wang X, Zhou L, Tong Y, Ren R, Leung KSM, Lau EHY, Wong JY, Xing X, Xiang N, Wu Y, Li C, Chen Q, Li D, Liu T, Zhao J, Liu M, Tu W, Chen C, Jin L, Yang R, Wang Q, Zhou S, Wang R, Liu H, Luo Y, Liu Y, Shao G, Li H, Tao Z, Yang Y, Deng Z, Liu B, Ma Z, Zhang Y, Shi G, Lam TTY, Wu JT, Gao GF, Cowling BJ, Yang B, Leung GM, Feng Z. Early transmission dynamics in Wuhan, China, of novel coronavirus-infected pneumonia. *N Engl J Med* 2020;**382**:1199–207.
5. Jalava K. First respiratory transmitted foodborne outbreak? *Int J Hyg Environ Health* 2020;**226**:113490.
6. World Health Organization (WHO). *2019-nCoV situation reports*. 2020. https://www.who.int/emergencies/diseases/novel-coronavirus-2019/situation-reports. [Accessed 15 October 2020].
7. WHO. *Critical preparedness, readiness and response actions for COVID-19: interim guidance*. March 19, 2020. https://www.who.int/publications-detail/critical-preparedness-readiness-and-response-actions-for-covid-19.
8. WHO. *World Health Organization. Report of the WHO-China joint mission on coronavirus disease 2019 (COVID-19)*. February 2020. p. 16–24 [Internet], Geneva, https://www.who.int/docs/default-source/coronaviruse/who-china-joint-mission-on-covid-19-final-report.pdf.
9. Naserghandi A, Allameh SF, Saffarpour R. All about COVID-19 in brief. *New Microb New Infect* 2020;**35**:100678.
10. Zhou P, Yang XL, Wang XG, Hu B, Zhang L, Zhang W, Si HR, Zhu Y, Li B, Huang CL, Chen HD, Chen J, Luo Y, Guo H, Jiang RD, Liu MQ, Chen Y, Shen XR, Wang X, Zheng XS, Zhao K, Chen QJ, Deng F, Liu LL, Yan B, Zhan FX, Wang YY, Xiao GF, Shi ZL. A pneumonia outbreak associated with a new coronavirus of probable bat origin. *Nature* 2020;**579**:270–3.
11. Wang LF, Shi Z, Zhang S, Field H, Daszak P, Eaton BT. Review of bats and SARS. *Emerg Infect Dis* 2006;**12**:1834–40.
12. Holmes KV, Lai MMC. Coronaviridae: the viruses and their replication. *Fields Virol* 1996;**3**:1075–93.
13. van der Hoek L, Pyrc K, Jebbink MF, Vermeulen-Oost W, Berkhout RJ, Wolthers KC, Wertheim-van Dillen PME, Kaandorp J, Spaargaren J, Berkhout B. Identification of a new human coronavirus. *Nat Med* 2004;**10**(4):368–73.
14. Monchatre-Leroy E, Boue F, Boucher JM, Renault C, Moutou F, Ar Gouilh M, Umhang G. Identification of alpha and beta coronavirus in wildlife species in France: bats, rodents, rabbits, and hedgehogs. *Virus* 2017;**9**(12):364.
15. Rodriguez-Morales AJ, Bonilla-Aldana DK, Balbin-Ramon GJ, Rabaan AA, Sah R, Paniz-Mondolfi A, Pagliano P, Esposito S. History is repeating itself: probable zoonotic spillover as the cause of the 2019 novel Coronavirus Epidemic. *Infez Med* 2020;**28**(1):3–5.
16. Dhama K, SharunK, Tiwari R, Sircar S, Bhat S, Malik YS, Singh KP, Chaikumpa W, Bonilla-Aldana DK, Rodriguez-Morales AJ. Coronavirus disease: COVID-19. *Clin Microbiol* 2019;**33**(4). e00028-20.
17. Zhu N, Zhang D, Wang W, Li X, Yang B, Song J, Zhao X, Huang B, Shi W, Lu R, Niu P, Zhan F, Ma X, Wang D, Xu W, Wu G, Gao GF, Tan W. A novel coronavirus from patients with pneumonia in China, 2019. *N Engl J Med* 2020;**382**(8):727–33.
18. Lai CC, Shih TP, Ko WC, Tang HJ, Hsueh PR. Severe acute respiratory syndrome coronavirus 2 (SARS-CoV-2) and coronavirus disease-2019 (COVID-19): the epidemic and the challenges. *Int J Antimicrob Agents* 2020;**55**:105924.

19. Huang C, Wang Y, Li X, Ren L, Zhao J, Hu Y, Zhang L, Fan G, Xu J, Gu X, Cheng Z, Yu T, Xia J, Wei Y, Wu W, Xie X, Yin W, Li H, Liu M, Xiao Y, Gao H, Guo L, Xie J, Wang G, Jiang R, Gao Z, Jin Q, Wang J, Cao B. Clinical features of patients infected with 2019 novel coronavirus in Wuhan, China. *Lancet* 2020;**395**(10223):497–506.
20. Baranwal A, Mahato K, Srivastava A, Maurya PK, Chandra P. Phytofabricated metallic nanoparticles and their clinical applications. *RSC Adv* 2016;**6**(107):105996–6010.
21. Centers for Disease Control and Prevention [CDC]. *How coronavirus spreads*. Atlanta: CDC; 2020.
22. Klein G. Spread of viruses through the food chain. *Dtsch Tierarztl Wochenschr* 2004;**111**:312–4.
23. European Food Safety Authority [EFSA]. *Coronavirus: no evidence that food is a source or transmission route*. EFSA: Parma; 2020.
24. Oakenfull RJ, Wilson AJ. *Qualitative risk assessment: what is the risk of food or food contact materials being A source or transmission route of SARS-CoV-2 for UK consumers?* London: Food Standards Agency; 2020.
25. Environmental Protection Agency List N. *Disinfectants for use against SARS-CoV-2*. Washington, DC: Environmental Protection Agency; 2020.
26. Harapan H, Itoh N, Yufika A, Winardi W, Keam S, Te H. Coronavirus disease 2019 (COVID-19): a literature review. *J Infect Public Health* 2020;**13**:667–73.
27. Djekic I, Tomasevic I, Radovanovic R. Quality and food safety issues revealed in certified food companies in three Western Balkans countries. *Food Control* 2011;**22**(11):1736–41.
28. Tomasevic I, Kuzmanović J, Anđelković A, Saračević M, Stojanović MM, Djekic I. The effects of mandatory HACCP implementation on microbiological indicators of process hygiene in meat processing and retail establishments in Serbia. *Meat Sci* 2016;**114**:54–7.
29. BSI Group. *Safe working during the covid-19 pandemic- general guidelines for organizations*. 2020. Version 1.
30. CAC. *Recommended international code of practice general principles of food hygiene*. 2003. CAC/RCP 1-1969.
31. Dzwolak W. HACCP in small food businesses – the Polish experience. *Food Control* 2014;**36**(1):132–7.
32. ISO Standard 22000. *Food safety management systems — requirements for any organization in the food chain*. Edition: 2 2018. p. 37.
33. Galimberti A, Cena H, Campone L, Ferri E, Dell'Agli M, Sangiovanni E. Rethinking urban and food policies to improve citizens safety after COVID-19 pandemic. *Front Nutr* 2020;**7**(181).
34. WHO. *Europe. WHO European Region – food and nutrition tips during self-quarantine*. 2020. www.euro.who.int/en/health-topics/health-emergencies/coronavirus-covid 19/publications-and-technical-guidance/noncommunicable-diseases/food-and-nutrition-tips-during-self-quarantine. [Accessed 14 November 2020].
35. Codex Alimentarius. *Committee on food hygiene, discussion paper on viruses in food. FAO/WHO document CX/FH99/11*. Rome, Italy: FAO/WHO; 1999.
36. Yekta R, Dastjerdi LV, Norouzbeigia S, Mortazavianb AM. *Food products as potential carriers of SARS-CoV-2*. Food Control; 2020. Published online on 11th November, 2020.
37. Ceylan Z, Meral R, Cetinkaya T. Relevance of SARS-CoV-2 in food safety and food hygiene: potential preventive measures, suggestions and nanotechnological approaches. *Virus Dis* 2020;**31**(2):154–60.

38. Codex Alimentarius. *Food hygiene: basic texts*. 2020. Accessed 10-Sep-20, http://www.fao.org/3/a1552e/a1552e00.pdf.
39. Mast EE, Alter MJ. Epidemiology of viral hepatitis. *Semin Virol* 1993;**4**:273–83.
40. Pintó RM, Saiz JC. Enteric hepatitis viruses. In: Zuckerman AJ, Mushahwar IK, editors. *Human Viruses in Water. Perspectives in Medical Virology Series*. **17**. Amsterdam, The Netherlands: Elsevier; 2007. p. 39–67.
41. The Lancet. *Infectious diseases challenges of coronavirus disease*. 2019. Lancet Infect. 2020.
42. Han J, Zhang X, He S, Jia P. Can the coronavirus disease be transmitted from food? A review of evidence, risks, policies and knowledge gaps. *Environ Chem Lett* 2021:5–16.
43. Thippareddi H, Balamurugan S, Patel J, Singh M, Brassard J. Coronaviruses – potential human threat from foodborne transmission? *Lebensm Wiss Technol (Food Sci)* 2020:134–110147.
44. Food and Drug Administration [FDA]. *Food safety and the coronavirus disease 2019 (COVID-19)*. White Oak Campus: FDA; 2020.
45. Kaul D. An overview of coronaviruses including the SARS-2 coronavirus - molecular biology, epidemiology and clinical implications. *Cur Med Res Pract* 2020;**10**:54–64.
46. Yu P, Zhu J, Zhang Z, Han Y. A familial cluster of infection associated with the 2019 novel coronavirus indicating possible person-to-person transmission during the incubation period. *J Infect* 2020;**221**(11):1757–61.
47. Pan X, Chen D, Xia Y, Wu X, Li T, Qu X, Zhou L, Liu J. Asymptomatic cases in a family cluster with SARS-CoV-2 infection. *Lancet Infect Dis* 2020;**20**(4):410–1.
48. Tong ZD, Tang A, Li KF, Li P, Wang HL, Yi JP, Zhang YL, Yan JB. Potential presymptomatic transmission of SARS-CoV-2, Zhejiang Province, China. *Emerg Infect Dis* 2020;**26**(5).
49. Kimball A, Hatfield KM, Arons M, James A, Taylor J, Spicer K, Bardossy AC, Oakley LP, Tanwar S, Chisty Z, Bell JM, Methner M, Harney J, Jacobs JR, Carlson CM, McLaughlin HP, Stone N, Clark S, Brostrom-Smith C, Page LC, Kay M, Lewis J, Russell D, Hiatt B, Gant J, Duchin JS, Clark TA, Honein MA, Reddy SC, Jernigan JA. Asymptomatic and Presymptomatic SARS-CoV-2 Infections in residents of a long-term care skilled nursing facility — King County, Washington. *Morb Mortal Wkly Rep (MMWR)* 2020;**69**(13):377–81.
50. Randazzo W, Fabra MJ, Falcó I, López-Rubio A, Sánchez G. Polymers and biopolymers with antiviral activity: potential applications for improving food safety. *Compr Rev Food Sci* 2018;**17**:754–68.
51. Warnes SL, Summersgill EN, Keevil CW. Inactivation of murine norovirus on a range of copper alloy surfaces is accompanied by loss of capsid integrity. *Appl Environ Microbiol* 2015;**81**(3):1085–91.
52. van Doremalen N, Bushmaker T, Morris D, Holbrook M, Gamble A, Williamson B, Lloyd-Smith J. Aerosol and surface stability of HCoV-19 (SARS-CoV-2) compared to SARS-CoV-1. *N Engl J Med* 2020;**382**:1564–7.
53. Sportelli MC, Longano D, Bonerba E, Tantillo G, Torsi L, Sabbatini L, Cioffi N, Ditaranto N. Electrochemical preparation of synergistic nanoantimicrobials. *Molecules* 2020;**25**:49.
54. WHO. *COVID-19 and food safety: guidance for food businesses: interim guidance*. April 07, 2020.
55. Foundation FSSC 22000. *Scheme documents, verson 5*. 2020. Issued on November, 2020, Netherlands.
56. Nakat Z, Bou-Mitri C. COVID-19 and the food industry: readiness assessment. *Food Control* 2021;**121**:107661.
57. Szymendera SD, OSHA. *Interim guidance for workers and employers of workers at increased risk of occupational exposure*. Occupational Safety and Health Administration, Department of Labor's; April 6, 2020.

58. Ong SW, Tan YK, Chia PY, Lee TH, Ng OT, Wong MS, Marimuthu K. Air, surface environmental, and personal protective equipment contamination by severe acute respiratory syndrome coronavirus 2 (SARS-CoV-2) from a symptomatic patient. *J Am Med Assoc* 2020;**323**(16):1610–2.
59. WHO. *Rational use of personal protective equipment for coronavirus disease (COVID-19) and considerations during severe shortages*. 2020.
60. WHO. *WHO save lives: clean your hands in the context of Covid 19*. Rome, Italy: World Health Organisation & Food and Agriculture Organization of the United Nations.
61. CDC. *Cleaning and disinfection for community facilities*. April 1, 2020.
62. *BRC global standard for food safety*. London, UK: BRC Trading Ltd; 2018. Issue: 8.
63. IFS. *IFS Food, version 6.1*. Berlin, Germany: IFS Management GmbH; 2017.
64. CISA. *Guidance on the essential critical infrastructure workforce*. March 19, 2020.
65. Duda-Chodak A, Lukasiewicz M, Zieć G, Florkiewicz A, Filipiak-Florkiewicz A. Covid-19 pandemic and food: present knowledge, risks, consumers fears and safety. *Trends Food Sci Technol* 2020;**105**:145–60.
66. CDC. *Businesses and employers responding to coronavirus disease 2019 (COVID-19): interim guidance*. May 2020.
67. CDC. *Check and record everyday (CARE)*. March 11, 2020.
68. Bilal M, Nazir MS, Parra-Saldivar R, Iqbal HM. 2019-nCoV/COVID-19 approaches to viral vaccine development and preventive measures. *J Pure Appl Microbiol* 2020;**14**(1):25–9.
69. CDC. *Hand hygiene recommendations*. May 17, 2020.
70. Kampf G, Todt D, Pfaender S, Steinmann E. Review: persistence of coronaviruses on inanimate surfaces and their inactivation with biocidal agents. *J Hosp Infect* 2020;**104**:246–51.
71. CDC. *Human coronavirus types*. 2020. Retrieved from: www.cdc.gov/coronavirus/types.html. Accessed 7-Oct-20.
72. Fraser AM, Pascall MA. *Cleaning and sanitization of food-contact surfaces in retail/food service establishments*. 2010. Available online at: http://www.foodsafetymagazine.com. Accessed 12-Nov-20.
73. *EPA frequent questions about disinfectants and coronavirus (COVID-19). United states environmental protection agency*. April 29, 2020.
74. EPA. *List N: disinfectants for use against SARS-CoV-2. United states environmental protection agency*. May 28, 2020.
75. CDC. *Cleaning and disinfecting your facility*. April 28, 2020.
76. Fukuzaki S. Mechanisms of actions of sodium hypochlorite in cleaning and disinfection processes. *Biocontrol Sci* 2006;**11**(4):147–57.
77. Yépiz-Gómez MS, Gerba CP, Bright KR. Survival of respiratory viruses on fresh produce. *Food Environ Virol* 2013;**5**:150–6.
78. Sapers GM. Efficacy of washing and sanitizing methods for disinfection of fresh fruit and vegetable products. *Food Technol Biotechnol* 2001;**39**(4):305–11.
79. Gov U. K. *Working safely during COVID-19 in factories, plants and warehouses*. May 11, 2020.
80. IDFA. *Emergency prevention measures to achieve physical (social) distancing in food manufacturing facilities as related to COVID-19*. March 31, 2020.
81. WHO. *COVID-19 and Food Safety: guidance for competent authorities responsible for national food safety control systems in Interim guidance*. April 22, 2020 [Rome, Italy].
82. FBIA. *Screening food industry employees for COVID-19 symptoms or exposure*. March 30, 2020.
83. ILO. International Labor Organization. *The six-step COVID-19 business continuity plan*. 2020. April 1, 2020.

84. FBIA. *FBIA's food industry recommended protocols when employee/visitor/customer test positive for COVID-19*. March 25, 2020.
85. WHO. *Contact tracing in the context of COVID-19*. May 10, 2020.
86. Zhang H, Chen R, Chen J, Chen B. COVID-19 Transmission within a family cluster in Yancheng, China. *Front Med* 2020;**7**:387.
87. Bienkov A. *Coronavirus: loss of smell and taste may be hidden symptom of COVID-19 – business insider*. 2020.
88. Olaimat AN, Shahbaz HM, Fatima N, Munir S, Holley RA. Food safety during and after the era of COVID-19 pandemic. *Front Microbiol* 2020;**11**:1854.
89. Rizou M, Galanakis IM, Aldawoud TMS, Galanakis CM. Safety of foods, food supply chain and environment within the COVID-19 pandemic. *Trends Food Sci Technol* 2020;**102**:293–9.
90. CDC. *Workplaces and businesses, considerations for restaurant and bar operators. USA department of health and human services*. 2020. Accessed on 26th December.
91. Sampath Kumar NS, Chintagunta AD, Jeevan Kumar SP, Roy S, Kumar M. Immunotherapeutics for Covid-19 and post vaccination surveillance. *3 Biotech* 2020;**10**(12):527.
92. Singh S, Prakash C, Ramakrishna S. Three-dimensional printing in the fight against novel virus COVID-19: technology helping society during an infectious disease pandemic. *Technol Soc* 2020;**62**:101305.
93. Djekic I, Nikolić A, Uzunović M, Marijke A, Liu A, Han J, Brnčić M, Knežević N, Papademas P, Lemoniati K. Covid-19 pandemic effects on food safety—multi-country survey study. *Food Control* 2021;**122**:107800.
94. Galanakis CM, Aldawoud TMS, Rizou M, Rowan NJ, Ibrahim SA. Food ingredients and active compounds against the coronavirus disease (COVID-19) pandemic: a comprehensive review. *Foods* 2020;**9**:1701.
95. Albin J. *Quarantine cuisine: easy meals to support a healthy immune system. COVID; prevention*. Dallas: UT Southwestern Medical Centre; 2020. https://utswmed.org/medblog/easy-immune-boosting-food-covid19/. [Accessed 29 December 2020].
96. Shakoor H, Feehan J, Al Dhaheri AS, Ali HI, Platat C, Ismail LC, Apostolopoulos V, Stojanovskaa L. Immune-boosting role of vitamins D, C, E, zinc, selenium and omega-3 fatty acids: could they help against COVID-19? *J Maturitus* 2021;**143**:1–9.
97. Kumar A, Kubota Y, Chernov M, Kasuya H. Potential role of zinc supplementation in prophylaxis and treatment of COVID-19. *Med Hypotheses* 2020;**144**:109848.
98. Hawkins J, Baker C, Cherry L. Black elderberry (*Sambucus nigra*) supplementation effectively treats upper respiratory symptoms: a meta-analysis of randomized, controlled clinical trials. *Compl Ther Med* 2019;**42**:361–5.
99. Arshad MS, Khan U, Sadiq A, Khalid W, Hussain M, Yasmeen A, Asghar Z, Rehana H. Coronavirus disease (COVID-19) and immunity booster green foods: a mini review. *Food Sci Nutr* 2020;**8**(8):3971–6.
100. Sahoo BM, Banik BK. Medicinal plants: source for immunosuppressive agents. *Immunol: Curr Res* 2018;**2**:106.
101. Anywar G, Kakudidi E, Byamukama R, Mukonzo J, Schubert A, Oryem-Origa H. Medicinal plants used by traditional medicine practitioners to boost the immune system in people living with HIV/AIDS in Uganda. *Eur J Integr Med* 2019;**35**:101011.
102. WHO. *COVID-19 and food safety: Guidance for food businesses: Interim guidance*. 07 April 2020.

CHAPTER 12

An overview of food safety and COVID-19 infection: nanotechnology and cold plasma applications, immune-boosting suggestions, hygienic precautions

Zafer Ceylan[1], Elvan Ocak[2], Yılmaz Uçar[3], Kadir Karakus[4], Turgay Cetinkaya[5]
[1]Van Yüzüncü Yıl University, Faculty of Tourism, Department of Gastronomy and Culinary Arts, Van, Turkey; [2]Van Yüzüncü Yıl University, Engineering Faculty, Food Engineering Department, Van, Turkey; [3]Fatsa Faculty of Marine Sciences, Ordu University, Ordu, Turkey; [4]Department of Animal Science, Faculty of Agriculture, Malatya Turgut Özal University, Malatya, Turkey; [5]Food Processing Department, Armutlu Vocational School, Yalova University, Yalova, Turkey

12.1 Introduction

Humankind struggles with many health matters, particularly those caused by bacteria and viruses. Bacteria-based illnesses have gained notoriety globally. Studies have mainly focused on pathogenic bacteria in food. In other words, other potential hazardous effects were mostly ignored. In this respect, today, humans are learning how to fight viruses.[1–3]

A typical virus has RNA or DNA and is mostly 300 nm in size. Unlike bacteria, viruses may not reproduce in a food environment because viruses are obligate intracellular parasites. In other words, they can only survive by moving their genetic material from one host cell to another. However, water and food may act as vectors for the transmission of viruses to humans via the fecal-oral route.[4–6] A virus can spread to food in several ways: (1) contamination during food processing through infected food personnel; (2) direct contamination of human feces without soil or water treatment; (3) ingestion of food animal origin with zoonotic viruses; and (4) inadequate treatment of wastewater.[7–11]

Ebola, avian influenzas, Nipah, MERS-CoV, SARS-CoV, and currently the most dangerous one, SARS-CoV-2, are the main zoonotic viruses that

Environmental Management of COVID-19
ISBN 978-0-323-85780-2
https://doi.org/10.1016/B978-0-323-85780-2.00004-4

affect human health directly.[12–14] CoVs can easily pass to humans from different sources like animals. According to previous studies, CoVs can transfer by consumption of infected animals.[15–21] Especially in the case of foods not being heated or cooked, these viruses can cause diseases in humans. Also, it has been observed that some food materials such as chicken or some fish samples are contaminated with SARS-CoV-2.[22–24] Pneumonia, diarrhea, fever, loss of odor and taste, and organ failures are common symptoms of SARS-CoV-2 and its infection.[25] Because of its long incubation stage, SARS-CoV-2 can have more infectious effects on society. In terms of food, the aforementioned viruses can be related with the consumption of wild animals. In China, in the first step, as can be seen from various news sources, the main source of spread of SARS-CoV-2 is also associated with some animals such as bats, pangolins, and civets. Furthermore, the relationship between animal-based food consumption and coronavirus disease (COVID-19) infections was associated with Wuhan's Seafood Wholesale Market, which was immediately closed at the beginning of January 2020.[8,26,27] A graphical illustration of the possible routes of SARS-CoV-2 transmission from foods to humans is presented in Fig. 12.1. Following the initial stages of COVID-19 spread in China, World Health Organization (WHO) and some leader organizations such as Food and Drug Administration (FDA) and European Food Safety Authority (EFSA) began preparing reports, which consisted of frequently asked questions related to the COVID-19 outbreak and food safety.[28,29] At that moment, consumers were thinking that foods could be contaminated by COVID-19. Restaurants, markets, and manufacturing facilities were seen as potential areas of risk.[11,30–34] Consumers have observed a lack of information regarding such risks. After the initial period of the COVID-19 outbreak, new studies related to food safety and COVID-19 have revealed the worry and awareness among consumers.[8] Also, it has emphasized the Hazard Analysis and Critical Control Point (HACCP) principles for the food industry.[35]

The International Commission on Microbiological Specifications for Foods (ICMSF) claimed that it is highly improbable that the ingestion of SARS-CoV-2 will result in illness, and there is still no evidence that food is a major source of transmission.[25] Oral transmission via food consumption has not been reported. One of the ways in which COVID-19 can be spread is through cross-contamination of surfaces. It has been reported that CoV particles may survive for hours to days on surfaces.[11,34,36,37] However, the possibility of transmission through inanimate surfaces has been observed to be lower.[25,38]

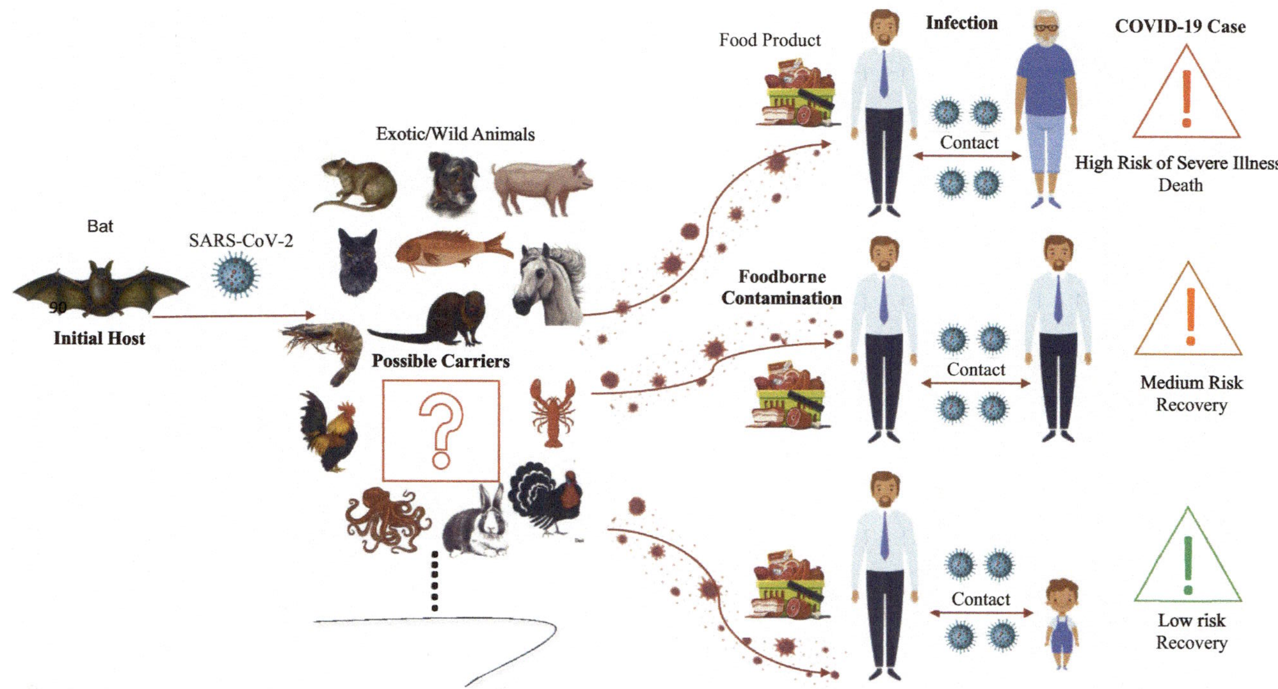

Figure 12.1 Graphical abstract of emerging SARS-CoV-2 that shows the probable main source, possible intermediate carriers, contamination to food products, infection in humans, and finally, COVID-19 cases, respectively.

Food preservation methods and hygiene practices, already in use for food safety, could be important in eliminating different microbiological risks at the onset of a COVID-19 outbreak. In this respect, to improve consumer safety, recently different certificate systems have been applied.[14,33,39] Besides novel approaches like nanotechnology applications or packaging applications, some minor preventions for industrial kitchens and households are gaining importance.[8,34,40–43] On the other hand, consumption of vegetables and dairy products, as well as foods rich in protein such as fish and red meat, will boost the human immune system. In addition, from their harvesting to the plate, food safety should be enhanced.[33,44]

This study aims to highlight the probable risks based on food material contaminated with potential COVID-19 infection from food surfaces, food packaging materials, food processing equipment, and other items in the food processing industry. With this study, novel perspectives including nanoscience and cold plasma treatment have been presented to society for the elimination or prevention of COVID-19.

12.2 Food safety and COVID-19

Considering risk management, there is currently no evidence that food is a source or route of transmission of SARS-CoV-2 food or food packaging, according to Centers for Disease Control and Prevention (CDC), U.S. Department of Agriculture (USDA), and EFSA.[23,45,46] However, there are a few reports that SARS-CoV-2 is found on food packaging materials. These reports show the presence of an RNA virus which may exist as a hazard to human health. As is known, the first case related to food-contact material (a salmon cutting board) with SARS-CoV-2 was determined in the wholesale market in Beijing.[11,23,24,47–49] Since July 2020, some cases of food contamination have been reported in China, where SARS-CoV-2 was detected on imported frozen food packaging materials such as shrimp.[23,24] SARS-CoV-2 was also detected in a food shipping container.[23] Food contamination that occurred with COVID-19 may be observed via the farming activity, processing, storage, transport, and retailing process in which foods might come in contact with staff and atmospheric conditions.[23,50] Ferretti et al. (2020) emphasized the importance of environmental conditions. In this sense, SAR-CoV-2 can transfer from symptomatic people at a 90% rate. On the other hand, 10% defines the environment, surfaces, and other sources.[51] Therefore, food safety related to COVID-19 should never be ignored.

12.2.1 Vegetables and fruits

According to Mullis et al., a kind of coronavirus called bovine on lettuce samples was stable for 14 days when it was stored at 4°C.[52] In addition, another type (CoV 229E) was found on lettuce after two days of storage at 4°C. On the other hand, this virus surprisingly could not survive on the surface because of the acidity in strawberry samples.[53]

Significant records related to COVID-19 in fruit and vegetables have not been directly detected yet. On the contrary, recently published articles have revealed that indirect contamination can affect vegetables and fruits. For example, irrigation with sewage is able to infect vegetables and fruit with COVID-19.[4] Also, if plants or fruit trees are irrigated by sewage constituting SARS-CoV-2, there would be a risk in terms of contamination in vegetables and fruits. In another study, to remove or decline the virus load on the surface of them, a good washing is suggested but the fruit or vegetable's surface structure would play a key role.[1] In this sense, the pH value of the fruit could play a significant role in the elimination of COVID-19.[54,55]

12.2.2 Animal resources

Breeders play a major role in animal production. With socioeconomic concerns, instead of small-sized family businesses, farms that are formed by the gathering of thousands of animals in closed areas have increased. Diseases transmitted through animal origin have increased because of industrial animal production. CoVs are responsible for respiratory and intestinal infections in animals.[56,57] For example, the host animal for MERS-CoV was dromedary camels. Foodborne transmission through consumption of unpasteurized camel milk and raw camel meat was an infection source for humans.[58] Like MERS-CoV, SARS-CoV-2 is also a zoonotic virus and has a genome similarity of 96% to a SARS-related bat CoV.[56] In fact, a zoonotic transmission from animals to humans has been reported in a Dutch mink farm during the COVID-19 pandemic.[56] Around 20 million mink in Europe have been culled in case of further zoonotic transmission possibility.[59,60] On November 12, 2020 the Food and Agriculture Organization (FAO) announced guidelines about the exposure of humans or animals to SARS-CoV-2 from wild, livestock, companion, and aquatic animals.[56]

Determining disease-resistant animal species by epigenetic studies may be beneficial for future applications. Genomic selection applications for animals that have resistance to diseases, will make a significant contribution to preventing the spread of COVID-19. In this way breeding studies could be carried out for the ones that are resistant to COVID-19.[61,62] Animals that are the sources of infection are a concern. To avoid the SARS-CoV-2 that causes disease, while planning animal shelters, attention should be paid to natural ventilation, adequate living space, controlled production and hygiene. All these data show that further investigation is needed for livestock, and precautions need to be taken against COVID-19 on an industrial scale.

12.2.3 Meat and meat products

CoVs can be transmitted from an infected person to meat samples by contaminated hands, or via sneezing or coughing if hygiene rules are not followed. Although the oral/alimentary transmission by consuming meat products does not play a role, according to new research, it was revealed that SARS-CoV-2 lost only a little of its infectivity on frozen meat even after three weeks.[63] Also, Liji (2020) has stated that SARS-CoV-2 may survive on frozen or cold storage salmon and meat samples for more than three weeks.[64] To the best of our knowledge, on August 12, 2020, the first known case of SARS-CoV-2 was detected on a chicken wing sample that had been imported from Brazil to China.[22,23]

The CDC requested data related to the number of poultry and meat plants (where beef, bison, lamb, pork, and poultry meats are processing) negatively affected, how many workers were affected in these plants, and the number of deaths based on COVID-19 in these facilities of the United States.[65] According to the report, 115 poultry or meat processing facilities, in which 4913 cases and 20 worker deaths were reported between April 9–27, 2020, were affected by this pandemic outbreak. It has been assumed that difficulties at workplaces regarding physical distancing, hygiene, and crowded living and transportation conditions were the factors that potentially affected infection risk. Other data revealed that a slaughterhouse in Germany was shut down on June 18, 2020 because COVID-19 test results of 1550 of 7000 workers were came out positive.[66] These data imply that the number of workers who have to work in crowded places in these facilities increased SARS-CoV-2 transmission risk.[65] Moreover, salmon from Europe was banned and Ecuadorean shrimp were suspended because of the SARS-CoV-2 outbreak.[24,64]

European Federation of Food Agriculture and Tourism Trade Unions stated that, according to a report from IUF consultation with Imperial College in London, the meat sector greatly relies on immigrants and cross-border workers in almost all European countries. In some meat factories, especially the old ones, the spread of the disease is 20 times more likely where there is insufficient ventilation. On the other hand, certain parts of meat processing plants are very cold and the virus spread faster in colder temperatures.[66]

All these data prove that during meat processing, SARS-CoV-2 transmission risk should be considered. The recognized risk to poultry and meat facility operation requires action plans to decrease risks to workers, preserve facility function, and maintain the food supply. Cleaning and disinfection, screening staff for symptoms, and providing multilingual disinfection training might slow the virus spread among workers in meat and poultry processing plants. Temperature monitoring and symptom screening are important to avoid the introduction of COVID-19 into a facility from infected persons.[65] Inactivation of SAR-CoV-2 in artificially contaminated chicken, pork, and salmon samples reveal that three weeks' storage at 4, −20, and −80°C, respectively, did not decrease infection level.[11,24] Above all, precautions in the meat sector should be improved and updated.

12.2.4 Dairy products

Some natural milk preservative components like lactoferrin (LCF), lactadherin and glycomacropeptide in milk serve an antiviral role against some viruses, particularly HIV,[67] and may also have antiviral activity for SARS-CoV−2. It is known that LCF can have an interaction with heparin glycosaminoglycan (HSPG) cell receptors. Also, the aforementioned receptors play a significant role in cells of the first term of the virus infection from CoV. These relationships contain the angiotensin-converting enzyme 2 (ACE2) receptor, a metallopeptidase that can bind virus terminals and also facilitate cell entry. All mechanisms mentioned above were revealed in the SARS-CoV epidemic in 2002. In this regard, similar interactions and mechanisms in ACE2 have been observed in SARS-CoV-2.[68]

Besides SARS-CoV, SARS-CoV-2 and MERS-CoV were determined in 41.9% of camel milk samples.[69] In this sense, the experiment indicated that MERS-CoV inoculated to camel milk could survive up to 72 h at 4 and 22°C, respectively. As stated by van Doremalen et al., the virus survived for more than two days in unpasteurized milk samples.[70] However, the consumption of milk by pasteurization prevents such foodborne contamination. In order to inactivate a virus like SARS-CoV, it is evaluated as sufficient to heat the milk to ~60°C for 30 min.[71] In addition to the heating processes, Ultraviolet (UV) light, alkaline pH (>12), or acidic pH (<3) are the other tools to inactive SARS-CoV.[72,73]

They may be recommended for the elimination of dairy products with SARS-CoV-2. Another widely consumed dairy product is yogurt, which is fermented by lactic acid bacteria following heat treatment. The last product possesses about 4.5 pH, but Chin et al. (2020) has reported that SARS-CoV-2 is evaluated as stable from 3 to 10 pH ranges.[54] Contrary to these aforementioned issues, some metabolites such as lactic acid, hydrogen peroxide (H_2O_2), and bacteriocins shaped by lactic acid bacteria at the end of fermentation possess antiviral properties.[74] So, the last product could provide a good opportunity to support the immune systems of humans in pandemic terms.

Yekta et al. noted that for dairy products, SARS-CoV-2 at −20°C for 2 years could be found as stable, so ice cream, frozen yogurt, and dairy desserts may be defined to be SARS-CoV-2 carriers.[1] Moreover, many more studies on dairy products should be revealed for potential contamination risk.

12.3 Food packaging applications and SARS-CoV-2

Food packaging applications are gaining importance. In the case of outbreaks, all worldwide consumers by time are getting much more worried because of the possible contamination risk of SARS-CoV-2. In this sense, we already know that packaging materials play a key role. Also, if possible, to decrease contamination risks with SARS-CoV-2 on food, different kinds of food products should be processed, preheated, and pasteurized or sterilized. Moreover, following the mentioned applications, food packaging material can be stored. For example, sous vide technology can provide a lower contamination risk for food samples because food material can be placed into the food packaging material then vacuum packaged, sealed, cooked, cooled, and stored (Fig. 12.2). When food products are cooked with sous vide technology, they could be directly consumed or heated, for instance, in a microwave oven. In addition to sous vide technology, intelligent and active packaging enriched with nanosensors could be used with basic food packaging materials used widely in the food industry.[75,76] The average risk of contamination with food or food packaging is evaluated to be lower.[77]

Zhang et al. revealed that fecal-oral contamination of the virus is an important issue.[78] In the case of handling food packages and containers, SARS-CoV-2 could be transmitted from the food packaging material surface to the eyes, nose, and mouth. But the aforementioned contamination way is not evaluated to be a separate route for this pandemic virus.[79] So, packaged food samples and canned foods may be preferred in this risky period.

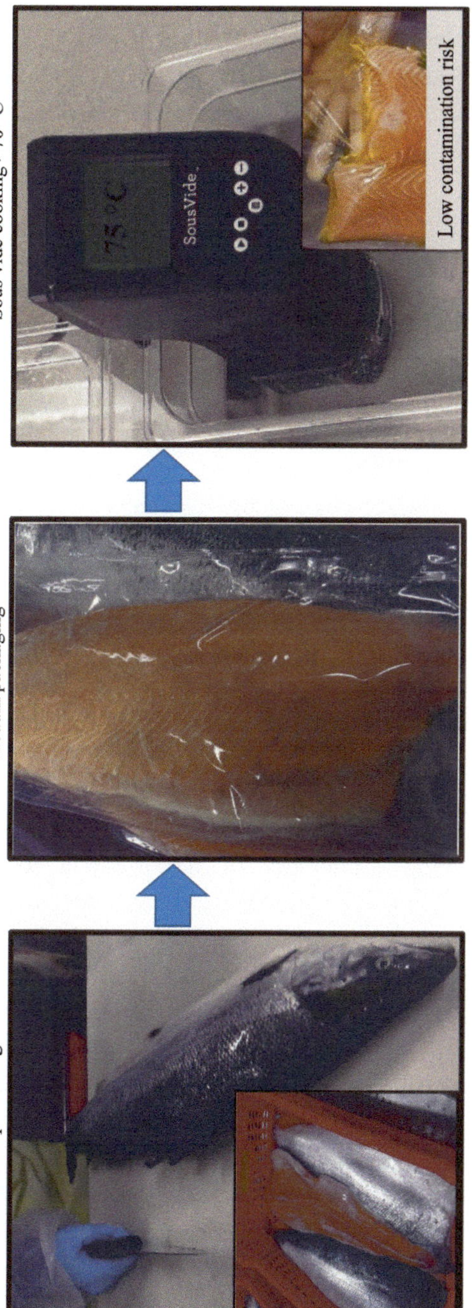

Figure 12.2 Main steps of sous vide cooking to lower viral contamination risks for food products.

12.4 Limitation of SARS-CoV-2 spread

12.4.1 Nanotechnological approach

The particle size of SARS-CoV-2 ranges from 50 to 200 nm.[77] Ignatov has reported the exact size as 120 nm.[80] The use of nanomaterials based on antiviral activity provided inactivation of some viruses in chicken embryo models and shrimp samples.[42,81,82] These studies have shown that nanotechnological approaches should be considered for the purpose of developing resistance against SARS-CoV-2.[41–43,83] Antimicrobial coatings by metal oxide nanostructures and nanopolymers may inhibit the transmission of COVID-19.[83,84] Silver (Ag) nanoparticles and zinc (Zn) nanomaterials indicated inhibition in viral replication, virus inactivation, viral protein translation, and polyprotein processing.[40,85] Nanoform of Ag and copper (Cu) cations could affectively damage the nucleotides of some viruses.[43] In this respect, TiO_2, ZnO, and Cu nanoparticles are suggested to control the disease by surface coating and disinfection. In one study, the use of nanosilver colloids was investigated for therapeutic purposes in the early stages of COVID-19. As a result of the study, it was determined that antimicrobial doses of nanosilver colloids also show antiviral effects. In the same study, it was stated that nanosilver colloids can be applied both at home and in the hospital through inhalation.[80] In another study, 2–15 nm size Ag nanoparticles provided a successful inhibition of SARS-CoV-2 at concentrations above 1 ppm.[86] In this regard, a relevance between the intake of SARS-CoV-2 contaminated food and the possibility of infections as well as the advancement of antiviral packaging using nanotechnology ought to be searched.[79] The use of nanomaterial coatings or films (containing Cu, Ag, and ZnO nanoparticles) has a potential against SARS-CoV-2 to avoid contamination of food packaging surfaces, and thus this approach may reduce its rapid transmission.[41] Also, nanosized materials can be integrated with other technologies for obtaining much safer materials, which can be contacted by foods. In this respect, packaging material enriched with nanoparticles in active and intelligent applications could be advised in order to provide food safety against SARS-CoV-2. Moreover, nanoparticles, nanofibers, nanoencapsulation technologies, nanoemulsions, and nanocomposites providing a larger contact area on the surface of the material could be used to increase food safety.

12.4.2 Cold plasma

Cold plasma has demonstrated great potential for inactivation of SARS-CoV-2.[10,87–90] Since it does not generate waste or toxic chemicals, this method is seen as environmentally friendly. This technique also has potential with extensive applications in the disinfection of food surfaces, equipment, and food packages.[88,91] Furthermore, the in-package plasma treatment can be combined with the bacteriostatic nanophotocatalysts onto food packing films.[91,92]

The mechanism is attributed to free radicals that diffuse through the capsid to reach and damage the RNA of the virus.[93] Although Reactive Oxygen Species (ROS) and Reactive Nitrogen Species (RNS) are considered the main contributors to the inactivation of viruses,[88] plasma does not fully damage the virus but it can delay its ability to infect.[94] Photons, positive and negative ions, neutral atoms, UV photons, and free radicals could play a key role.[95,96] The occurrence and role of ozone and UV photons in cold plasma species can affect the protein structure and damage RNA in a virus.[97] The electrostatic forces related to plasma technology may affect the disruption of cell membranes that lead to cell death. In this respect, gas composition, flow rate, voltage parameters, and food composition are the other main factors to inactivate microbial load in food.[98,99]

In terms of SARS-CoV-2, cold plasma applications have been discussed to degrade the RNA virus.[100,101] Cold plasma with Argon gas to deactivate SARS-CoV-2 on various surfaces, including plastic, metal, and cardboard, was also employed. These studies showed that cold plasma could be an effective method to avoid transmission and infections on various surfaces.[89,101] Also, Verma (2020) has reported that application parameters such as duration should be optimized for the elimination of viruses on the contaminated material.[94] In terms of food products, inactivation of the Tulane virus on spinach,[102] murine norovirus-1 and hepatitis A virus on fresh meats,[95] norovirus surrogates on blueberries[97] and some other food systems,[103] Tulane virus on romaine lettuce,[104] and Newcastle disease virus on poultry[105] was effectively applied by cold plasma treatment.

Major foodborne viruses on surfaces such as stainless steel[106] were also studied. These studies demonstrated that cold plasma is effective for reducing microbial load in surface areas such as stainless steel or other industry materials that may have close contact with food products.[107] For these reasons, cold plasma application can be considered for decontamination of food processing equipment. Furthermore, cold plasma usage can decrease the rapid contamination of COVID-19 in water samples.[88,108] It is known that most food or food products, such as a different kind of seafood and meat, contain high levels of water. So, in further studies, cold plasma treatment can be combined with food samples that are widely consumed.

12.5 Suggestions for consumers

Suggestions will be evaluated from two perspectives. The main aim is to explain how to decrease the risk in food, food packaging, or kitchen equipment contaminated with SARS-CoV-2. After purchasing foods, packaging materials including food should be washed or cleaned by using different materials having higher acidic pH value. For example, canned sample surfaces should be washed with disinfectants. Cooking temperature or heating processes are highly important to eliminate the potential risk. If possible, as can be seen from the present study, the midpoint temperature of food products such as fish and meat should be controlled before consumption. Also, especially fruits with lower acidity should be washed. Recently, studies reported that freezing or chilling is not meant to eliminate the virus; therefore, it is not thought that they are safer for COVID-19. Moreover, this sort of sample must be well cooked like fresh food samples. Kitchen equipment (knives, cutting board, etc.) must be continually disinfected by using hot water.

CoVs are thermolabile and do not withstand normal cooking temperatures.[109] SARS-CoV was found to be inactivated after 15 min incubation at >75°C. Similarly, SARS-CoV-2 can be inactivated after incubation for 5 min at 70°C.[54] These findings recommend that cooking at >70°C is adequate for virus inactivation, but transmission from frozen food should still be considered; therefore, thorough hand washing after handling raw food is necessary.[33,73,110]

Besides the aforementioned hygienic precautions, in order to support the immune system in humans, some recently published articles have presented some suggestions. Ashraf et al. has suggested that consumption of honey (1 g/kg per person/day) will gain a powerful resistance toward COVID-19.[111] Adequate and balanced nutrition is important for boosting the immune system. In this respect, consuming foods rich in vitamins, minerals, and ω-3 fatty acids may have a potential role to help fight COVID-19. Vitamin B plays a role in cell functioning and energy metabolism. Vitamin D decreases rates of infection. The use of vitamin C can help minimize the symptoms caused by COVID-19, while vitamin E and some minerals like selenium may assist in the recovery stage.[48,112–114] Also, an increase in immunomodulatory effects can be provided by using ω-3 fatty acids.[112,113] According to recent studies, polyphenols such as resveratrol and flavonoids have potential inhibitors.[48,115,116] There is also indefinite proof supporting the future use of β-glucan to address COVID-19 due in part to variability in the immune response resulting from heterogeneity in the polysaccharide branch.[44] Nanoencapsulation improves bioavailability through controlled-release properties of substances in food with more health-promoting properties. This contribution summarizes the development of effective antioxidants such as coenzyme Q10, quercetin, curcumin, and probiotics that influences body immune systems in a positive way. In addition to these, micronutrients, functional food ingredients like probiotics, herbs, flavonoids, and carotenoids are also evaluated to be natural immune boosters.[48,116–118] As could be seen in several studies, a balanced diet enriched with different vitamins, minerals, and also probiotics may be playing a key role in this fight.

12.6 Conclusion

The potential effect of COVID-19 related to the safety of food is an important issue for the food industry. Further investigation is needed for livestock, and precautions need to be taken against COVID-19 on an industrial scale. Because SARS-CoV-2 is a zoonotic virus, in regards to animal resources, identifying disease-resistant animal genotypes and including them in breeding programs through genetic mapping is necessary. Furthermore, potential viral contamination risks on meat and dairy food products should be further investigated. Novel approaches, such as nanoscience (nanofibers, nanocapsules, and nanocomposites), cold plasma, and ozone applications can be considered as safe and effective methods toward SARS-CoV-2 prevention. The mentioned applications could be combined or integrated with sous vide, active, and intelligent packaging applications to improve food safety against future viral outbreaks. Consumers must take hygienic precautions to decrease the SARS-CoV-2 spread risk for frozen or chilled foods. The heating/cooking process, pH value, and temperature duration for all foods must be controlled. Packaging materials and food-contact surfaces should be cleaned to prevent contamination. Also, a balanced diet with intake of vitamins, minerals, phenolic compounds, and ω-3 fatty acids from various sources, consumption of dairy products, and honey-based products such as bee propolis and other supplementary materials play key roles in supporting the immune system.

References

1. Yekta R, Vahid-Dastjerdi L, Norouzbeigi S, Mortazavian AM. Food products as potential carriers of SARS-CoV-2. *Food Contr* 2020:107754. https://doi.org/10.1016/j.foodcont.2020.107754.
2. Knorr D, Khoo C-SH. COVID-19 and food: challenges and research needs. *Front Nutr* 2020;**7**:598913. https://doi.org/10.3389/fnut.2020.598913.
3. WHO and FAO. *Viruses in Food: scientific Advice to Support Risk Management Activities: meeting Report. Viruses in food: scientific advice to support risk management activities: meeting report (who.int)*. 2020. Accessed 25.12.20.
4. Rusiñol M, Hundesa A, Cárdenas-Youngs Y, et al. Microbiological contamination of conventional and reclaimed irrigation water: evaluation and management measures. *Sci Total Environ* 2020;**710**:136298. https://doi.org/10.1016/j.scitotenv.2019.136298.
5. Miranda RC, Schaffner DW. Virus risk in the food supply chain. *Curr Opin Food Sci* 2019;**30**:43–8. https://doi.org/10.1016/j.cofs.2018.12.002.
6. Öksüztepe G, Demir P. Gıda Güvenliği ve Virüsler. *Turkiye Klinikleri J Food Hyg Technol-Special Topics*. 2016;**2**(3):49–55.
7. Alp D, Kuleasan H. Gıda kaynaklı viral gastroenteritler. *TURJAF* 2018;**6**:1592–8. https://doi.org/10.24925/turjaf.v6i11.1592-1598.2054.
8. Ceylan Z, Meral R, Cetinkaya T. Relevance of SARS-CoV-2 in food safety and food hygiene: potential preventive measures, suggestions and nanotechnological approaches. *Virus Dis* 2020;**31**:154–60. https://doi.org/10.1007/s13337-020-00611-0.
9. Duizer E, Koopmans M. Emerging food-borne viral diseases. In: Marion PGK, Dean OC, Albert B, Michael PD, editors. *Food-borne viruses progress and challenges*. Washington DC: ASM Press; 2008. p. 117–45.
10. Roos YH. Water and pathogenic viruses inactivation—food engineering perspectives. *Food Eng Rev* 2020:1–17. https://doi.org/10.1007/s12393-020-09234-z.
11. Mardones FO, Rich KM, Boden LA, et al. The COVID-19 pandemic and global food security. *Front Vet Sci* 2020;**7**. https://doi.org/10.3389/fvets.2020.578508.
12. Chen Y, Liu Q, Guo D. Emerging coronaviruses: genome structure, replication, and pathogenesis. *J Med Vir* 2020;**92**:418–23. https://doi.org/10.1002/jmv.25681.
13. Koonin E v, Senkevich TG, Dolja V. The ancient Virus World and evolution of cells. *Biol Direct* 2006;**1**:29. https://doi.org/10.1186/1745-6150-1-29.
14. Galanakis CM. The food systems in the era of the coronavirus (COVID-19) pandemic crisis. *Foods* 2020;**9**:1–10. https://doi.org/10.3390/foods9040523.
15. Wang LF, Shi Z, Zhang S, Field H, Daszak P, Eaton BT. Review of bats and SARS. *Emerg Infect Dis* 2006;**12**:1834–40. https://doi.org/10.3201/eid1212.060401.
16. Chhikara BS, Rathi B, Singh J. Corona virus SARS-CoV-2 disease COVID-19: infection, prevention and clinical advances of the prospective chemical drug therapeutics. *Chem Biol Let* 2020;**7**(1):63–72.
17. Yuan J, Lu Y, Cao X, Cui H. Regulating wildlife conservation and food safety to prevent human exposure to novel virus. *Ecosyst Health Sustain* 2020;**6**:1741325. https://doi.org/10.1080/20964129.2020.1741325.
18. Monchatre-Leroy E, Boué F, Boucher J-M, et al. Identification of alpha and beta coronavirus in wildlife species in France: bats, rodents, rabbits, and hedgehogs. *Viruses* 2017;**9**:364. https://doi.org/10.3390/v9120364.
19. Dhama K, Pawaiya RVS, Chakrabort S, Tiwari R, Saminathan M, Verma AK. Coronavirus infection in equines: a review. *Asian J Anim Vet Adv* 2014;**9**:164–76. https://doi.org/10.3923/ajava.2014.164.176.
20. van der Hoek L, Pyrc K, Jebbink MF, et al. Identification of a new human coronavirus. *Nat Med* 2004;**10**:368–73. https://doi.org/10.1038/nm1024.

21. Oakenfull RJ, Wilson AJ. *Qualitative risk assessment: what is the risk of food or food contact materials being a source or transmission route of SARS-CoV-2 for UK consumers?* 2020. https://www.food.gov.uk/sites/default/files/media/document/web-version-qualitative-risk-assessment-risk-of-food-or-food-contact-materials-as-transmission-route-of-sars-cov-2-002.pdf. Accessed 28.11.20.
22. Shenzhen Municipal Health Commission (SMHC). *Detection of SARS-CoV-2 on an imported chicken wing sample.* 2020. http://wjw.sz.gov.cn/yqxx/content/post_7998108.html. Accessed 26.11.20.
23. Han J, Zhang X, He S, Jia P. Can the coronavirus disease be transmitted from food? A review of evidence, risks, policies and knowledge gaps. *Environ Chem Lett* 2020:1–12. https://doi.org/10.1007/s10311-020-01101-x.
24. Fisher D, Reilly A, Kang A, Zheng E, Cook AR, Anderson DE. *Seeding of outbreaks of COVID-19 by contaminated fresh and frozen food.* bioRxiv; 2020. https://doi.org/10.1101/2020.08.17.255166. 2020.08.17.255166.
25. ICMSF. *ICMSF opinion on SARS-CoV-2 and its relationship to food safety.* 2020. https://www.icmsf.org/publications/papers/. Accessed 25.11.20.
26. Li X, Song Y, Wong G, Cui J. Bat origin of a new human coronavirus: there and back again. *Sci China Life Sci* 2020;**63**:461–2. https://doi.org/10.1007/s11427-020-1645-7.
27. Hui DS, I Azhar E, Madani TA, et al. The continuing 2019-nCoV epidemic threat of novel coronaviruses to global health — the latest 2019 novel coronavirus outbreak in Wuhan, China. *Int J Infect Dis* 2020;**91**:264–6. https://doi.org/10.1016/j.ijid.2020.01.009.
28. WHO. *Novel coronavirus (2019-NCoV) SITUATION REPORT - 3.* 2020. https://apps.who.int/iris/bitstream/handle/10665/330762/nCoVsitrep23Jan2020-eng.pdf. Accessed 25.11.20.
29. EFSA. *COVID-19 and food safety-questions and answers.* 2020. Accessed 26.11.20.
30. Jalava K. First respiratory transmitted food borne outbreak? *Int J Hyg Environ Health* 2020;**226**:113490. https://doi.org/10.1016/j.ijheh.2020.113490.
31. Song HJ, Yeon J, Lee S. Impact of the COVID-19 pandemic: evidence from the U.S. restaurant industry. *Int J Hosp Manag* 2021;**92**:102702. https://doi.org/10.1016/j.ijhm.2020.102702.
32. de Freitas RSG, Stedefeldt E. COVID-19 pandemic underlines the need to build resilience in commercial restaurants' food safety. *Food Res Int* 2020;**136**:109472. https://doi.org/10.1016/j.foodres.2020.109472.
33. Rizou M, Galanakis IM, Aldawoud TMS, Galanakis CM. Safety of foods, food supply chain and environment within the COVID-19 pandemic. *Trends Food Sci Tech* 2020;**102**:293–9. https://doi.org/10.1016/j.tifs.2020.06.008.
34. Zuber S, Brüssow H. Covid 19: challenges for virologists in the food industry. *Microb Biotechnol* 2020;**13**:1689–701. https://doi.org/10.1111/1751-7915.13638.
35. Tayar M. Gıda Güvenliği ve COVID-19. *Bull Vet Pharm Tox* 2020;**11**:61–71. https://doi.org/10.38137/vetfarmatoksbulten.765700.
36. van Doremalen N, Bushmaker T, Morris DH, et al. Aerosol and surface stability of SARS-CoV-2 as compared with SARS-CoV-1. *N Engl J Med* 2020;**382**:1564–7. https://doi.org/10.1056/NEJMc2004973.
37. Akram MZ. Inanimate surfaces as potential source of 2019-nCoV spread and their disinfection with biocidal agents. *Virus Dis* 2020;**31**:94–6. https://doi.org/10.1007/s13337-020-00603-0.
38. Goldman E. Exaggerated risk of transmission of COVID-19 by fomites. *Lancet Infect Dis* 2020;**20**:892–3. https://doi.org/10.1016/S1473-3099(20)30561-2.
39. Djekic I, Nikolić A, Uzunović M, et al. Covid-19 pandemic effects on food safety - multi-country survey study. *Food Contr* 2021;**122**:107800. https://doi.org/10.1016/j.foodcont.2020.107800.

40. Imani SM, Ladouceur L, Marshall T, Maclachlan R, Soleymani L, Didar TF. Antimicrobial nanomaterials and coatings: current mechanisms and future perspectives to control the spread of viruses including SARS-CoV-2. *ACS Nano* 2020;**14**:12341–69. https://doi.org/10.1021/acsnano.0c05937.
41. Sportelli MC, Izzi M, Kukushkina EA, et al. Can nanotechnology and materials science help the fight against SARS-CoV-2? *Nanomaterials* 2020;**10**:802. https://doi.org/10.3390/nano10040802.
42. Vazquez-Munoz R, Lopez-Ribot JL. Nanotechnology as an alternative to reduce the spread of COVID-19. *Challenges* 2020;**11**:15. https://doi.org/10.3390/challe11020015.
43. Ruiz-Hitzky E, Darder M, Wicklein B, et al. Nanotechnology responses to COVID-19. *Adv Healthcare Mater* 2020;**9**:2000979. https://doi.org/10.1002/adhm.202000979.
44. Galanakis CM, Aldawoud TMS, Rizou M, Rowan NJ, Ibrahim SA. Food ingredients and active compounds against the coronavirus disease (COVID-19) pandemic: a comprehensive review. *Foods* 2020;**9**:1701. https://doi.org/10.3390/foods9111701.
45. U.S. Food & Drug Administration. *COVID-19 frequently asked questions*. 2020. https://www.fda.gov/emergency-preparedness-and-response/coronavirus-disease-2019-covid-19/covid-19-frequently-asked-questions. Accessed 27.11.20.
46. FDA. *Food safety and the coronavirus disease 2019 (COVID-19)*. 2020. https://www.fda.gov/food/food-safety-during-emergencies/food-safety-and-coronavirus-disease-2019-covid-19. Accessed 27.11.20.
47. Aday S, Aday MS. Impact of COVID-19 on the food supply chain. *Food Qual Saf* 2020;**4**:167–80. https://doi.org/10.1093/fqsafe/fyaa024.
48. Ayseli YI, Aytekin N, Buyukkayhan D, Aslan I, Ayseli MT. Food policy, nutrition and nutraceuticals in the prevention and management of COVID-19: advice for healthcare professionals. *Trends Food Sci Technol* 2020;**105**:186–99. https://doi.org/10.1016/j.tifs.2020.09.001.
49. Global Times. *Beijing supermarkets stop selling salmon after wholesalers test positive for coronavirus - global Times*. 2020. https://www.globaltimes.cn/content/1191462.shtml. Accessed 26.11.20.
50. Lacombe A, Quintela I, Liao Y, Wu VCH. Food safety lessons learned from the COVID-19 pandemic. *J Food Saf* 2020:e12878. https://doi.org/10.1111/jfs.12878.
51. Ferretti L, Wymant C, Kendall M, et al. Quantifying SARS-CoV-2 transmission suggests epidemic control with digital contact tracing. *Science* 2020;**368**:eabb6936. https://doi.org/10.1126/science.abb6936.
52. Mullis L, Saif LJ, Zhang Y, Zhang X, Azevedo MSP. Stability of bovine coronavirus on lettuce surfaces under household refrigeration conditions. *Food Microbiol* 2012;**30**:180–6. https://doi.org/10.1016/j.fm.2011.12.009.
53. Yépiz-Gómez MS, Gerba CP, Bright KR. Survival of respiratory viruses on fresh produce. *Food Environ Virol* 2013;**5**:150–6. https://doi.org/10.1007/s12560-013-9114-4.
54. Chin AWH, Chu JTS, Perera MRA, et al. Stability of SARS-CoV-2 in different environmental conditions. *Lancet Microbe* 2020;**1**:e10. https://doi.org/10.1016/S2666-5247(20)30003-3.
55. O'Brien B, Goodridge L, Ronholm J, Nasheri N. Exploring the potential of foodborne transmission of respiratory viruses. *Food Microbiol* 2021;**95**:103709. https://doi.org/10.1016/j.fm.2020.103709.
56. el Masry I, von Dobschuetz S, Plee L, et al. Exposure of Humans or Animals to SARS-CoV-2 from wild, livestock, Companion and aquatic animals: qualitative exposure assessment. *FAO Animal Product Health* 2020:181. https://doi.org/10.4060/ca9959en.
57. Hernández M, Abad D, Eiros JM, Rodríguez-Lázaro D. Are animals a neglected transmission route of SARS-CoV-2? *Pathogens* 2020;**9**:480. https://doi.org/10.3390/pathogens9060480.

58. Gossner C, Danielson N, Gervelmeyer A, et al. Human–dromedary camel interactions and the risk of acquiring zoonotic Middle East respiratory syndrome coronavirus infection. *Zoonoses Public Health* 2016;**63**:1–9. https://doi.org/10.1111/zph.12171.
59. Azkur A. Covid-19 and animals. *Bull Vet Pharm Tox* 2020;**11**:49–60. https://doi.org/10.38137/vetfarmatoksbulten.768811.
60. Koopmans M. SARS-CoV-2 and the human-animal interface: outbreaks on mink farms. *Lancet Infect Dis* 2020:1–2. https://doi.org/10.1016/S1473-3099(20)30912-9.
61. Koyun H, Karakus K, Tas A. Impacts of climite changes in animal genetic diversity. *TJSR* 2016;**9**:01–2. https://doi.org/10.1098/rspb.2011.2363.
62. Ekinci MS, Akyol İ, Karaman M, Özköse E. Recent Advances in Applications of Animal Biotechnology. *KSU J Sci Eng* 2005;**8**(2):89–95.
63. BfR. German Federal Institute for Risk Assessment. *Can the new type of coronavirus Be transmitted via food and objects?* 2020. https://doi.org/10.1056/NEJMc2004973?url_ver=Z39.88.
64. Liji T. *Fresh and frozen food can harbor SARS-CoV-2 for at least 21 days*. 2020. https://www.news-medical.net/news/20200819/Fresh-and-frozen-food-can-harbor-SARS-CoV-2-for-at-least-21-days.aspx. Accessed 27.11.2020.
65. Dyal JW, Grant MP, Broadwater K, et al. COVID-19 among workers in meat and poultry processing facilities – 19 states, April 2020. *MMWR Morb Mortal Wkly Rep* 2020;**69**:557–61. https://doi.org/10.15585/mmwr.mm6918e3.
66. EFFAT. *Covid-19 outbreaks in slaughterhouses and meat processing plants state of affairs and proposals for policy action at EU level.* 2020. EFFAT-Report-Covid-19-outbreaks-in-slaughterhouses-and-meat-packing-plants-State-of-affairs-and-proposals-for-policy-action-at-EU-level-30.06.2020.pdf; Accessed 26.11.20.
67. Giansanti F, Panella G, Leboffe L, Antonini G. Lactoferrin from milk: nutraceutical and pharmacological properties. *Pharmaceuticals* 2016;**9**:61. https://doi.org/10.3390/ph9040061.
68. Peroni DG, Fanos V. Lactoferrin is an important factor when breastfeeding and COVID-19 are considered. *Acta Paediatr* 2020;**109**:2139–40. https://doi.org/10.1111/apa.15417.
69. Reusken CB, Farag EA, Jonges M, et al. Middle East respiratory syndrome coronavirus (MERS-CoV) RNA and neutralising antibodies in milk collected according to local customs from dromedary camels, Qatar, April 2014. *Euro Surveill* 2014;**19**:20829. https://doi.org/10.2807/1560-7917.ES2014.19.23.20829.
70. van Doremalen N, Bushmaker T, Karesh WB, Munster VJ. Stability of Middle East respiratory syndrome coronavirus in milk. *Emerg Infect Dis* 2014;**20**:1263–4. https://doi.org/10.3201/eid2007.140500.
71. Rabenau HF, Cinatl J, Morgenstern B, Bauer G, Preiser W, Doerr HW. Stability and inactivation of SARS coronavirus. *Med Microbiol Immunol* 2005;**194**:1–6. https://doi.org/10.1007/s00430-004-0219-0.
72. Darnell MER, Subbarao K, Feinstone SM, Taylor DR. Inactivation of the coronavirus that induces severe acute respiratory syndrome, SARS-CoV. *J Virol Met* 2004;**121**:85–91. https://doi.org/10.1016/j.jviromet.2004.06.006.
73. Pressman P, Naidu AS, Clemens R. COVID-19 and food safety. *Nutr Today* 2020;**55**:125–8. https://doi.org/10.1097/NT.0000000000000415.
74. Tachedjian G, Aldunate M, Bradshaw CS, Cone RA. The role of lactic acid production by probiotic Lactobacillus species in vaginal health. *Res Microbiol* 2017;**168**:782–92. https://doi.org/10.1016/j.resmic.2017.04.001.
75. Zafer C, Ünal Sengör GF. Determination of some quality changes of sous vide-sea bass fillets (*Dicentrarchus labrax*, linnaeus, 1758) treated with dried basil, fresh garlic, and dill weed. *Acta Aquatica Turcica* 2019;**15**:126–34. https://doi.org/10.22392/actaquatr.577330.

76. Gonzalez-Fandos E, Laorden AM. Sous vide technology. In: Ozogul Y, editor. *Innovative technologies in seafood processing*. Boca Raton, Florida: CRC Press; 2019 [chapter 13].
77. Anelich LECM, Lues R, Farber JM, Parreira VR. SARS-CoV-2 and risk to food safety. *Front Nutr* 2020;**7**. https://doi.org/10.3389/fnut.2020.580551.
78. Zhang W, Du R-H, Li B, et al. Molecular and serological investigation of 2019-nCoV infected patients: implication of multiple shedding routes. *Emerg Microb Infect* 2020;**9**:386–9. https://doi.org/10.1080/22221751.2020.1729071.
79. Olaimat AN, Shahbaz HM, Fatima N, Munir S, Holley RA. Food safety during and after the era of COVID-19 pandemic. *Front Microbiol* 2020;**11**. https://doi.org/10.3389/fmicb.2020.01854.
80. Ignatov I. Antiviral effects of nano colloidal silver, water catholyte, oxidal with methylene blue. Possible effects of influence over coronavirus SARS-CoV and SARS-CoV-2 with disease COVID-19. In: *Global congress on infectious diseases. Sofia, Bulgaria*; 2020.
81. Nazaktabar A, Lashkenari MS, Araghi A, Ghorbani M, Golshahi H. In vivo evaluation of toxicity and antiviral activity of polyrhodanine nanoparticles by using the chicken embryo model. *Int J Biol Macromol* 2017;**103**:379–84. https://doi.org/10.1016/j.ijbiomac.2017.05.069.
82. Ufaz S, Balter A, Tzror C, et al. Anti-viral RNAi nanoparticles protect shrimp against white spot disease. *Mol Syst Des Eng* 2018;**3**:38–48. https://doi.org/10.1039/C7ME00092H.
83. Abd Elkodous M, El-Sayyad GS, Abdel-Daim MM. Engineered nanomaterials as fighters against SARS-CoV-2: the way to control and treat pandemics. *Environ Sci Pollut Res* 2020:1–7. https://doi.org/10.1007/s11356-020-11032-3.
84. Rai PK, Usmani Z, Thakur VK, Gupta VK, Mishra YK. Tackling COVID-19 pandemic through nanocoatings: Confront and exactitude. *Curr Res Green Sustain Chem* 2020;**3**:100011. https://doi.org/10.1016/j.crgsc.2020.100011.
85. Lara HH, Ayala-Nuñez N v, Ixtepan-Turrent L, Rodriguez-Padilla C. Mode of antiviral action of silver nanoparticles against HIV-1. *J Nanobiotechnol* 2010;**8**:1. https://doi.org/10.1186/1477-3155-8-1.
86. Jeremiah SS, Miyakawa K, Morita T, Yamaoka Y, Ryo A. Potent antiviral effect of silver nanoparticles on SARS-CoV-2. *Biochem Biophys Res Commun* 2020;**533**:195–200. https://doi.org/10.1016/j.bbrc.2020.09.018.
87. Guo L, Yao Z, Yang L, et al. Plasma-activated water: An alternative disinfectant for S protein inactivation to prevent SARS-CoV-2 infection. *Chem Eng J* 2020:127742. https://doi.org/10.1016/j.cej.2020.127742.
88. Filipić A, Gutierrez-Aguirre I, Primc G, Mozetič M, Dobnik D. Cold plasma, a new hope in the field of virus inactivation. *Trends Biotechnol* 2020. https://doi.org/10.1016/j.tibtech.2020.04.003.
89. Bekeschus S, Kramer A, Suffredini E, von Woedtke T, Colombo V. Gas plasma technology—an asset to healthcare during viral pandemics such as the COVID-19 crisis? *IEEE Trans Radiat Plasma Med Sci* 2020;**4**:391–9. https://doi.org/10.1109/trpms.2020.3002658.
90. Mohan SV, Hemalatha M, Kopperi H, Ranjith I, Kumar AK. SARS-CoV-2 in environmental perspective: Occurrence, persistence, surveillance, inactivation and challenges. *Chem Eng J* 2021;**405**:126893. https://doi.org/10.1016/j.cej.2020.126893.
91. Peng P, Chen P, Zhou N, et al. Packed food and packaging materials disinfected by cold plasma. In: Daniela BA, editor. *Advances in cold plasma applications for food safety and preservation*. Richland, WA: Academic Press; 2020. p. 269–86.
92. Ostrikov K, Neyts EC, Meyyappan M. Plasma nanoscience: from nano-solids in plasmas to nano-plasmas in solids. *Adv Phys* 2013;**62**:113–224. https://doi.org/10.1080/00018732.2013.808047.

93. Pradeep P, Chulkyoon M. Non-thermal plasmas (NTPs) for inactivation of viruses in abiotic environment. *Res J Biotechnol* 2016;**11**(6):91–6.
94. Verma SS. Fighting COVID-19 with non-thermal plasma technology. *Indian Pract* 2020;**73**(7):42–5.
95. Bae SC, Park SY, Choe W, Ha S do. Inactivation of murine norovirus-1 and hepatitis A virus on fresh meats by atmospheric pressure plasma jets. *Food Res Int* 2015;**76**:342–7. https://doi.org/10.1016/j.foodres.2015.06.039.
96. Bermudez-Aguirre D. Advances in the inactivation of microorganisms and viruses in food and model systems using cold plasma. In: Daniela BA, editor. *Advances in cold plasma applications for food safety and preservation.* Richland, WA: Academic Press; 2020. p. 49–91.
97. Lacombe A, Niemira BA, Gurtler JB, et al. Nonthermal inactivation of norovirus surrogates on blueberries using atmospheric cold plasma. *Food Microbiol* 2017;**63**:1–5. https://doi.org/10.1016/j.fm.2016.10.030.
98. Feizollahi E, Misra N, Roopesh MS. Factors influencing the antimicrobial efficacy of dielectric barrier discharge (DBD) atmospheric cold plasma (ACP) in food processing applications. *Crit Rev Food Sci Nutr* 2020:1–24. https://doi.org/10.1080/10408398.2020.1743967.
99. Zhao YM, de Alba M, Sun DW, Tiwari B. Principles and recent applications of novel non-thermal processing technologies for the fish industry—a review. *Crit Rev Food Sci Nutr* 2019;**59**:728–42. https://doi.org/10.1080/10408398.2018.1495613.
100. Bisag A, Isabelli P, Laurita R, et al. Cold atmospheric plasma inactivation of aerosolized microdroplets containing bacteria and purified SARS-CoV-2 RNA to contrast airborne indoor transmission. *Plasma Process Polym* 2020;**17**:2000154. https://doi.org/10.1002/ppap.202000154.
101. Chen Z, Garcia G, Arumugaswami V, Wirz RE. Cold atmospheric plasma for SARS-CoV-2 inactivation. *Phys Fluids* 2020;**32**:111702. https://doi.org/10.1063/5.0031332.
102. Ailavadi S, Davidson PM, Morgan MT, D'Souza DH. Thermal inactivation kinetics of Tulane virus in cell-culture medium and spinach. *J Food Sci* 2019;**84**:557–63. https://doi.org/10.1111/1750-3841.14461.
103. Ahmed H, Maunula L, Korhonen J. Reduction of norovirus in foods by nonthermal treatments. *J Food Protect* 2020;**83**:2053–73. https://doi.org/10.4315/JFP-20-177.
104. Min SC, Roh SH, Niemira BA, Sites JE, Boyd G, Lacombe A. Dielectric barrier discharge atmospheric cold plasma inhibits *Escherichia coli* O157:H7, Salmonella, Listeria monocytogenes, and Tulane virus in Romaine lettuce. *Int J Food Microbiol* 2016;**237**:114–20. https://doi.org/10.1016/j.ijfoodmicro.2016.08.025.
105. Schiappacasse C, Peng P, Zhou N, et al. Inactivation of aerosolized Newcastle disease virus with non-thermal plasma. *Appl Eng Agric* 2020;**36**:55–60. https://doi.org/10.13031/aea.13699.
106. Park SY, Ha S do. Assessment of cold oxygen plasma technology for the inactivation of major foodborne viruses on stainless steel. *J Food Eng* 2018;**223**:42–5. https://doi.org/10.1016/j.jfoodeng.2017.11.041.
107. Yepez Xv, Misra NN, Keener KM. Nonthermal plasma technology. In: Demirci A, Feng H, Krishnamurthy K, editors. *Food safety engineering. Food engineering series.* Springer; 2020. p. 607–28.
108. Ghernaout D, Elboughdiri N. Disinfecting water: plasma discharge for removing coronaviruses. *OALib* 2020;**07**:1–29. https://doi.org/10.4236/oalib.1106314.
109. FAO. *Food safety in the time of COVID-19.* 2020. https://doi.org/10.4060/ca8623en.
110. Nakat Z, Bou-Mitri C. COVID-19 and the food industry: readiness assessment. *Food Contr* 2021;**121**:107661. https://doi.org/10.1016/j.foodcont.2020.107661.

111. Ashraf S, Ashraf S, Ashraf M, et al. *Honey and Nigella sativa against COVID-19 in Pakistan (HNS-COVID-PK): a multi-center placebo-controlled randomized clinical trial.* medRxiv; 2020. https://doi.org/10.1101/2020.10.30.20217364. 2020.10.30.20217364.
112. Shakoor H, Feehan J, al Dhaheri AS, et al. Immune-boosting role of vitamins D, C, E, zinc, selenium and omega-3 fatty acids: could they help against COVID-19? *Maturitas* 2021;**143**:1−9. https://doi.org/10.1016/j.maturitas.2020.08.003.
113. Junaid K, Qasim S, Yasmeen H, et al. Potential inhibitory effect of vitamins against COVID-19. *Comput Mater Continua (CMC)* 2020;**66**:707−14. https://doi.org/10.32604/cmc.2020.012976.
114. Shakoor H, Feehan J, Mikkelsen K, et al. Be well: a potential role for vitamin B in COVID-19. *Maturitas* 2021. https://doi.org/10.1016/j.maturitas.2020.08.007.
115. Annunziata G, Sanduzzi Zamparelli M, Santoro C, et al. May polyphenols have a role against coronavirus infection? An Overview of in vitro evidence. *Front Med* 2020;**7**. https://doi.org/10.3389/fmed.2020.00240.
116. Iddir M, Brito A, Dingeo G, et al. Strengthening the immune system and reducing inflammation and oxidative stress through diet and nutrition: considerations during the COVID-19 crisis. *Nutrients* 2020;**12**:1562. https://doi.org/10.3390/nu12061562.
117. Singh P, Tripathi MK, Yasir M, Khare R, Tripathi MK, Shrivastava R. Potential inhibitors for SARS-CoV-2 and functional food components as nutritional supplement for COVID-19: a review. *Plant Foods Hum Nutr* 2020;**75**:458−66. https://doi.org/10.1007/s11130-020-00861-9.
118. Jampilek J, Kralova K. Potential of nanonutraceuticals in increasing immunity. *Nanomaterials* 2020;**10**:2224. https://doi.org/10.3390/nano10112224.

CHAPTER 13

Public health management during COVID-19 and applications of point-of-care based biomolecular detection approaches

Monalisha Ghosh Dastidar[1], Sharmili Roy[2]
[1]Research School of Electrical, Energy, and Materials Engineering, College of Engineering and Computer Science, Australian National University, ACT, Canberra, Australia; [2]Division of Oncology, School of Medicine, Stanford University, Palo Alto, CA, United States

13.1 Overview of human coronavirus infections

Coronaviruses are phylogenetically classified under the family of coronaviridae (a subfamily of orthocoronavirinae) in order Nidovirales, which carries positive-sense single-stranded RNA (ss+RNA) as genetic material. In a study, it has been reported to have the biggest genome among all other types of RNA viruses.[1,2] Naturally, these viruses can be found in a wide range of species including animals as well as humans. The human coronaviruses (HCoVs) include alphacoronavirus (HCoV-229E and HCoV-NL63) and betacoronavirus (HCov-OC43 and HCoV-HKU1).[3] The first HCoVs were discovered in the 1960s from the upper respiratory infections of a patient.[4–6] In 2002, a strain of betacoronavirus (Subgenus Sarbecovirus) originated from bats and later spread to humans via civets in the province of southern China, which caused a severe acute respiratory syndrome (SARS) affecting the lungs, and thus the virus was named as the SARS-CoV.[7] After 10 years, in 2012 a betacoronavirus strain (Subgenus Merbecovirus) has been reported to be transmitted in humans via camels in the Middle East Asian region and was found to have similar symptoms as SARS and thus named as Middle East respiratory syndrome-related coronavirus (MERS-CoV).[8,9] The current outbreak of coronavirus in humans originated from Hubei province of China in December 2019. Studies have shown that it has also been originated sharing similar characteristics as the previous outbreak, thus later named SARS-CoV-2.[10] The genomic

Environmental Management of COVID-19
ISBN 978-0-323-85780-2
https://doi.org/10.1016/B978-0-323-85780-2.00009-3

sequencing of this virus revealed it is a group of betacoronavirus 79% distinctive from SARS-CoV.[11,12] Not only this, but it also shares structurally similar features such as the viral capsids proteins, containing S (Spike), E (Envelope), M (Membrane), and N (Nucleocapsid).[13]

13.1.1 Symptomatic characterization of SARS-CoV-2 patients

Infection of the SARS-CoV-2 causes a range of symptoms similar to the previous outbreaks, where apart from abnormal respiratory functioning, mild fever, cough, abnormal gastrointestinal behavior are reported as the major symptoms. In addition, there are asymptomatic cases have also been reported in various studies. A study directed toward asymptomatic infections was conducted by Kim et al., which shows that one-fifth of the patients develop either very mild or no symptoms, which were later considered as asymptomatic cases.[14,15] The most common symptoms of SARS-CoV-2 infection were slight headache and occurrence of cough with a sore throat.[16] The schematic characterization of symptoms in different cases has been represented in Fig. 13.1. Clinical symptoms are generally present with body fever (~102°F), fatigue, diarrhea, nausea, continuous dry cough, complete loss of smell and taste, severe respiratory problems, thoracic pain, and breathing apnea.

In addition, a study showed that patients with a premedical conditions such as hypertension, heart illness, chronic bronchitis, diabetes, and cerebral injuries were getting more severe upon contraction of the virus.[17] In severe cases, this virus caused organ damage along with other complications in the lungs, and sometimes death.[18–21]

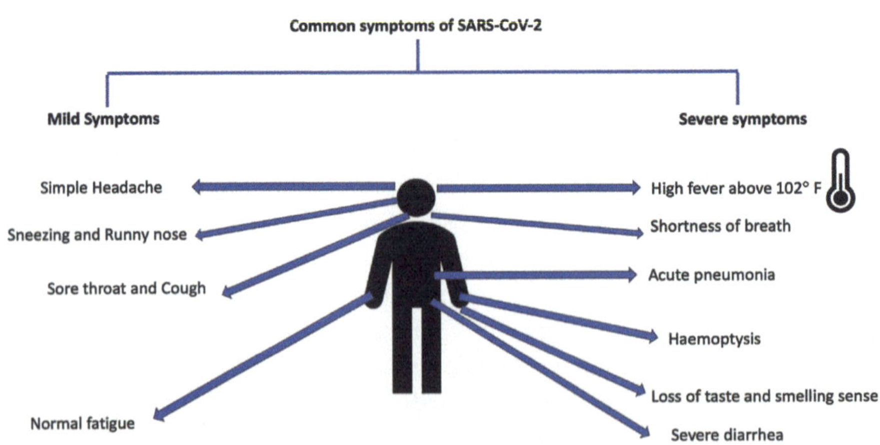

Figure 13.1 Representation of symptoms by COVID-19 patients. Two types of symptoms were observed: (1) mild, and (2) severe symptoms.

13.1.2 Transmission of SARS-CoV-2 agents and pathogenesis

The first epicenter of SARS-CoV-2 was in Wuhan, China. To reduce the rapid spreading of this virus, the Chinese administration had imposed a lockdown in Wuhan and nearby cities to break the transmission from one individual to another.[22,23] The transmission is reportedly air and surface borne, which resulted in rapid spread of the virus. Within four months of its origin, the virus had spread more than 100 countries across the world and thus, the World Health Organization (WHO) declared a pandemic on March 11, 2020.[24] Initially, the major outbreaks were observed in Europe, Iran, South Korea, and the United States. Few reports stated that the initial transmission occurred due to in-person contacts with close ones and overseas traveling, which created the patient zero for the region.[25] Virologists and epidemiologists have suggested that the transmission can be controlled by quick identifications of the carrier and the isolation of infected patients.[26] Another major cause of massive spread was attributed to the latency period of the virus which resulted in delayed symptoms even through the infection was quickly transmitting from one individual to another. The latency period of SARS-CoV-2 was reported for three days, which has greatly impacted the domestic transmission even after the complete lockdown and social distancing measures. In addition, "asymptomatic" cases have played a major role in connecting these domestic spreads. Studies reported that the transmission of this virus is not only human-to-human close contacts, but it is also spreading through human to air in close proximity areas[27] and fomite transmission[28] such as solid objects like plastics, copper, steels, used masks, and gloves. In these cases, when water droplets from infected individuals are expelled onto these solid surfaces,[29] the virus stays viable for a few hours. Another mode of transmission is the contamination of wastewater by fecal and excretory wastes.[30]

Until now, the pathogenesis of SARS-CoV-2 in humans is not fully known. On the basis of various studies and reports of genetic behavior of SARS-CoV-2 virus, to date, researchers are able to understand the mechanisms about the pathophysiology of this virus (Fig. 13.2).[31] It is also very essential to screen the viral loads, which is different from children to adults to elder individuals. Reports suggested that the amount of viral loads helps to understand whether a patient is having pre- or mild symptoms.[32] Further, in the critical case of a critical patient, the reliability of the epithelial-endothelial wall was interrupted rigorously.[33] Additionally, upon

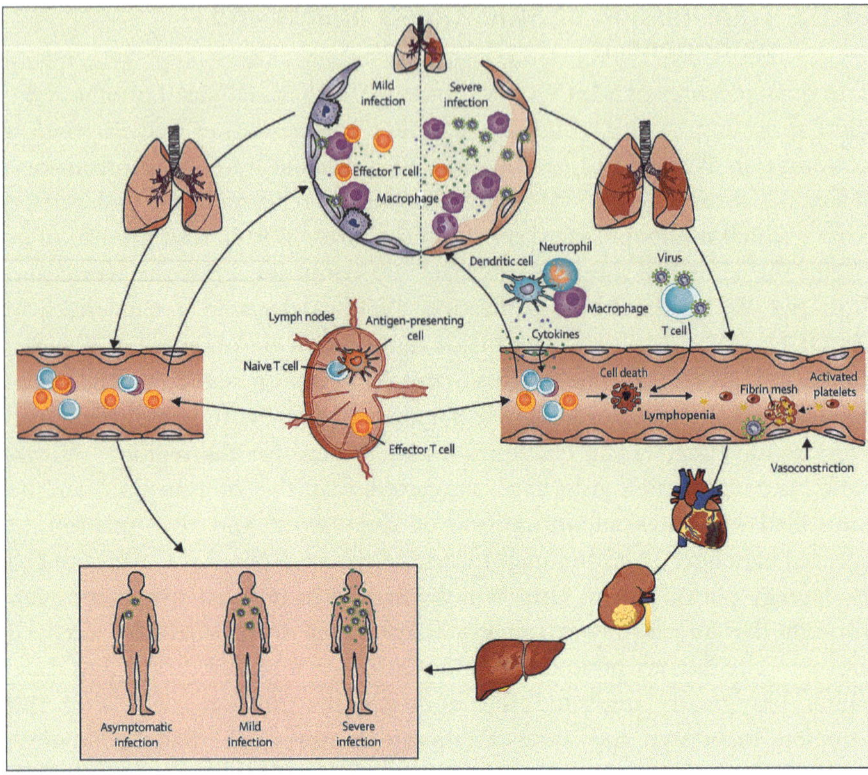

Figure 13.2 Pathophysiology of SARS-CoV-2 in humans. *(This figure is reproduced with permission from Li H, Liu L, Zhang D, et al. SARS-CoV-2 and viral sepsis: observations and hypotheses.* Lancet *2020;**395**(10235):1517–20.)*

the rupture of epithelial cells, SARS-CoV-2 can have an immense effect on the capillary system of the lung endothelial cells, which will cause a diffusion of a large amount of plasma components in the alveolar cavity.[34] On the other hand, the unrestrained virus infection caused by macrophage penetration which will lead to more injuries in the lungs. Simultaneously, the direct infection on other organs by this virus can cause systemic cytokine disturbance in the flow of immune pathogenesis. Hence, the effectiveness of antiviral therapies, immune responses, and study of the new adaptive immune responses are very much important for breaking the vicious cycles and improving the outcome of patients.[31]

Based on several studies and reports, investigators have hypothetically drawn an idea about the pathogenesis of SARS-CoV-2 infections.[35] According to several reports, the virus passes through nasal and larynx mucosa, after which it enters the lungs via the respiratory tract. During that time patients get a fever, dry cough, and shortness of breath.[36] Investigators explained that once the virus enters the body, it travels through the blood from the lungs to the intestines.[31,37,38] Upon infection, the count of white blood cells (WBC) get reduced to less than the marginal requirements in the body.[5] Hence, lymphopenia has been diagnosed in patients. Researchers have also predicted that B lymphocyte may reduce during this infection, which causes declined antibody production inside the body.[38] However, to confirm this theory more investigations are needed. Immune-related studies are necessary to comprehend and to know the insights of pathogenesis, and treatment of COVID-19 disease.[35]

13.2 Challenges for public health management during the COVID-19 pandemic

The health management of public health is very crucial during the time of a pandemic. Many restrictions are imposed for controlling the transmission within the population. However, there are several challenges associated with following those restrictions which makes it difficult to control the spread of SARS-CoV-2 virus. These challenges will not be mitigated completely even after arrival of vaccines, because postvaccination surveillance is also important, due to which the health workers need to go through numerous undefined challenges.[39] The major challenge to this situation is the asymptomatic cases, in which an individual or a group of individuals spread the virus without even knowing that they have contracted close with the virus. By the time these individuals realize their condition and start to experience the symptoms, they have already become superspreaders. For controlling this situation, a quick plan of action and proper guidelines have been delivered by WHO and the Centers for Disease Control and Prevention (CDC), where the front liners, essential professionals, and investigators have been instructed to use personal protective equipment (PPE) to manage the spread of COVID-19 in the public domain.[40] These organizations have also mentioned the risks factors, symptoms, and preventive measures as well as have suggested when to travel, quarantine, and get tested for COVID-19.

The management of a COVID-19 patient is extremely difficult in terms of contagiousness such as carrying patient's samples regularly for laboratory tests, regular investigations of various diagnostic detection processes, cleaning the bathrooms and surfaces used by COVID-19 patients, and so forth.[41] Since the majority of transmission of the virus comes from water droplets released by the patient during coughing, sneezing, and talking, therefore, covering the mouth and nose using a surgical mask or N95 mask are extremely necessary.[42] Essential workers need to wear PPE kits all the time when they are in contact with the patients or any suspected individual who came in close contact with the patients. Although wearing PPE is a vital safety measurement for avoiding droplets and airborne transmissions in close vicinity of an infected individual, it is difficult to wear the same for a long period of time due to excessive perspiration. Another important key factor is multidisciplinary management between intensive care units (ICUs), infectious disease (ID) departments, and infection control specialists.[43] After exploring the characteristics of this virus, investigators have drawn few management procedures to implement in the field.

13.2.1 Patient management

Health workers at hospitals need to take care of a COVID-19 patient starting from their admission until their recovery or ICU admissions or in a later-stage ventilation process. During this whole process, the hospital's administrative workers need to collaborate with ID physicians. COVID-19 suspected patients need to be kept in an isolated hygienic single room.[44] All CDC and the infection prevention control (IPC) guidelines should be followed by each personnel.[40] Later, on a regular basis those patients need to be tested by RT-PCR and CT scanning to confirm the case and to know the viral loads inside the patient's body.[45] In that case, samples are taken from upper and lower respiratory tracts (mostly from nasal and oropharyngeal swabs specimens). Additionally, a few days later blood, urine, and fecal samples should be collected for further analysis.[43,46,47] Thereafter, if the patient is recovering well, then they need to have regular checkups for testing the viral loads and when it will be confirmed by RT-PCR testing that the patient does not have any trace of SARS-CoV-2 elements then they will be discharged with proper hospital guidelines. On the other hand, if the patient's condition worsens, they need to be shifted to ventilators in ICUs with various life supporting systems. Thus, the overall process of an infected patient from SARS-CoV-2 is maintained through good health management decided by the health sector's expert professionals.[48]

13.2.2 Precision medicine management (PM)

Precision medicine (PM) management is a budding area in pathogen disease management, which develops with various advances in molecular biology, immunology sciences, bioinformatics, etc. for the treatment of health diseases.[49] Four major approaches that mainly occur under this management consist of prediction, prevention, personalization, and participation.[50] The first step is "prediction," which mainly focuses on the expectation of the manifestations of diseases by analyzing the risk factors, symptoms, and lifestyle, spreading social awareness with proper information to the public to prepare themselves to face the pandemic situation.[51] The next step is "prevention," where this management is applied for offering proper programs that will direct the progression of the pathology before the appearances occur. This process will also convey secondary prevention when the disease will be stable. The third step is "personalization," where various kinds of biological testing will occur, such as molecular, serological, and genetic analysis to know the behaviors of SARS-CoV-2 virus within infected patients. The last step of this management is called "participation," which includes a detailed treatment process and the involvement of academic institutes, health workers, biomedical professionals, and the patients themselves.[52] Fundamentally, PM involves the integration of a broad array of community and distinctive data that embraces clinical, molecular, genetic, and more improved biomarkers that can be used for the detection. This study mainly emphasizes how and why it is necessary to detect COVID-19 at an early stage, with a major focus on the strategies adopted for screening and detection of SARS-CoV-2. Hence, depending on the knowledge of detection and management of the virus, we aim toward contributing in controlling the COVID-19 pandemic.

13.3 Importance of early diagnosis and treatment of SARS-CoV-2

Because scientific studies have reported that SARS-CoV-2 is spreading rapidly, quick detection of this virus is a very crucial step to stop the spread. The rapid replication of this virus increase the viral loads. The very first step is to ban all kind of travel process and local transportations except essentials belongings The very first step is to restrict all kinds of traveling processes and local transportations except essentials belongings.[53] The next step is to isolate all the immediate travelers and test them by quick diagnostic procedures with 14 days of the quarantine course. Already, many countries have implemented these steps to evaluate COVID-19 patients.[54,55] Hence, early diagnosis is important to isolate patients quickly from healthy people.[56] Researchers are focusing on more POC-based diagnostic processes to detect the virus at earliest. The gold standards for detection SARS-CoV-2 is reverse transcriptase polymerase chain reaction (RT-PCR).[57] Many other crucial POC strategies with advanced biomolecular bioanalysis processes have been implemented for early diagnostic purposes. Numerous rapid detection kits are available in the market to control the spread.[58] Presently, nucleic acid-based and genetic sequencing are widely being used to detect this contagious infection.[59] Nucleic acid amplification tests like isothermal amplification diagnostic processes are rapid and straightforward as compared to the conventional processes such as cell culture based, ELISA-based techniques.[60] Thus, these techniques are playing a major role in initial diagnosis of SARS-CoV-2. These test specimens are collected from upper and lower respiratory tract.[61,62] Serological testing is also important. In this process lateral flow assays (LFA), ELISA techniques have been used for population screening, identification of plasma donors, and evaluation of immunity power.[63,64] The third essential part is biomedical monitoring by advanced POC devices, and the major role of these strategies are to check the severity of the virus, evaluation, and therapeutic monitoring of patients. Many biomarkers like cardiac,[65] renal,[66] inflammatory,[67] and hepatic[68] markers have been implicated in the COVID-19 pandemic.

Additionally, early-stage sample collection is a vital factor in terms of reducing the viral loads for patients. To know the acute phase of this infection time of repeated testing is needed. In serological testing, monitoring the IgA response is also important. By this response, it will be easy to know when patients are exposed to the virus. Generally, the peak time is for IgM antibody is around 18–21 days from contracting the virus to get a positive response.[69] Although detailed study of antibody testing is not yet confirmed and many factors are unknown. In addition the serological processes have poor sensitivity issues. Henceforward, the biosensors approaches come into the picture where diagnosis process is very fast and affordable to the general health public.

On the other hand, medical diagnosis of SARS-CoV-2 at an early stage is also very much essential, so that the increase of viral loads can be reduced. Following all the preventive and control measures can reduce the risk of exposure with an infected person but at early stage of infection proper medication is a critical issue. According to studies, COVID-19 patients with pneumonia, high fever, and chest congestion require oxygen therapy with a flow rate of 5 L/min. The concentration of oxygen obtained for adults is >90% SpO_2, for children and pregnant women is 95% SpO_2. Patients with acute respiratory problems and hypoxemic respiratory failure require invasive mechanical ventilation.[68,70] For the mild symptoms cases noninvasive and high-flow nasal oxygen therapy can be implemented at an early stage. Further, if the condition of patients get worse within 1 h then endotracheal intubation and mechanical lung ventilations must be applied.[71] However, specific pharmacological treatment is not yet confirmed in the case of COVID-19. In many places, doctors have used empirical antiviral with lopinavir, ribavirin, nelfinavir, or other same derivatives.[72] In Japan, doctors have used inhibitor enzyme transmembrane protease, serine 2 (TMPRSS2).[73,74] Antimalaria drugs such as chloroquine and its derivatives hydroxychloroquine are used in many countries.[75] In a few cases, therapeutics strategies are also used like steroids, immunoglobulin, Janus kinase inhibition, and cytokine blockade.[76] These strategies are not only used for quick relief for patients, but these treatments are curable for COVID-19. More accurate medicines and vaccines are yet to come. Thus, in the current situation, development of technologies for early detection is crucial. Later in this chapter the various advances in sensor technology to improve COVID-19 situations are discussed.

13.4 Strategies adopted for the screening and detection of SARS-CoV-2

The screening, early detection, and public health management of the SARS-CoV-2 infected individuals have been done in various ways, where the preliminary ways for the screening have been performed by biomolecular technologies. Currently the gold standards technology is RT-PCR based process. Based on RT-PCR tests, many advanced technologies have been developed. Many commercial kits are available based on RT-PCR and nucleic acid testing in the market. Apart from these, recently diagnostics-based screening, biosensor, and nanomaterial-based POC detections are also employed for early diagnosis of SARS-CoV-2 virus. This section describes all such advanced in technologies used for SARS-CoV-2 preliminary screening and monitoring in daily bases.

13.4.1 Technology empowered screening of patients

Infected individuals by SARS-CoV-2 commonly show a mild fever at first and then it develops severity at a later stage. Targeting these mild symptoms, temperature based screening have been initiated in public domains such as at airports, public parks, and various public transportations.[77] Recently in Switzerland, dual functional plasmonic biosensor combined with plasmonic photo thermal effect and localized surface plasmon resonance (LSPR) sensing process had a very advantageous solution for SARS-CoV-2 diagnoses.[78] This study showed high sensitivity and rapid reliable assay for this virus detection. In this study, researchers have used two-dimensional gold nanoislands, which were functionalized with complementary DNA (cDNA) receptors. This principle worked to sense the selected sequences from SARS-CoV-2 strain via nucleic acid hybridization. To improve the sensitivity, researchers employed thermoplasmonic heat on gold nanoislands and illuminated at particular plasmonic resonance frequency (Fig. 13.3). The detection limit (LOD) of this process is 0.22 pM.[78] Previously developed body temperature sensor and thermal imaging based infrared thermography (IRT) technology can be used for controlling recent

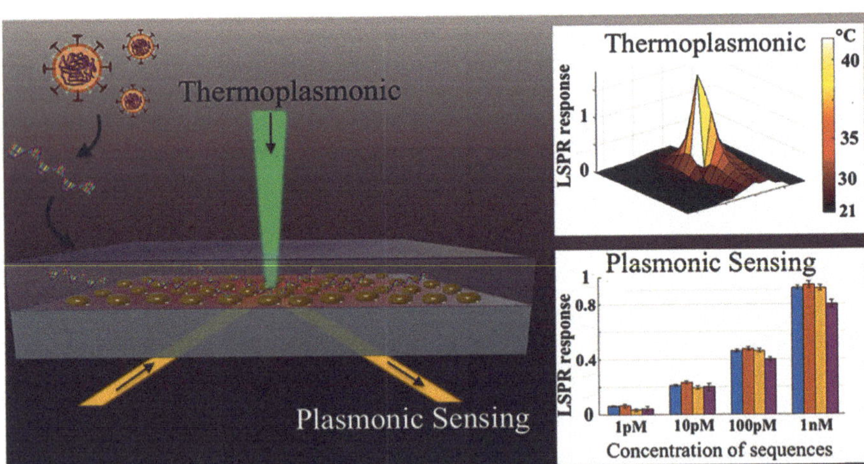

Figure 13.3 Schematic representation of dual-functional plasmonic photothermal biosensors for SARS-CoV-2 detection. Two-dimensional gold nanoislands functionalized with cDNA receptors for sensing the selected sequences from SARS-CoV-2 via nucleic acid hybridization. *(This figure is reproduced with the permission from Qiu G, Gai Z, Tao Y, Schmitt J, Kullak-Ublick GA, Wang J. Dual-functional plasmonic photothermal biosensors for highly accurate severe acute respiratory syndrome coronavirus 2 detection.* ACS Nano. *2020;**14**(5):5268–77.)*

outbreaks by SARS-CoV-2 screening.[79] Apart from these technologies, recent geotagging-based cellphone screening is also helpful for separating the infected people from healthy populations.[80] In this process, when a person is infected by the SARS-CoV-2 virus, then the individual will be detected by geotagging and cellular code. This is linked with the device the person carries. Hence, this process gives alerts and signals to healthy individuals to separate themselves from the infected ones.[81] These important technologies are gaining in demand to help further detect SARS-CoV-2 in early stages.

In several cases, it has been reported that the infected personnel who is asymptomatic is difficult to be diagnosed and isolated those individuals from healthy population. In these cases, thermal sensors fail. For these incidents, screening at a large scale for each individual is necessary by biomolecular, colorimetric sensor, electrochemical sensor, or any clinical assisted diagnostics methods. Therefore, the aforementioned techniques empower the screening and quick detection schemes that can evaluate healthy people from infected ones. In the following section, we have described such frequently used techniques for the screening/detection SARS-CoV-2.

13.4.2 Biomolecular techniques for SARS-CoV-2 detection

In clinics and diagnostic centers, many gold-standard biomolecular techniques have been used, where they examine the samples from the infected symptomatic/asymptomatic individual using thermal cycler by RT-PCR, enzyme-linked immunoassays (ELISA), and many next generation sequencing (NGS) platforms. For this, the sample from an infected patient is collected either in the normal laboratory settings or in the clinical settings. These techniques have mainly been targeted for identifying the specific signature sequence from the viral genome. The most common biomolecular technique is based on the reverse transcriptase quantitative polymerase chain reaction (RT-qPCR).[82] Usually, viral RNA is converted into cDNA by RT-PCR/qRT-PCR amplification. Infected samples such as swabs from respiratory tracts and bronchoalveolar lavage fluid were collected and transferred into sterile tubes that contais universal transport medium for RT-PCR reactions (Fig. 13.4.1). Most of the commercial kits target four different viral genes such as RdRp genes for RNA dependent RNA polymerase, ORF1a/b gene for human RNA polymerase protein, N genes for nucleocapsid protein, and E gene for envelope protein[83] (Table 13.1). However, there are limitations to this process, which needs to be addressed

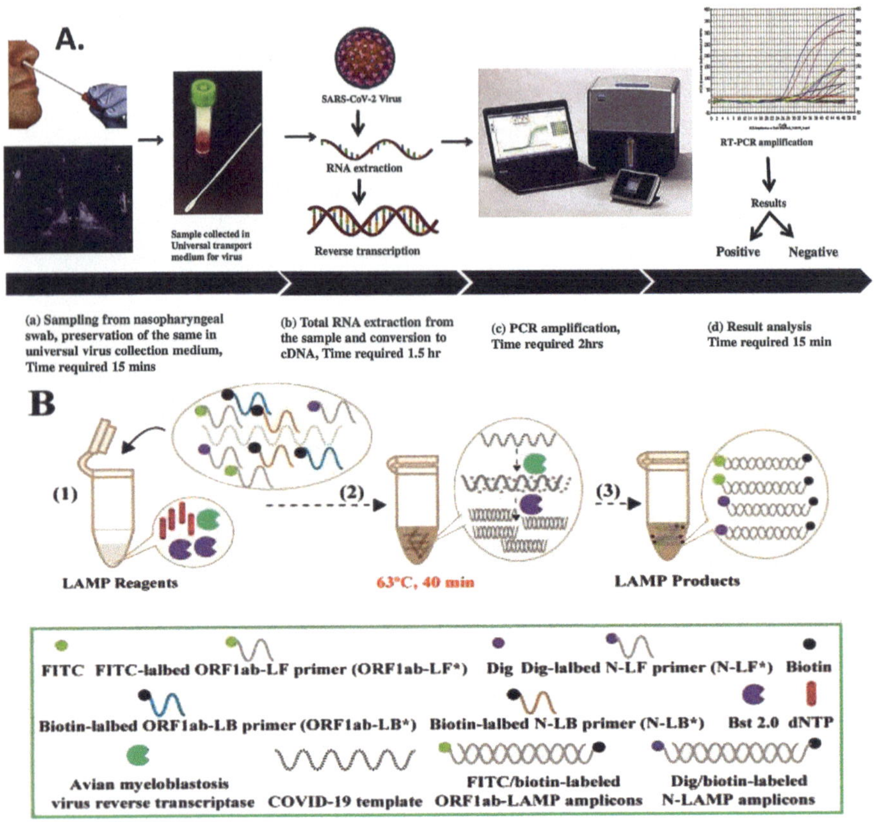

Figure 13.4 Schematic representation of biomolecular detection of SARS-CoV-2: (1) workflow of RT-PCR based detection; (2) mechanisms of RT-LAMP reaction for the detection of SARS-CoV-2. *(This figure is reproduced with permission from Das Mukhopadhyay C, Sharma P, Sinha K, Rajarshi K. Recent trends in analytical and digital techniques for the detection of the SARS-Cov-2.* Biophys Chem. *2021;**270**:106538 and Zhu X, Wang X, Han L, et al. Reverse transcription loop-mediated isothermal amplification combined with nanoparticles-based biosensor for diagnosis of COVID-19 166. medRxiv; 2020. p. 112437. 2020.2003.2017.20037796.)*[85]

further. RT-PCR method uses more time for expertized sample handling, post-PCR analysis. This process also has cross-contaminating issues and the sensitivity is inferior compared to other POC-based detection processes.[84] In a few reports it is stated that the sampling process (RNA samples) is very complicated. Thus, the sample preparation for the RT-qPCR is one of the tedious part especially when it comes to the larger scale of examinations.

To address such limitations, researchers have developed isothermal amplification processes like loop-mediated isothermal amplification (LAMP),[86–88]

Table 13.1 Diverse biomolecular processes used for detection of SARS-CoV-2.

Diagnostic platforms	Sample type	Detection limit	Analytes	Detection time	Advantages	Limitations	References
RT-PCR	Fecal sample	3.9 copies/reaction	Nucleic acids	Not mentioned	High sensitivity and high specificity	High-end laboratory procedures with trained labors	104
qRT-PCR	Respiratory samples (swabs)	M Gene 10^6 copies/reaction	RNA	Not mentioned	Good sensitivity	High-end procedures and Carry over issues	105
Real-time RT-PCR	Respiratory swabs	11.2 copies/reaction	Genomic RNA	Not mentioned	Good sensitivity and specificity	Different COVID-19 gene assays were performed with different commercial kits which created problem in determination of sensitivity	59
RT-PCR	Nasal swab and sputum	20 folds of diluted samples	Viral RNA	3–4 h	Moderate sensitivity	High-end laboratory process	106
ELISA	Blood samples	92.37% sensitive	Antigen and antibodies (IgG)	1–5 h	Good sensitivity	ELISA is a very conventional laboratory based procedure with high risk of contamination	107

Continued

Table 13.1 Diverse biomolecular processes used for detection of SARS-CoV-2.—cont'd

Diagnostic platforms	Sample type	Detection limit	Analytes	Detection time	Advantages	Limitations	References
ELISA	Plasmid	IgG = 0.95, IgA = 0.975 IgM = 0.916	Antigen and antibody	Limited time frame	Quantitative ELISA method for SARS-CoV-2 detection with good sensitivity	High risk of contamination, need high-end instruments for sample analysis	108
RT-LAMP	Serum, saliva and nasal swab, urine	1.02 fg	Nucleic acids	30 min	Very high sensitivity with high specificity	This study was performed with nucleotide oligos of COVID-19 from GeneBank not directly test related samples	109
RT-LAMP	Clinical samples from direct COVID-19 patients	3 copies/ reaction	RNA	30 −50 min	Very good specificity and high accuracy with high sensitivity on-spot biomolecular assay	Not mentioned	110

RCA	DNA Oligo nucleotides	~0.96 p.m.	Circular DNA	10 min	Rapid, robust and highly sensitive assay	RCA has limitation for the length of circular template	101
One step RT-LAMP	Clinical samples	500 copies/mL	Nucleocapsid protein	60 min	Good specificity, and straight forward process	This study is not as good as other quantitative process and sensitivity is also poor	111
NASBA	Saliva samples	10 copies/reaction	Viral RNA	60 min	Portable, scalable, high accuracy, rapid and affordable	NASBA has poor RNA integrity issues and NASBA needs to improve the specificity of the reactions	100

rolling circle amplification (RCA),[89] nucleic acid sequence-based amplification (NASBA) (Table 13.1).[90] These isothermal amplifications, especially LAMP reaction, reduces time, multiple steps, costs, and development on-site based on rapid detection (Fig. 13.4.2).[87,91–93] Based on this principle, researchers have developed a LAMP-assisted colorimetric rapid detection method for SARS-CoV-2 in clinical samples. This method demonstrated excellent results and qualitative visual detection of the SARS-CoV-2 virus.[94] Likewise, using the same LAMP principle, one group of researchers have detected the SARS-CoV-2 virus from the unpurified clinical samples. Due to the versatile nature of LAMP assay, it is easy to detect target analytes from unpurified samples. As a result, this process leads to the whole reaction quickly and more cost-effectively using less reagents and high-end instruments.[95] Since LAMP reaction is an isothermal process, which means for the whole reaction, it needs only one single temperature; thus this process required only one heat block to perform the whole experiment.[96] Based on this concept LAMP assay can be used as a noninstrumental and POC system.[97,98] Accordingly, paper-based LAMP reaction, based on a colorimetric process provides better sensitivity and time of response. Researchers developed a LAMP-based method for the rapid colorimetric detection of COVID-19; the sensitivity was ~97.6% (42/43) and read-out time was ~30 min[99] based on the isothermal concept. NASBA is also used for SARS-CoV-2 RNA detection. Reported sensitivity by NASBA was 10 copies/reaction within 10 min.[100] RCA isothermal process is also one of the helpful methods for rapid detection of SARS-CoV-2 at this current situation. One of the reports stated that RCA can detect SARS-CoV-2 circular DNA within 10 min and the sensitivity of the assay was ~0.96 pM.[101] On the other hand, for a detailed study of SARS-CoV-2 behavior and determination of mutations, NGS technology is playing a major role.[102] Additionally, advanced molecular biology techniques based on gene editing have also been used to develop the assay for the detection of the SARS-CoV-2 in clinical samples. In recent evolution, a clustered regularly interspaced short palindromic repeat (CRISPR)-based technique has been employed to achieve higher sensitivity and prompt discovery in molecular assay.[103]

13.5 Nanobiosensing techniques for SARS-CoV-2 detection

Even though biomolecular technologies are considered as gold-standard procedures due to their constant, efficient, and reliable outcomes, but these processes have many drawbacks, such as requiring high-end instruments, taking too much time, and consuming the use of many reagents and skilled personnel to perform the experiments.[112,113] Furthermore, their requirement for these reliable instruments has many restrictions, which are very challenging in rural areas with less-developed facilities. To conquer such drawbacks and to expand the prospect for the prompt detection of the SARS-CoV-2 at on-site for rapid assessment, nanomaterial-based biosensing methods have been employed to improve the unsophisticated devices that can be handled by anyone as a POC system.[114] These devices are established using nanomaterial-based processes, which help to detect pathogens and are robust, rapid, and with very good sensitivity.[115] Nanomaterials provide outstanding transducing capabilities to the device due to their astonishing optical and electronic properties.[114] In recent years, these nanomaterial-based applications in biological sciences are becoming an emerging technology due to their outstanding analytical performance, lower LOD determination, wide detection range, and less involvement of high-end instruments.[116] Apart from these advantages, the main benefits of nanomaterial-based technologies are reduction of reagent volumes, detection and response time, user-friendly applications, and high stability. The major application of nanomaterials is based on different kinds of biosensor devices, which include, for example, colorimetric, fluorescence, plasmonic, and electrochemical devices.[117,118]

Biosensor contains a bioreceptor as a sensing component and a signal transducer to transform any kind of biological signal into a physical, electrochemical, optical, or any other quantifiable signal. Transformation of a biological signal into a measurable signal, biosensor is performed by a physicochemical transducer, such as an optical, thermal, and mass process.[119] To amplify the signal, a signal processor is necessary, including a bioreceptor such as DNA, RNA, antibodies, and enzymes, which is linked to the target analytes (Fig. 13.5).

Based on these simple principles, researchers are applying this concept for early diagnosis of SARS-CoV-2 screening. Many nanobiosensing-based detection methods have been developed for on-site monitoring of SARS-CoV-2.[121] However, these techniques are also having some limitations such

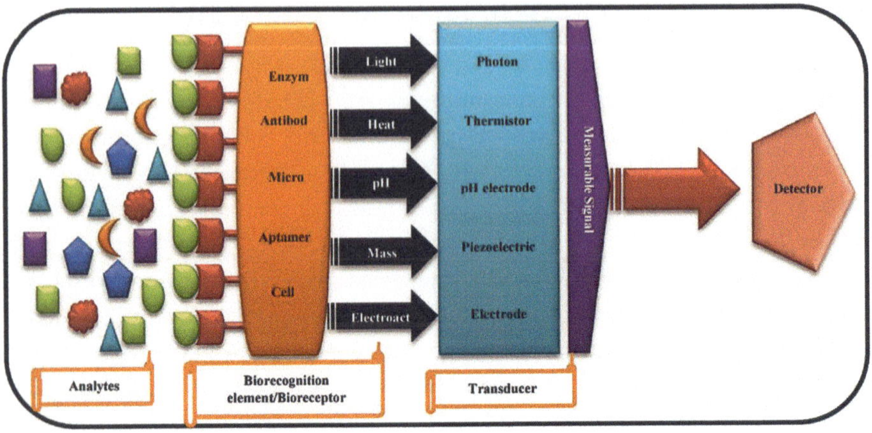

Figure 13.5 Schematic representation of a typical nanobiosensor principle. *(This figure is adopted from Girigoswami K, Akhtar N. Nanobiosensors and fluorescence based biosensors: an overview.* Int J Nano Dimens *2019;**10**(1):1–17.)*[120]

as the designing of sensor platforms to obtain the lowest LOD and to improve the signal-to-noise ratio. To overcome these problems, researchers are attempting to improvise more by incorporating different unique surface modifications to the biosensor platforms using advanced nanomaterials. Among all the sensor strategies, electrochemical-based biosensing platforms are used widely. In a later section we have summarized studies that have been conducted using electrochemical biosensors for SARS-CoV-2 detection.

13.5.1 Electrochemical-based biosensors for the detection of SARS-CoV-2

Fundamentally, an electrochemical biosensor is a biosensor device with an electrochemical transducer. It consists of a chemically modified electrode with conducting or semiconducting material. These materials are coated with a biochemical film.[122] Based on this concept, several electrochemical based biosensor studies were employed in detection of SARS-CoV-2 (Table 13.2). In general, electrochemical-based colorimetric detection was used in viral detection such as LFA, lateral flow immunochromatographic strips (LFICS), fluorescence resonance energy transfer (FRET), Field effect transistors (FET), etc.[123] These processes are not only used before the clinical/real sample but also allows for the controlled release of samples to the detection zone, which assist a device to obtain a better sensitivity. Employing this strategy, Lin et al.

Table 13.2 Summary table of available electrochemical biosensing strategies for detection SARS-CoV-2.

Detection principles	Samples source	Nanomaterial used in probe	LOD	Linear range	Signal process	Response time	Recognition element	References
Functionalized TiO_2 Nanotube-based electrochemical biosensor	Nasal and saliva	Co-TNT [Cobalt functionalized TiO_2 nanotubes]	~0.7 nM	14–1400 nM	Amperometry	~30s	Spike protein of SARS-CoV-2 (Receptor-binding domain, RBD)	133
Paper-based electrochemical biosensors	Nasopharyngeal swab specimens	ssDNA capped with AuNPs on graphene-coated filter paper	6.9 copies/μL	585.4–5.854×10^7 copies/μL	CV and EIS	Less than 5 min	Nucleocapsid phosphoprotein (N genes)	134
Ultrasensitive super sandwich electrochemical sensor using smartphone	Sputum, throat swabs, urine, plasma, feces, oral swabs and serum samples	Calixarene functionalized graphene oxide targeting RNA	200 copies/μL	10-fold serial dilution	EIS and DPV	Not reported	ORF1ab gene	135
Paper-based electrochemical biosensor	Human Serum	Redox conversation $[Fe(CN)_6]^{3-/4-}$ between captured immunoglobulins	~1 ng/mL	1–1000 ng/mL	EIS, CV	30 min	Spike protein	136
Impedance-based	Serum samples from positive	RBD-Monoclonal	Not reported	Not reported	EIS	Less than 5 min	RBD	137

Continued

Table 13.2 Summary table of available electrochemical biosensing strategies for detection SARS-CoV-2.—cont'd

Detection principles	Samples source	Nanomaterial used in probe	LOD	Linear range	Signal process	Response time	Recognition element	References
electrochemical sensor	COVID-19 patients	antibody CR3022						
Electrochemical immunosensor	Nasal swabs	Carbon nanofiber-modified screen printed	0.8 pg/mL	1–1000 ng/mL	SWV	Not reported	Nucleocapsid N protein	138
Plasmonic electrochemical biosensor detection	Not reported	Functionalized AuNPs conjugated specific monoclonal antibody	~4.2 fmol	Not reported	Plasmonic metasensors	~80 min	Spike protein	139
FET-based electrochemical sensor	Nasal swabs	Graphene-SiO_2/Si substrate	2.42×10^2 copies/mL	Not reported	FET	Not reported	Spike protein	125
Microfluidic chip based electrochemical sensor	Saliva and nasal swabs	Gold nanospikes in a microfluidic chip	~0.5 p.m.	Not reported	Surface plasmon resonance	30 min	Spike protein	140

have developed an LFA-based device for rapid and sensitive determination of the anti-SARS-CoV-2 antibody, which is a biomarker that originates in the infected individuals.[124] Simultaneously, Seo et al. developed FET-based detection process of SARS-CoV-2 spike protein detection with high sensitivity and specificity. The sensitivity of this process was discovered as 1 fg/mL in phosphate-buffered saline and 100 fg/mL in clinical medium from nasal swabs (Fig. 13.6A).[125] The detection of the antibody is the indirect format and carries reliability issues; thus to obtain improved detection, the sensing is based on highly precise techniques, where CRISPR-Cas12 based lateral flow system has been adopted to effectively detect the specific RNA from the viral genome.[126] Other than these types of biosensor systems, few biosensors were made using conventional biological processes such as ELISA-LFA techniques for preliminary detection of SARS-CoV-2 virus. This process has also successfully served as an efficient tool for screening a wide range of populations for managing the public health.[127] Besides, LFICS based detection has been developed by Huang et al. for analysis of SARS-CoV-2. This knowledge incorporates gold nanoparticles colloid-based procedures, which can detect IgM and IgG antibodies against a SARS-CoV-2 viral strain in human blood within a very short period of time, ~15 min.[128] In addition, another report by Li et al., who discovered that the testing sensitivity of LFICS procedure is 88.66% and the specificity is 90.63%, which is also an outstanding outcome (Fig. 13.6C and D).[129] Further, detecting and destroying this virus at an early stage is extremely important. Hence, targeting this issue, researchers suggested a strategy based on metallic-nanomaterials in disinfectants such as silver, titanium dioxide, and copper nanoparticles by emitting toxic ions, which will destroy the viral particles.[130]

Due to the rapid and sensitive detection capability, the electrochemical-based biosensing platforms have widely been accepted worldwide and were impressively discovered for the detection and screening of various disease biomarkers.[131] In the current circumstances of SARS-CoV-2 detection, these electrochemical procedures have been significantly exploited, which includes voltammetry, amperometry, impedance-based, and FET-assisted detection for screening the infected individuals. For example, Mahari et al. have developed a system to detect the SARS-CoV-2-specific antigen, using the differential potential voltametric technique, where the antibody against SARS-CoV-2 has been immobilized on the sensing probe surface. Here, the analytical signal has been recorded using Zobell's solution and obtained a wide range of the antigen concentration between one fM and 1M with the sensitivity of 90 fM.[132]

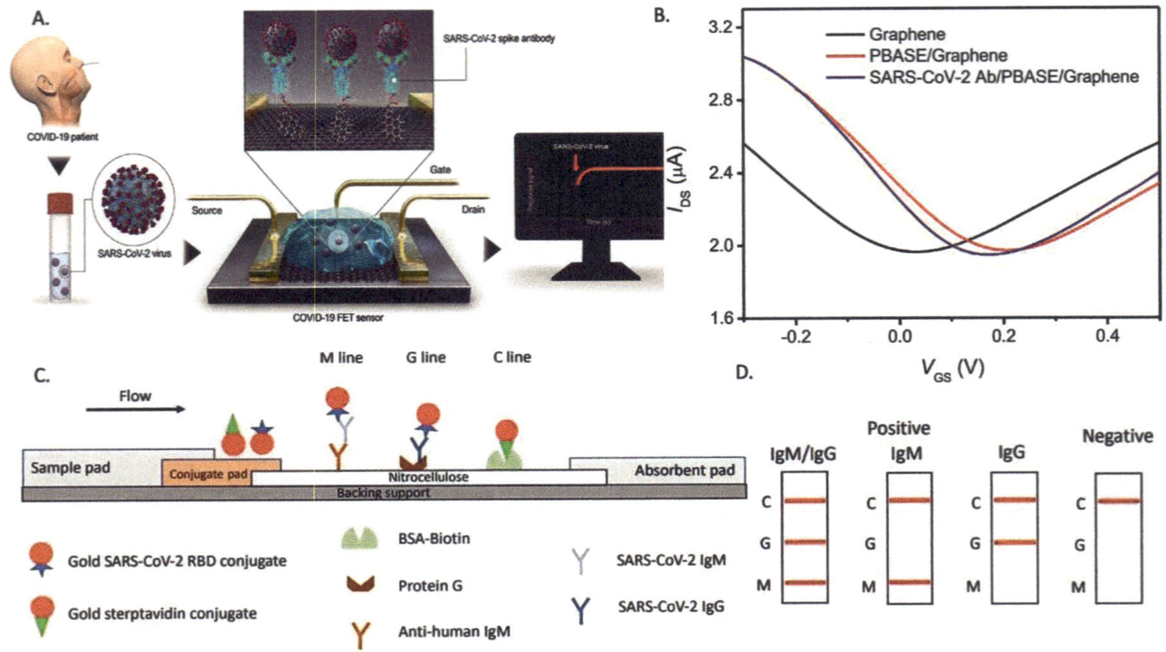

Figure 13.6 A schematic diagram of FET sensor for SARS-CoV-2: (A) working principle; (B) measurement of SARS-CoV-2 signals through FET sensors in steps of the antibody conjugation; (C) schematic representation of SARS-CoV-2 IgM-IgG antibody test diagram of detection device; (D) diagram of different testing outcomes where C indicates control line, G indicates IgG line, M indicates IgM line, and G means IgG immunoglobulin. *(These figures are adopted with the permission from Seo G, Lee G, Kim MJ, et al. Rapid detection of COVID-19 causative virus (SARS-CoV-2) in human nasopharyngeal swab specimens using field-effect transistor-based biosensor.* ACS Nano *2020;***14***(4):5135–42. and Li Z, Yi Y, Luo X, et al. Development and clinical application of a rapid IgM-IgG combined antibody test for SARS-CoV-2 infection diagnosis.* J Med Virol *2020;***92***(9):1518–24.)*

13.5.2 Colorimetric-based biosensors for the detection of SARS-CoV-2

Colorimetric biosensors are one of the best options for SARS-CoV-2 detection and random screening due to its simple, rapid outcome and on field experiments. This kind of setup is much more important at this current scenario in terms of isolating infected individuals from healthy ones. Using the colorimetric principle, several biosensor devices have been proposed for on-site monitoring of SARS-CoV-2 detection (Table 13.3). In contrast, conventional processes and electrochemical-based detection colorimetric detection can be more portable since these processes do not require a laboratory facility for the whole experiment. For example, Moitra et al. developed naked eye detection of SARS-CoV-2 DNA by using colorimetric assay. In this case, the group has used gold nanoparticles capped with thiol-based antisense oligonucleotides for N gene from SARS-CoV-2. This report said that this process can detect positive SARS-CoV-2 strain within 10 min with the sensitivity of 0.18 ng/μL.[141] Another group of researchers have developed colorimetric biosensors based on gold nanoparticles functionalized with antibodies targeting three different proteins of SARS-CoV-2. The remarkable feature of this biosensor is to detect the quantification of viral load which is similar to threshold cycle 36.5 targeting RdRp and N genes. The application of this method is claimed as POC testing.[142] Considering these facts, we can conclude that, in the field of lab-on-chip diagnostic sensors for SARS-CoV-2, colorimetric biosensors play major roles in terms of rapidity and cost.

13.6 Conclusion and future direction

To address COVID-19 encounters, proper health management is much needed in terms of controlling the spread of the virus. The isolation process and the infrastructure of patient units such as ICU rooms, ventilators, operating rooms, and washrooms need to be maintained in a hygienic way using disinfectants, sanitization, and continuous hand washing. The demand for PPE has increased significantly, since more patients are getting admitted to hospitals. So, distributing PPE to health workers at the correct time is very much needed. In addition, to protect patients, medical professions, health care systems, virologists, and other investigators who are associated with COVID-19 from unnecessary infections, health care management is very much essential at this moment. Additionally, it is important to perform

Table 13.3 Various colorimetric biosensors available for SARS-CoV-2 detection.

Testing approaches	Testing type	Sample source	Gene/region detection	LOD	Detection time	Target analyte	References
Colorimetric biosensor based on gold nanoparticle (AuNP)	POC	Nasal swabs,	RdRp and N genes	Threshold cycle (Ct) = 36.5	3 min	Spike protein, membrane protein, envelope proteins	142
Colorimetric biosensor based on botulinum neurotoxin receptor	POC	Blood, urine, Stool and Nasopharyngeal swabs	N genes	1a.m.	1–2 h	Recombinant plasmid	143
RT-PCR colorimetric assay	Laboratory based	Upper respiratory specimens	ORF1ab and N genes	10^5 dilution factors	Total ~2 –3 h But observable color change found within 10 min	Nucleic acids template	144
Colorimetric detection using CRISPER/dCas9	Laboratory based	Nasopharyngeal aspirates and sputum specimens	N1, N2, and N3 genes	30 pM	~2 h	Nucleic acids	145
Colorimetric detection using DNAzyme sensor	Laboratory and POC-based	Swabs from nasopharyngeal and oropharyngeal	N genes	10^1 to 10^7 copies	Total duration ~1 h and visual observation ~5 min	RNA	146

Paper-based colorimetric biosensor using electrochemical assay	Laboratory and POC-based	Not mentioned	Immunoglobulins IgG and IgM	~1 ng/mL	30 min	Spike protein	136
Colorimetric biosensor using LAMP assay	POC	Nasal swabs	N gene	~62.5 viral copies/LAMP reaction	50–60 min	Plasmids	147
Colorimetric detection using RT-LAMP	POC	Nasal swabs	N gene	10^2 RNA copies	30 min	RNA	148
Colorimetric detection using paper-based ELISA	POC	Not mentioned	SARS-CoV-2 antibody	100 ng/μL	20 min	Recombinant nucleocapsid	149
Colorimetric detection using binding chemistry	POC	Cells infected with SARS-CoV-2	N gene	0.18 ng/μL	10 min	Nucleic acids	141

regular self-monitoring and self-care by using easily accessible diagnostic support, which could improve the outcomes for many individuals and health care professionals without having close contact with infected patients. Due to lack of space in the hospital for 14 days isolation is difficult, rather the isolation process can be possibly done at a patient's home instead if the symptoms of COVID-19 is mild. Instead of admitting into the hospital and occupying the hospital beds that can be used for those severely affected or who are having heavy symptoms, patients could monitor themselves using rapid and on-site diagnostic approaches. Therefore, advancements in biosensor systems are helping patients for quick, accurate, and easy identification of their condition. The biosensor-based quick methodologies have the potential to reduce the transmission of the respiratory virus. The rapid and early detection of SARS-CoV-2 is very much essential in terms of stopping the infections. Biosensor-based POC systems are not only useful for on-the-spot diagnostics, but it is also helpful for environmental monitoring in public areas such as airports, ports, railways stations, and various working institutes. Scaling up the number of preliminary tests across the world, biosensor-based facilities would be the best way to stop the virus. In the near future, advancements in biosensor-based technology integrated with microfluidic sensors and the establishment of nanotechnologies such as nanosized sensors, nanovaccines, and antiviral nanocomposites will help to bring some solution in terms of stopping this virus. Henceforth, the development of biosensor-based technologies and health management will have great potential to combat the present scenario and to overcome any other pandemic situations in the future.

References

1. Giwa A, Desai A. Novel coronavirus COVID-19: an overview for emergency clinicians. *Emerg Med Pract* 2020;**22**(2 Suppl. 2):1–21.
2. Baranwal A, Mahapatra S, Purohit B, Roy S, Chandra P. Insights into novel coronavirus and COVID-19 outbreak. In: Chandra P, Roy S, editors. *Diagnostic strategies for COVID-19 and other coronaviruses*. Singapore: Springer Singapore; 2020. p. 1–17.
3. Loeffelholz MJ, Tang Y-W. Laboratory diagnosis of emerging human coronavirus infections – the state of the art. *Emerg Microb Infect* 2020;**9**(1):747–56.
4. Su S, Wong G, Shi W, et al. Epidemiology, genetic recombination, and pathogenesis of coronaviruses. *Trends Microbiol* 2016;**24**(6):490–502.
5. Fontana F, Alfano G, Mori G, et al. COVID-19 pneumonia in a kidney transplant recipient successfully treated with tocilizumab and hydroxychloroquine. *Am J Transplant* 2020;**20**(7):1902–6.
6. Hamre D, Procknow JJ. A new virus isolated from the human respiratory tract. *Proc Soc Exp Biol Med* 1966;**121**(1):190–3.

7. Peiris JSM, Lai ST, Poon LLM, et al. Coronavirus as a possible cause of severe acute respiratory syndrome. *Lancet* 2003;**361**(9366):1319–25.
8. Zaki AM, van Boheemen S, Bestebroer TM, Osterhaus ADME, Fouchier RAM. Isolation of a novel coronavirus from a man with pneumonia in Saudi Arabia. *N Engl J Med* 2012;**367**(19):1814–20.
9. Assiri A, Al-Tawfiq JA, Al-Rabeeah AA, et al. Epidemiological, demographic, and clinical characteristics of 47 cases of Middle East respiratory syndrome coronavirus disease from Saudi Arabia: a descriptive study. *Lancet Infect Dis* 2013;**13**(9):752–61.
10. Coronaviridae Study Group of the International Committee on Taxonomy of V. The species Severe acute respiratory syndrome-related coronavirus: classifying 2019-nCoV and naming it SARS-CoV-2. *Nat Microbiol* 2020;**5**(4):536–44.
11. Lu R, Zhao X, Li J. Genomic characterisation and epidemiology of 2019 novel coronavirus: implications for virus origins and receptor binding. *Lancet* January 30, 2020:30251–8. published online ahead of print.
12. Zhu N, Zhang D, Wang W, et al. A novel coronavirus from patients with pneumonia in China, 2019. *N Engl J Med* 2020;**382**(8):727–33.
13. Walls AC, Park Y-J, Tortorici MA, Wall A, McGuire AT, Veesler D. Structure, function, and antigenicity of the SARS-CoV-2 spike glycoprotein. *Cell* 2020;**181**(2):281–292.e286.
14. Gu K, Kim MJ, Ra SH, et al. Clinical characteristics of asymptomatic and symptomatic patients with mild COVID-19. *Clin Microbiol Infect* 2020;**26**(7):948.e941–3.
15. Report on the epidemiological features of coronavirus disease 2019 (COVID-19) outbreak in the Republic of Korea from January 19 to March 2, 2020. *J Kor Med Sci* 2020;**35**(10):e112.
16. Abd El-Aziz TM, Stockand JD. Recent progress and challenges in drug development against COVID-19 coronavirus (SARS-CoV-2) - an update on the status. *Infect Genet Evol* 2020;**83**:104327.
17. Rottnek F. Exhibit B.
18. Soufi GJ, Hekmatnia A, Nasrollahzadeh M, et al. SARS-CoV-2 (COVID-19): new discoveries and current challenges. *Appl Sci* 2020;**10**(10):3641.
19. Chan JF-W, Yuan S, Kok K-H, et al. A familial cluster of pneumonia associated with the 2019 novel coronavirus indicating person-to-person transmission: a study of a family cluster. *Lancet* 2020;**395**(10223):514–23.
20. Wang L, Wang Y, Ye D, Liu Q. Review of the 2019 novel coronavirus (SARS-CoV-2) based on current evidence. *Int J Antimicrob Agents* 2020;**55**(6):105948.
21. Menni C, Valdes A, Freydin MB, et al. *Loss of smell and taste in combination with other symptoms is a strong predictor of COVID-19 infection.* medRxiv; 2020. 2020.2004.2005.20048421.
22. Wu JT, Leung K, Leung GM. Nowcasting and forecasting the potential domestic and international spread of the 2019-nCoV outbreak originating in Wuhan, China: a modelling study. *Lancet* 2020;**395**(10225):689–97.
23. Yuen K-S, Ye ZW, Fung S-Y, Chan C-P, Jin D-Y. SARS-CoV-2 and COVID-19: the most important research questions. *Cell Biosci* 2020;**10**(1):40.
24. Rockett RJ, Arnott A, Lam C, et al. Revealing COVID-19 transmission in Australia by SARS-CoV-2 genome sequencing and agent-based modeling. *Nat Med* 2020;**26**(9):1398–404.
25. Ghinai I, McPherson TD, Hunter JC, et al. First known person-to-person transmission of severe acute respiratory syndrome coronavirus 2 (SARS-CoV-2) in the USA. *Lancet* 2020;**395**(10230):1137–44.
26. Anderson RM, Heesterbeek H, Klinkenberg D, Hollingsworth TD. How will country-based mitigation measures influence the course of the COVID-19 epidemic? *Lancet* 2020;**395**(10228):931–4.

27. Morawska L, Cao J. Airborne transmission of SARS-CoV-2: the world should face the reality. *Environ Int* 2020;**139**:105730.
28. Augenbraun BL, Lasner ZD, Mitra D, et al. Assessment and mitigation of aerosol airborne SARS-CoV-2 transmission in laboratory and office environments. *J Occup Environ Hyg* 2020;**17**(10):447−56.
29. van Doremalen N, Bushmaker T, Morris DH, et al. Aerosol and surface stability of SARS-CoV-2 as compared with SARS-CoV-1. *N Engl J Med* 2020;**382**(16):1564−7.
30. Di Maria F, Beccaloni E, Bonadonna L, et al. Minimization of spreading of SARS-CoV-2 via household waste produced by subjects affected by COVID-19 or in quarantine. *Sci Total Environ.* 2020;**743**:140803.
31. Li H, Liu L, Zhang D, et al. SARS-CoV-2 and viral sepsis: observations and hypotheses. *Lancet* 2020;**395**(10235):1517−20.
32. Jones TC, Mühlemann B, Veith T, et al. *An analysis of SARS-CoV-2 viral load by patient age*. medRxiv; 2020. 2020.2006.2008.20125484.
33. Zeng H-L, Chen D, Yan J, et al. Proteomic characteristics of bronchoalveolar lavage fluid in critical COVID-19 patients. *FEBS J* 2020. https://doi.org/10.1111/febs.15609 [n/a(n/a)].
34. Briguglio M, Porta M, Zuffada F, et al. SARS-CoV-2 aiming for the heart: a multicenter Italian perspective about cardiovascular issues in COVID-19. *Front Physiol* 2020;**11**. 571367-571367.
35. Lin L, Lu L, Cao W, Li T. Hypothesis for potential pathogenesis of SARS-CoV-2 infection—a review of immune changes in patients with viral pneumonia. *Emerg Microb Infect* 2020;**9**(1):727−32.
36. Guan W-j, Ni Z-y, Hu Y, et al. *Clinical characteristics of 2019 novel coronavirus infection in China*. medRxiv; 2020. 2020.2002.2006.20020974.
37. Guan W-j, Ni Z-y, Hu Y, et al. Clinical characteristics of coronavirus disease 2019 in China. *N Engl J Med* 2020;**382**(18):1708−20.
38. Wang D, Hu B, Hu C, et al. Clinical characteristics of 138 hospitalized patients with 2019 novel coronavirus-infected pneumonia in Wuhan, China. *J Am Med Assoc* 2020;**323**(11):1061−9 [Internet]. 2020 Feb [cited 2020 Mar 23].
39. Sampath Kumar NS, Chintagunta AD, Jeevan Kumar SP, Roy S, Kumar M. Immunotherapeutics for Covid-19 and post vaccination surveillance. *3 Biotech* 2020;**10**(12):527.
40. Organization WH. *Infection prevention and control during health care when novel coronavirus (nCoV) infection is suspected: interim guidance*, vol. 25; 2020. January 2020.
41. Chia PY, Coleman KK, Tan YK, et al. Detection of air and surface contamination by SARS-CoV-2 in hospital rooms of infected patients. *Nat Commun* 2020;**11**(1):2800.
42. Asadi S, Cappa CD, Barreda S, Wexler AS, Bouvier NM, Ristenpart WD. Efficacy of masks and face coverings in controlling outward aerosol particle emission from expiratory activities. *Sci Rep* 2020;**10**(1):15665.
43. Bouadma L, Lescure F-X, Lucet J-C, Yazdanpanah Y, Timsit J-F. Severe SARS-CoV-2 infections: practical considerations and management strategy for intensivists. *Intensive Care Med* 2020;**46**(4):579−82.
44. Zhang G-Q, Pan H-Q, Hu X-X, et al. The role of isolation rooms, facemasks and intensified hand hygiene in the prevention of nosocomial COVID-19 transmission in a pulmonary clinical setting. *Infect Dis Povert* 2020;**9**(1):104.
45. Li W, Su YY, Zhi SS, et al. Virus shedding dynamics in asymptomatic and mildly symptomatic patients infected with SARS-CoV-2. *Clin Microbiol Infect* 2020;**26**(11):1556.e1551−6.
46. Thompson JR, Nancharaiah YV, Gu X, et al. Making waves: wastewater surveillance of SARS-CoV-2 for population-based health management. *Water Res* 2020;**184**:116181.

47. Long Q-X, Tang X-J, Shi Q-L, et al. Clinical and immunological assessment of asymptomatic SARS-CoV-2 infections. *Nat Med* 2020;**26**(8):1200–4.
48. Hick JL, Hanfling D, Wynia MK, Pavia AT. *Duty to plan: health care, crisis standards of care, and novel coronavirus SARS-CoV-2.* NAM Perspectives; 2020.
49. Hodson R. Precision medicine. *Nature* 2016;**537**(7619). S49-S49.
50. Crisci CD, Ardusso LRF, Mossuz A, Müller L. A precision medicine approach to SARS-CoV-2 pandemic management. *Curr Treat Opt Aller* 2020;**7**(3):422–40.
51. Carta MG, Romano F, Orrù G. The true challenges of the covid-19 epidemics: the need for essential levels of care for all. *Open Respir Med J* 2020;**14**:8–9.
52. Hou Y-CC, Yu H-C, Martin R, et al. Precision medicine integrating whole-genome sequencing, comprehensive metabolomics, and advanced imaging. *Proc Natl Acad Sci USA* 2020;**117**(6):3053.
53. Nicola M, Alsafi Z, Sohrabi C, et al. The socio-economic implications of the coronavirus pandemic (COVID-19): a review. *Int J Surg* 2020;**78**:185–93.
54. Whitelaw S, Mamas MA, Topol E, Van Spall HGC. Applications of digital technology in COVID-19 pandemic planning and response. *Lancet Dig Health* 2020;**2**(8):e435–40.
55. Chandra P, Roy S. *Diagnostic strategies for COVID-19 and other coronaviruses.* Springer; 2020.
56. Roy S, Baranwal A. Diverse molecular techniques for early diagnosis of COVID-19 and other coronaviruses. In: Chandra P, Roy S, editors. *Diagnostic strategies for COVID-19 and other coronaviruses.* Singapore: Springer Singapore; 2020. p. 135–59.
57. Chau CH, Strope JD, Figg WD. COVID-19 Clinical Diagnostics and Testing Technology. *Pharmacotherapy: The Journal of Human Pharmacology and Drug Therapy* 2020;**40**(8):857–68.
58. Afzal A. Molecular diagnostic technologies for COVID-19: limitations and challenges. *J Adv Res* 2020;**26**:149–59.
59. Chan JF-W, Yip CC-Y, To KK-W, et al. Improved molecular diagnosis of COVID-19 by the novel, highly sensitive and specific COVID-19-RdRp/Hel real-time reverse transcription-PCR assay Validated <em>In Vitro</em> and with clinical specimens. *J Clin Microbiol* 2020;**58**(5). e00310-00320.
60. Roy S, Hossain MM, Safavieh M, Lubis HN, Zourob M, Ahmed MU. CHAPTER 16 isothermal DNA amplification strategies for food biosensors. In: *Food biosensors.* The Royal Society of Chemistry; 2017. p. 367–92.
61. Carter LJ, Garner LV, Smoot JW, et al. Assay techniques and test development for COVID-19 diagnosis. *ACS Cent Sci* 2020;**6**(5):591–605.
62. World Health O. *Laboratory testing for coronavirus disease (COVID-19) in suspected human cases: interim guidance, 19 March 2020.* World Health Organization; 2020.
63. Mary Kathryn B, Giuseppe L, Andrea H, et al. Molecular, serological, and biochemical diagnosis and monitoring of COVID-19: IFCC taskforce evaluation of the latest evidence. *Clin Chem Lab Med* 2020;**58**(7):1037–52.
64. Sethuraman N, Jeremiah SS, Ryo A. Interpreting diagnostic tests for SARS-CoV-2. *J Am Med Assoc* 2020;**323**(22):2249–51.
65. Lippi G, Lavie CJ, Sanchis-Gomar F. Cardiac troponin I in patients with coronavirus disease 2019 (COVID-19): evidence from a meta-analysis. *Prog Cardiovasc Dis* 2020;**63**(3):390–1.
66. Chu KH, Tsang WK, Tang CS, et al. Acute renal impairment in coronavirus-associated severe acute respiratory syndrome. *Kidney Int* 2005;**67**(2):698–705.
67. Yang X, Yu Y, Xu J, et al. Clinical course and outcomes of critically ill patients with SARS-CoV-2 pneumonia in Wuhan, China: a single-centered, retrospective, observational study. *Lancet Resp Med* 2020;**8**(5):475–81.
68. Huang C, Wang Y, Li X, et al. Clinical features of patients infected with 2019 novel coronavirus in Wuhan, China. *Lancet* 2020;**395**(10223):497–506.

69. Padoan A, Sciacovelli L, Basso D, et al. IgA-Ab response to spike glycoprotein of SARS-CoV-2 in patients with COVID-19: a longitudinal study. *Clin Chim Acta* 2020;**507**:164–6.
70. Dzieciatkowski T, Szarpak L, Filipiak K, Jaguszewski M, Ladny J, Smereka J. COVID-19 challenge for modern medicine. *Cardiol J* 2020;**27**(2):175–83.
71. MacLaren G, Fisher D, Brodie D. Preparing for the most critically ill patients with COVID-19: the potential role of extracorporeal membrane Oxygenation. *J Am Med Assoc* 2020;**323**(13):1245–6.
72. Zhang L, Liu Y. Potential interventions for novel coronavirus in China: a systematic review. *J Med Virol* 2020;**92**(5):479–90.
73. Shen LW, Mao HJ, Wu YL, Tanaka Y, Zhang W. TMPRSS2: a potential target for treatment of influenza virus and coronavirus infections. *Biochimie* 2017;**142**:1–10.
74. Yamaya M, Shimotai Y, Hatachi Y, et al. The serine protease inhibitor camostat inhibits influenza virus replication and cytokine production in primary cultures of human tracheal epithelial cells. *Pulm Pharmacol Therapeut* 2015;**33**:66–74.
75. Cortegiani A, Ingoglia G, Ippolito M, Giarratano A, Einav S. A systematic review on the efficacy and safety of chloroquine for the treatment of COVID-19. *J Crit Care* 2020;**57**:279–83.
76. Richardson P, Griffin I, Tucker C, et al. Baricitinib as potential treatment for 2019-nCoV acute respiratory disease. *Lancet* 2020;**395**(10223):e30–1.
77. Filonchyk M, Hurynovich V, Yan H, Gusev A, Shpilevskaya N. Impact assessment of COVID-19 on Variations of SO_2, NO_2, CO and AOD over east China. *Aerosol Air Qual Res* 2020;**20**(7):1530–40.
78. Qiu G, Gai Z, Tao Y, Schmitt J, Kullak-Ublick GA, Wang J. Dual-functional plasmonic photothermal biosensors for highly accurate severe acute respiratory syndrome coronavirus 2 detection. *ACS Nano* 2020;**14**(5):5268–77.
79. Lahiri BB, Bagavathiappan S, Jayakumar T, Philip J. Medical applications of infrared thermography: a review. *Infrared Phys Technol* 2012;**55**(4):221–35.
80. Vashist SK, Mudanyali O, Schneider EM, Zengerle R, Ozcan A. Cellphone-based devices for bioanalytical sciences. *Anal Bioanal Chem* 2014;**406**(14):3263–77.
81. Altman SH, Sivo NT, Pernikoff TB, Deshpande A, Pernikoff RM. Geo-tagged journal system for location-aware mobile communication devices. In: *Google Patents*; 2011.
82. Najioullah F, Viron F, Césaire R. Evaluation of four commercial real-time RT-PCR kits for the detection of dengue viruses in clinical samples. *Virology Journal* 2014;**11**(1):164.
83. Das Mukhopadhyay C, Sharma P, Sinha K, Rajarshi K. Recent trends in analytical and digital techniques for the detection of the SARS-Cov-2. *Biophys Chem* 2021;**270**:106538.
84. van Elden LJR, Anton MAM, van Alphen F, et al. Frequent detection of human coronaviruses in clinical specimens from patients with respiratory tract infection by use of a novel real-time reverse-transcriptase polymerase chain reaction. *J Infect Dis* 2004;**189**(4):652–7.
85. Zhu X, Wang X, Han L, et al. *Reverse transcription loop-mediated isothermal amplification combined with nanoparticles-based biosensor for diagnosis of COVID-19* **166**. medRxiv; 2020. p. 112437. 2020.2003.2017.20037796.
86. Munirah H, Roy S, Ying JL, Rahman IA, Ahmed MU. Rapid detection of pork DNA in food samples using reusable electrochemical sensor. *Sci Bruneiana* 2016;**15**.
87. Azam NFN, Roy S, Lim SA, Uddin Ahmed M. Meat species identification using DNA-luminol interaction and their slow diffusion onto the biochip surface. *Food Chem* 2018;**248**:29–36.

88. Roy S, Ahmed MU. System and Method for Immobilization free electro-chemiluminescence DNA detection using a luminophore dye for multi-species detection. In: *Google Patents*; 2018.
89. Lee I-H, Park JY, Chai Y-G, Zhang B-T. RCA-based detection methods for resolution refutation. In: *Paper presented at: DNA computing; 2004*; 2004 [Berlin, Heidelberg].
90. Widjojoatmodjo MN, Borst A, Schukkink RAF, et al. Nucleic acid sequence-based amplification (NASBA) detection of medically important Candida species. *J Microbiol Methods* 1999;**38**(1):81–90.
91. Roy SXWS, Abd Rahman I, Ahmed MU. Based visual detection of Salmonella bacteria using isothermal DNA amplification and magnetic bead aggregation. *Malay J Microbiol* 2016;**12**(5):332–8.
92. Roy S, Wei SX, Ying JLZ, Safavieh M, Ahmed MU. A novel, sensitive and label-free loop-mediated isothermal amplification detection method for nucleic acids using luminophore dyes. *Biosens Bioelectron* 2016;**86**:346–52.
93. Roy S, Rahman IA, Ahmed MU. Electrochemical DNA biosensor using graphene biochip for species identification. In: *Google Patents*; 2019.
94. Zhang Y, Odiwuor N, Xiong J, et al. *Rapid molecular detection of SARS-CoV-2 (COVID-19) virus RNA using colorimetric LAMP*. medRxiv; 2020. 2020.2002.2026.20028373.
95. Park G-S, Ku K, Baek S-H, et al. Development of reverse transcription loop-mediated isothermal amplification assays targeting severe acute respiratory syndrome coronavirus 2 (SARS-CoV-2). *J Mol Diagn* 2020;**22**(6):729–35.
96. Roy S, Rahman IA, Santos JH, Ahmed MU. Meat species identification using DNA-redox electrostatic interactions and non-specific adsorption on graphene biochips. *Food Contr* 2016;**61**:70–8.
97. Roy S, Rahman IA, Ahmed MU. Paper-based rapid detection of pork and chicken using LAMP–magnetic bead aggregates. *Analy Methods* 2016;**8**(11):2391–9.
98. Roy S, Mohd-Naim NF, Safavieh M, Ahmed MU. Colorimetric nucleic acid detection on paper microchip using loop mediated isothermal amplification and crystal Violet dye. *ACS Sens* 2017;**2**(11):1713–20.
99. Yu L, Wu S, Hao X, et al. *Rapid colorimetric detection of COVID-19 coronavirus using a reverse tran-scriptional loop-mediated isothermal amplification (RT-LAMP) diagnostic plat-form: iLACO*. medRxiv; 2020. 2020.2002.2020.20025874.
100. Wu Q, Suo C, Brown T, Wang T, Teichmann SA, Bassett AR. *INSIGHT: a scalable isothermal NASBA-based platform for COVID-19 diagnosis*. bioRxiv; 2020. 2020.2006.2001.127019.
101. Jiao J, Duan C, Xue L, Liu Y, Sun W, Xiang Y. DNA nanoscaffold-based SARS-CoV-2 detection for COVID-19 diagnosis. *Biosens Bioelectron* 2020;**167**:112479.
102. Yelagandula R, Bykov A, Vogt A, et al. *SARSeq, a robust and highly multiplexed NGS assay for parallel detection of SARS-CoV2 and other respiratory infections*. medRxiv; 2020. 2020.2010.2028.20217778.
103. Bai H, Cai X, Zhang X. *Landscape Coronavirus Disease 2019 test (COVID-19 test) in vitro–A comparison of PCR vs Immunoassay vs Crispr-Based test*. 2020.
104. Corman VM, Landt O, Kaiser M, et al. Detection of 2019 novel coronavirus (2019-nCoV) by real-time RT-PCR. *Euro Surveill* 2020;**25**(3):2000045.
105. Toptan T, Hoehl S, Westhaus S, et al. *Optimized qRT-PCR approach for the detection of intra-and extracellular SARS-CoV-2 RNAs*. bioRxiv; 2020.
106. Younes N, Al-Sadeq DW, Al-Jighefee H, et al. Challenges in laboratory diagnosis of the novel coronavirus SARS-CoV-2. *Viruses* 2020;**12**(6):582.
107. Sapkal G, Shete-Aich A, Jain R, et al. Development of indigenous IgG ELISA for the detection of anti-SARS-CoV-2 IgG. *Indian J Med Res* 2020;**151**(5):444–9.

108. Roy V, Fischinger S, Atyeo C, et al. SARS-CoV-2-specific ELISA development. *J Immunol Methods* 2020:484–5. 112832.
109. Lamb LE, Bartolone SN, Ward E, Chancellor MB. *Rapid detection of novel coronavirus (COVID-19) by reverse transcription-loop-mediated isothermal amplification*. medRxiv; 2020. 2020.2002.2019.20025155.
110. Lu R, Wu X, Wan Z, et al. Development of a novel reverse transcription loop-mediated isothermal amplification method for rapid detection of SARS-CoV-2. *Virol Sin* 2020;**35**(3):344–7.
111. Jiang M, Pan W, Aratehfar A, et al. *Development and validation of a rapid, single-step reverse transcriptase loop-mediated isothermal amplification (RT-LAMP) system potentially to be used for reliable and high-throughput screening of COVID-19*. medRxiv; 2020. 2020.2003.2015.20036376.
112. Purohit B, Kumar A, Mahato K, Roy S, Chandra P. Chapter 9 - cancer cytosensing approaches in miniaturized settings based on advanced nanomaterials and biosensors. In: Maurya PK, Singh S, editors. *Nanotechnology in modern animal biotechnology*. Elsevier; 2019. p. 133–47.
113. Kumar A, Roy S, Srivastava A, et al. Chapter 10 - nanotherapeutics: a novel and powerful approach in modern Healthcare system. In: Maurya PK, Singh S, editors. *Nanotechnology in modern animal biotechnology*. Elsevier; 2019. p. 149–61.
114. Roy S, Malode SJ, Shetti NP, Chandra P. Modernization of biosensing strategies for the development of lab-on-chip integrated systems. *Bioelectrochem Inter Eng* 2019:325–42.
115. Shetti NP, Malode SJ, Roy S, Chandra P, Reddy KR, Chatterjee S. 11 - electroanalytical techniques for investigating biofilms: applications in biosensing and biomolecular interfacing. In: Kanchi S, Sharma D, editors. *Nanomaterials in diagnostic tools and devices*. Elsevier; 2020. p. 293–329.
116. Pérez-López B, Merkoçi A. Nanomaterials based biosensors for food analysis applications. *Trends Food Sci Technol* 2011;**22**(11):625–39.
117. Rawtani D, Khatri N, Tyagi S, Pandey G. Nanotechnology-based recent approaches for sensing and remediation of pesticides. *J Environ Manag* 2018;**206**:749–62.
118. Pundir CS, Deswal R, Kumar P. Quantitative analysis of sarcosine with special emphasis on biosensors: a review. *Biomarkers* 2019;**24**(5):415–22.
119. Sharifi M, Avadi MR, Attar F, et al. Cancer diagnosis using nanomaterials based electrochemical nanobiosensors. *Biosens Bioelectron* 2019;**126**:773–84.
120. Girigoswami K, Akhtar N. Nanobiosensors and fluorescence based biosensors: an overview. *Int J Nano Dimens* 2019;**10**(1):1–17.
121. Forssén P, Samuelsson J, Lacki K, Fornstedt T. Advanced analysis of biosensor data for SARS-CoV-2 RBD and ACE2 interactions. *Anal Chem* 2020;**92**(17):11520–4.
122. Kutner W, Wang J, L'Her M, Buck RP. Analytical aspects of chemically modified electrodes: classification, critical evaluation and recommendations (IUPAC Recommendations 1998). *Pure Appl Chem* 1998;**70**(6):1301–18.
123. Khan MZH, Hasan MR, Hossain SI, Ahommed MS, Daizy M. Ultrasensitive detection of pathogenic viruses with electrochemical biosensor: state of the art. *Biosens Bioelectron* 2020;**166**:112431.
124. Chen Z, Zhang Z, Zhai X, et al. Rapid and sensitive detection of anti-SARS-CoV-2 IgG, using lanthanide-doped nanoparticles-based lateral flow immunoassay. *Anal Chem* 2020;**92**(10):7226–31.
125. Seo G, Lee G, Kim MJ, et al. Rapid detection of COVID-19 causative virus (SARS-CoV-2) in human nasopharyngeal swab specimens using field-effect transistor-based biosensor. *ACS Nano* 2020;**14**(4):5135–42.
126. Margolin E, Burgers WA, Sturrock ED, et al. Prospects for SARS-CoV-2 diagnostics, therapeutics and vaccines in Africa. *Nat Rev Microbiol* 2020;**18**(12):690–704.

127. Hoste ACR, Venteo A, Fresco-Taboada A, et al. Two serological approaches for detection of antibodies to SARS-CoV-2 in different scenarios: a screening tool and a point-of-care test. *Diagn Microbiol Infect Dis* 2020;**98**(4):115167.
128. Huang L, Tian S, Zhao W, Liu K, Ma X, Guo J. Multiplexed detection of biomarkers in lateral-flow immunoassays. *Analyst* 2020;**145**(8):2828–40.
129. Li Z, Yi Y, Luo X, et al. Development and clinical application of a rapid IgM-IgG combined antibody test for SARS-CoV-2 infection diagnosis. *J Med Virol* 2020;**92**(9):1518–24.
130. Talebian S, Wallace GG, Schroeder A, Stellacci F, Conde J. Nanotechnology-based disinfectants and sensors for SARS-CoV-2. *Nat Nanotechnol* 2020;**15**(8):618–21.
131. Rusling JF, Kumar CV, Gutkind JS, Patel V. Measurement of biomarker proteins for point-of-care early detection and monitoring of cancer. *Analyst* 2010;**135**(10):2496–511.
132. Mahari S, Roberts A, Shahdeo D, Gandhi S. *eCovSens-ultrasensitive novel in-house built printed circuit board based electrochemical device for rapid detection of nCovid-19 antigen, a spike protein domain 1 of SARS-CoV-2*. bioRxiv; 2020. 2020.2004.2024.059204.
133. Vadlamani BS, Uppal T, Verma SC, Misra M. Functionalized tio2 nanotube-based electrochemical biosensor for rapid detection of SARS-CoV-2. *Sensors* 2020;**20**(20):5871.
134. Alafeef M, Dighe K, Moitra P, Pan D. Rapid, ultrasensitive, and quantitative detection of SARS-CoV-2 using antisense oligonucleotides directed electrochemical biosensor chip. *ACS Nano* 2020;**14**(12):17028–45.
135. Zhao H, Liu F, Xie W, et al. Ultrasensitive supersandwich-type electrochemical sensor for SARS-CoV-2 from the infected COVID-19 patients using a smartphone. *Sensor Actuator B Chem* 2021;**327**:128899.
136. Yakoh A, Pimpitak U, Rengpipat S, Hirankarn N, Chailapakul O, Chaiyo S. Paper-based electrochemical biosensor for diagnosing COVID-19: detection of SARS-CoV-2 antibodies and antigen. *Biosens Bioelectron* 2020;**176**:112912. https://doi.org/10.1016/j.bios.2020.112912.
137. Rashed MZ, Kopechek JA, Priddy MC, et al. Rapid detection of SARS-CoV-2 antibodies using electrochemical impedance-based detector. *Biosens Bioelectron* 2021;**171**:112709.
138. Eissa S, Zourob M. Development of a low-cost cotton-tipped electrochemical immunosensor for the detection of SARS-CoV-2. *Anal Chem* 2020.
139. Ahmadivand A, Gerislioglu B, Ramezani Z, Kaushik A, Manickam P, Ghoreishi SA. *Femtomolar-level detection of SARS-CoV-2 spike proteins using toroidal plasmonic metasensors*. 2020. arXiv preprint arXiv:200608536.
140. Funari R, Chu K-Y, Shen AQ. Detection of antibodies against SARS-CoV-2 spike protein by gold nanospikes in an opto-microfluidic chip. *Biosens Bioelectron* 2020;**169**:112578.
141. Moitra P, Alafeef M, Dighe K, Frieman MB, Pan D. Selective naked-eye detection of SARS-CoV-2 mediated by N gene targeted antisense oligonucleotide capped plasmonic nanoparticles. *ACS Nano* 2020;**14**(6):7617–27.
142. Ventura BD, Cennamo M, Minopoli A, et al. Colorimetric test for fast detection of SARS-CoV-2 in nasal and throat swabs. *ACS Sens.* 2020;**5**(10):3043–8.
143. Song F, Shen Y, Wei Y, et al. Botulinum toxin as an ultrasensitive reporter for bacterial and SARS-CoV-2 nucleic acid diagnostics. *Biosens Bioelectron* 2021;**176**:112953.
144. Karami A, Hasani M, Azizi Jalilian F, Ezati R. Conventional PCR assisted single-component assembly of spherical nucleic acids for simple colorimetric detection of SARS-CoV-2. *Sensor Actuator B Chem* 2021;**328**:128971.
145. Moon J, Kwon H-J, Yong D, et al. Colorimetric detection of SARS-CoV-2 and drug-resistant pH1N1 using CRISPR/dCas9. *ACS Sens* 2020;**5**(12):4017–26.

146. Anantharaj A, Das SJ, Sharanabasava P, et al. Visual detection of SARS-CoV-2 RNA by conventional PCR-induced generation of DNAzyme sensor. *Front Molec Biosci* 2020;**7**:444.
147. González-González E, Lara-Mayorga IM, García-Rubio A, et al. *Scaling diagnostics in times of COVID-19: rapid prototyping of 3D-printed water circulators for Loop-mediated Isothermal Amplification (LAMP) and detection of SARS-CoV-2 virus*. medRxiv; 2020. 2020.2004.2009.20058651.
148. Baek YH, Um J, Antigua KJC, et al. Development of a reverse transcription-loop-mediated isothermal amplification as a rapid early-detection method for novel SARS-CoV-2. *Emerg Microb Infect* 2020;**9**(1):998−1007.
149. Kasetsirikul S, Umer M, Soda N, Sreejith KR, Shiddiky MJ, Nguyen N-T. Detection of the SARS-CoV-2 humanized antibody with paper-based ELISA. *Analyst* 2020;**145**(23):7680−6.

CHAPTER 14

Prescription, over-the-counter (OTC), herbal, and other treatments and preventive uses for COVID-19

Jaime A. Yáñez[1,2], Sun Ah Chung[3], Brenda Rojas Román[3], Palmer J. Hernández-Yépez[4], Franko O. Garcia-Solorzano[5], Shyla Del-Aguila-Arcentales[6], Fiorella Inga-Berrospi[4], Christian R. Mejia[7], Aldo Alvarez-Risco[8]

[1]Universidad Peruana de Ciencias Aplicadas. Facultad de Educación, Carrera de Educación y Gestión Del Aprendizaje, Lima, Peru; [2]Teoma Global. Gerencia Corporativa de Asuntos Científicos y Regulatorios, Lima, Peru; [3]Universidad Cristiana de Bolivia. Comunidad Científica de Estudiantes de Medicina, Santa Cruz, Bolivia; [4]Universidad Privada Norbert Wiener. Grupo de Investigación en Gestión y Salud Pública, Lima, Peru; [5]Universidad Científica del Sur. Facultad de Ciencias de la Salud, Carrera de Medicina Humana, Lima, Peru; [6]Escuela Nacional de Marina Mercante "Almirante Miguel Grau", Callao, Peru; [7]Universidad Continental, Lima, Peru; [8]Universidad de Lima, Facultad de Ciencias Empresariales y Económicas. Carrera de Negocios Internacionales. Lima, Peru

14.1 Introduction

Since November 2019, a new type of coronavirus infected patients in China.[1,2] This virus has spread rapidly to 213 countries and territories around the world causing a pandemic that at the time of writing (end of September 2020) has taken the lives of over 1 million people.[3] The COVID-19 pandemic migrated from its initial epicenter in Asia to Europe and is currently causing a tremendous burden in Latin America. This situation has exposed fragile health care systems, vulnerable socioeconomic conditions,[4] higher risk factors because of obesity and increasing rates of undernutrition,[5] lack of procurement of medical supplies,[6] and an inadequate reporting of COVID-19 cases and deaths.[7] For instance, the Ministry of Health of Peru has recognized that its health care system was fragmented before the pandemic, and the current situation has made its fragility become more evident.[8] It has been reported that Peru preferentially uses rapid serological tests (75%) to report active confirmed COVID-19 cases, which could lead to an overestimation of positive cases.[9] Latin America has not implemented public policies to control self-medication that it is currently occurring at alarming rates as speculative preventive measures against

Environmental Management of COVID-19
ISBN 978-0-323-85780-2
https://doi.org/10.1016/B978-0-323-85780-2.00001-9

COVID-19 start to flourish.[10] This event has challenged many health care systems, causing the collapse of many hospitals around the globe.[7,10–13] It also impacted the global economy, leaving it vulnerable and probably prone to lose an estimated 2.4% of the GDP of the major economies.[14,15] Unemployment rates have remarkably increased, leading to a possible recession.[16] The COVID-19 pandemic was accompanied by conspiracy theories, rumors, and infodemic, aggravated because of the availability to access social media, and the development of mental health issues on the population due to the lockdown.[17–21]

Currently, with over six months of the outbreak of the pandemic, the COVID-19 crisis is expected to markedly affect people's well-being and mental health as reported in China,[22–24] Singapore,[25] Iran,[26,27] Italy,[28,29] France,[30] United Kingdom,[31] Spain,[32,33] Chile,[34] Bolivia,[35,36] Ecuador,[36–38] Brazil,[39–41] Peru,[36,42] and indigenous populations.[43] This disruption in people's work and life has been accompanied by an unprecedented infodemic of fake news.[44] This plethora of misinformation and false reports have spread faster because of social media.[44] The entire world has been affected by this infodemic, and Latin America has been no exception. However, Peru was the only country in Latin America that took a different approach by implementing prison sentences to the creators and disseminators of fake news.[17]

The infodemic has been accompanied by a surge of unproven religious and herbal treatments for COVID-19 prevention.[45] Herbal remedies or phytomedicinal self-medication use is common in developing countries[46,47] and patients often do not properly report it to their physicians.[48,49] It has been reported in a cross-sectional survey that patients do not disclose this information since they are afraid of their doctor's disagreement or negative response because most of them followed advice from a nonmedical source (family, friend, internet, or social media).[48] Other patients stated that their doctor did not ask them, while others considered that it was not necessary to inform their doctor.[48] However, possible drug-herbal interactions are discussed in various reputable pharmacopeias that detail herbal use, efficacy and safety.[50–52] Furthermore, drug-herbal interaction databases allow physicians to check possible interactions online.[53,54] This is a confounding factor since people typically self-medicate regardless of the effort of regulatory agencies to educate the population.[55] Unproven prescription drugs have been falsely promoted for COVID-19 prevention and treatment.[56] Health literacy has been defined as the individuals' capacity to obtain, process, and understand basic health information and services needed to make appropriate health decisions,[57] and to address or solve a health-related problem.[58] Latin America is a region with a low literacy rate that affects directly the drug utilization research (DUR), which encompasses the use, efficacy, and safety of medicines based on local research.[59] This is primarily caused by the existence of a very limited generation of scientific publications by local investigators.[59,60] On the available studies, data collection bias has been identified since they present methodological issues such as limited validation, small sample size, and unavailability of information and involvement of the public sector.[60] All this causes a significant impact in the clinical and regulatory decisions that impact directly the public health decision making.[61] In general, public health decisions in Latin America are adaptations of public health policies of other countries,[11] rather than an actual country-based decision justified by local research.[59] A clear example of this was the implementation of supervised walks for children and adolescents during the COVID-19 lockdown in Peru,[11] which was based on the same measure that was implemented in Spain.[62] However, this decision in Peru was implemented on May 18, 2020 when data from Spain already indicated that the number of COVID-19 cases increased in this age group,[63] but importantly occurred at the time that the Multisystem Inflammatory Syndrome in Children (MIS-C) cases kept increasing in COVID-19 children around the world.[64] This resulted in the report of the first case of concomitant MIS-C and COVID-19 in a three-year-old child in Peru, the first case in Latin America.[11]

The U.S. Food and Drug Administration (FDA) declared that there are not any approved drugs or therapeutics to treat or prevent COVID-19 yet.[65] That doesn't mean that health professionals don't know how to treat a patient with COVID-19, the management consists of symptomatic care and individualized treatment depending on the type of patient we have.[65] However, this lack of knowledge has created a pathway leading to a growth in the tendency of self-medication, self-treatment, and self-care within the society.[18,66] People put their health at risk due to peer pressure, medication availability, previous experience self-medicating, among other correlated factors.[10,13,67] For example, a person can take dietary supplements such as silver instead of receiving professional care to treat or prevent coronavirus without any sustainable evidence of the effects of those substances against the SARS-CoV-2 or the adverse effects of those products just because someone recommended it on social media or TV.[68]

This chapter intends to present the safety, efficacy, and toxicology profiles of different products that were promoted by the people during the COVID-19 pandemic promising to treat or prevent the disease, such as prescription drugs, over-the-counter (OTC) drugs, herbal products, and unproven chemicals.[13,69–72] It also shows the need to implement strategies to guide the population and health care professionals in order to take safe and effective measures to fight against this virus.[73,74] The COVID-19 pandemic has shown us that the population needs to be urgently correctly informed about the protocols, ways of prevention, unsafe medications and treatments for this disease, and what to do if a person suspects of having the disease.[75]

14.2 Health literacy and drug utilization

The Institute of Medicine[76] defined health literacy as the individuals' capacity to obtain, process, and understand basic health information and services needed to make appropriate health decisions. In this sense, Sorensen et al.[77] reported that about 50% of people in eight European countries had limited knowledge of health. This could be explained because during formal education, given at elementary and high schools, health issues are not a fundamental component of the current curricula. This lack of proper health education would prevent people from taking wise decisions in health promotion for themselves and the ones surrounding them. Health literacy has a direct influence on the ability of people to filter the health information they receive. So, lack of proper health literacy is one of the reasons why people tend to receive and/or distribute news without proper verification.[78]

Digital health literacy not only requires the ability to navigate efficiently searching for trusted sources on the Internet; most importantly, it requires the ability to evaluate the quality of the information on the Internet to discern the true information from the wrong one. Digital health literacy also can be defined as the ability to search, find, understand, and evaluate health information from electronic sources and apply knowledge acquired to address or solve a health problem.[58] Likewise, to be able to communicate efficiently with health professionals through the various technologies available for making health decisions.[79] It needs to be realized that the COVID-19 global pandemic has showed us that in-person and face-to-face doctor consultations are not possible and that there are still gaps for us to understand and use the full potential of digital tools.[80] However, this will require adequate training for health professionals and users, accompanied with more funding, research, and policy changes.[80]

The progress of the countries of the American continent[81] in relation to digital health is varied. On the one hand, 61% of countries have a national digital health strategy, but several of them have yet to move from the phase of formulating digital health policies and strategies to the actual implementation of services to patients.[81] Thus, it is reported that telehealth is still in the initial stages, which could explain that the population has serious limitations to access useful information through the Internet and that they can also filter the correct information.[81] It has also reported that 52.6% have an electronic health information system but only 26.3% have legislation that supports its use in national systems, also explained by the lack of financing.[81] Even though some initial efforts for digital health ecosystem for Latin America have been implemented, large-scale or nationwide coverage of digital health interventions to support health workforce development is still rarely reported in the literature.[82–84] Because of the earlier onset of the COVID-19 pandemic in Europe, eHealth services were implemented earlier in countries such as Spain[85] and were quickly replicated in the United States.[86] As the pandemic moved toward Latin America, various countries implemented telemedicine systems with Peru,[84] Brazil,[87] and Mexico[88] taking the lead in March, while Colombia[89] implemented it in April, and Argentina[90] in May 2020. However, other countries in the region have been more resilient in this aspect, such as Uruguay, which just approved telemedicine regulations,[91] Paraguay that just launched its first official telemedicine app,[92] and Bolivia started a telemedicine pilot test to evaluate its implementation.[93]

To provide high-quality eHealth services there are certain requirements that we consider need to be followed (Table 14.1). First, it is important to empower health care professionals for them to develop sufficient digital communication skills, which starts by acknowledging that the patient might perceive eHealth by a lower quality of clinical care.[94] Second, patients need to be empowered by increasing their digital health literacy, which consists of the ability to search, find, understand, and evaluate health information

Table 14.1 Requirements to ensure the quality of eHealth programs.

Requirement 1: Empower health care professionals
Develop sufficient digital communication skills
Acknowledge and circumvent the innate limitations of eHealth
Provide information to patient ahead of time
Requirement 2: Empower patients
Increase their digital health literacy
Adequate preparedness ahead of time

from electronic sources and apply knowledge acquired to address or solve a health problem.[58] Recently, a masterclass was published that exemplifies the actual wording to use for an eHealth consultation considering the critical components that need to be present such as greeting, introduction, courtesy, equipment check, establish remote experience, consent, signpost, preparation check, ID check, reason for call, agenda-setting, and key clinical questions.[95]

Low levels of health literacy and specifically digital health literacy are the right field to start an apparently inexhaustible spread of false information. This lack of literacy is relevant because rumors had three times more share than verified stories.[96] Furthermore, it has been reported that there is a lack of general knowledge about COVID-19 symptoms[97] and lack of information on what drugs to take.[98]

14.3 Prescription drugs

Self-medication is more common in countries where health care systems tend to be less effective because of long waiting times in health care facilities, difficulty in obtaining physicians' appointments, insufficient stock of essential medicines, delay in attention, and insufficient amount of available beds/space in health care facilities.[99] This has resulted in the self-medication of various prescription drugs that have no confirmed clinical efficacy against SARS-CoV-2,[100] which correlated with the Google trend data for web search query for the terms "hydroxychloroquine,"[101] "ivermectin,"[102] and "azithromycin"[103] when potential COVID-19 related benefits were reported in the media (Fig. 14.1). Worldwide Google trend data has been reported previously for the term "self-medication," "self-care," and "self-administration."[70]

Some of them include the antimalaria drug hydroxychloroquine,[104] the antibiotic azithromycin,[105] and the antiretrovirals lopinavir and ritonavir.[106] Because of the lack of monitoring, these drugs could cause a shortage of these drugs for patients that need them for approved conditions,[107] a direct impact in the price of these drugs,[108] and jeopardize peoples' health due to their known adverse events.[100,109]

This self-medication trend has been reported to have increased worldwide based on the number of Google searches since the pandemic started.[70] This global trend has caused a tremendous medical challenge[110,111] because the various prescription drugs currently approved for COVID-19 symptoms carry adverse drug reactions.[72] Furthermore, there is

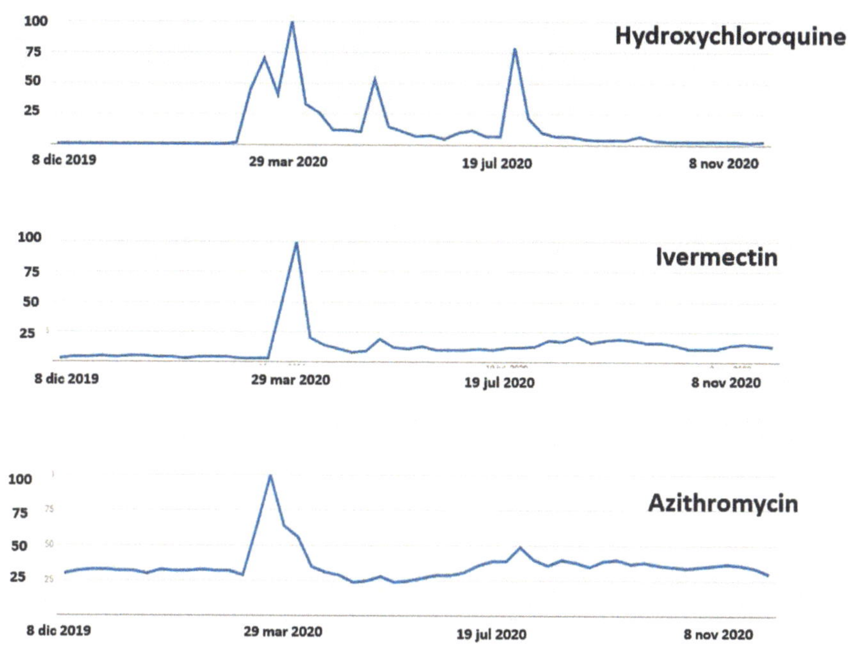

Figure 14.1 Worldwide Google trend data for web search query for the terms "hydroxychloroquine,"[101] "ivermectin,"[102] and "azithromycin."[103]

high risk of incorrect dosage, improper route of administration, longer use than intended, improper storage, risk of dependency and abuse, and increased prevalence of pathogenic resistance to drugs.[72,112] Even though self-medication intentions are common worldwide, it has only been reported in Saudi Arabia[113] and Kenya,[72] while actual statistics of prescription drug use has been reported in Peru.[13]

Regarding the consumption of the antibiotic azithromycin, an in vitro study assessed the combination of azithromycin and hydroxychloroquine showing a synergistic effect against SARS-CoV-2.[114,115] The possible antiinflammatory properties of azithromycin, which could improve the disease progression, was also reported.[114,115] However, its administration in combination therapy with hydroxychloroquine has been implicated in the elevation of the QT interval.[116] Given the limited data available to ensure the efficacy of combination therapy, the American Society for Infectious Diseases recommends that the hydroxychloroquine/chloroquine plus azithromycin combination should be limited to clinical trials.[116,117] Regarding the use of hydroxychloroquine in combination with azithromycin was publicly endorsed by President Trump, which caused self-medication causing several reports of severe poisoning in Nigeria and the United States.[108] This public endorsement triggered various in vitro and in vivo studies that reported a decrease in viral load and mitigation in the cytokine storm in critically ill patients with SARS-CoV-2.[118] Similarly, various studies reported the positive use of hydroxychloroquine and azithromycin for treatment in hospitalized COVID-19 patients.[118–121] However, the consumption of hydroxychloroquine with or without azithromycin caused an increase in cardiotoxic risk such as QT prolongation, torsades de pointes, and sudden death in hospitalized patients with COVID-19.[116,122,123] Even though the literature alerts to the possible risks with the combination of azithromycin and hydroxychloroquine, it is recommended in Peru for moderate and severe COVID-19 cases in a clinical setting since May 2020.[109,124] For mild COVID-19 patients, hydroxychloroquine is recommended at a dose of 400 mg orally every 12 h for the first day, and 200 mg orally every 12 h for 6 days.[124] In the case of moderate-to-severe COVID-19 patients, hydroxychloroquine is recommended at a dose of 200 mg orally every 8 h for 7–10 days, or hydroxychloroquine + azithromycin at a dose regimen of 200 mg orally every 8 h for 7–10 days (hydroxychloroquine) + 500 mg orally on the first day and then 250 mg every 24 h for 5 days.[124] A case report in Peru of a 13-year-old girl with encephalitis and COVID-19 serological diagnosis presented an unfavorable clinical evolution despite treatment with hydroxychloroquine, azithromycin, and corticosteroids, dying on the third day of hospitalization.[125] This published case report accompanied by the observed lack of efficacy and increased risk triggered that ivermectin, hydroxychloroquine, and azithromycin were removed from the official COVID-19 treatment in Peru.[126]

Regarding the consumption of antiretrovirals in order to prevent a viral disease such as COVID-19 it is risky because liver damage has been observed in COVID-19 patients.[127] Considering that antiretrovirals list liver damage as a common adverse effect,[128,129] it would make them inappropriate for use in COVID-19 patients. Their use is relevant for other pathologies, but a recent review has determined that there is no clear evidence of the beneficial effects of antiretrovirals in the prevention of COVID-19.[130,131] More research is still needed to determine the cost and benefit of antiretrovirals for COVID-19.

COVID-19 treatment has been a widely discussed aspect throughout 2020, with the appearance of various drugs that showed properties to control symptoms and even improve survival. As the weeks went by, these perspectives changed, leaving medicines aside, changing doses, and replacing medicines. First, it can be mentioned that there is no antiviral treatment and there is not yet a vaccine available, therefore the initial approach is based on symptom relief and oxygen therapy. In the case of patients with an oxygen saturation lower than 95%, they require high oxygen concentrations. Silent hypoxemia has been able to be detected with the use of an oximeter, which determines if the patient requires to be taken to the emergency services immediately.[132,133] Corticosteroids gained an important role, as described in the recovery trial, which shown that dexamethasone reduced deaths by a third among critical COVID-19 patients.[134] The lopinavir/ritonavir combination appeared as an alternative treatment,[135] but then studies emerged that showed superior treatments such as arbidol.[136] However, the lack of efficacy of lopinavir/ritonavir[137] and both lopinavir/ritonavir and arbidol[138] has been evidenced in COVID-19 patients. Chloroquine and hydroxychloroquine were proposed as immunomodulatory agents,[139] including their concomitant use with azithromycin.[140] The possible adverse event in the alteration of the QT interval was also reported due to concomitant use;[122] however, such adverse events were later discarded.[141] Furthermore, the use for chloroquine and hydroxychloroquine as a preventive drug for healthy and asymptomatic COVID-19 patients still needs a proper double-blind clinical trial.[142] It is important to mention that each country has adapted different pharmaceutical treatment options that are constantly updated.

14.4 Over-the-counter (OTC) drugs

OTC drugs are medicines that can be sold without a prescription directly to the consumer according to the specific regulations of each country. In the context of the COVID-19 pandemic, OTC drugs can be used as symptomatic treatment for mild COVID-19 cases without the need for the patient to go to the hospital. It is important to notice that due to the ongoing situation regarding COVID-19 pandemic, hospitals regularly are saturated. As a consequence, patients who have mild symptoms of coronavirus can be treated at home. The mild symptoms of COVID-19 include fever, dry cough, and tiredness.[143,144] It is important to note that loss of smell (anosmia) has been recognized as a prominent clinical symptom in COVID-19 patients without any other significant signs.[145,146] Since there is no available treatment for COVID-19 related anosmia[145] it is important to have a proper olfactory evaluation.[147,148] Other symptoms that are less common but still considered mild are body aches, headache,

Table 14.2 OTC drugs as COVID-19 symptomatic treatment.

Type of drug	Generic name	Symptom
Analgesic	Acetaminophen[270]	Fever, mild pain
Antiinflammatory	Ibuprofen[270]	Fever, moderate pain
	Naproxen[271]	
Cough suppressant	Dextromethorphan[272]	Cough
Antihistamines	Loratadine[273]	Nasal congestion, Sneezing
	Cetirizine[273]	
Sympathomimetic	Pseudoephedrine[274,a]	
	Phenylephrine[275]	
Opioid	Loperamide[276]	Diarrhea

[a]Pseudoephedrine is an OTC drug when combined with an antihistamine, such as loratadine or dexbrompheniramine.

sore throat, nasal congestion, and diarrhea.[143,144] We present the OTC drugs that have been recommended for mild COVID-19 symptoms. Table 14.2 shows the OTC drugs that have been used for COVID-19 symptomatic treatment.

14.4.1 Fever and body aches

In the management of fever due to COVID-19, and in general to any pathology, acetaminophen is recommended. The antipyretic effect and its safety at recommended doses make acetaminophen one of the most-used OTC drugs during the COVID-19 pandemic.[10,149] However, the increasing use of acetaminophen for managing the COVID-19 associated fever is generating concern because patients can take high doses, which can lead to acute liver injury (ALI) or acute liver failure (ALF) due to overdose.[150] Acetaminophen standard therapeutic oral dose is 0.5–1 g every 4–6 h to a maximum of 4 g/day, but more importantly it has a dose-dependent toxicity.[151] Acetaminophen can cause hepatoxicity after major overdose,[152] and severe liver damage has been observed with long-term use even at therapeutic doses in patients with alcoholic liver disease or viral infections.[153] Furthermore, it has been reported that long-term consumption of acetaminophen carries a potential risk factor for chronic renal failure,[154] cardiovascular and gastrointestinal diseases, and even mortality.[151] This is corroborated in recent studies that have reported an increase in alanine aminotransferase, aspartate aminotransferase, bilirubin, and creatinine in patients with confirmed COVID-19.[155] In addition, it has been reported that more than half of patients with ALI and ALF induced by acetaminophen have undetectable levels of acetaminophen, which is concerning.[156] Therefore, clinicians should not rule out the possibility of acetaminophen toxicity and should pay attention to patients with a history of suspected acetaminophen poisoning or associated biochemical profile.[156] This is important now with a potential significant increase in acetaminophen use due to the COVID-19 pandemic.[149]

Nonsteroidal antiinflammatory drugs (NSAIDs) play an important role in the treatment of fever and muscle aches. Ibuprofen is the most widely used OTC NSAID and has good properties for fever and pain. However, in the context of COVID-19 pandemic, it should be noted that the use of ibuprofen had a controversy due to statements in France, where it was stated that ibuprofen could worsen the clinical state in COVID-19 patients.[157,158] Nevertheless, both the World Health Organization (WHO) and the European Medicines Agency (EMA) recommended not avoiding NSAIDs when clinically indicated due to lack of scientific evidence. Recent studies suggest that there are two phases in the immune response induced by the SARS-CoV-2.[159] The first corresponds to the incubation and nonsevere stages, in which an immune response is required to eliminate the SARS-CoV-2 and prevent progression to severe stages of the disease.[159] Thus, the defense mechanism in the initial stage might be blocked by NSAIDs. The second phase corresponds to the severe stage, in which lung damage appears to be related to a cytokine storm due to an acute immune reaction.[159] It is in this second stage where the use of NSAIDs might be of more importance. In addition, the strong antipyretic efficacy of ibuprofen, which is more potent to reduce fever compared to acetaminophen, may be interfering with the benefits of a fever response.[160] Nonetheless, there is no strong evidence that can support the worsening of COVID-19 symptoms due to ibuprofen, and more research is needed such as case-control studies, cohorts, and randomized clinical trials.[161,162] Therefore, acetaminophen continues to be the first option in treating fever during the SARS-CoV-2 infection. NSAIDs are clinically recommended when the fever is high and the aches are really strong. In the case it is required to take an NSAID, although there is no significant evidence to remove ibuprofen from the list, people can opt for naproxen, which is also an OTC drug, and its effects last longer than ibuprofen. Likewise, it is important to keep in mind the presence of fever as a symptom in areas of prevalence of malaria and dengue, because it can be a factor that is confused with COVID-19 when in reality it denotes the possibility of contagion by these two diseases.[163]

14.4.2 Dry cough

In the treatment of dry cough, dextromethorphan has been widely used.[164] In the context of COVID-19, it has been reported that dextromethorphan has a pro-viral activity because it stimulated the growth of the virus in monkey epithelial cells.[165] A cellular stress coping process appears to be started by dextromethorphan, and this mechanism is also used by the SARS-CoV-2 for its replication.[165] Consequently, its use should merit caution and further study in the context of COVID-19 treatment.[165] In contrast, the TAS2R gene is believed to play an important role in host defense pathways, and has been reported that dextromethorphan is an agonist of this gene and could improve immunity especially in the treatment of dry cough.[166] However, both studies are not conclusive enough to promote or restrict the use of dextromethorphan for the management of dry cough in SARS-CoV-2 infection.

14.4.3 Nasal congestion

Despite the fact that nasal congestion is a not-so-frequent symptom during COVID-19 infection, it is a symptom that should not be ignored. There are some OTC drugs for this symptom such as antihistamines, phenylephrine, and pseudoephedrine in combination with antihistamines: pseudoephedrine and loratadine, or dexbrompheniramine and pseudoephedrine. In an experimental study, it was reported that pseudoephedrine had a protective effect in mice infected with influenza A virus because it could mitigate the cytokine storm and improved lung pathological damage.[167] However, a clinical study is still necessary to support its use in humans.

14.4.4 Diarrhea

Diarrhea is a not-so-frequent symptom in COVID-19 infection, but its management is important.[168] It has been suggested that diarrhea may be related to the use of large amounts of either antibacterial or antiviral drugs and should be considered as an adverse reaction, and consequently the first option in those cases is to suspend the use of these drugs.[169] Another proposed mechanism is that there is a direct infection of gastrointestinal cells by SARS-CoV-2, which uses ACE2 (angiotensin-converting enzyme 2) as receptor to enter the cells, and this enzyme plays an important role in regulating intestinal inflammation and diarrhea.[169] Although loperamide, an OTC drug, is widely used for milder diarrhea, there is no evidence that can support the efficacy of this drug as well as other antidiarrheal drugs. Therefore, it is suggested that adequate rehydration and electrolytes monitoring should be performed as in all patients with diarrhea.[170]

14.5 Herbals

The lack of evidence to recommend a specific treatment for COVID-19 and the absence of an available vaccine to prevent this disease, lead us to consider the supportive care and the symptomatic treatment as good strategies to deal with COVID-19. Historically, herbal medicine has been used in several epidemics of acute respiratory infectious diseases, including severe acute respiratory syndrome (SARS) and influenza.[171] This knowledge is valuable to inspire possible treatments for COVID-19.[277] In this section, we present the use of several herbal medicines, especially Traditional Chinese Medicine (TCM), that have been evaluated in randomized controlled trials (RCT) and some other herbal compounds with an antiviral effect that have evaluated in preclinical studies.

14.5.1 Prevention

Several studies showed good experience using TCM for the prevention of SARS cases during the outbreak of 2003.[172] Due to structural and genetic similarities between SARS-CoV and SARS-CoV-2, TCM is proving also to prevent COVID-19 cases currently. A prospective randomized study in 22,065 subjects, evaluated the use of *Huoxiang Zhengqi* oral liquid and *Jinhao Jiere* granules to prevent COVID-19; however, no suspected or confirmed COVID-19 case occurred in the control and intervention group, which did not make it possible to assess their effectiveness against COVID-19.[173] There is a controversial study in Israel that reported that people with vitamin D deficiency in plasma was associated with a higher risk of COVID-19 infection,[174] whereas other studies in the United Kingdom did not find any association between the Vitamin D plasma level and the risk of COVID-19 infection[175] or severe COVID-19.[176]

14.5.2 Treatment

There is a large number of studies, and they were summarized in two systematic reviews and meta-analysis of RCTs by Xiong et al.[177] and Wang et al.[178] that compared the use of TCM plus standard care, in comparison with standard care only, in COVID-19 patients. For these studies, the standard care included antiviral drugs, antibiotics, corticosteroids, supportive oxygen, and others, whereas TCM included more than a 100 herbals. The top five most common were licorice root, baical skullcap root, pinellia rhizome, forsythia fruit, and bitter apricot seed. Potential positive benefits have been published, however, most of them have a high risk of bias and small sample size,[177,178] which do not allow us to truly recommend its use in COVID-19 patients.

14.5.2.1 Clinical improvement

It has been reported that TCM reduce the length of hospitalization in terms of clinical symptoms score showing positive results for COVID-19 symptom improvement in patients that received TCM compared to standard care.[177,178] Wang et al.[178] included studies with different scales to assess this outcome and it may explain the high heterogeneity ($I^2 = 94\%$). However, in the Xiong et al.[177] study this outcome was measured in just two studies not generating heterogeneity, and the pooled result maintained its direction and significance. It was also observed that TCM reduced the number of cases of cough and fatigue and the duration of fever and fatigue in COVID-19 patients.[177] In terms of biomarkers, the C-reactive protein was reported to be significantly lower in the intervention group in the pooled result in both meta-analyses.[177,178] However, no significant differences were found in white cell count and lymphocytes between the intervention and the control group. According to CT findings, COVID-19 patients who were treated with TCM plus standard care had significantly better improvement than the control group, in both meta-analyses.[177,178]

14.5.2.2 Mortality and adverse events

The overall result from the meta-analysis showed no significant differences in the number of deaths. No severe or moderate adverse effects were reported in any of the systematic reviews; no significant differences were found between intervention and control group. The most common type of adverse event was gastrointestinal, but the recovery was rapid.[177,178]

14.5.2.3 TCM in a guidelines of treatment for COVID-19

The sixth edition of the Guidelines of Diagnosis and Treatment for COVID-19 for China[179] recommended the use of TCM according to the stage of the disease. For the patients in the medical observation period with fatigue with gastrointestinal discomfort: *Huo Xiang Zheng Qi Shui*, and for fatigue with fever: *Lian Hua Qing Wen capsule*, *Shu Feng Jie Du capsule*, and *Jin Hua Qing Gan granule*. For confirmed COVID-19 case with mild disease they recommend *Qing Fei Pai Du Tang*. For severe COVID-19 they recommended *Xi Yan Ping injection*, *Xue Bi Jing injection*, Re *Du Ning injection,* Tan *Re Qing injection*, *Xing Nao Jing injection*, and *Qing Fei Pai Du Tang*. For COVID-19 patients who required intense care they recommended *Shen Fu injection*, *Sheng Mai injection*, *Shen Mai injection*, *Su He Xiang pill*, and *An Gong Niu Huang pill*.

In July 2020, the number of the registered clinical trials for COVID-19 were 75, of them 26 aimed to evaluate TCM and only 19 aim to evaluate the combinations of TCM and Western medicine.[180] Unfortunately, most of them have several limitations since the design protocol that may not permit to validate their results and conclusions.

14.5.3 Potential candidates for clinical trials

Several herbal isolated compounds have been proved in preclinical studies as a potential inhibitor for SARS-CoV-2. Several phytochemicals were assessed in silico using molecular docking that showed that oleanolic acid extracted from *Allium cepa* was a superior inhibitor for SARS-CoV-2 than remdesivir, whereas *Moringa olifera* did not show any antiviral effect.[181] However, these findings need to be assessed in in vitro and in vivo future studies. Regarding TCM, baicalin and baicalein extracted from *Shuanghuanglian* were described as potent antiviral agents against SARS-CoV-2 in an in vitro study.[182]

14.5.4 Dietary supplements

14.5.4.1 Zinc

Zinc has proved its effectiveness against common flu and immunological prevention among some risk patients like older adults.[183] For instance, high oral dose of zinc salts has been reported to improve clinical indicators such as oxygen saturation and fever in COVID-19 adult patients.[183] However, this study was performed only in four (4) patients and further investigation in a larger group of patients is necessary. Currently, four registered clinical trials are being performed to evaluate the comparative effectiveness of zinc supplements against vitamin C and usual care for the treatment of mild-to-moderate symptoms of SARS-CoV-2 community acquired.[184]

14.5.4.2 Vitamin C

For many years, the administration of ascorbic acid has been prescribed to enhance the immune system in patients who are susceptible to viral diseases such as common flu.[185] Thus, a study performed in mice showed that vitamin C (ascorbic acid) combined with red ginseng juice enhanced T-cell immune response and stimulated NK cells.[186] In the same study, the interruption of the influenza A viral lytic cycle was observed accompanied with an improved toleration to lung inflammation.[186] A systematic review of ongoing randomized clinical trials related to the use of vitamin C in

COVID-19 patients identified that a high dose of vitamin C improved biomarker levels such as ferritin and D-dimer, and caused a trend to decrease oxygen requirement after the infusion of vitamin C.[187] Similarly, a case report illustrated the clinical improvement of a patient after five days of infusion with vitamin C.[187] Thus, vitamin C has been proposed a prophylactic and adjunctive medical treatment for COVID-19.[188]

14.5.4.3 Omega-3

A Cochrane meta-analysis of 10 studies examined the effects of omega-3 fatty acids, γ-linolenic acid and antioxidants in acute respiratory distress syndrome (ARDS) patients.[189] Quality of evidence was identified exhibiting no benefits in mortality and uncertain effects in reducing intensive care unit (ICU) length of stay, ventilator days, or oxygenation.[189]

14.6 Unproven chemicals

The COVID-19 pandemic has caused the disruption in people's work and life, which has been accompanied by an unprecedented infodemic of fake news.[44] This plethora of misinformation and false reports have spread faster because of social media.[44] The infodemic has been matched with a low health literacy condition across the world. Health literacy has been defined as the individuals' capacity to obtain, process, and understand basic health information and services needed to make appropriate health decisions,[57] and to address or solve a health-related problem.[58] All this has caused a surge and revamping of unproven chemicals that has been promoted as preventives and even miracle cures for COVID-19.[45] In this section we present the reported efficacy, safety, and toxicology profile of some of these speculative preventive and treatment options against COVID-19 such as Miracle Mineral Solution (MMS),[190] chlorine dioxide solution (CDS),[191] colloidal silver,[192] and hydrogen peroxide,[193] which is summarized in Table 14.3.

14.6.1 Chlorine dioxide solution (CDS) and Miracle Mineral Solution (MMS)

CDS is a chemical that has previously been portrayed as a miracle cure for multiple diseases,[226] and has been revamped in the current pandemic as an unproven cure and preventive for COVID-19.[191] Multiple health and sanitary regulatory agencies around the world do not support nor recommend the prophylactic use of chlorine dioxide and its chemical predecessor Miracle Mineral Solution (MMS) in humans.[194] The CDS defenders cite the unpublished research work of its most representative promoter Andreas Ludwig Kalcker and his pending patent applications.[227–229] They also cite

Table 14.3 Summary of the efficacy, safety, and toxicology profile of unproven chemicals used as speculative preventive and treatment options against COVID-19.

Unproven chemical	Efficacy	Safety	Toxicology profile	References
Chlorine dioxide solution (CDS)	NEE[a]	NEE[a]	Nausea, vomiting, diarrhea, severe dehydration, anuria, unconsciousness, cyanosis and chocolate brown serum.	194–203
Miracle Mineral solution (MMS)	NEE[a]	NEE[a]	Nausea, vomiting, diarrhea, severe dehydration, fever, chills or rigors, tachycardia, and dry cough	194–203
Colloidal silver	NEE[a]	NEE[a]	Argyria or jaundice, dehydration, fever, confusion, hypotension, hypoxia, lethargy, slurred speech, organ damage, profound disfiguration, hypochloremia, hyponatremia, cell apoptosis or necrosis, oxidative stress, and cholelithiasis	204–221
Hydrogen peroxide	NEE[a]	Not safe	Vomiting, gastric injury, skin burning or mechanical pressure injury, oxygen embolism, unconsciousness, and respiratory paralysis	193,222–225

[a]NEE = Not Enough Evidence.

the unpublished online testimonies, misinterpreted research animal and human studies, and its presumed safety based on its use in water treatment and blood transfusion bags.[190,230] CDS supporters have drafted 25 protocols to use CDS (alone or in combination with dimethyl sulfoxide [DMSO]) via the oral, intravenous, rectal, vaginal, topical, and ocular route in adults and children.[230] These *protocols* do not have efficacy nor safety data nor regulatory agency approval anywhere in the world. The FDA first warned consumers about the dangers of MMS and CDS in 2010,[194] however, people are still ingesting it, and their numbers appear on the rise, especially since the COVID-19 pandemic.

14.6.1.1 Efficacy

The promoters of CDS and MMS are making false and dangerous claims by presenting them as a remedy for autism, cancer, HIV/AIDS, hepatitis, flu, and recently, COVID-19.[194,231] Their promoters declare that vomiting and diarrhea are evidence that CDS or MMS is working; however, this lacks any scientific evidence that supports its safety or efficacy.[194,197,199]

14.6.1.2 Toxicology profile

This chemical can cause nausea, vomiting, diarrhea, and symptoms of severe dehydration.[195–197] There is a case report of life-threatening complications from poisoning with sodium chlorite, where a 55-year-old man arrived to the ICU in a cyanotic state with lowered consciousness displaying anuria and chocolate brown serum.[200] The patient required renal replacement therapy, treatment with methylene blue, and red blood cells transfusion.[200] Another case report indicated that a 41-year-old woman developed Kikuchi-Fujimoto disease (KFD)[201] after consuming MMS, which resulted in fever spiked to 40°C, chills, tachycardia, rigors, and dry cough.[202] She recovered after 16 days of treatment for her symptoms and continuous follow-up.[202] The use of these unproven chemicals is dangerous, and extreme caution needs to be implemented.[203]

14.6.2 Colloidal silver

Colloidal silver is a substance that consists of silver nanoparticles suspended on a liquid in very low quantities, similar as in homeopathy.[192] In humans, silver is absorbed through inhalation, parenteral route, ingestion, or topically, and most of it is metabolized by binding with other proteins in the liver and kidneys.[232]

14.6.2.1 Efficacy

Researchers have found that colloidal silver nanoparticles have antifungal, antibacterial, and antiviral activity.[205–207,232] Negatively charged silver nanoparticles showed a significantly higher antibacterial activity than regular colloidal silver.[208] It has been reported that silver can inhibit the replication of the following viruses: HIV-1, Hepatitis B virus, Tacaribe virus, recombinant respiratory syncytial virus, Influenza A/H1N1 virus, monkey pox virus, herpes simplex virus-1, human parainfluenza virus, vaccinia virus, murine norovirus-1,[209,210] and H3N2 Influenza virus.[211] However, further research must be performed to assess the effect of colloidal silver against SARS-CoV-2.

Another study done with the transmissible gastroenteritis virus (TGEV, a type of coronavirus) has shown that pure silver nanoparticles and silver nanowires do have inhibitory action against TGEV coronavirus, but colloidal silver does not.[212] Colloidal silver as people consumed today (orally and topically), might not help to prevent or treat COVID-19, but inhaling 3–7 nm colloidal silver particles could have an effect against bacteria and viruses (such as the SARS-CoV-2) without the risk of developing drug resistance in the lungs.[213,214] It has been reported that silver could help in the treatment of different types of cancer, such as breast cancer, leukemia, lung carcinoma, and skin carcinoma, among others.[209] However, the results are inconclusive and more studies are necessary.

14.6.2.2 Safety

Contradictory literature is found about the use of silver and its interaction within the organism. Some researchers say that silver nanoparticles are safe to use, and that is why many products used today contain silver, such as creams for burned skin and dressings for open wounds.[215,233] Other researchers discourage the use of this substance, claiming that it can cause leukocytoclastic vasculitis, argyria, neurological symptoms, and problems in different organs after long-term consumption.[216,217]

14.6.2.3 Toxicological profile

In high concentrations or long-term ingestion, silver can cause argyria, an irreversible brown-black pigmentation on tissues caused by the accumulation of silver granules, which seems to not be harmful to the patient.[204,233] Other scientists reference silver's toxicity because in high quantity or over a prolonged time, it can cause neurological problems (confusion), hypochloremia, hyponatremia, profound disfiguration, damage to mitochondria reducing the cellular ATP content, and increases the generation of reactive oxygen species (hydrogen peroxide and superoxide production) leading to oxidative stress.[209,218–220] Furthermore, it can cause damage in the DNA-promoting apoptosis, recruitment of E-cadherin to the junction between cultured keratinocytes, irregularities in the cycle of mitosis or changes in the chromosomal structure, changes in the structure of the cellular cytoskeleton and increase in cell migration in cultured keratinocytes.[209,218–220] This could cause hypotension, jaundice, obscure fingers, dry mucous membranes, cholelithiasis, lethargy, fever, hypoxia, slurred speech, and accumulation of silver in different organs (liver, spleen, lungs, colon, and kidneys) that can lead to long-term damage.[209,218–220] Another study reported that silver nanoparticles are four times less toxic than silver ions, but they still cause an increase in the levels of apoptosis and necrosis in THP-1 monocytic cells as the concentration and time increase.[234] It also needs to be mentioned that silver nitrate causes damage, inhibits metabolic activity, and reduces or suppresses the proliferation activity in keratinocytes, diminishes fibroblasts activity, and attacks feeder cells of keratinocytes.[221]

14.6.2.4 Legal issues

During the COVID-19 pandemic, many businesses, politicians, and anti-vaccination groups have taken advantage of the misinformation and desperation of the people to take unacceptable measures, such as selling unauthorized products like colloidal silver to treat and prevent coronavirus.[69,235] Due to this, the FDA's Office of Criminal Investigators and the Department of Justice of the United States of America have taken legal actions against companies, such as *My Doctor Suggests LLC* and *GP Silver LLC,* that were accused of promoting and selling silver-based products to cure and prevent COVID-19.[236–238]

14.6.3 Hydrogen peroxide

Intentional ingestion of high concentration hydrogen peroxide for health purposes has gained popularity in certain patient populations, purported benefits are due to the increased oxygen released into the bloodstream.[193] Hydrogen peroxide is also known as oxydol or dihydrogen dioxide, which is a chemical that appears as a colorless liquid, used in the production of inorganic and organic chemicals.[239,240] It is also used in a wide range of cleaning and personal care products, including surgical disinfectant, toothpaste and mouthwashes, bathroom cleaners and laundry stain removers, hair dyes, and bleaches.[239,240] Hydrogen peroxide can also be found in OTC first-aid antiseptics; it is used as a bleaching agent in some food products, and it has other consumer and industrial uses as well including water treatment.[239,240] Hydrogen peroxide is available in dilute form (3%–10%) for household use and OTC and in concentrated form (greater than 30%) for industrial use.[222]

Hydrogen peroxide injected into the bloodstream is considered an alternative treatment that claims to cure cancer, called hyperoxygenation therapy. This theory is based on the erroneous concept that cancer is caused by oxygen deficiency and can be cured by exposing cancer cells to more oxygen than they can tolerate. There is no evidence that this treatment can cure any serious disease, and it has been reported to be potentially harmful.[223] At the same time, there is also no evidence that gargling diluted hydrogen peroxide eliminates SARS-CoV-2, which was another theory that hydrogen peroxide can help prevent COVID-19.[224]

14.6.3.1 Efficacy

The SARS-CoV-2 can be persistent on inanimate surfaces, and this is why there is a list of biocidal chemicals such as hydrogen peroxide. This product is known to clean areas that might be contaminated with virus such as SARS-CoV-2. The efficacy of this chemical with a concentration of 0.5% is effective within 1 min.[241] However, its use is limited to inanimate surfaces, while no evidence exists for oral ingestion.

14.6.3.2 Safety

The growing naturopathic health industry has promoted the use of hydrogen peroxide in treating a wide variety of medical conditions. Ingestion of hydrogen peroxide can cause poisoning and results in morbidity through three main mechanisms: direct caustic injury, oxygen gas formation, and lipid peroxidation. Severe toxicity has resulted from the use of hydrogen peroxide solutions to irrigate wounds within closed body cavities or under pressure as oxygen gas has resulted in embolism.[193,225]

14.6.3.3 Toxicological profile

Hydrogen peroxide ingestion, inhalation, or contact with skin and eyes can be toxic. The severity of the effect depends on the concentration. Low concentrations can cause vomiting, mild gastrointestinal irritation, and gastric distension. Higher concentrations (>35%) can burn the exposed skin entering the adjacent tissues and blood vessels, consequently causing mechanical pressure injury and causing gastrointestinal erosion, oxygen embolism, and loss of consciousness followed by respiratory paralysis.[193,222,224]

14.7 Medication therapy management

In the health care system, a series of steps are established to ensure that patients who attend health care centers can achieve medical attention to their ailments. So, when it comes to chronic diseases, patients manage to get medical diagnosis and receive a prescription. However, the medical literature shows different results in relation to the control of diseases, which is directly related to the level of health literacy of patients.[242] In this way, it has been possible to recognize problems in adherence to drug treatment, especially in chronic diseases as diabetes,[243] tuberculosis,[244] glaucoma,[245] hypertension,[246] and others. Nonadherence is a drug-related problem.

14.7.1 Drug-related problem

Pharmaceutical Care Network Europe (PCNE) stablished the definition for drug-related problem (DRP) as an event or circumstance involving drug therapy that actually or potentially interferes with desired health outcomes.[247] DRPs have been reported in different types of patients such as ischemic stroke,[248] epilepsy,[249] rheumatoid arthritis,[250] and other illnesses.

14.7.2 Medication therapy management/pharmaceutical care

To identify, solve, and prevent DRPs, a clear methodology is required, and for this there is the medication therapy management, also known as pharmaceutical care, which seeks to contribute to achieving therapeutic goals as in the case of patients with diabetes,[251–253] hypertension,[254,255] and other diseases. However, some barriers to provide pharmaceutical care have been detected.[256,257] It seeks to reduce the discrepancies between what is indicated by the doctor and the drugs that are finally used by the patient. In this way, it is possible to reduce DRPs, which achieves three specific impacts: reduction of workload, reduction of morbidity and mortality, and reduction of (re)hospitalizations (Fig. 14.2).[278,279]

14.7.3 Pharmaceutical care during the COVID-19 pandemic

COVID-19 is an infectious disease of pandemic proportions, with more than 34 million cases and over 1 million deaths reported worldwide (at the time of writing).[3] The current pandemic has generated many problems with the use of drugs for prevention and treatment, influenced by fake news[17] and other aspects. During COVID-19, people have needed to receive pharmaceutical care and it has been shown that pharmacists have been able to provide different services to patients. The care of patients with COVID-19 within hospital settings presented different limitations that has

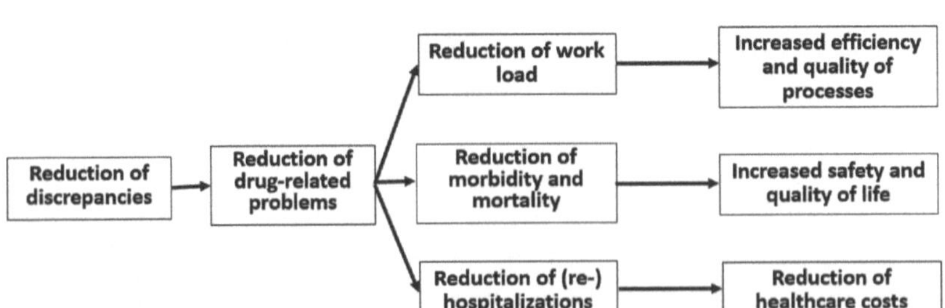

Figure 14.2 Impact of reduction of discrepancies in the use of medication.

generated negative impact in the mental health and job satisfaction of health care professionals.[12,26,35,37,258] Despite this, it has been possible to report clinical interventions to optimize the use of medications and avoid self-medication. There is a need for health professionals to guide patients and their families to take the necessary actions to prevent further infections, especially the most vulnerable.[9,11,259–261] Thus, Cadogan and Hughes[262] recognized that pharmaceutical care can contribute to the following actions:

1. Provide objective and reliable information on the disease and associated symptoms.
2. Educate the public about infection control and preventive measures to reduce transmission (e.g., hand hygiene, social distancing, and self-isolation).
3. Implement infection control measures (e.g., cleaning and disinfection of the pharmacy environment, limiting public access to the pharmacy).
4. Maintain continuity of pharmacy services, including supplies of essential medications and other products (e.g., hand sanitizers, protective masks).
5. Facilitate continued supply of OTC and prescription medications to patients (including emergency supply of repeat medications where necessary).

Likewise, Li et al.[263] were able to recognize innovative strategies to deal with COVID-19 such as establish evidence-based drug evaluation and guidelines; remote inpatient order review and dispensing tele-pharmaceutical care, telehealth counseling, and patient education; and multimedia health education. At the hospital level, Song et al.[264] has also evidenced the contribution in medical rounds of pharmacists to optimize pharmacotherapy. Many community pharmacies have been on the frontline of health service to fight against COVID-19, dispensing products, and providing counseling about drugs to patients,[265] and fighting against misinformation.[266] Specifically, there is evidence from Germany that shows that pharmaceutical care in the pharmacy, telephone, and video call was provided from community pharmacies with 44.2% of pharmacies that conducted medication reviews during the COVID-19 epidemic. The contribution from the rural pharmacy in actions against COVID-19 has even been described.[267] In China expert teams were created to focus on issues related to pharmaceuticals, pharmacokinetics, and pharmacotherapy, as well as treatment alternatives due to drug shortages.[268] Finally, it has been reported that telepharmacy services have continued to guide patients.[269]

Fig. 14.3 shows that medication therapy management/pharmaceutical care must consider the different kind of information that the patients

Figure 14.3 Information sources for health decision making.

receive to work closely with them to make good health decisions to prevent adverse events and the patient can reach the therapeutic goals during COVID-19 care.

Health care systems need to empower health professionals who have direct access to the population as in the case of community pharmacists, whom the population had available to make consultations during the COVID-19 pandemic. Community pharmacists have served as source of information for the general public[280,281] to clearly understand the messages from the authorities and prescribers, to avoid self-medication and to report adverse reactions by the use of prescription and OTC drugs, herbals, and unproven chemicals for the treatment of COVID-19.

14.8 Conclusion

COVID-19 is a new disease and we are still understanding its pathophysiology and symptomatology, and are in the process of producing an effective and safe vaccine and specific treatments. Currently, various prescription and OTC drugs, herbal products, and unproven chemicals have been used in an effort to manage the COVID-19-related symptoms. The efficacy, safety, and toxicology of these has been presented, and caution needs to be taken with all of them. Medication therapy management or pharmaceutical care is critical as well as consultation with community pharmacists. The sense of urgency and desperation against a pandemic is evident, but the caution against the preventive and COVID-19-related symptom treatment is warranted. Future studies and therapeutic strategies will continue to be produced in order to counter a pandemic that has affected so many people and lives across the globe.

References

1. Ma J. *Coronavirus: China's first confirmed Covid-19 case traced back to November 17*. 28 September 2020. Available from: https://www.scmp.com/news/china/society/article/3074991/coronavirus-chinas-first-confirmed-covid-19-case-traced-back.
2. Davidson H. *First Covid-19 case happened in November, China government records show - report*. 28 September 2020. Available from: https://www.theguardian.com/world/2020/mar/13/first-covid-19-case-happened-in-november-china-government-records-show-report.
3. Dong E, Du H, Gardner L. An interactive web-based dashboard to track COVID-19 in real time. *Lancet Infect Dis* 2020;**20**(5):533–4.
4. Almeida F. Exploring the impact of COVID-19 on the sustainability of health critical care systems in South America. *Int J Health Pol Manag* 2020. In press.
5. Cuevas A, Barquera S. COVID-19, obesity and undernutrition: a major challenge for Latin American countries. *Obesity* 2020. In press.
6. Rubin R, Abbasi J, Voelker R. Latin America and its global partners toil to procure medical supplies as COVID-19 pushes the region to its limit. *J Am Med Assoc* 2020;**324**(3):217–9.
7. Yáñez JA, et al. *Demographic and geographic COVID-19 death risk factors in Peru. A nationwide analysis*. EClinicalMedicine; 2020. Available at: SSRN, https://ssrn.com/abstract=3648543.
8. Panamericana. *COVID-19 sí habría ocasionado más muertes en el país que casos reconocidos por el gobierno*. 05 July 2020. Available from: https://panamericana.pe/nacionales/295269-covid-19-ocasionado-muertes-pais-casos-reconocidos-gobierno.
9. Yáñez JA, Alvarez-Risco A, Delgado-Zegarra J. *Rapid Response: Does Peru really have that high number of COVID-19 confirmed cases? The deception of combining RT-PCR and rapid test results*. 01 July 2020. Available from: https://www.bmj.com/content/369/bmj.m2518/rr-4.
10. Rojas Román B, et al. Tratamiento de la COVID-19 en Perú y Bolivia y los riesgos de la automedicación. *Rev Cubana Farmac* 2020;**53**(2):1–20.
11. Yáñez JA, Alvarez-Risco A, Delgado-Zegarra J. Covid-19 in Peru: from supervised walks for children to the first case of Kawasaki-like syndrome. *BMJ* 2020;**369**:m2418.
12. Yáñez JA, et al. Anxiety, distress, and turnover intention of healthcare workers in Peru by their distance to the epicenter during the COVID-19 crisis. *Am J Trop Med Hyg* 2020;**103**(4):1614–20.
13. Quispe-Cañari JF, et al. Self-medication practices during the COVID-19 pandemic among the adult population in Peru: a cross-sectional survey. *Saudi Pharmaceut J* 2021;**29**(1):1–11.
14. Duffin E. *Impact of the coronavirus pandemic on the global economy - Statistics & Facts*. 28 September 2020. Available from: https://www.statista.com/topics/6139/covid-19-impact-on-the-global-economy/.
15. Yan J, et al. Hospitality workers' COVID-19 risk perception and depression: a transactional theory of stress model. *Int J Hospit Manag* 2021;**95**:102935.
16. Lones L, Palumbo D, Coronavirus DB. *A visual guide to the economic impact*. 28 September 2020. Available from: https://www.bbc.com/news/business-51706225.
17. Alvarez-Risco A, et al. The Peru approach against the COVID-19 infodemic: insights and strategies. *Am J Trop Med Hyg* 2020;**103**(2):583–6.
18. Matias T, Dominski FH, Marks DF. Human needs in COVID-19 isolation. *J Health Psychol* 2020;**25**(7):871–82.
19. Datta R, et al. The infodemics of COVID-19 amongst healthcare professionals in India. *Med J Armed Forces India* 2020;**76**(3):276–83.

20. Pfefferbaum B, North CS. Mental health and the covid-19 pandemic. *N Engl J Med* 2020;**383**(6):510–2.
21. Tasnim S, Hossain MM, Mazumder H. Impact of rumors and misinformation on COVID-19 in social media. *J Prev Med Public Health* 2020;**53**(3):171–4.
22. Tang PM, et al. Geographical identification of the vulnerable groups during COVID-19 crisis: the typhoon eye effect and its boundary conditions. *Psychiatr Clin Neurosci* 2020;**74**(10):562–3.
23. Lai J, et al. Factors associated with mental health outcomes among health care workers exposed to coronavirus disease 2019. *JAMA Netw Open* 2020;**3**(3):e203976.
24. Kang L, et al. Impact on mental health and perceptions of psychological care among medical and nursing staff in Wuhan during the 2019 novel coronavirus disease outbreak: a cross-sectional study. *Brain Behav Immun* 2020;**87**:11–7.
25. Tan BYQ, et al. Psychological impact of the COVID-19 pandemic on health care workers in Singapore. *Ann Intern Med* 2020:M20–1083. Online ahead of print.
26. Zhang SX, et al. At the height of the storm: healthcare staff's health conditions and job satisfaction and their associated predictors during the epidemic peak of COVID-19. *Brain Behav Immun* 2020;**87**:144–6.
27. Jahanshahi AA, et al. The distress of Iranian adults during the Covid-19 pandemic - more distressed than the Chinese and with different predictors. *Brain Behav Immun* 2020;**87**:124–5.
28. Moccia L, et al. Affective temperament, attachment style, and the psychological impact of the COVID-19 outbreak: an early report on the Italian general population. *Brain Behav Immun* 2020;**87**:75–9.
29. Mazza C, et al. A nationwide survey of psychological distress among Italian people during the COVID-19 pandemic: immediate psychological responses and associated factors. *Int J Environ Res Publ Health* 2020;**17**(9):3165.
30. Maher A, et al. COVID-19 outbreak situation and its psychological impact among surgeon in training in France. *World J Urol* 2020:1–2. Epub ahead of print.
31. Bacon AM, Corr PJ. Coronavirus (COVID-19) in the United Kingdom: a personality-based perspective on concerns and intention to self-isolate. *Br J Health Psychol* 2020;**25**(4):839–48 [Epub ahead of print].
32. Ozamiz-Etxebarria N, et al. Stress, anxiety, and depression levels in the initial stage of the COVID-19 outbreak in a population sample in the northern Spain. *Cad Saúde Pública* 2020;**36**(4). e00054020.
33. Gonzalez-Sanguino C, et al. Mental health consequences during the initial stage of the 2020 coronavirus pandemic (COVID-19) in Spain. *Brain Behav Immun* 2020;**87**:172–6.
34. Caqueo-Urízar A, et al. Mental health and the COVID-19 pandemic in Chile. *Psychol Trauma* 2020;**12**(5):521–3.
35. Zhang SX, et al. Succumbing to the COVID-19 pandemic e healthcare workers not satisfied and intend to leave their jobs. *JMIR Prep* 2021. https://doi.org/10.1007/s11469-020-00418-6.
36. Zhang SX, et al. Developing and testing a measure of COVID-19 organizational support of healthcare workers – results from Peru, Ecuador, and Bolivia. *Psychiatr Res* 2020;**291**:113174.
37. Chen X, et al. Belief in conspiracy theory about COVID-19 predicts mental health and well-being: a study of healthcare staff in Ecuador. *JMIR Public Health Surveill* 2020;**6**(3):E20737.
38. Paz C, et al. Anxiety and depression in patients with confirmed and suspected COVID-19 in Ecuador. *Psychiatr Clin Neurosci* 2020;**74**(10):554–5.
39. Dal'Bosco EB, et al. Mental health of nursing in coping with COVID-19 at a regional university hospital. *Rev Bras Enferm* 2020;**73**(Suppl. 2):e20200434.

40. Forlenza OV, Stella F. Impact of SARS-CoV-2 pandemic on mental health in the elderly: perspective from a psychogeriatric clinic at a tertiary hospital in São Paulo, Brazil. *Int Psychogeriatr* 2020:1–5.
41. Castro-de-Araujo LFS, Machado DB. Impact of COVID-19 on mental health in a low and middle-income country. *Ciência Saúde Coletiva* 2020;**25**(Suppl. 1):2457–60.
42. Yáñez JA, et al. Anxiety, distress, and turnover intention of healthcare workers in Peru by their distance to the epicenter during the COVID-19 crisis. *Am J Trop Med Hyg* 2020;**103**(4):1614–20.
43. Júnior JG, et al. The mental health of those whose rights have been taken away: an essay on the mental health of indigenous peoples in the face of the 2019 Coronavirus (2019-nCoV) outbreak. *Psychiatr Res* 2020;**289**:113094.
44. Hua J, Shaw R. Corona virus (COVID-19) "infodemic" and emerging issues through a data lens: the case of China. *Int J Environ Res Publ Health* 2020;**17**(7).
45. Kadam AB, Atre SR. Negative impact of social media panic during the COVID-19 outbreak in India. *J Trav Med* 2020;**27**(3).
46. Shafie M, et al. Prevalence and determinants of self-medication practice among selected households in Addis Ababa community. *PloS One* 2018;**13**(3). e0194122-e0194122.
47. Gore PR, Madhavan S. Consumers' preference and willingness to pay for pharmacist counselling for non-prescription medicines. *J Clin Pharm Therapeut* 1994;**19**(1):17–25.
48. Alqathama A, et al. Herbal medicine from the perspective of type II diabetic patients and physicians: what is the relationship? *BMC Complement Med Ther* 2020;**20**(1):65.
49. Clement YN, et al. A gap between acceptance and knowledge of herbal remedies by physicians: the need for educational intervention. *BMC Compl Alternative Med* 2005;**5**:20.
50. Alamgir ANM. Pharmacopoeia and herbal monograph, the aim and use of WHO's herbal monograph, WHO's guide lines for herbal monograph, pharmacognostical research and monographs of organized, unorganized drugs and drugs from animal sources. In: *Therapeutic use of medicinal plants and their extracts: volume 1: pharmacognosy*. Cham: Springer International Publishing; 2017. p. 295–353.
51. Robertson J. American herbal pharmacopoeia. Botanical pharmacognosy – microscopic characterization of botanical medicines. *Aust J Forensic Sci* 2016;**48**(3):359–61.
52. Abourjaily P. American Herbal Pharmacopoeia and Therapeutic Compendium (A botanical supplement monograph series). *Nutr Clin Care* 2001;**4**(4):221–2.
53. Medscape. *Drug Interaction Checker*. 27 July 2020. Available from: https://reference.medscape.com/drug-interactionchecker.
54. WebMD. *Drug Interaction Checker*. 27 July 2020. Available from: https://www.webmd.com/interaction-checker/default.htm.
55. Barros GAMd, et al. The use of analgesics and risk of self-medication in an urban population sample: cross-sectional study. *Brazilian J Anesthesiol* 2019;**69**(6):529–36.
56. Tapia L. COVID-19 and fake news in the Dominican Republic. *Am J Trop Med Hyg* 2020;**102**(6):1172–4.
57. PeruSalem. *Urgente Dos ministras están pasando el test Coronavirus x sospecha a estar infectadas.Una d ellas es la ministra d economía María Antonieta Alva quien hoy salió en medios y estaba triste y débil*. 26 March 2020. Available from: https://twitter.com/SalemPeru/status/1237822141598322688.
58. Galán-Rodas E, Zamora A. Alfabetización Digital en Salud para fortalecer los sistemas de salud en Centroamérica. *Revista Hispanoamericana de Ciencias de la Salud* 2015;**1**(1):29–33.
59. Salas M, et al. Challenges facing drug utilization research in the Latin American region. *Pharmacoepidemiol Drug Saf* 2020;**29**(11):1353–63 [n/a(n/a)].

60. Durán CE, et al. Systematic review of cross-national drug utilization studies in Latin America: methods and comparability. *Pharmacoepidemiol Drug Saf* 2016;**25**(1):16–25.
61. Ali A. Methodological challenges in observational research: a pharmacoepidemiological perspective. *Br J Pharmaceut Res* 2013;**3**:161–75.
62. Sanidad Md, editor. *Guía de buenas prácticas en las salidas de la población infantil durante el estado de alarma*. Gobierno de España; 2020.
63. Castro C. *Sanidad reconoce preocupación por el aumento de contagios en niños. El Independiente*. 18 May 2020. Available from: https://www.elindependiente.com/vida-sana/salud/2020/05/13/sanidad-reconoce-preocupacion-por-el-aumento-de-contagios-en-ninos/.
64. CDC. *Multisystem Inflammatory Syndrome in Children (MIS-C) Associated with Coronavirus Disease 2019 (COVID-19)*. 03 June 2020. Available from: https://emergency.cdc.gov/han/2020/han00432.asp.
65. CDC. *Therapeutic Options. Coronavirus Disease 2019 (COVID-19)*. 28 September 2020. Available from: https://www.cdc.gov/coronavirus/2019-ncov/hcp/therapeutic-options.html.
66. Blenkinsopp A, Bradley C. Patients, society, and the increase in self medication. *BMJ* 1996;**312**:629–32.
67. Kassie AD, Bifftu BB, Mekonnen HS. Self-medication practice and associated factors among adult household members in Meket district, Northeast Ethiopia, 2017. *BMC Pharmacol Toxicol* 2018;**19**(1). 15-15.
68. Adams KK, Baker WL, Sobieraj DM. Myth busters: dietary supplements and COVID-19. *Ann Pharmacother* 2020;**54**(8):820–6.
69. FDA. *Fraudulent Coronavirus Disease 2019 (COVID-19) Products. Health Fraud Scams*. 24 May 2020. Available from: https://www.fda.gov/consumers/health-fraud-scams/fraudulent-coronavirus-disease-2019-covid-19-products.
70. Onchonga D. A Google Trends study on the interest in self-medication during the 2019 novel coronavirus (COVID-19) disease pandemic. *Saudi Pharm J* 2020;**28**(7):903–4.
71. Molento MB. COVID-19 and the rush for self-medication and self-dosing with ivermectin: a word of caution. *One Health* 2020;**10**:100148.
72. Onchonga D, Omwoyo J, Nyamamba D. Assessing the prevalence of self-medication among healthcare workers before and during the 2019 SARS-CoV-2 (COVID-19) pandemic in Kenya. *Saudi Pharmaceut J* 2020;**28**(10):1149–54.
73. AphA. *Medication Therapy Management (MTM) Services*. 28 September 2020. Available from: https://www.pharmacist.com/medication-therapy-management-services.
74. AphA. *Principles of Practice for Pharmaceutical Care*. 28 September 2020. Available from: https://www.pharmacist.com/principles-practice-pharmaceutical-care.
75. Webster RK, et al. How to improve adherence with quarantine: rapid review of the evidence. *Public Health* 2020;**182**:163–9.
76. Nielsen-Bohlman L, Panzer AM, Kindig DA, editors. *Health literacy: a prescription to end confusion*; 2004 [Washington (DC)].
77. Sorensen K, et al. Health literacy in Europe: comparative results of the European health literacy survey (HLS-EU). *Eur J Public Health* 2015;**25**(6):1053–8.
78. Chou WS, Oh A, Klein WMP. Addressing health-related misinformation on social media. *J Am Med Assoc* 2018;**320**(23):2417–8.
79. Dunn P, Conard S. Improving health literacy in patients with chronic conditions: a call to action. *Int J Cardiol* 2018;**273**:249–51.
80. Torous J, et al. Digital mental health and COVID-19: using technology today to accelerate the curve on access and quality tomorrow. *JMIR Ment Health* 2020;**7**(3):e18848.

81. Mariscal J, Herrera Rosado F, Varela Castro S. *Estudio sobre TIC y salud pública en América Latina: la perspectiva de e-salud y m-salud*. 2018. Available from: https://itu.int/en/ITU-D/ICT-Applications/Documents/Publications/Estudio_sobre_TICS_y_salud_publica_en_America_latina_S.PDF.
82. Curioso WH. Building capacity and training for digital health: challenges and opportunities in Latin America. *J Med Internet Res* 2019;**21**(12):e16513.
83. Walter HC, Eden G-R. The role of telehealth in the fight against COVID-19 and the evolution of the Peruvian regulatory framework. *Acta Méd Peru* 2020;**37**(3).
84. Alvarez-Risco A, Del-Aguila-Arcentales S, Yanez JA. Telemedicine in Peru as a result of the COVID-19 pandemic: perspective from a country with limited internet access. *Am J Trop Med Hyg* 2021. https://doi.org/10.4269/ajtmh.21-0255.
85. ESA. *Telemedicina en primera línea*. 16 November 2020. Available from: https://www.esa.int/Space_in_Member_States/Spain/Telemedicina_en_primera_linea.
86. Univadis. *Estados Unidos redescubre la telemedicina en la pandemia de coronavirus*. 16 November 2020. Available from: https://www.univadis.es/viewarticle/estados-unidos-redescubre-la-telemedicina-en-la-pandemia-de-coronavirus-715070.
87. Carreño I. *Brasil regula el uso de la telemedicina ante pandemia del COVID-19*. 16 November 2020. Available from: https://digitalpolicylaw.com/brasil-regula-el-uso-de-la-telemedicina-ante-pandemia-del-covid-19/.
88. Mexico. *Telemedicina satelital conecta 35 hospitales públicos y centros de salud en zonas urbanas y rurales que atienden COVID-19*. 16 November 2020. Available from: https://www.gob.mx/salud/prensa/telemedicina-satelital-conecta-35-hospitales-publicos-y-centros-de-salud-en-zonas-urbanas-y-rurales-que-atienden-covid-19-243918.
89. MinSalud. *Telesalud y Telemedicina para la Prestación de Servicios de Salud en la Pandemia por COVID-19*. 16 November 2020. Available from: https://www.minsalud.gov.co/Ministerio/Institucional/Procesos%20y%20procedimientos/PSSS04.pdf.
90. Argentina. *Lanzamiento de Tele-Covid: el servicio de telemedicina destinado a personas con cobertura estatal exclusiva*. 16 November 2020. Available from: https://www.argentina.gob.ar/noticias/lanzamiento-de-tele-covid-el-servicio-de-telemedicina-destinado-personas-con-cobertura.
91. Salud. *Telemedicina: salud y tecnología al servicio de la ciudadanía*. 16 November 2020. Available from: https://www.gub.uy/agencia-gobierno-electronico-sociedad-informacion-conocimiento/comunicacion/noticias/telemedicina-salud-tecnologia-servicio-ciudadania.
92. SaludDigital. *Lanzan en Paraguay primera aplicación de telemedicina*. 16 November 2020. Available from: https://saluddigital.com/plataformas-digitales/lanzan-en-paraguay-primera-aplicacion-de-telemedicina/.
93. SaludDigital. *Servicios de Salud en La Paz, Bolivia* presentaron una chatbot de consultas y diagnósticos en línea para personas con síntomas de COVID-19. 16 November 2020. Available from: https://saluddigital.com/plataformas-digitales/servicios-de-salud-en-la-paz-bolivia-presentaron-una-chatbot-de-consultas-y-diagnosticos-en-linea-para-personas-con-sintomas-de-covid-19/.
94. Ackerman SL, Gleason N, Shipman SA. Comparing patients' experiences with electronic and traditional consultation: results from a multisite survey. *J Gen Intern Med* 2020;**35**(4):1135−42.
95. Roberts LC, Osborn-Jenkins L. Delivering remote consultations: talking the talk. *Musculoskelet Sci Pract* 2020:102275.
96. Sommariva S, et al. Spreading the (fake) news: exploring health messages on social media and the implications for health professionals using a case study. *Am J Health Educ* 2018;**49**(4):246−55.
97. Gomez Tejeda JJ, et al. Evaluación del nivel de conocimiento sobre COVID-19 durante la pesquisa en la población de un consultorio. *Inicio* 2020;**59**(278):1−2.

98. Huaroto F, et al. Intervenciones farmacológicas para el tratamiento de la Enfermedad por Coronavirus (COVID-19). *An Fac Med* 2020;**81**(1):71–9.
99. Meena P, et al. Self-medication in developing countries a systematic review. *J Pharm Technol, Res Manag* 2016;**4**(2).
100. Mallhi TH, et al. Drug repurposing for COVID-19: a potential threat of self-medication and controlling measures. *Postgrad Med* 2020. In press.
101. Google. *Worldwide Google trend data for web search query for the term 'hydroxychloroquine'*. 04 December 2020. Available from: https://trends.google.es/trends/explore?q=hydroxychloroquine.
102. Google. *Worldwide Google trend data for web search query for the term 'ivermectin'*. 04 December 2020. Available from: https://trends.google.es/trends/explore?q=ivermectin.
103. Google. *Worldwide Google trend data for web search query for the term 'azithromycin'*. 04 December 2020. Available from: https://trends.google.es/trends/explore?q=azithromycin.
104. Hasan S, Kow CS, Merchant HA. Is it worth the wait? Should Chloroquine or Hydroxychloroquine be allowed for immediate use in CoViD-19? *British J Pharm* 2020;**5**(1):1–5.
105. Molina JM, et al. No evidence of rapid antiviral clearance or clinical benefit with the combination of hydroxychloroquine and azithromycin in patients with severe COVID-19 infection. *Med Maladies Infect* 2020;**50**(4). 384-384.
106. Lim J, et al. Case of the index patient who caused tertiary transmission of COVID-19 infection in Korea: the application of lopinavir/ritonavir for the treatment of COVID-19 infected pneumonia monitored by quantitative RT-PCR. *J Kor Med Sci* 2020;**35**(6). e79-e79.
107. Jaffe S. Regulators split on antimalarials for COVID-19. *Lancet* 2020;**395**(10231):1179.
108. Busari S, Adebayo B. *Nigeria records chloroquine poisoning after Trump endorses it for coronavirus treatment.* 29 August 2020. Available from: https://edition.cnn.com/2020/03/23/africa/chloroquine-trump-nigeria-intl/index.html.
109. Rojas B, et al. Tratamiento contra el COVID-19 en Perú y Bolivia y el riesgo de automedicarse. *Rev Cubana Farmac* 2020;**53**(2):e435.
110. Hughes CM, McElnay JC, Fleming GF. Benefits and risks of self medication. *Drug Saf* 2001;**24**(14):1027–37.
111. Mandal S. Can over-the-counter antibiotics coerce people for self-medication with antibiotics? *Asian Pac J Trop Dis* 2015;**5**:S184–6.
112. Menary KR, et al. The prevalence and clinical implications of self-medication among individuals with anxiety disorders. *J Anxiety Disord* 2011;**25**(3):335–9.
113. Mansuri FMA, et al. Estimating the public response to mitigation measures and self-perceived behaviours towards the COVID-19 pandemic. *J Taibah Univ Med Sci* 2020;**15**(4):278–83.
114. Damle B, et al. Clinical pharmacology perspectives on the antiviral activity of azithromycin and use in COVID-19. *Clin Pharmacol Ther* 2020;**108**(2):201–11.
115. Andreani J, et al. In vitro testing of combined hydroxychloroquine and azithromycin on SARS-CoV-2 shows synergistic effect. *Microb Pathog* 2020;**145**:104228.
116. Mercuro NJ, et al. Risk of QT interval prolongation associated with use of hydroxychloroquine with or without concomitant azithromycin among hospitalized patients testing positive for coronavirus disease 2019 (COVID-19). *JAMA Cardiol* 2020;**5**(9):1036–41.
117. Bhimraj A, et al. Infectious diseases society of America guidelines on the treatment and management of patients with COVID-19. *Clin Infect Dis* 2020. In press.
118. Gautret P, et al. Hydroxychloroquine and azithromycin as a treatment of COVID-19: results of an open-label non-randomized clinical trial. *Int J Antimicrob Agents* 2020:105949.

119. Yao X, et al. In vitro antiviral activity and projection of optimized dosing design of hydroxychloroquine for the treatment of severe acute respiratory syndrome coronavirus 2 (SARS-CoV-2). *Clin Infect Dis* 2020;**71**(15):732–9.
120. Gautret P, et al. Clinical and microbiological effect of a combination of hydroxychloroquine and azithromycin in 80 COVID-19 patients with at least a six-day follow up: a pilot observational study. *Travel Med Infect Dis* 2020;**34**:101663.
121. Million M, et al. Early treatment of COVID-19 patients with hydroxychloroquine and azithromycin: a retrospective analysis of 1061 cases in Marseille, France. *Trav Med Infect Dis* 2020;**35**:101738.
122. Chorin E, et al. The QT interval in patients with COVID-19 treated with hydroxychloroquine and azithromycin. *Nat Med* 2020;**26**(6):808–9.
123. Ramireddy A, et al. Experience with hydroxychloroquine and azithromycin in the coronavirus disease 2019 pandemic: implications for QT interval monitoring. *J Am Heart Assoc* 2020;**9**(12):e017144.
124. MINSA. *Resolución Ministerial N° 270-2020-MINSA - prevención, Diagnóstico y Tratamiento de personas afectadas por COVID-19.* Ministerio de Salud del Perú; 2020.
125. Conto-Palomino NM, et al. Encephalitis associated with COVID-19 in a 13-year-old girl: a case report. *Medwave* 2020;**20**(7):e7984.
126. RPP. *Zamora plantea el retiro de la ivermectina e hidroxicloroquina: "No han mostrado ningún beneficio".* 07 September 2020. Available from: https://rpp.pe/peru/actualidad/coronavirus-en-peru-victor-zamora-plantea-el-retiro-de-la-ivermectina-e-hidroxicloroquina-no-han-mostrado-ningun-beneficio-noticia-1291101?ref=rpp.
127. Velarde-Ruiz Velasco JA, García-Jiménez ES, Remes-Troche JM. Manifestaciones hepáticas y repercusión en el paciente cirrótico de COVID-19. *Rev Gastroenterol México* 2020;**85**(3):303–11.
128. Alonso-Bello CD, et al. Lesión hepática inducida por antirretrovirales. *Rev Hosp Jua Mex* 2018;**85**(1):49–56.
129. Mallolas J, et al. Hepatotoxicidad asociada al tratamiento antirretroviral. *Enf Emerg* 2003;**5**(2):97–104.
130. Ford N, et al. Systematic review of the efficacy and safety of antiretroviral drugs against SARS, MERS or COVID-19: initial assessment. *J Int AIDS Soc* 2020;**23**(4):e25489.
131. Dong L, Hu S, Gao J. Discovering drugs to treat coronavirus disease 2019 (COVID-19). *Drug Discov Ther* 2020;**14**(1):58–60.
132. Tobin MJ, Laghi F, Jubran A. Why COVID-19 silent hypoxemia is baffling to physicians. *Am J Respir Crit Care Med* 2020;**202**(3):356–60.
133. Quaresima V, Ferrari M. COVID-19: efficacy of prehospital pulse oximetry for early detection of silent hypoxemia. *Crit Care* 2020;**24**(1):501.
134. Ledford H. Coronavirus breakthrough: dexamethasone is first drug shown to save lives. *Nature* 2020;**582**(7813):469.
135. Bimonte S, et al. Potential antiviral drugs for SARS-cov-2 treatment: preclinical findings and ongoing clinical research. *In Vivo* 2020;**34**(3 Suppl. l):1597–602.
136. Zhu Z, et al. Arbidol monotherapy is superior to lopinavir/ritonavir in treating COVID-19. *J Infect* 2020;**81**(1):e21–3.
137. Cao B, et al. A trial of lopinavir–ritonavir in adults hospitalized with severe covid-19. *N Engl J Med* 2020;**382**(19):1787–99.
138. Wen CY, et al. [Real-world efficacy and safety of lopinavir/ritonavir and arbidol in treating with COVID-19 : an observational cohort study]. *Zhonghua Nei Ke Za Zhi* 2020;**59**(0):E012.
139. Catteau L, et al. Low-dose hydroxychloroquine therapy and mortality in hospitalised patients with COVID-19: a nationwide observational study of 8075 participants. *Int J Antimicrob Agents* 2020;**56**(4). 106144-106144.

140. Arshad S, et al. Treatment with hydroxychloroquine, azithromycin, and combination in patients hospitalized with COVID-19. *Int J Infect Dis* 2020;**97**:396–403.
141. Skipper CP, et al. Hydroxychloroquine in nonhospitalized adults with early COVID-19: a randomized trial. *Ann Intern Med* 2020;**173**(8):623–31.
142. Li X, et al. Is hydroxychloroquine beneficial for COVID-19 patients? *Cell Death Dis* 2020;**11**(7):512.
143. Kim GU, et al. Clinical characteristics of asymptomatic and symptomatic patients with mild COVID-19. *Clin Microbiol Infect* 2020;**26**(7):948.e1–3.
144. Esakandari H, et al. A comprehensive review of COVID-19 characteristics. *Biol Proced Online* 2020;**22**. 19-19.
145. Meng X, et al. COVID-19 and anosmia: a review based on up-to-date knowledge. *Am J Otolaryngol* 2020;**41**(5). 102581-102581.
146. Heidari F, et al. Anosmia as a prominent symptom of COVID-19 infection. *Rhinology* 2020;**58**(3):302–3.
147. Vaira LA, et al. Validation of a self-administered olfactory and gustatory test for the remotely evaluation of COVID-19 patients in home quarantine. *Head Neck* 2020;**42**(7):1570–6.
148. Meng X, et al. Smartphone-enabled wireless otoscope-assisted online telemedicine during the COVID-19 outbreak. *Am J Otolaryngol* 2020;**41**(3):102476.
149. Romano S, et al. Time-trend analysis of medicine sales and shortages during COVID-19 outbreak: data from community pharmacies. *Res Soc Adm Pharm* 2021;**17**(1):1876–81.
150. Rodríguez-Morales AJ, Cardona-Ospina JA, Murillo-Muñoz MM. Gastroenterologists, hepatologists, COVID-19 and the use of acetaminophen. *Clin Gastroenterol Hepatol* 2020;**18**(9):2142–3.
151. Roberts E, et al. Paracetamol: not as safe as we thought? A systematic literature review of observational studies. *Ann Rheum Dis* 2016;**75**(3):552–9.
152. Prescott LF. Therapeutic misadventure with paracetamol: fact or fiction? *Am J Therapeut* 2000;**7**(2):99–114.
153. Day RO, Graham GG, Whelton A. The position of paracetamol in the world of analgesics. *Am J Therapeut* 2000;**7**(2):51–4.
154. McLaughlin JK, et al. Analgesic use and chronic renal failure: a critical review of the epidemiologic literature. *Kidney Int* 1998;**54**(3):679–86.
155. Rodriguez-Morales AJ, et al. Clinical, laboratory and imaging features of COVID-19: a systematic review and meta-analysis. *Trav Med Infect Dis* 2020;**34**:101623.
156. Leventhal TM, et al. Acetaminophen is undetectable in plasma from more than half of patients believed to have acute liver failure due to overdose. *Clin Gastroenterol Hepatol* 2019;**17**(10):2110–6.
157. Day M. Covid-19: ibuprofen should not be used for managing symptoms, say doctors and scientists. *BMJ* 2020;**368**:m1086.
158. Micallef J, Soeiro T, Jonville-Béra A-P. Non-steroidal anti-inflammatory drugs, pharmacology, and COVID-19 infection. *Therapies* 2020;**75**(4):355–62.
159. Qin C, et al. Dysregulation of immune response in patients with coronavirus 2019 (COVID-19) in Wuhan, China. *Clin Infect Dis* 2020;**71**(15):762–8.
160. Jamerson BD, Haryadi TH. The use of ibuprofen to treat fever in COVID-19: a possible indirect association with worse outcome? *Med Hypoth* 2020;**144**. 109880-109880.
161. Yousefifard M, et al. Non-steroidal anti-inflammatory drugs in management of COVID-19; A systematic review on current evidence. *Int J Clin Pract* 2020;**74**(9):e13557.
162. Rinott E, et al. Ibuprofen use and clinical outcomes in COVID-19 patients. *Clin Microbiol Infect* 2020;**26**(9):1259.e5–7.

163. Verduyn M, et al. Co-infection of dengue and COVID-19: a case report. *PLoS Negl Trop Dis* 2020;**14**(8). e0008476-e0008476.
164. Smith SM, Schroeder K, Fahey T. Over-the-counter (OTC) medications for acute cough in children and adults in community settings. *Coch Datab Syst Rev* 2014;**2014**(11):Cd001831.
165. Gordon DE, et al. A SARS-CoV-2 protein interaction map reveals targets for drug repurposing. *Nature* 2020;**583**(7816):459−68.
166. Li X, et al. Existing bitter medicines for fighting 2019-nCoV-associated infectious diseases. *Faseb J* 2020;**34**(5):6008−16.
167. Deng L, et al. Pseudoephedrine protects mice from infection of H1N1 virus. *Int J Comput Vis* 2020;**4**(1):014−20.
168. Villamizar-PeÑa R, Gutiérrez-Ocampo E, Rodríguez-Morales AJ. Pooled prevalence of diarrhea among COVID-19 patients. *Clin Gastroenterol Hepatol* 2020;**18**(10):2385−7.
169. Ye Q, et al. The mechanism and treatment of gastrointestinal symptoms in patients with COVID-19. *Am J Physiol Gastrointest Liver Physiol* 2020;**319**(2):G245−g252.
170. D'Amico F, et al. Diarrhea during COVID-19 infection: pathogenesis, epidemiology, prevention, and management. *Clin Gastroenterol Hepatol* 2020;**18**(8):1663−72.
171. Liu X, et al. Chinese herbs combined with Western medicine for severe acute respiratory syndrome (SARS). *Coch Database Syst Rev* 2006;(1):Cd004882.
172. Luo H, et al. Can Chinese medicine Be used for prevention of corona virus disease 2019 (COVID-19)? A review of historical classics, research evidence and current prevention programs. *Chin J Integr Med* 2020;**26**(4):243−50.
173. Yan BH, et al. Large- scale prospective clinical study on prophylactic intervention of COVID-19 in community population using Huoxiang Zhengqi Oral Liquid and Jinhao Jiere Granules. *Zhongguo Zhongyao Zazhi* 2020;**45**(13):2993−3000.
174. Merzon E, et al. Low plasma 25(OH) vitamin D level is associated with increased risk of COVID-19 infection: an Israeli population-based study. *FEBS J* 2020;**287**(17):3693−702.
175. Hastie CE, et al. Vitamin D concentrations and COVID-19 infection in UK Biobank. *Diabetes Metab Syndr* 2020;**14**(4):561−5.
176. Raisi-Estabragh Z, et al. Greater risk of severe COVID-19 in Black, Asian and Minority Ethnic populations is not explained by cardiometabolic, socioeconomic or behavioural factors, or by 25(OH)-vitamin D status: study of 1326 cases from the UK Biobank. *J Public Health* 2020;**42**(3):451−60.
177. Xiong X, et al. Chinese herbal medicine for coronavirus disease 2019: a systematic review and meta-analysis. *Pharmacol Res* 2020;**160**:105056.
178. Wang X, et al. A systematic review and meta-analysis of the efficacy and safety of arbidol in the treatment of coronavirus disease 2019. *Medicine (Baltim)* 2020;**99**(30):e21402.
179. Yang Y, et al. Traditional Chinese medicine in the treatment of patients infected with 2019-new coronavirus (SARS-CoV-2): a review and perspective. *Int J Biol Sci* 2020;**16**(10):1708−17.
180. Zhu RF, et al. Systematic review of the registered clinical trials for coronavirus disease 2019 (COVID-19). *J Transl Med* 2020;**18**(1):274.
181. Fitriani I, et al. *In silico approach of potential phytochemical inhibitor from Moringa oleifera, cocos nucifera, Allium cepa, psidium guajava, and Eucalyptus globulus for the treatment of COVID-19 by molecular docking*. 2020.
182. Su H, et al. *Discovery of baicalin and baicalein as novel, natural product inhibitors of SARS-CoV-2 3CL protease in vitro*. 2020.
183. Finzi E. Treatment of SARS-CoV-2 with high dose oral zinc salts: a report on four patients. *Int J Infect Dis* 2020;**99**:307−9.

184. Arentz S, et al. Zinc for the prevention and treatment of SARS-CoV-2 and other acute viral respiratory infections: a rapid review. *Adv Integ Med* 2020;**7**(4):252–60. p. 10.1016/j.aimed.2020.07.009.
185. Gorton HC, Jarvis K. The effectiveness of vitamin C in preventing and relieving the symptoms of virus-induced respiratory infections. *J Manip Physiol Ther* 1999;**22**(8):530–3.
186. Kim H, et al. Red ginseng and vitamin C increase immune cell activity and decrease lung inflammation induced by influenza A virus/H1N1 infection. *J Pharm Pharmacol* 2016;**68**(3):406–20.
187. Baladia E, et al. Vitamin C for COVID-19: a living systematic review. *Medwave* 2020;**20**(6):e7978.
188. Feyaerts AF, Luyten W. Vitamin C as prophylaxis and adjunctive medical treatment for COVID-19? *Nutrition* 2020;**79–80**:110948.
189. Dushianthan A, et al. Immunonutrition for adults with ARDS: results from a Cochrane systematic review and meta-analysis. *Respir Care* 2020;**65**(1):99–110.
190. Kalcker AL. *How to make CDS? with Andreas Kalcker.* 01 August 2020. Available from: https://lbry.tv/@Kalcker:7/How_to_make_CDS:a.
191. AEMEMI. *Dioxido de cloro una terapeutica efectiva para el tratamiento del SARS-CoV2 (COVID-19).* 31 July 2020. Available from: https://drive.google.com/file/d/1AJXO_GnBl2o4kJgltQi2XRr4Am42TGQh/view.
192. NIH. *Colloidal Silver.* 28 September 2020. Available from: https://www.nccih.nih.gov/health/colloidal-silver.
193. Watt BE, Proudfoot AT, Vale JA. Hydrogen peroxide poisoning. *Toxicol Rev* 2004;**23**(1):51–7.
194. FDA. *FDA Warns Consumers of Serious Harm from Drinking Miracle Mineral Solution (MMS).* 28 September 2020. Available from: https://www.prnewswire.com/news-releases/fda-warns-consumers-of-serious-harm-from-drinking-miracle-mineral-solution-mms-99656679.html.
195. INCHEM. *Material Safety Datasheet for Chlorine Dioxide.* 31 July 2020. Available from: http://www.inchem.org/documents/icsc/icsc/eics0127.htm.
196. MSDS-Europe. *Safety data sheet knowledge base – R phrases and S phrases.* 01 August 2020. Available from: https://www.msds-europe.com/r-phrases-s-phrases/.
197. FDA. *Danger: Don't Drink Miracle Mineral Solution or Similar Products.* 28 September 2020. Available from: https://www.fda.gov/consumers/consumer-updates/danger-dont-drink-miracle-mineral-solution-or-similar-products.
198. Medsafe. *Miracle Mineral Solution – dangerous and potentially life-threatening side effects.* 28 September 2020. Available from: https://medsafe.govt.nz/safety/Alerts/MiracleMineralSolution.asp.
199. TGA. *Safety advisory - false and misleading claims relating to COVID-19.* 28 September 2020. Available from: https://www.tga.gov.au/alert/miracle-mineral-solution-mms-0.
200. Gebhardtova A, et al. A case of severe chlorite poisoning successfully treated with early administration of methylene blue, renal replacement therapy, and red blood cell transfusion: case report. *Medicine (Baltim)* 2014;**93**(9):e60.
201. Mahajan T, Merriman RC, Stone MJ. Kikuchi-Fujimoto disease (histiocytic necrotizing lymphadenitis): report of a case with other autoimmune manifestations. *Proceedings* 2007;**20**(2):149–51.
202. Loh JMR, Shafi H. Kikuchi-Fujimoto disease presenting after consumption of 'Miracle Mineral Solution' (sodium chlorite). *BMJ Case Rep* 2014. bcr2014205832.
203. Lardieri A, et al. Harmful effects of chlorine dioxide exposure. *Clin Toxicol* 2020:1–2.

204. Lansdown ABG. A pharmacological and toxicological profile of silver as an antimicrobial agent in medical devices. *Adv Pharmacol Sci* 2010;**2010**:910686.
205. Pandoli O, et al. Colloidal silver nanoparticles: an effective nano-filler material to prevent fungal proliferation in bamboo. *RSC Adv* 2016;**6**(100):98325–36.
206. Valsalam S, et al. Biosynthesis of silver and gold nanoparticles using Musa acuminata colla flower and its pharmaceutical activity against bacteria and anticancer efficacy. *J Photochem Photobiol B Biol* 2019;**201**:111670.
207. Sharma V, et al. Green synthesis of silver nanoparticles from medicinal plants and evaluation of their antiviral potential against chikungunya virus. *Appl Microbiol Biotechnol* 2019;**103**(2):881–91.
208. Salvioni L, et al. Negatively charged silver nanoparticles with potent antibacterial activity and reduced toxicity for pharmaceutical preparations. *Int J Nanomed* 2017;**12**:2517–30.
209. Wei L, et al. Silver nanoparticles: synthesis, properties, and therapeutic applications. *Drug Discov Today* 2015;**20**(5):595–601.
210. Galdiero S, et al. Silver nanoparticles as potential antiviral agents. *Molecules* 2011;**16**(10):8894–918.
211. Xiang D, et al. Inhibition of A/Human/Hubei/3/2005 (H3N2) influenza virus infection by silver nanoparticles in vitro and in vivo. *Int J Nanomed* 2013;**8**:4103–13.
212. Lv X, et al. Inhibitory effect of silver nanomaterials on transmissible virus-induced host cell infections. *Biomaterials* 2014;**35**(13):4195–203.
213. Zachar O. *Nanomedicine formulations for respiratory infections by inhalation delivery - covid-19 and beyond.* 2020.
214. Gurunathan S, et al. Antiviral potential of nanoparticles-can nanoparticles fight against coronaviruses? *Nanomaterials* 2020;**10**(9).
215. Vlachou E, et al. The safety of nanocrystalline silver dressings on burns: a study of systemic silver absorption. *Burns* 2007;**33**(8):979–85.
216. Mohan N, et al. Colloidal silver ingestion associated with leukocytoclastic vasculitis in an adolescent female. *Am J Case Rep* 2019;**20**:730–4.
217. Griffith RD, et al. Colloidal silver: dangerous and readily available. *JAMA Dermatol* 2015;**151**(6):667–8.
218. Stepien KM, et al. Unintentional silver intoxication following self-medication: an unusual case of corticobasal degeneration. *Ann Clin Biochem* 2009;**46**(6):520–2.
219. Montano E, et al. Colloidal silver induces cytoskeleton reorganization and E-cadherin recruitment at cell-cell contacts in HaCaT cells. *Pharmaceuticals* 2019;**12**(2):72.
220. Tobarran N, Hieger MA. Acute silver toxicity from colloidal silver overdose. *Am J Therapeut* 2020;**27**(6):e682–4.
221. Poon VKM, Burd A. In vitro cytotoxity of silver: implication for clinical wound care. *Burns* 2004;**30**(2):140–7.
222. ATSDR. *Medical Management Guidelines for Hydrogen Peroxide.* 28 September 2020. Available from: https://www.atsdr.cdc.gov/MMG/MMG.asp?id=304&tid=55.
223. Questionable methods of cancer management: hydrogen peroxide and other 'hyper-oxygenation' therapies. *CA Cancer J Clin* 1993;**43**(1):47–56.
224. CIMLAC. *Mitos y verdades sobre la COVID-19.* 28 September 2020. Available from: http://www.fasgo.org.ar/images/mitos-y-verdades-sobre-covid-redcimlac.pdf.
225. Pritchett S, Green D, Rossos P. Accidental ingestion of 35% hydrogen peroxide. *Canadian J Gastroenterol* 2007;**21**(10):665–7.
226. Kalcker AL. *Forbidden health: incurable was yesterday.* Amazon Digital Services LLC - KDP Print US; 2019.

227. Kalcker AL. *Pharmaceutical composition for treating acute intoxication, I.u.E. Schweizer Zentrum für wissenschaftliche Forschung*. 2017. Editor.
228. Kalcker AL. *Pharmaceutical composition for the treatment of internal inflammations, I.u.E. Schweizer Zentrum für wissenschaftliche Forschung*. 2017. Editor.
229. Kalcker AL. *Pharmaceutical composition for treating infectious diseases, I.u.E. Schweizer Zentrum für wissenschaftliche Forschung*. 2017. Editor.
230. Kalcker AL. *CDS Protocols*. 01 August 2020. Available from: https://andreaskalcker.com/en/protocolos/.
231. Medsafe: potentially fatal adverse effects with Miracle Mineral Solution. *React Wkly* 2020;**1808**(1). 3-3.
232. Lansdown AB. Silver in health care: antimicrobial effects and safety in use. *Curr Probl Dermatol* 2006;**33**:17−34.
233. Okan D, Woo K, Sibbald RG. So what if you are blue? Oral colloidal silver and argyria are out: safe dressings are in. *Adv Skin Wound Care* 2007;**20**(6):326−30.
234. Foldbjerg R, et al. PVP-coated silver nanoparticles and silver ions induce reactive oxygen species, apoptosis and necrosis in THP-1 monocytes. *Toxicol Lett* 2009;**190**(2):156−62.
235. The Lancet Infectious D. The COVID-19 infodemic. *Lancet Infect Dis* 2020;**20**(8). 875-875.
236. Doheny K. *FDA Warns of Silver, Other Bogus COVID-19 Cures*. 28 September 2020. Available from: https://www.webmd.com/lung/news/20200730/fda-warns-of-silver-other-bogus-covid-cures.
237. FDA. *Court Orders Halt to Sale of Silver Product fraudulently Touted as COVID-19 Cure*. 28 September 2020. Available from: https://www.fda.gov/inspections-compliance-enforcement-and-criminal-investigations/press-releases/court-orders-halt-sale-silver-product-fraudulently-touted-covid-19-cure.
238. Olding R, Hughes S. *Fake Utah Doc Peddled 'Ingestible Silver' as a Bogus COVID Cure: Feds*. 28 September 2020. Available from: https://www.thedailybeast.com/fake-utah-doctor-gordon-pedersen-peddled-ingestible-silver-as-a-covid-19-cure-feds-say?ref=scroll.
239. ChemicalSafetyFacts. *Hydrogen Peroxide*. 28 September 2020. Available from: https://www.chemicalsafetyfacts.org/hydrogen-peroxide/.
240. PubChem. *PubChem Compound Summary for CID 784, Hydrogen peroxide*. 28 September 2020. Available from: https://pubchem.ncbi.nlm.nih.gov/compound/Hydrogen-peroxide.
241. Kampf G, et al. Persistence of coronaviruses on inanimate surfaces and their inactivation with biocidal agents. *J Hosp Infect* 2020;**104**(3):246−51.
242. Alvarez-Risco A, et al. Health literacy, pharmaceutical care, and population health. In: A.-R.A., et al., editors. *Building sustainable cities*; 2020. p. 83−95.
243. Guerci B, et al. Lack of treatment persistence and treatment nonadherence as barriers to glycaemic control in patients with type 2 diabetes. *Diabetes Ther* 2019;**10**(2):437−49.
244. Zegeye A, et al. Prevalence and determinants of anti-tuberculosis treatment non-adherence in Ethiopia: a systematic review and meta-analysis. *PloS One* 2019;**14**(1):e0210422.
245. Wolfram C, Stahlberg E, Pfeiffer N. Patient-reported nonadherence with glaucoma therapy. *J Ocul Pharmacol Therapeut* 2019;**35**(4):223−8.
246. Ruzicka M, et al. Use of directly observed therapy to assess treatment adherence in patients with apparent treatment-resistant hypertension. *JAMA Intern Med* 2019;**179**(10):1433−4.

247. da Costa FA, van Mil JWF, Alvarez-Risco A. *The pharmacist guide to implementing pharmaceutical care*. Springer International Publishing; 2018.
248. Chen Q, et al. Characteristics of drug-related problems among hospitalized ischemic stroke patients in China. *Int J. Clin Pharm* 2020;**42**(4):1237–41.
249. Vikash L, et al. Evaluation of Drug utilization and Analysis of Drug related problems of Anti-Epileptic drugs at tertiary care teaching hospital. *Drug Invent Today* 2020;**14**(6):910–5.
250. Ma SN, Zaman Huri H, Yahya F. Drug-related problems in patients with rheumatoid arthritis. *Therapeut Clin Risk Manag* 2019;**15**:505–24.
251. Shareef J, et al. *Evaluating the effect of pharmacis" s delivered counseling on medication adherence and glycemic control in patients with diabetes mellitus*. 2016.
252. Lim PC, et al. Study investigating the impact of pharmacist involvement on the outcomes of diabetes medication therapy adherence program Malaysia. *Pak J Pharm Sci* 2016;**29**(2):595–601.
253. Shao H, et al. Effect of pharmaceutical care on clinical outcomes of outpatients with type 2 diabetes mellitus. *Pat Pref Adher* 2017;**11**:897–903.
254. Alvarez-Risco A, Quiroz Delgado D, Del-Aguila-Arcentales S. Pharmaceutical care in hypertension patients in a Peruvian hospital. *Indian J Public Health Res Devel* 2016;**7**:198–202.
255. Omboni S, Tenti M, Coronetti C. Physician–pharmacist collaborative practice and telehealth may transform hypertension management. *J Hum Hypertens* 2019;**33**(3):177–87.
256. Wen S, et al. Pharmaceutical care in Asia. In: Alves da Costa F, Van Mil JWF, Alvarez-Risco A, editors. *The pharmacist guide to implementing pharmaceutical care*. Springer; 2019. p. 191–7.
257. Wong YX, et al. Perception of community pharmacists in Malaysia about mental healthcare and barriers to providing pharmaceutical care services to patients with mental disorders. *Community Ment Health J* 2020;**56**(1):88–98.
258. Zhang SX, et al. Unprecedented disruption of lives and work: health, distress and life satisfaction of working adults in China one month into the COVID-19 outbreak. *Psychiatr Res* 2020;**288**:112958.
259. Gonzales-Tamayo L, Arevalo-Oropeza M, Yáñez JA. COVID-19 physician deaths in Peru: a result of an underfunded and fragmented healthcare system. *F1000Res* 2020. Available at: SSRN: Submitted for publication, https://ssrn.com/abstract=3676849.
260. Gonzales-Tamayo L, et al. Primary immunodeficiency during COVID-19: can colchicine play a role in this overlooked population? *F1000Res* 2020. Available at: SSRN: Submitted for publication, https://ssrn.com/abstract=3676857.
261. Yáñez JA, Alvarez-Risco A, Delgado-Zegarra J. *Rapid Response: Clearing the path for COVID-19 in Peru? The decision of supervised walks for children and adolescents*. 03 June 2020. Available from: https://www.bmj.com/content/369/bmj.m1918/rr-9.
262. Cadogan CA, Hughes CM. On the frontline against COVID-19: community pharmacists' contribution during a public health crisis. *Res Soc Adm Pharm* 2020;**17**(1):2032–5.
263. Li H, et al. Fighting against COVID-19: innovative strategies for clinical pharmacists. *Res Soc Adm Pharm* 2020;**17**(1):1813–8.
264. Song Z, et al. Hospital pharmacists' pharmaceutical care for hospitalized patients with COVID-19: recommendations and guidance from clinical experience. *Res Soc Adm Pharm* 2020;**17**(1):2027–31.
265. Carico Jr RR, Sheppard J, Thomas CB. Community pharmacists and communication in the time of COVID-19: applying the health belief model. *Res Social Adm Pharm* 2021;**17**(1):1984–7.

266. Erku DA, et al. When fear and misinformation go viral: pharmacists' role in deterring medication misinformation during the 'infodemic' surrounding COVID-19. *Res Soc Adm Pharm* 2021;**17**(1):1954–63.
267. Adunlin G, Murphy PZ, Manis M. COVID-19: how can rural community pharmacies respond to the outbreak? *J Rural Health* 2021;**37**(1):153–5.
268. Kamimura A, et al. Perceptions of mental health and mental health services among college students in Vietnam and the United States. *Asian J Psychiatr* 2018;**37**:15–9.
269. Yemm KE, Arnall JR, Cowgill NA. Necessity of pharmacist-driven nonprescription telehealth consult services in the era of COVID-19. *Am J Health Syst Pharm* 2020;**77**(15):1188.
270. Quispe-Cañari JF, et al. Self-medication practices during the COVID-19 pandemic among the adult population in Peru: a cross-sectional survey. *Saudi Pharmaceut J* 2021;**29**(1):1–11.
271. Chiou WC, Hsu MS, Chen YT, Yang JM, Tsay YG, Huang HC, Huang C. Repurposing existing drugs: identification of SARS-CoV-2 3C-like protease inhibitors. *J Enzyme Inhib Med Chem,* 2021;**36**(1):147–53.
272. Sarkar I, Sen A. In silico screening predicts common cold drug Dextromethorphan along with Prednisolone and Dexamethasone can be effective against novel Coronavirus disease (COVID-19). *J Biomol Struct Dyn,* 2020:1–5. https://doi.org/10.1080/07391102.2020.1850528.
273. Morán Blanco JI, Alvarenga Bonilla JA, Homma S, Suzuki K, Fremont-Smith P, Villar Gómez de Las Heras K. Antihistamines and azithromycin as a treatment for COVID-19 on primary health care – A retrospective observational study in elderly patients. *Pulm Pharmacol Ther,* 2021;**67**:101989. https://doi.org/10.1016/j.pupt.2021.101989.
274. Lv Y, Wang S, Liang P, Wang Y, Zhang X, Jia Q, He L. Screening and evaluation of anti-SARS-CoV-2 components from Ephedra sinica by ACE2/CMC-HPLC-IT-TOF-MS approach. *Anal Bioanal Chem,* 2021;**413**(11):2995–3004. https://doi.org/10.1007/s00216-021-03233-7.
275. Addar A, Al Fraidi O, Nazer A, Althonayan N, Ghazwani Y. Priapism for 10 days in a patient with SARS-CoV-2 pneumonia: a case report. *J Surg Case Rep,* 2021;**2021**(4):rjab020. https://doi.org/10.1093/jscr/rjab020.
276. Pasquereau S, Nehme Z, Haidar Ahmad S, Daouad F, Van Assche J, Wallet C, Herbein G. Resveratrol inhibits HCoV-229E and SARS-CoV-2 coronavirus replication in vitro. *Viruses,* 2021;**13**(2). https://doi.org/10.3390/v13020354.
277. Villena-Tejada M, Vera-Ferchau I, Cardona-Rivero A, Zamalloa-Cornejo R, Quispe-Florez M, Frisancho-Triveño Z, Abarca-Melendez RC, Alvarez-Sucari SG, Mejia CR, Yáñez JA. Use of medicinal plants for COVID-19 prevention and respiratory symptom treatment during the pandemic in Cusco, Peru: a cross-sectional survey. *PLOS ONE* 2021. Submitted for publication.
278. Wen S, et al. Pharmaceutical care in Asia. In: Alves da Costa F, Van Mil JWF, Alvarez-Risco A, editors. The pharmacist guide to implementing pharmaceutical care. Springer; 2019:191–7.
279. Wong YX, et al. Perception of community pharmacists in Malaysia about mental healthcare and barriers to providing pharmaceutical care services to patients with mental disorders. *Community Ment Health J* 2020;**56**(1):88–98.
280. Hess K, Bach A, Won K, Seed SM. Community Pharmacists Roles During the COVID-19 pandemic. *J Pharm Pract,* 2020. 897190020980626, https://doi.org/10.1177/0897190020980626.
281. Agomo CO, Ogunleye J, Portlock J. Enhancing the public health role of community pharmacists – a qualitative research utilising the theoretical domains framework. Innovations in pharmacy, 2020;**11**(4). 10.24926/iip.v2491 1i24924.23393, https://doi.org/10.24926/iip.v11i4.3393.

SECTION IV

Challenges and opportunities faced due to COVID-19

CHAPTER 15

Addressing the global challenge of access to supplies during COVID-19: mask reuse and local production of alcohol-based hand rub

Alexandra Peters, Chloé Guitart, Didier Pittet
Infection Control Programme, University of Geneva Hospitals and Faculty of Medicine, Geneva, Switzerland

15.1 Introduction/background

In light of the current COVID-19 pandemic, many health care facilities worldwide are lacking a steady source of supplies. Access to alcohol-based hand rub (ABHR), and personal protective equipment (PPE), especially masks, is a cornerstone for infection prevention and control (IPC) practices during a pandemic. The pandemic caused major disruptions to the global shipping infrastructure, when coupled with a drastic, almost simultaneous global increase in demand for these products. This emergency situation warrants the taking of extraordinary measures to minimize the negative health impact of a lack of supplies; numerous health care facilities found themselves without enough appropriate equipment to safely take care of their patients. Two areas where new approaches are rising to meet the challenges that COVID-19 has posed for global supply chains are the local production of ABHR and masks, and their decontamination and reuse. The crisis has led to new supply chains and changes to the distribution in manufacturing systems.[1]

Many countries have been working together with local industries to make it easier to produce the necessary supplies locally. This includes measures such as capital investment, help with the procurement of raw materials and distribution, as well as help from a legal standpoint. Because medical equipment and alcohol are two types of products that are rather heavily regulated, countries are trying to deregulate these laws and at least

Environmental Management of COVID-19
ISBN 978-0-323-85780-2
https://doi.org/10.1016/B978-0-323-85780-2.00008-1

temporarily modify them to allow for a more efficient production during the pandemic.[2–4]

Local production for ABHR has already been implemented in a number of places around the globe, following the World Health Organization (WHO) guidelines.[5] This is usually done in order to help make hand rub more accessible in low- and middle-resource settings. During the first few months of the pandemic, we observed this model of local production shifting to high-income countries (HICs). Though local production started at a small scale in hospitals and pharmacies, production at an industrial level soon became essential to assure sufficient quality and quantity of ABHR. Accordingly, certain perfume, cosmetics, and alcohol manufacturers shifted production to make hand rub at a reasonable price, mainly for health care settings.[1–4] The ABHR formulation, manufacturing process, quality control, and challenges to local production are four themes that will be explored in more detail later in the chapter.

Local production of masks also increased globally in response to the initial widespread shortage of PPE, and the realization that most of the world was principally dependent on China for its masks. However, this took time to implement, and even though the initial crisis and mask shortage is over, many low resource countries still do not have the infrastructure to make local production of masks a viable option.[6] Therefore, during the shortage, the decontamination and reuse of health care workers' N95/FFP2 face masks seemed a promising solution and was increasingly studied by institutions all over the world. A significant number of studies in the last six months have focused on this issue, and this chapter will review how such a program would have to be implemented, what factors must be taken into account if attempting to reprocess masks locally, and what methods seem to work best.[7] It is important to state that decontaminating respirators should only be done in emergency situations where better options are not available.

This chapter will explore both of these phenomena. First, because access to sufficient and high-quality supplies still is an important challenge faced in this pandemic, and secondly because they can teach us lessons for future emergencies. In the age of globalization, global pandemics will undoubtedly become a more regular occurrence. A growing and highly mobile human population encroaching on ecosystems that they have never been in contact with, and changes in land use, are known to influence the frequency of outbreak events of emerging or reemerging pathogens as well as the speed of their spread and distance they cover.[8,9]

For governments and politicians, preparing for pandemics is a double-edged sword. On the one hand, if a pandemic does not occur, preparing a country for one is seen as a waste of public money; if it does arrive, then whatever preparation was put in place will be seen as inadequate. Global supply chains involve huge infrastructures and multiple high-level stakeholders that are often entrenched in legal and political structures. It is therefore important for national and global health systems and supply chain systems to try to be exactly what they are traditionally not: agile and adaptable to whatever global emergency may present itself. To work toward this technically unachievable goal, studying what measures can be implemented constructively in the face of supply chain failure is valuable.[6–9]

Both local production and reprocessing of N95 masks have been implemented in numerous areas around the world. By analyzing how the implementation of these measures worked and the challenges that were faced, we can analyze what measures worked in which geographic regions, how they were implemented, and the aspects that determined their relative success or failure. Studying this means that not only will countries or regions be better prepared if faced with the same challenges, but also that they can undoubtedly apply some of the broader aspects of their experiences to address novel future challenges that affect the global supply chain.[7–9]

15.2 Evidence in the literature of hand hygiene and masking for preventing transmission of SARS-CoV-2

The main source of transmission of severe acute respiratory syndrome coronavirus 2 (SARS-Cov-2) is droplets of symptomatic or pauci-symptomatic people directly or indirectly by hands and environmental surfaces.[10–12] That is why both wearing masks and performing hand hygiene are crucial for controlling the pandemic. To limit the virus propagation, public health institutions recommend basic prevention precautions including social distancing, wearing a mask when people are in close contact, hand hygiene, and environmental cleaning.[13–15]

Masks are a key tool to slowing the pandemic both in the community and in health care settings.[16,17] There is now sufficient evidence in the literature (including a systematic review published in the Lancet) that masks work well for stopping COVID-19.[18,19] When someone coughs or sneezes, they emit the virus particles inside droplets of mucus of different sizes. Using masks directly decreases the virus droplet numbers that can reach people and will prevent the droplets produced by coughing or sneezing from entering the nose and mouth of another person, or from contaminating a nearby surface. At the very least, a mask will greatly reduce the number of viruses that are able to reach a nearby person. This will not only help keep people from being infected but will also increase the probability that they only get mild symptoms, since the severity of COVID-19 infection is also dependent on the viral load a person is exposed to.[20]

A randomized case-control study during the 2009 influenza pandemic determined the value of hand hygiene for the prevention of influenza or influenza-like illnesses that are also transmitted by droplets.[21] It concluded that "the face mask with hand hygiene group showed a significant reduction in respiratory illness compared to the group with the mask only."[21] Hand hygiene is a crucial measure to prevent the cross-transmission as SARS-CoV viruses can survive on surfaces for prolonged periods.[22]

Hand hygiene may be performed by two different methods: handwashing with soap and water or hand rubbing with an ABHR. The latter is preferred in most clinical situations due to its greater microbiological efficacy, better skin tolerance, more practical application, and possible availability in almost all situations especially with the possibility of a pocket-sized dispenser.[23,24] Enveloped viruses, including SARS-CoV-2, have the lowest level of resistance to ABHR compared to other microorganisms.[25] This means that they are generally the easiest type of virus to kill on surfaces and hands, and die very quickly while in contact with alcohol.[26–29] The evidence of the efficacy of various disinfectants against SARS-CoV-2 is not easy to prove mainly because it is a new virus and microbiological efficacy of in-vitro research methods are more difficult to test on viruses compared to bacteria.[28,30]

ABHR remains indispensable for preventing the spread of infections.[31] It is well known that ABHR can be successfully used as an effective infection prevention measure during viral outbreaks. Zika, Ebola, SARS-CoV-1, Middle East respiratory syndrome coronavirus (MERS), and other enveloped viruses are all efficiently killed by the two WHO ABHR formulations.[26] A recent study shows that both the commercial and WHO formulations of ABHR can effectively inactivate SARS-CoV-2,[32] and that ethanol and 2-propanol at a concentration of >30% inactivated the virus in 30 s. This provides strong evidence supporting the use of ABHR in the context of the COVID-19 pandemic, and it is recommended by the WHO, Centers for Disease Prevention and Control (CDC), and European Center for Disease Prevention and Control.[13,33–35]

15.3 The global situation since the beginning of the pandemic

The COVID-19 pandemic resulted in significant overconsumption of PPE, including masks, ABHR, and certain medical devices. The demand for some protective equipment has increased by several thousand percent. In the United States, ABHR sales soared 470% in the first week of March 2020, compared to the same period in 2019. Likewise, in Italy, sales of ABHR in supermarkets increased by 561% in the first three weeks of the pandemic compared to the year before.[36] Around the world, many supermarkets, pharmacies, hospitals, and other health care facilities all ran out of their stock of ABHR. In Italy, these shortages have been reported to have contributed to the high burden of infections and deaths among hospital staff.[37] It is logical to imagine that this also happened in other areas where health care facilities did not have appropriate supplies.

A lack of foresight coupled with the dramatic increase in need have led to the organizational or even the collapse of supply chains. Lockdowns further exacerbated the problem, causing production shutdowns and disruptions to transportation and logistics.[38] This is especially pertinent because the countries most affected by COVID-19 at the start of the pandemic are among the countries most involved in global trade. The G7 countries, which include China, Japan, the United States, France, Italy, United Kingdom, and Germany are responsible for 65% of world manufacturing and 41% of world manufacturing exports.[38]

At the beginning of the pandemic, being able to procure supplies was only the first major challenge; the next one was being able to pay for them. In the absence of a national or international price control program, the prices of PPE skyrocketed.[39] According to the WHO, the prices of surgical masks increased sixfold, tripled for N95 respirators, and doubled for gowns.[40] Market manipulation was a global phenomenon, with stocks of supplies often being sold to the highest bidder, and stories of governments buying supplies off the tarmac destined for other countries.[41] This competition disproportionately benefited economically powerful countries and could potentially leave those with smaller budgets in need of equipment, thus further widening inequalities. This has resulted in approaches such as accepting donations of homemade PPE and reusing materials. It goes without saying that these strategies are far from ideal; if the quality of PPE is inadequate, or a protocol for reuse is not well implemented, it jeopardizes the safety of the people that rely on it to protect them.[42]

15.4 Strategies to adapt

Never since the beginning of the globalized world economy have so many countries needed the same resources at the same time. For both masks and ABHR, various countries have been loosening restrictions on manufacturing to encourage local production and meet the demand for these.[2]

For ABHR, the options for addressing shortages are limited to ameliorating supply chain/distribution networks and increasing local production. The formula can be made principally from either ethanol or isopropyl alcohols, but there is not really a way around their availability, as the composition is quite limited by the specific molecules needed. During a pandemic, it is crucial for local governments to reinforce the already existing ABHR production and distribution networks. If these measures are not adequate to cover the increasing demand, it is essential for them to create new networks by working with companies that are already well equipped for production of similar products at large scale. Small-scale production in pharmacies or health care facilities (HCF)s can also be implemented. For masks, the fabric used is not as limited, in the sense that different materials can function as a mask, and therefore a number of materials can be tested and considered.[4]

Local regulations have to be adapted to the extraordinary situation in order to facilitate local production. Japan is a good example: in order to facilitate access to priority equipment, the government established cooperation between private companies for PPE such as surgical masks, ABHR, and medical devices including ventilators. It has put in place facilitative measures such as subsidies for investments in the factories producing the masks and PPE, support for the purchase of raw materials and the deregulation of alcohol consumption. The Japanese government was thus able to provide their hospitals with the necessary PPE, including 45 million surgical masks.[3]

In Switzerland, the solidarity of many companies and individuals brought about the implementation of a national plan for local ABHR production increased the availability of ABHR and avoided major shortages, as well as excessive pricing.[43] In certain countries including France and United States, they passed specific laws during the crisis to cap the price of masks and ABHR in order to maintain a reasonable cost of these essential products.[44,45]

In response to the challenge of inequality faced to this global shortage, the WHO and several partners have set up a procurement portal through which countries can order PPE.[46] Although members of this procurement portal can advise on prioritization between countries, there is no formal international mechanism for allocating PPE to regions that are most in need. Ideally countries would prepare plans for how to handle shortage, otherwise situations where some countries have no access to supplies and others have a surplus, are more likely.[47]

15.5 ABHR local production

15.5.1 WHO and the global implementation of ABHR

Studies have shown that as many as 50%—70% of infections are transmitted through hands due to a lack of proper hand hygiene during daily patient care.[48] Proper hand hygiene can be achieved by using a product that has proven antimicrobial efficacy and when users performed it at the right time and in the right way. According to the WHO multimodal strategy for hand hygiene improvement,[49] five elements are considered essential for achieving best practices. The first of these is "system change," i.e., the implementation of a system making ABHR available wherever it is needed in sufficient quantity and quality.[24,49] Over the past decades, this strategy has been largely implemented all around the world in order to enhance patient safety in HCF.[23,50]

Commercially available ABHR is produced mainly in HICs, and "system change" has been fully implemented in all HCF in these countries. However, in low- and middle-income countries (LMICs), these products are seldom commercially available and their cost can exceed the cost of ABHR in HICs mainly due to transportation and importation fees, making it completely unaffordable.[51]

To overcome this problem and to facilitate the development of a global approach to hand hygiene improvement in health care, the WHO developed two alcohol-based formulations for hand rubbing, which were tested and validated for antimicrobial efficacy and skin tolerability.[23,52,53] These were made patent-free to facilitate production and affordability to all institutions globally. These ABHR formulations are so vital to health care that they are classified as Essential Medicines by the WHO.[54] Access to ABHR is therefore also a central component of the WHO's overall objective of universal health coverage.[55,56]

WHO formulations of ABHR are recommended by the WHO Guidelines on Hand Hygiene in Health Care settings,[23] and the specific methodology adapted for their local production were tested in pilot sites mostly located in LMICs. Alcohol for hand rub can be manufactured from a distillation process using the byproducts of crops such as manioc, bananas, corn, and sugarcane, many of which are cultivated in the majority of LMICs.[5] The WHO formulation for ABHR, little known to HICs before 2020, has now become very popular. The views of the e-learning video developed by the University Hospitals of Geneva pharmacy, which details the process of the production of the WHO formulation production increased from 5000 to 52,000 since the start of the epidemic (as of September 9, 2020).[57]

15.5.2 ABHR formulation

The formulation of ABHR product is important for the microbial efficacy, tolerance, and acceptability of users.[23,58] For safety and practical reasons, it is better to keep it simple and to use well-known active ingredients and excipients. The WHO formulation is made up of alcohol (active ingredient), glycerol (emollient), peroxide hydrogen (sporicide), and water. All these ingredients are ubiquitous, inexpensive, and have been well-known for many years.[5,23]

The active ingredient is the alcohol. The efficacy of alcohol is affected by type of alcohol, its concentration, contact time, and skin conditions.[23] Ethanol, isopropanol, n-propanol, or a combination of both are commonly used as active agents.[59] They are active against bacteria, all enveloped and most nonenveloped viruses.[60] Contrary to other antiseptics, alcohol used in this way does not have the potential to develop bacterial resistance.[27,61] Though there is limited evidence for bacterial tolerance of certain strains after long-term exposure to low concentrations of alcohol, there is no evidence of any resistance to any alcohol at concentrations of anywhere near 60%, which is the minimum required concentration for ABHR.[62–64] The reason is that alcohol's mechanism of action is aspecific. Alcohols are rapidly germicidal on hands through protein denaturation.[25,60,65] The effective and safe concentration of alcohol for disinfection is between 60% and 95% (v/v) according to the CDC and the WHO,[14,59,66] including for use against SARS-CoV-2. The WHO recommended to use a concentration of ethanol at 80% (v/v) (formulation 1) or a concentration of isopropanol at 75% (v/v) (formulation 2). Recent studies showed that increasing concentration of alcohol in ABHR by 5% (v/v) could reduce the contact time necessary to be effective.[67]

Adding an emollient to the formulation is crucial to improve skin tolerability, but it can also reduce the effectiveness of the alcohol, thus an adequate balance should be found between the two to ensure both efficacy and acceptability.[68] Alcohols are not effective against spores, making a sporicidal agent necessary to avoid the presence of viable bacterial spores in the final product.[69] The solution could be a filtered method (0.22 μm) during the filling bottle process, or a sporicidal such as hydrogen peroxide could be added at low concentration, which is active against spores after 72 h.[70] Other potential additives besides the emollients and the aforementioned sporicidal agents include viscosity enhancers, buffers, preservatives, colorants, and fragrances, though not all products have all of these.[5,59] Many nonessential additives can influence the antimicrobial activity, which is why the efficacy of the final product must be carefully evaluated for every formulation. Though nonessential additives are often used to enhance the acceptability of the product, they can decrease the tolerability and increase the risk of allergenic properties.[59,68–70]

15.5.3 Manufacturing process of ABHR

Industrial manufacturing is the best way to provide ABHR in quality and quantity. On the one hand, for safety reasons, it is essential to have premises suitable for the preparation of alcohol and storage of the finished product. One needs a spacious facility that is ventilated and not accessible to outsiders. The automated production chain of large-scale industrial production means more quick and efficient filling of bottles and printing and application of labels with less staff necessary. Still, certain situations make local, small-scale production necessary in pharmacies, hospitals, or others strategic places which has been the case in sanitary crisis situations.[71] This type of emergency manufacturing can adversely affect the quality of the final product.[36] The selected components and the manufacturing process are the variables that ultimately influence the quality of the final product, so production and quality control are crucial.

All parameters of such production including the facility, the operators, the documentation, the production process and quality control must be in accordance with good manufacturing practices (GMP).[72] The GMP are the minimum standards that a product manufacturer must meet in their production processes. The GMP limits the risks of cross-contamination of products (by an internal or external contaminant) and the risks of confusion, in particular with regard to labeling and the identification of ingredients.

These standards insist that hygiene and organizational practices must be implemented at all levels. The level of GMP depends on local regulations and volume of hand rub produced.[73]

Because the expected efficacy of the preparation will be developed only if their quality and quantity corresponds to the procedure, it is important to maintain a high standard of quality control of both the raw materials, especially active ingredients, and of the final product.[73] Regarding the concentration and dosage of alcohol, gas chromatography is the gold standard for production, but simpler procedures such as alcoholmeters can be used as recommended by the WHO.[5] According to the WHO guidelines of local ABHR production, simplified methods could be implemented to production and quality control processes in difficult contexts such as a pandemic. Other active ingredients such as hydrogen peroxide should also be analyzed in the final preparation.[5,57] In theory, even "inactive" components, such as emollients or perfumes, should be controlled since the effectiveness of ABHR is also dependent on these components.[5,57]

An e-learning about the production process at a local level is available freely online to increase access.[57] No guide or training for the ABHR manufacturing process exists at this time.

15.5.4 ABHR production challenges

Although analysis of the feasibility testing, quality control, and costs of local ABHR production have led to very encouraging results, particularly in terms of cost reduction compared to commercial products in LMICs, there are still challenges.[5,51] Some of these have always existed and are inherent to the process of local production, especially in LMICs, where there are both common and widespread. One of the main challenges lies in the procurement of the main ingredients and containers as all the necessary raw materials are not produced locally in most cases. There are also a number of constraints link to distribution, due to the flammability of alcohols. The flash points of ethanol 80% (v/v) and of isopropanol 75% (v/v) are 17.5 and 19°C, respectively, meaning that adapted transportation methods are necessary to move alcohols or ABHR between locations.[5] Furthermore, disruptions of regular budget allocations for buying ABHR are frequent in LMICs. Governments, institutions, HCWs, or patients themselves could be the buyer. This question remains unresolved in low resources settings, disrupting the procurement of ABHR needed for the continuity of quality care.[51,74]

With the arrival of the pandemic, these issues, once confined to LMICs, became a global problem. Although low resources settings faced the brunt of challenges and shortages, COVID-19 exacerbated the existing issues with local ABHR production for everyone. Even HICs were faced with difficulty in supplying alcohol and appropriate containers that were both easily transportable and compatible with alcohol. Help often came from the private sector; cosmetics, perfume, and beverage packaging helped in production, packaging, and distribution of ABHR. The solidarity and creativity shown by industries during this time helped to successfully compensate for this lack of ABHR, and ensure that remains available at reasonable prices.[43,75]

Regulation of alcohol sales and distribution is also often problematic since these laws are often not adapted to this extraordinary situation. In certain European countries, such as in Switzerland, WHO formulations do not meet the standards for biocidal products for use in HCFs.[76] This is due to the fact that commercially products are usually used in these countries. Due to the high standards of European norms, the WHO formulation in its current incarnation does not pass the European norms, and thus cannot be sold commercially to health care facilities within Europe (publication EN and WHO formulation). The same challenge regarding the WHO formulation is also present in other HICs. However, translating the WHO formulations from v/v to w/w and reducing its glycerol content to 0.5% resolves this issue. Because alcohol consumption is harmful to human health, some countries including Japan or the United States implemented local regulations to limit the importation, distribution, and sale of the substance over the country without distinguishing between drinkable alcohol and pharmaceuticals.[3,77] If this regulatory aspect is not revised during a pandemic, regulations can block the production and prevent the use of the final product.

15.6 Masks/filtering face piece respirators

It is important to note that we will mostly be covering surgical and N95 face masks in this chapter. Cloth face masks were often recommended for use in the community as a stopgap during the time period where there was a drastic lack of supply until surgical mask importation or production could increase to the point where there were enough for use in both health care facilities and in the community.[4]

Surgical masks and even some cloth masks are able to stop most of the droplets in which the viruses reside, but N95 masks are the only ones that can actually physically stop the SARS-CoV-2 virion. Mask filters collect particles through a variety of mechanisms: inertial impaction, diffusion, interception (which some authors argue is a form of diffusion), and electrostatic attraction (Fig. 15.1).[78–80] Disposable surgical masks are recommended for the general population because they are less expensive, and easier and more comfortable to wear. Because COVID-19 is mostly transmitted by hands or droplets, surgical masks are sufficient for protection in the community, and in most clinical activities in health care settings. N95 masks are recommended for situations when health care workers are taking care of patients known to be infected with COVID, especially if they are performing procedures such as tracheal aspirations that could cause the aerosolization of the virus. N95 masks are designed to create a seal around the face, and people need training to wear them correctly. Furthermore, because they are uncomfortable, a person who has not been trained on their use or in hand hygiene will be more likely to touch or adjust the mask without thinking, potentially contaminating it and rendering it useless.[4,78–80]

Before exploring the possibility of reusing masks, it is important to state that no matter the method of decontamination, this process is not recommended if other options are available. Not only do the decontamination methods have very little evidence for clinical efficacy and safety outside of a laboratory setting, but as the process is relatively complex, there are many opportunities for mistakes to be made.[78–80]

15.7 Logistics

Although much attention has recently been given in the literature to the different methods of mask decontamination, there is little in the way of recommendations concerning the collection and redistribution systems for those masks. If the masks cannot be collected, stored, reprocessed, and returned in a safe and orderly way, even a potentially excellent decontamination method will fail. It is important to note that each method of decontamination is quite labor intensive and involves numerous people besides the health care workers that need to wear the mask.[40,78–81]

Because masks cannot be washed or disinfected before being reprocessed, the safety of the process cannot be guaranteed. Therefore, it is important that the collection and distribution of the masks is individualized,

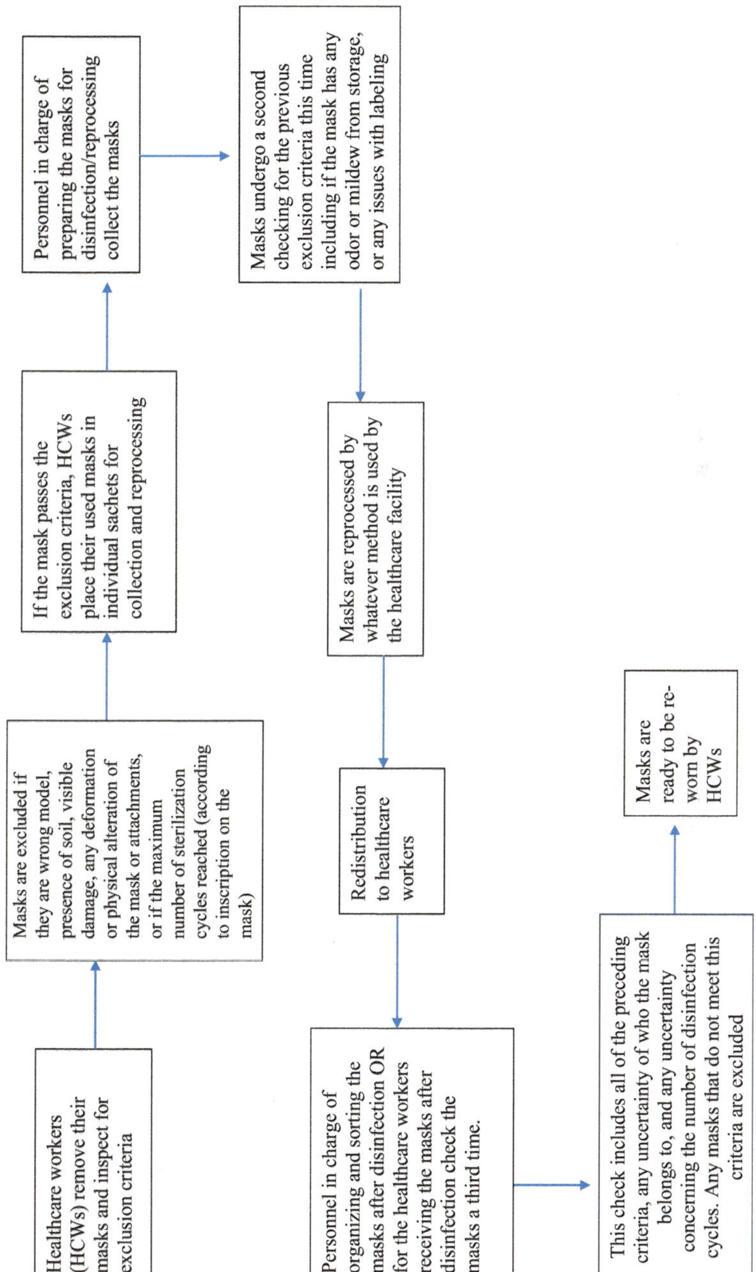

Figure 15.1 Flow chart for the logistics of mask reprocessing.

meaning that a specific mask must go back to the person that wore it the first time. Even if the decontamination process is 100% effective on SARS-CoV-2, this is a very easy virus to destroy, and other more tenacious microorganisms such as spores or mycobacteria might not be as easy to kill. This is especially an issue if the masks are soiled, and though they would be inspected for any soil before reprocessing, a visual inspection will inevitably miss some residues.[80–83]

Nonetheless, this visual inspection is the starting point of any mask reprocessing, and only eligible masks should undergo decontamination. Any visible soil or wetness (including mucus or lipstick stains), damage to the mask's integrity, or problem with the elastic bands mean that the mask should automatically be discarded.[7,40,81] It is also important that each mask is reused by the same health care worker, if one does not know who a mask belongs to, it should be discarded as well.[40,82] This will prevent and transmission of microorganisms between the health care workers if the decontamination procedure fails, or in the case that the method not effective enough to kill the microorganisms that are more tenacious than SARS-CoV-2 (Fig. 15.2).[83]

15.7.1 Decontamination and reuse

The minimum requirement for a method of decontamination is disinfection, meaning a minimum of a 3-log reduction in the pathogen tested.[84] If a method reaches sterilization, this means that a 6-log reduction of the microorganism has been reached. It is crucial that the masks are not damaged by the decontamination process. If the physical integrity, filtration capacity, or the electrostatic charge of the masks (which greatly increases the mechanical filtration capacity) is damaged, the mask is rendered useless.[85]

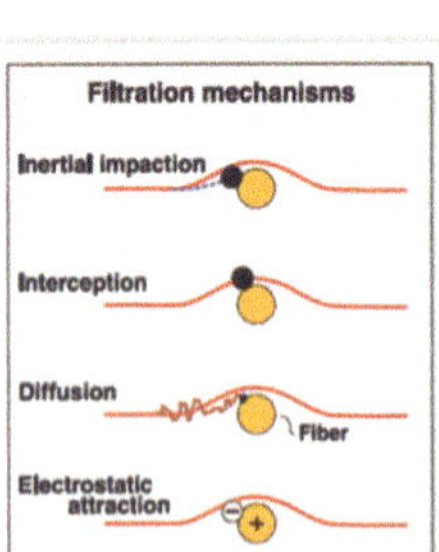

- **Inertial impaction:** With this mechanism, particles having too much inertia due to size or mass cannot follow the airstream as it is diverted around a filter fiber. This mechanism is responsible for collecting larger particles.
- **Interception:** As particles pass close to a filter fiber, they may be intercepted by the fiber. Again, this mechanism is responsible for collecting larger particles.
- **Diffusion:** Small particles are constantly bombarded by air molecules, which causes them to deviate from the airstream and come into contact with a filter fiber. This mechanism is responsible for collecting smaller particles.
- **Electrostatic attraction:** Oppositely charged particles are attracted to a charged fiber. This collection mechanism does not favor a certain particle size.

Figure 15.2 How masks work. *(Source: N95 respirators and surgical masks | Blogs | CDC. Available from: https://blogs.cdc.gov/niosh-science-blog/2009/10/14/n95/.https://blogs.cdc.gov/niosh-science-blog/2009/10/14/n95/).*

In order to be recommended for an emergency situation such as a shortage of masks, decontamination methods need to be both microbiologically effective (with a minimum of 3-log reduction in microbial contamination), and be able to be performed without any physical damage to the masks themselves. Not all methods that have been tested were able to reach these standards. Notably, liquids such as ethanol, sodium hypochlorite, detergent wipes, and benzalkonium chloride wipes, as well as UVA rays performed insufficiently.[86–89] There are some methods of decontamination that are able to effectuate disinfection and sterilization. It is important to note that the success of an experiment may vary depending on what model of mask they are testing on. When reviewing the literature, we found quite a high degree of variation, even within the same study if the models of the masks tested were different. Gasses such as ethylene oxide, gaseous hydrogen peroxide (with or without peracetic acid), peracetic acid dry-fogging system, and ozone all performed quite well,[90] although ozone tended to deform the rubber band that attaches the face mask, which potentially disqualifies that method for some masks.[91–93] Heat-based disinfection includes dry heat, moist heat, microwave-generated moist heat, and steam (under pressure). In terms of physical degradation of the masks, all of these methods performed well in terms of fit and filtration. Microbiologically, these methods varied in efficacy, with dry heat being the least effective.[90] Still, because these methods are among the simplest to implement, they will no doubt be important for use in low resource environments. Ultraviolet germicidal irradiation (UVGI) using UVC rays could also be quite effective in some experiments, but the results seemed to be less consistent.[90]

It is important to note that most of these experiments were not performed on SARS-CoV-2, which is probably good, since the virus is relatively easy to kill. One could argue that a number of the methods that performed well on microorganisms that are far more difficult to kill than SARS-CoV-2 inherently allows for a margin of error. This is meant in the sense that even if the method performed significantly less well in clinical settings than in laboratory settings, there is still a high chance of it being effective against SARS-CoV-2.[84–90]

When analyzing which method of decontamination to implement, it is important for institutions to review a number of factors. Each of the methods have different requirements in terms of the microbiological efficacy of the method, the time needed to complete the decontamination, the cost, risks for staff, the complexity of the procedure, the infrastructure or equipment needed, and whether the decontamination method can be used repeatedly on the same mask without degrading the material. Some of these methods require very specialized machinery or can be toxic if not deployed properly or if the masks are not allowed to off-gas. Other methods are incredibly simple, with some experiments even seeing how masks could be decontaminated effectively with widely available equipment such as microwaves and electric rice cookers. Ultimately, if an institution is forced to resort to decontaminating N95 masks, the method that they should use is the one that they can implement correctly and use safely.

15.7.2 Local production

Before resorting to decontamination, many countries have been working on increasing their own capacity to produce masks. Although the local production of face masks is now widespread, it is a very new phenomenon, and thus has not been extensively discussed in the literature. One important difficulty in the process is the nonwoven materials from which the mask is manufactured; the most popular material is melt-blown polypropylene. This material, which consists of very thin plastic nanofibers, is made by complex machines that cost almost €4 million.[87] The quality of the masks depends, in part, on the precision of these machines. One crucial part of local production is to find cheaper viable alternatives to these machines.[84–90]

Electrospinning is another production method that can produce the nanofibers needed for face masks, and it is generally a more simple process, though productivity is generally higher with melt blowing.[88] Polymers with higher molecular weights are more likely to make thicker fibers with electrospinning than with melt blowing. Generally, either additives or injected fluids are needed for both production methods.[88] One study looked at simplifying this process to increase the local production of polypropylene masks by modifying cotton candy machines to produce nanofilaments of polypropylene. It also explored the possibility of using a car battery as a source of electricity.[78]

Locally produced N95 masks are currently being produced, used, and tested. One study, which looked at a number of different criteria including breathing resistance, virus filtration efficacy, NaCl particle penetration, and fit, analyzed locally produced masks made of various layers different types of nonwoven polypropylene fabric. They found that the locally produced masks were able to achieve similar filtering capacities as FFP2/N95 masks, which is encouraging for local production.[93]

15.7.3 Other methods and materials

Another strategy that is not discussed here in detail is the extended use of surgical/N95 face masks instead of disposing them as often as recommended.[7,94] This strategy is more difficult to assess, because of how it was implemented often varied from one institution to another, and was directly dependent on what masks were available at any given time. Generally, this strategy includes guidelines of when a mask is to soiled to reuse, wearing a cleanable face shield over the mask to protect it from contamination, and how to fold and store masks for reuse as safely as possible.[94]

In some instances, strategies have also been implemented to change the manufacturing materials used in production or for repurposing other PPE for making masks. One study looked at repurposing surgical wrapping for creating masks that performed close to the filtration capacity of N95 masks, and with a single layer of the material, worked well to replace surgical masks.[95] Some others looked at using commonly available materials such as polyester, silk, nylon, etc., what their filtration and breathability is, and in one case, whether these materials could be triboelectrically charged to increase filtration.[96,97] Another study looked at the effect of combining different fabrics such as silk, chiffon, and cotton to create hybrid masks. They seemed to have a higher filtration than masks made with a single type of fabric, possibly in part due to the combined effect of both mechanical and electrostatic filtration.[98]

15.7.4 Looking forward

To optimize local production in the future, it is important to work with supranational organizations, governments, and nongovernmental actors. Often, public—private partnerships work well for such endeavors and help to ensure their sustainability. Concerning the reuse of masks, it is far better to focus on a steady supply and reducing the need for them to be reused, and the safety and these procedures are difficult to ensure, and the financial cost of implementations can be higher than buying new disposable masks.[94–98]

15.8 Conclusion

The current COVID-19 pandemic has challenged the world in unprecedented ways, and forced countries and international organizations to rethink the globalized economy as we know it. Implementing aspects of deglobalization and ensuring sufficient local production and access to the most crucial products for community health has become a high priority for everyone. Access to masks and ABHR are crucial for health care facilities and for the community. The success of infection prevention programs depends on them. The pandemic has shifted how the world thinks about its resources and production, and countries around the world have tackled the issues of locally producing masks and ABHR, producing masks with different materials and equipment and decontaminating and reusing them if there are no other viable options.

In this unprecedented global crisis, local production is crucial. It has a number of potential advantages over products imported from other countries. It is more affordable and therefore more accessible at the point of care/need. It could help the development of the local economy and create local jobs, which in turn could ensure sustainability of ABHR availability. Such production could positively influence health policies and increase community awareness about the importance of IPC measures. Lastly, it could make current IPC measures and programs more sustainable, and prevent cross-transmission in the long term, preparing the health system for a potential future pandemics.

Humans are resourceful creatures in times of adversity, but some of the measures taken during this time should be lasting. Being more self-sufficient in infection prevention during a pandemic is an issue of both national and international security. It will be in countries' best interest to continue to explore the avenues of self-sufficiency; not least because this will simultaneously contribute to addressing issues of access to basic health supplies in low resource areas around the world.

References

1. Shokrani A, Loukaides EG, Elias E, Lunt AJG. Exploration of alternative supply chains and distributed manufacturing in response to COVID-19; a case study of medical face shields. *Mater Design* 2020;**192**:108749.
2. *Face masks and surgical masks for COVID-19: "manufacturing, purchasing, importing, and donating masks during the public health emergency"*. US Food and Drug Administration; 2020. https://www.fda.gov/medical-devices/coronavirus-covid-19-and-medical-devices/face-masks-including-surgical-masks-and-respirators-covid-19.
3. *Current status of production and supply of face masks, antiseptics and toilet paper*. METI Ministry of Economy, Trade and Industry; 2020. https://www.meti.go.jp/english/covid-19/mask.html.
4. *Masks for all: using purchase guarantees and targeted deregulation to boost production of essential medical equipment*. Mercatus Center; 2020. https://www.mercatus.org/publications/covid-19-crisis-response/masks-all-using-purchase-guarantees-and-targeted-deregulation.
5. *Guide to local production: WHO-recommended handrub formulations*. World Health Organization; 2010. https://www.who.int/gpsc/5may/Guide_to_Local_Production.pdf.

6. Bradsher K. China dominates medical supplies. In: *This Outbreak and the Next.* The New York Times; 2020.
7. *Recommended guidance for extended use and limited reuse of N95 filtering facepiece respirators in healthcare settings.* NIOSH Centers for Disease Control and Prevention; 2020. https://www.cdc.gov/niosh/topics/hcwcontrols/recommendedguidanceextuse.html.
8. Garrett L. *The coming plague: newly emerging diseases in a World out of balance.* New York, NY: Picador; 1994. https://www.lauriegarrett.com/the-coming-plague.
9. Peters A, Vetter P, Guitart C, Lotfinejad N, Pittet D. Understanding the emerging coronavirus: what it means for health security and infection prevention. *J Hosp Infect* 2020;**104**:440–8.
10. Liu, J. et al. Community transmission of severe acute respiratory syndrome Coronavirus 2, Shenzhen, China, 2020. Emerging Infect Dis 26, 1320–1323 .
11. Chan JFW, et al. A familial cluster of pneumonia associated with the 2019 novel coronavirus indicating person-to-person transmission: a study of a family cluster. *Lancet* 2020;**395**:514–23.
12. *Modes of transmission of virus causing COVID-19: implications for IPC precaution recommendations.* World Health Organization; 2020. https://www.who.int/news-room/commentaries/detail/modes-of-transmission-of-virus-causing-covid-19-implications-for-ipc-precaution-recommendations.
13. *Infection prevention and control and preparedness for COVID-19 in healthcare settings - fifth update.* European Centre for Disease Prevention and Control; 2020. https://www.ecdc.europa.eu/en/publications-data/infection-prevention-and-control-and-preparedness-covid-19-healthcare-settings.
14. *Communities, schools, workplaces, & events.* Centers for Disease Control and Prevention; 2020. https://www.cdc.gov/coronavirus/2019-ncov/community/reopen-guidance.html.
15. *COVID-19 latest information on the outbreak in the Western Pacific.* World Health Organization; 2020. https://www.who.int/westernpacific.
16. *When and how to use masks.* World Health Organization; 2020. https://www.who.int/emergencies/diseases/novel-coronavirus-2019/advice-for-public/when-and-how-to-use-masks.
17. *Advice on the use of masks in the community, during home care and in healthcare settings in the context of the novel coronavirus (COVID-19) outbreak.* World Health Organization; 2020. www.who.int/publications-detail-redirect/advice-on-the-use-of-masks-in-the-community-during-home-care-and-in-healthcare-settings-in-the-context-of-the-novel-coronavirus-(2019-ncov)-outbreak.
18. Eikenberry SE, et al. To mask or not to mask: Modeling the potential for face mask use by the general public to curtail the COVID-19 pandemic. *Infect Dis Model* 2020;**5**:293–308.
19. Chu DK, et al. Physical distancing, face masks, and eye protection to prevent person-to-person transmission of SARS-CoV-2 and COVID-19: a systematic review and meta-analysis. *Lancet* 2020;**395**:1973–87.
20. Magleby R, et al. Impact of SARS-CoV-2 viral load on risk of intubation and mortality among hospitalized patients with Coronavirus disease 2019. *Clin Infect Dis* June 30, 2020.
21. Aiello AE, et al. Mask use, hand hygiene, and seasonal influenza-like illness among young adults: a randomized intervention trial. *J Infect Dis* 2010;**201**:491–8.
22. Otter JA, et al. Transmission of SARS and MERS coronaviruses and influenza virus in healthcare settings: the possible role of dry surface contamination. *J Hosp Infect* 2016;**92**:235–50.
23. *WHO guidelines on hand hygiene in health care : a summary.* World Health Organization; 2009. https://www.who.int/gpsc/information_centre/hand-hygiene-summary/en/.

24. Pittet D, Boyce JM, Allegranzi B. *System change. Hand hygiene: a handbook for medical professionals*. Willey Blackwell; 2017. p. 144–51.
25. *Disinfection & sterilization guidelines*. Centers for Disease Control and Prevention; 2019. https://www.cdc.gov/infectioncontrol/guidelines/disinfection/index.html.
26. Siddharta A, et al. Virucidal activity of World Health Organization–recommended formulations against enveloped viruses, including Zika, Ebola, and Emerging Coronaviruses. *J Infect Dis* 2017;**215**:902–6.
27. Kampf G. Efficacy of ethanol against viruses in hand disinfection. *J Hosp Infect* 2018;**98**:331–8.
28. *How we know disinfectants should kill the COVID-19 coronavirus*. Chemical & Engineering News; 2020. https://cen.acs.org/biological-chemistry/infectious-disease/How-we-know-disinfectants-should-kill-the-COVID-19-coronavirus/98/web/2020/03.
29. List N. *Disinfectants for Coronavirus (COVID-19)*. US Environmental Protection Agency; 2020. https://www.epa.gov/pesticide-registration/list-n-disinfectants-coronavirus-covid-19.
30. *Chemical disinfectants and antiseptics - Quantitative suspension test for the evaluation of virucidal activity in the medical area - Test method and requirements (Phase 2/Step 1)*. 2019. EN 14476:2013+A2:2019 Standards.
31. Pittet D, et al. Effectiveness of a hospital-wide programme to improve compliance with hand hygiene. Infection Control Programme. *Lancet* 2000;**356**:1307–12.
32. Kratzel A, et al. Inactivation of severe acute respiratory syndrome Coronavirus 2 by WHO-recommended hand rub formulations and alcohols. *Emerg Infect Dis J* 2020;**26**(7). Centers for Disease Control and Prevention.
33. Pradhan D, Biswasroy P, Kumar Naik P, Ghosh G, Rath G. A review of current interventions for COVID-19 prevention. *Arch Med Res* 2020;**51**:363–74.
34. *Hand hygiene recommendations guidance for healthcare providers about hand hygiene and COVID-19*. Centers for Disease Control and Prevention; 2020. https://www.cdc.gov/coronavirus/2019-ncov/hcp/hand-hygiene.html.
35. *Interim recommendations on obligatory hand hygiene against transmission of COVID-19*. World Health Organization; 2020. https://www.who.int/publications/m/item/interim-recommendations-on-obligatory-hand-hygiene-against-transmission-of-covid-19.
36. Berardi A, et al. Hand sanitisers amid CoViD-19: A critical review of alcohol-based products on the market and formulation approaches to respond to increasing demand. *Int. J. Pharm.* 2020;**584**:119431.
37. Lazzerini M, Putoto G. COVID-19 in Italy: momentous decisions and many uncertainties. *Lancet Global Health* 2020;**8**:e641–2.
38. Barua S. Understanding Coronanomics: the economic implications of the Coronavirus (COVID-19) Pandemic. *SSRN J* 2020. https://papers.ssrn.com/sol3/papers.cfm?abstract_id=3566477.
39. *States forced into bidding war for medical equipment*. US News & World Report; 2020. https://www.usnews.com/news/best-states/articles/2020-04-07/states-compete-in-global-jungle-for-personal-protective-equipment-amid-coronavirus.
40. *Shortage of personal protective equipment endangering health workers worldwide*. World Health Organization; 2020. https://www.who.int/news/item/03-03-2020-shortage-of-personal-protective-equipment-endangering-health-workers-worldwide.
41. Willsher K, Holmes O, McKernan B, Tondo L. *US hijacking mask shipments in rush for coronavirus protection*. The Guardian; 2020. http://www.theguardian.com/world/2020/apr/02/global-battle-coronavirus-equipment-masks-tests.
42. Atkinson C, et al. *Supply chain manipulation, misrepresentation, and magical thinking during the COVID-19 pandemic*. The American Review of Public Administration; 2020. https://journals.sagepub.com/doi/full/10.1177/0275074020942055.

43. Dupont K. *La formule genevoise du gel hydro-alcoolique explose sur YouTube*. May 6, 2020. https://www.heidi.news/geneva-solutions/la-formule-genevoise-du-gel-hydro-alcoolique-explose-sur-youtube.
44. Combatting Price Gouging & Hoarding. *United States Department of Justice*. 2020. https://www.justice.gov/coronavirus/combattingpricegouginghoarding.
45. *Article 1 - Décret n° 2020-858 du 10 juillet 2020 relatif aux prix de vente des gels et solutions hydro-alcooliques et des masques de type ou de forme chirurgicale à usage unique*. Légifrance; 2020. https://www.legifrance.gouv.fr/loda/article_lc/LEGIARTI000042107726.
46. *COVID-19 partners platform*. World Health Organization; 2020. https://covid19partnersplatform.who.int/.
47. Burki T. Global shortage of personal protective equipment. *Lancet Infect Dis* 2020;**20**:785–6.
48. Pittet D, et al. Evidence-based model for hand transmission during patient care and the role of improved practices. *Lancet Infect Dis* 2006;**6**:641–52.
49. *A guide to the implementation of the WHO multimodal hand hygiene improvement strategy*. World Health Organization; 2009. https://www.who.int/gpsc/5may/Guide_to_Implementation.pdf?ua=1.
50. Allegranzi B, et al. Global implementation of WHO's multimodal strategy for improvement of hand hygiene: a quasi-experimental study. *Lancet Infect Dis* 2013;**13**:843–51.
51. Bauer-Savage J, et al. Local production of WHO-recommended alcohol-based handrubs: feasibility, advantages, barriers and costs. *Bull World Health Organ* 2013;**91**:963–9.
52. Pittet D, et al. Double-blind, randomized, crossover trial of 3 hand rub formulations: fast-track evaluation of tolerability and acceptability. *Infect Control Hosp Epidemiol* 2007;**28**:1344–51.
53. Steinmann J, et al. Virucidal activity of 2 alcohol-based formulations proposed as hand rubs by the World Health Organization. *Am J Infect Control* 2010;**38**:66–8.
54. *WHO model lists of essential medicines*. World Health Organization; 2020. http://www.who.int/medicines/publications/essentialmedicines/en/.
55. Sustainable development goals. Department of Economic and Social Affairs Sustainable Development. United Nations; 2021. https://sdgs.un.org/.
56. Wirtz VJ, et al. Essential medicines for universal health coverage. *Lancet* 2017;**389**:403–76.
57. von Grünigen S. *Fabrication de la solution hydro-alcoolique pour l'hygiène des mains. Pharm-Ed*. University Hospitals of Geneva; 2017. https://pharmed.datapharma.ch/lessons/fabrication-de-la-solution-hydro-alcoolique-pour-lhygiene-des-mains/.
58. Sickbert-Bennett EE, et al. Comparative efficacy of hand hygiene agents in the reduction of bacteria and viruses. *Am J Infect Control* 2005;**33**:67–77.
59. Boyce J, et al. Guideline for hand hygiene in health-care settings. Recommendations of the healthcare infection control practices advisory committee and the HICPAC/ SHEA/ APIC/ IDSA Hand Hygiene Task Force. Society for Healthcare Epidemiology of America/Association for Professionals in Infection Control/Infectious Diseases Society of America. *MMWR Recomm Rep* 2002;**51**:1–45. Quiz CE1-4.
60. Gold NA, et al. *Alcohol Sanitizer. in StatPearls*. StatPearls Publishing; 2020. https://www.ncbi.nlm.nih.gov/books/NBK513254/.
61. Kampf G, Kramer A. Epidemiologic background of hand hygiene and evaluation of the most important agents for scrubs and rubs. *Clin Microbiol Rev* 2004;**17**:863–93.
62. Kampf G. Biocidal agents used for disinfection can enhance antibiotic resistance in gram-negative species. *Antibiotics (Basel)* 2018;**7**. https://pubmed.ncbi.nlm.nih.gov/30558235/.
63. Russell AD. Bacterial resistance to disinfectants: present knowledge and future problems. *J Hosp Infect* 1999;**43**(Suppl):S57–68.

64. Maillard JY. Bacterial resistance to biocides in the healthcare environment: should it be of genuine concern? *J Hosp Infect* 2007;**65**(Suppl 2):60–72.
65. *How does ethyl alcohol kill bacteria?* Science Line. University of California Santa Barbara UCSB; 2009. http://scienceline.ucsb.edu/getkey.php?key=2160.
66. *Healthcare providers: clean hands count logo clean hands count for healthcare providers.* Centers for Disease Control and Prevention; 2020. https://www.cdc.gov/handhygiene/providers/index.html.
67. Suchomel M, et al. Testing of the World Health Organization recommended formulations in their application as hygienic hand rubs and proposals for increased efficacy. *Am J Infect Control* 2012;**40**:328–31.
68. Suchomel M, et al. Glycerol significantly decreases the three hour efficacy of alcohol-based surgical hand rubs. *J Hosp Infect* 2013;**83**:284–7.
69. Weber DJ, Sickbert-Bennett E, Gergen MF, Rutala WA. Efficacy of selected hand hygiene agents used to remove *Bacillus atrophaeus* (a surrogate of *Bacillus anthracis*) from contaminated hands. *JAMA* 2003;**289**:1274–7.
70. Lee M, Hunt P, Weir. The use of hydrogen peroxide as a sporicide in alcohol disinfectant solutions. *EHP* 1996:203–6.
71. Allegranzi B, et al. Successful implementation of the World Health Organization hand hygiene improvement strategy in a referral hospital in Mali, Africa. *Infect Control Hosp Epidemiol* 2010;**31**:133–41.
72. *Essential medicines and health products: production.* World Health Organization; 2020. http://www.who.int/medicines/areas/quality_safety/quality_assurance/production/en/.
73. Bonnabry P, Voss A. Hand hygiene agents. In: *Hand hygiene: a handbook for medical professionals.* Willey Blackwell; 2017. p. 51–7.
74. Allegranzi B, Sax H, Pittet D. Hand hygiene and healthcare system change within multi-modal promotion: a narrative review. *J. Hosp. Infect.* 2013;**83**(Suppl 1):S3–10.
75. Coronavirus. *Coronavirus : un bateau-usine doit produire des bouteilles de solution hydro-alcoolique à Sète.* France 3 Occitanie, https://france3-regions.francetvinfo.fr/occitanie/herault/sete/coronavirus-bateau-usine-doit-produire-bouteilles-solution-hydro-alcoolique-sete-1809714.html.
76. *Authorisation biocidal products.* Swiss Federal Council; 2017. https://www.anmeldestelle.admin.ch/chem/en/home/themen/pflicht-hersteller/zulassung-biozidprodukte.html.
77. *Trade practice enforcement during COVID-19 Pandemic. TTB industry circular.* US Department of Treasury; May 2020. https://www.ttb.gov/industry-circulars/ttb-industry-circular-2020-3.
78. *N95 respirators, surgical masks, and face masks.* US Food and Drug Administration; 2020. https://www.fda.gov/medical-devices/personal-protective-equipment-infection-control/n95-respirators-surgical-masks-and-face-masks.
79. *Reusability of facemasks during an influenza pandemic: facing the flu. Institute of Medicine.* Washington, DC: The National Academies Press; 2006. https://doi.org/10.17226/11637.
80. Bandi M. *N95-electrocharged filtration principle based face mask design using common materials.* Okinawa Institute of Science and Technology Graduate University; 2020. https://groups.oist.jp/nnp/diy-face-mask.
81. Ney H. *Modalités de retraitement des masques N95 / FFP2 en cas de pénurie.* Société Suisse de Stérilisation; 2020. http://www.sssh.ch/covid-19/masques-ffp2n95/.
82. *Recommendation for healthcare facilities in Switzerland for sterilizing protection FFP masks following the norm EN149 (FFP1-3, N95, or equivalent) with vapor hydrogen peroxide. National COVID-19 Science Task Force (NCS-TF).* 2020. https://sciencetaskforce.ch/en/policy-brief/recommendation-for-healthcare-facilities-in-switzerland-for-sterilizing-protection-ffp-masks-3/.

83. Brosseau L, Berry Ann R. *N95 respirators and surgical masks*. Blogs. Centers for Disease Control and Prevention; 2020. https://blogs.cdc.gov/niosh-science-blog/2009/10/14/n95/.
84. Choudhary A. Sterility Assurance Level (SAL) and Log Reduction (D Value) in sterilization. *Pharm Guidelines* December 2020. https://www.pharmaguideline.com/2013/12/log-reduction-and-sterility-assurance-level-sal.html.
85. Pourdeyhimi, B. Facemasks — challenges during pandemics. The Nonwovens Institute. NC State University; 2021. https://thenonwovensinstitute.com/facemasks-challenges-during-pandemics/.
86. Heimbuch BK, et al. Cleaning of filtering facepiece respirators contaminated with mucin and *Staphylococcus aureus*. *Am J Infect Control* 2014;**42**:265—70.
87. Lin T-H, Tang F-C, Hung P-C, Hua Z-C, Lai C-Y. Relative survival of Bacillus subtilis spores loaded on filtering facepiece respirators after five decontamination methods. *Indoor Air* 2018.
88. Fischer RJ, et al. Effectiveness of N95 respirator decontamination and reuse against SARS-CoV-2 virus. *Emerg Infect Dis J* September 2020;**26**(9). Centers for Disease Control and Prevention.
89. Smith JS, et al. Effect of various decontamination procedures on disposable N95 mask integrity and SARS-CoV-2 infectivity. *medRxiv* 2020. https://www.medrxiv.org/content/10.1101/2020.04.11.20062331v2.
90. Peters, A. Decontaminating N95/FFP2 masks for reuse during the COVID-19 epidemic: a systematic review. In submission.
91. Feng, E., Cheng, A. COVID-19 has caused a shortage of face masks. But they're surprisingly hard to make. National Public Radio; 2020. https://www.npr.org/sections/goatsandsoda/2020/03/16/814929294/covid-19-has-caused-a-shortage-of-face-masks-but-theyre-surprisingly-hard-to-make.
92. Nayak R. *Polypropylene nanofibers: melt electrospinning versus meltblowing*. Springer International Publishing; 2017. https://doi.org/10.1007/978-3-319-61458-8.
93. Boogaard B, et al. Efficacy assessment of newly-designed and locally-produced filtering facemasks during the SARS-CoV-2 pandemic. *medRxiv* 2020. https://www.medrxiv.org/content/10.1101/2020.09.04.20188185v1.
94. Extended use or re-use of single-use surgical masks and filtering facepiece respirators: a rapid evidence review. CEBM; 2020. https://www.cebm.net/covid-19/extended-use-or-re-use-of-single-use-surgical-masks-and-filtering-facepiece-respirators-a-rapid-evidence-review/.
95. Vrielink T, Meijer J. Using surgical wrapping material for the fabrication of respirator masks. *PLos One* 2020. https://journals.plos.org/plosone/article?id=10.1371/journal.pone.0236239.
96. Zhao M, et al. Household materials selection for homemade cloth face coverings and their filtration efficiency enhancement with triboelectric charging. *Nano Lett* 2020;**20**:5544—52.
97. Parlin AF, et al. Silk fabric as a protective barrier for personal protective equipment and as a functional material for face coverings during the COVID-19 pandemic. *medRxiv* 2020. https://www.medrxiv.org/content/10.1101/2020.06.25.20136424v1.
98. Konda A, et al. Aerosol filtration efficiency of common fabrics used in respiratory cloth masks. *ACS Nano* 2020;**14**:6339—47.

CHAPTER 16

Challenges, opportunities, and future perspectives

Amirhossein Takian[1,2,3], Hajar Haghighi[1,a], Azam Raoofi[1,a]
[1]Department of Health Management, Policy & Economics, School of Public Health, Tehran University of Medical Sciences (TUMS), Tehran, Iran; [2]Department of Global Health & Public Policy, School of Public Health, Tehran University of Medical Sciences (TUMS), Tehran, Iran; [3]Health Equity Research Center (HERC), Tehran University of Medical Sciences (TUMS), Tehran, Iran

Success is not final, failure it is not fatal; it's the courage to continue that counts
Winston Churchill

Key messages:
- Resilient and strong health systems are the foundation of an effective and timely response to pandemics.
- Cooperation and partnership must be at the heart of the response to tackle the pandemic and achieve a peaceful world free of disease and injustice, and built upon solidarity, cohesion, and convergence.
- Health is the precondition, outcome, and indicator of a sustainable society, and it has an inevitable relationship with the economic, political, and social aspects of human life.

16.1 Introduction

The emergence of COVID-19 in Hubei province of China in late 2019 has become a great shock and the worst challenge of the century in many areas, including in health care systems, economics, business, transportation, etc. in roughly 223 countries.[1] From 1918, when the H1N1 epidemic broke out, to the emergence of COVID-19, the world has not experienced such a widespread pandemic. Although there are major similarities and differences between the two outbreaks, COVID-19 is far more dangerous and severe than influenza.[2] The COVID-19 outbreak has imposed a global health emergency and its global impact have caused growing concern to the extent

[a] These two authors have contributed equally to this work. Their names have been ordered alphabetically.

Environmental Management of COVID-19
ISBN 978-0-323-85780-2
https://doi.org/10.1016/B978-0-323-85780-2.00011-1

that the World Health Organization (WHO) declared the crisis on January 30, 2020 a "Public Health Emergency of International Concern" (PHEIC).[3]

The prevalence of COVID-19 in various countries has been changing frequently. At the beginning of the global outbreak, China, South Korea, Italy, and Iran were ranked as the top four countries in terms of total confirmed cases. On March 4, 2021, while the United States, India, Brazil, and Russia were ranked the top four countries with confirmed COVID-19 cases, the number of new cases in China had dropped dramatically since early March 2020, ranking China as the 84th on that date.[4]

Coping with this global health crisis is a complex issue that requires vision, planning, and flexibility at the individual, organizational, and international levels. Policy makers need to adopt interdisciplinary and multidisciplinary approaches to succeed in their battle against the pandemic.[5] The COVID-19 experience has shown us that countries' preparedness is the most fundamental factor to prevent more costly consequences and governments to respond to the crisis more appropriately.[1,5]

For instance, the government of China, the first country that reported the spread of severe acute respiratory syndrome coronavirus 2 (SARS-CoV-2), established a central committee headed by the prime minister as well as provincial counterparts commanded by governors to combat COVID-19 outbreak. In the United States (US), for coordinating the administration's efforts to battle this pandemic, the White House Coronavirus Taskforce was established. Federal, state, and local governments have been given the authority to make decisions for their mitigation strategies based on guidelines, mostly prepared by the Centers for Disease Control and Prevention (CDC). Some European countries such as Italy and Spain also delegated authority and decision-making to their states to combat against COVID-19 independently. On the other hand, learning from the experiences previous SARS pandemic in 2003, the Southeast Asian countries including Taiwan, South Korea, Hong Kong, and Singapore showed more preparedness to control the pandemic and decrease the number of new positive cases. The government of Iran established the National COVID-19 committee, led by the president and coordinated through the Ministry of Health and Medical Education (MoHME). The provincial COVID-19 committees were also established in each province, headed by governors with the aim of improving intersectional cooperation to combat the pandemic.[6]

Most notably, community participation in the implementation of the adopted strategies is an essential need that many governments may encounter some challenges for its implementation during this pandemic.[7] Governments need to know that overcoming this challenge requires a greater understanding of public perceptions regarding these strategies, as well as they need to appreciate the context, where public adherence behaviors may increase or decrease.[8] For instance, governments need to understand whether encouraging people to observe protective measures is enough or detention measures such as fines against any violations needs to be enforced to determine people's attitudes and perceptions.[7]

Delayed execution of these strategies will lead to public uncertainty and diminishing trust in the governments. Previous epidemics' experiences have shown that although significant cooperation is needed to succeed in fighting the virus, it is not clear how people will react to unexpected events such as outbreaks. Although international organizations and governments always endeavor to provide appropriate and timely guidelines to deal with outbreaks and make them available to the public, such emergency guidelines are not enough on their own to anticipate and identify barriers and enablers in relation to people's responses and reactions to the crises.[9] Many factors including access to resources, socioeconomic status, health literacy, and public awareness of the consequences of noncompliance with recommendations, will affect the level of acceptance of guidelines and recommendations by people in the community.[7] Therefore, realistic evaluation of proposed public health intervention strategies, along with awareness of what increases public acceptance of the adopted solutions and strategies are among the significant factors that public health, the scientific community, and governments may face during this pandemic.[10,11]

16.2 The coronaviruses epidemic and the challenges and opportunities

Despite all challenges, COVID-19 has also provided health officials and governments with a variety of opportunities for policy learning.[1] In this chapter, we will review global challenges and risks due to the COVID-19 pandemic, along with the opportunities, consequences, and impact on different sectors of societies.

In this section, first, we will discuss the importance of resilient and strong health care systems, the value of coordination and collaboration among and within countries, and some significant consequences of this pandemic on the world economy. This will be followed by highlighting the importance of social media, digital technology, and psychological outlook that governments have been facing during the COVID-19 pandemic.

16.2.1 Health care systems

Although some countries performed better than others, COVID-19 revealed fundamental shortages in this regard worldwide. The countries' ability to deal with the COVID-19 pandemic largely depends on their health system resilience and strength. Such resilience is related to the extent and speed to which health systems are able to diagnose, detect, and control the spread of the virus; the capacity of laboratories for performing required number of COVID-19 diagnosis tests per day (particularly polymerase chain reaction [PCR]); the process of health care screening; provision of necessary infrastructure for contact tracing; mobilization of a sufficient workforce through employment of medical students and retired health professionals to help the health system during crisis, and preparing hospitals to be sufficiently ready to deal with the pandemic, among other measures.[5]

The COVID-19 pandemic is revealing inadequate infrastructure and insufficient facilities, including personal protective equipment (PPE) such as masks, gowns, shields, gloves, disinfectant solutions; intensive care unit (ICU) beds; prepared human resources; and ventilators in many low and middle income countries (LMICs) as well as some high-income countries (HICs) that were not properly equipped to deal with this crisis in a timely manner.[12] The magnitude of problem multiplies in countries like Iran as a result of unilateral sanctions[13] that has brought overwhelming challenges to the health system,[14] particularly for people in the lower socioeconomic status.[15] In addition, many citizens with chronic conditions, e.g., non-communicable diseases (NCDs)[16] or in need of acute care, did not seek required care due to their fear and anxiety to get infected with COVID-19.[17–19] These situations will impose greater challenges and exacerbate the COVID-19 consequences for the health systems worldwide.[12,20]

Ever since the COVID-19 pandemic was declared, several elective procedures and nonemergency care have been suspended in both private and public medical centers in many countries. This has been inevitable, aiming to reduce the spread of virus and preserve PPEs, ventilators supply, and ICU beds for acute patients[21,22] and has threatened the ability of many health organizations to pay their expenditure. A study conducted in 2011 revealed that more than 48% of hospitals' revenue in the United States came from elective surgeries. Due to the COVID-19 pandemic, many hospitals, particularly private centers, which are mostly financed through elective procedures, faced financial challenges and had to lay off or reshuffle their staff.[21,22]

Worse still, almost one-quarter of the world population reside in slum areas or among refugees and displaced people, for whom protective measures, i.e., social distancing, mask wearing, and hand washing might not be feasible all times. Furthermore, about three-quarters of the workforce in the LMICs work in the informal sectors, where lockdown and social distancing might be a selection between hunger and danger. Consequently, these people are more exposed to COVID-19 and may impose more burden on the health system.[12] Indeed, the most vulnerable countries that are at the risk of natural and humanitarian crises are three times more likely to be exposed to COVID-19, while six times less likely to have access to needed health care.[23] Despite numerous sacrifices by frontline health workers, the extensive spread of COVID-19 and continued increase in symptomatic and asymptomatic patients that are referred to in health care centers present signs of inadequate global response to COVID-19. For instance, insufficient health care providers, high burden of health care personnel, inadequate PPE, fear of being infected with COVID-19, as well as governments' inability to properly incentivize and compensate health care workers can be mentioned, which will create the vicious cycle of shortage of health workforce and imposes an additional burden on health systems.[24]

16.2.2 Cooperation between and within countries

COVID-19 has affected all aspects of public life in almost all countries.[25] Following the declaration of pandemic by WHO on March 11, 2020, governments in many countries began to enforce a number of preventive measures, including mandatory quarantine and lockdown, borders closing, and imposing international travel restrictions. In the light of globalization and the so-called "global village," where governments are interdependent in governing their countries, this has become a big challenge.[12] On the other hand, due to the fundamental knowledge gap about this emerging pandemic, known as a "global emergency," to combat the crisis, countries need cooperation, instead of competition, in several aspects.[26] These include, but are not limited to collaborative research on vaccine and medication, innovative technologies for production of more suitable PPE, respirators and ventilators, and more importantly the exchange of experiences and policy lessons. Comparative policy and management learning is crucial to ensure the accuracy of the information provided to policy makers, as well as to prevent infodemics and the spread of fake news.[27] Last but not least, intercountries collaboration at the highest political level will pave the way for meaningful ground root cooperation between universities and research institutes, which is essential to combat the ongoing pandemic[27] and turn this challenge to an opportunity for rebuilding the societies when the crisis ends. The ultimate goal of this cooperation is to establish a resilient, sustainable, and fair world based on solidarity, cohesion, and convergence.[12] The COVID-19 pandemic has led to greater social solidarity within communities. In Nigeria, some organizations and companies provide food for those who lost their jobs as a result of lockdown. Some restaurants provide meals for hospital staff who cannot leave their workplace due to high workload in the UK. In China, volunteers from across the country traveled to Wuhan to assist medical staff. In France, people who lost their jobs in the cities were directed to rural areas to help farmers. In some countries, emergency teams have been established to help elderly, homeless, and disabled citizens.[12] In Iran, some hotels participated in the campaign to accommodate recovering COVID-19 patients. In addition, Tehran Exhibition Center and one of the largest shopping malls in the capital city of Tehran were equipped with 2000 and 3000 beds, respectively, and were converted into mobile hospitals.[6,28]

This pandemic has highlighted the importance of intersectional collaboration among relevant sectors such as military forces, ministry of road and transportation, religious groups, ministry of higher education, research centers, ministry of health, and legislative authorities for countries to better manage such crises. Improving the policy dialogue between miscellaneous stakeholders provides an opportunity for policy makers to make decisions based on the comments of various actors and can implement them with greater acceptance and compliance consequently.[12] Worse still, other disasters are likely to occur simultaneously with outbreaks, for instance about 1 billion people are exposed to floods. Expanded cooperation in the health, social, and humanitarian sectors is therefore the key to deal with simultaneous crises, especially during the current global economic downturn.[29]

16.2.3 Economics

COVID-19 is not just a health issue. The pandemic is also a political, economic, and social challenge for the global community. The crisis led all governments to spend more for providing extra health care services as well as to support affected businesses and families. Despite these aids, the deteriorating economic, social and health situation of almost all countries is obvious.[27,30,31] These challenges resulted from various reasons including lockdown, business shutting down, travel bans and massive unemployment.[27] In other words, COVID-19 pandemic has led to direct and indirect costs. Direct costs refer to the resources needed for prevention, treatment and rehabilitation, while indirect costs refer to the costs that are imposed on society due to the disability of individuals and the reduction of labor productivity.[32]

Depending on their economic make up, the impact of the pandemic on economies has been diverse. Nations with heavy reliance on international industries like tourism, manufacturing, agriculture, and international trade have been affected the most, both directly and indirectly.[33–35] Extreme escalation of job losses also influenced individuals' livelihoods and doubled the challenge.[36] In the shadow of inadequate government support, most workforce in the informal sectors, e.g., garbage collectors and child laborers, have been left with no choice than working in high-risk environment to avoid starving to death.[27]

According to the World Bank report, global economy was estimated to have shrunk by 5% due to the COVID-19 pandemic until June 2020 (Figs. 16.1–16.4).[31,38] The Chinese government, for example, announced a significant reduction of its production in February 2020.[24] Some

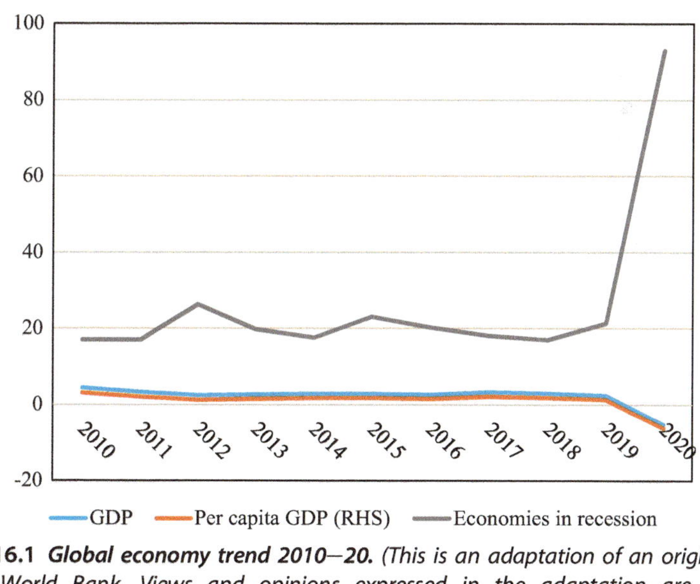

Figure 16.1 ***Global economy trend 2010–20.*** *(This is an adaptation of an original work by the World Bank. Views and opinions expressed in the adaptation are the sole responsibility of the author or authors of the adaptation and are not endorsed by the World Bank. World Bank.* Global economic prospects. *Washington, DC: World Bank; June 2020.)*[37]

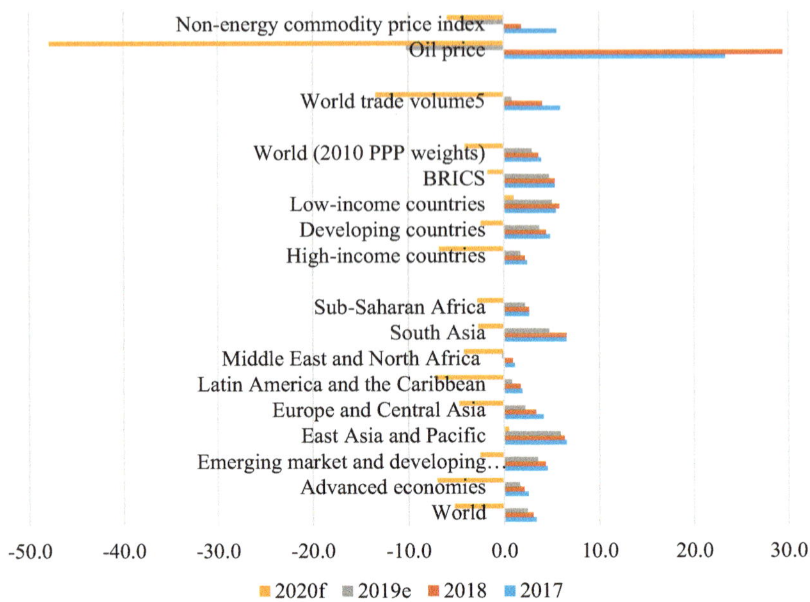

Figure 16.2 ***Real gross domestic product (GDP) (percent change from previous year).*** *(This is an adaptation of an original work by the World Bank. Views and opinions expressed in the adaptation are the sole responsibility of the author or authors of the adaptation and are not endorsed by the World Bank. World Bank.* Global economic prospects. *Washington, DC: World Bank; June 2020.)*[37]

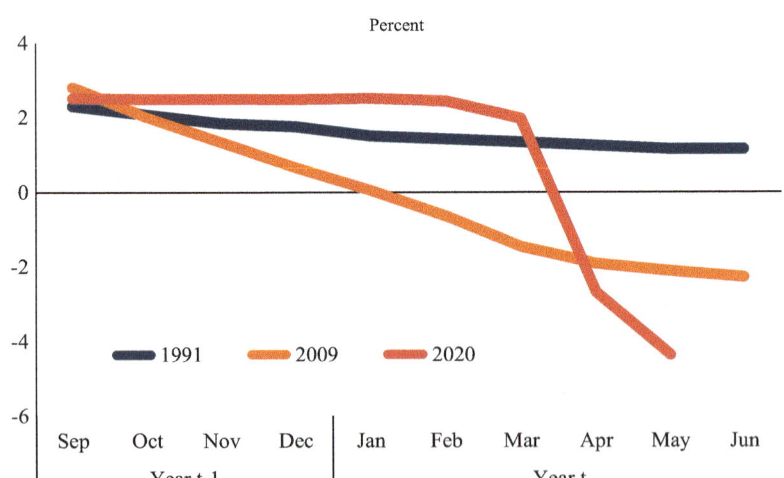

Figure 16.3 ***Consensus forecasts of global gross domestic product (GDP) growth.*** *(Source from: World Bank.* Global economic prospects. *Washington, DC: World Bank; June 2020. https://doi.org/10.1596/978-1-4648-1553-9. License: Creative Commons Attribution CC BY 3.0 IGO.)*[37]

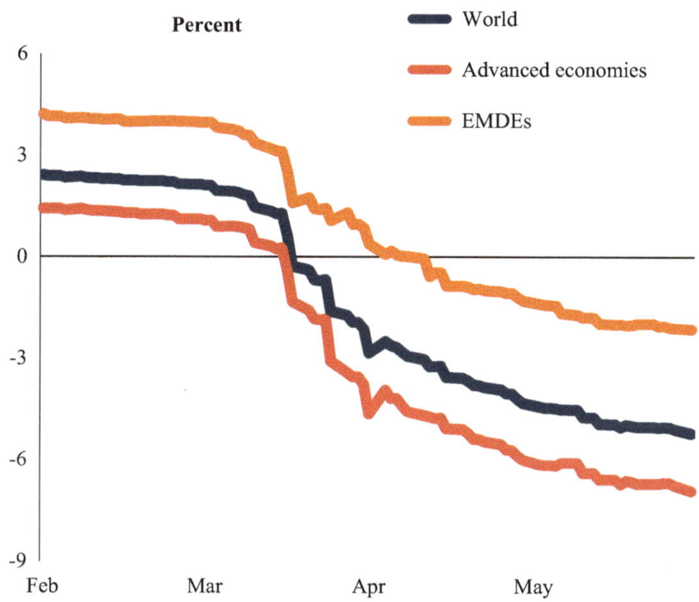

Figure 16.4 ***Consensus forecasts of gross domestic product (GDP) growth for 2020, February–June 2020.*** *(Source from: World Bank.* Global economic prospects. *Washington, DC: World Bank; June 2020. https://doi.org/10.1596/978-1-4648-1553-9. License: Creative Commons Attribution CC BY 3.0 IGO.)*[37]

governments have allocated considerable additional budgets to deal with this crisis. Nevertheless, some LMICs are concerned to face financial crisis as well as their increasing gap with the HICs.[27] Worse still, many countries have been experiencing sharp declines in their foreign investment due to the pandemic. "This is a crisis that the world has never seen before" the International Monetary Fund had reported. It is estimated that COVID-19 outbreak can decrease the economic growth of HICs for one or two years, while this damage in the LMICs may last up to 10 years.[12]

In this regard, as of September 23, 2020 the United States has set the largest stimulus package at $2.6 trillion to support its domestic economy, as well as for individuals, hospitals, and medical care, and the insurance benefits intended to prevent the economic impact of the COVID-19 outbreak (Fig. 16.2).[39] Let alone that the United States withdrew from the WHO membership amid the ongoing crisis.[38]

16.2.4 Social media

Social media, while being an appropriate tool for disseminating information and raising public awareness to prevent the spread of disease, has been the cause of fear and anxiety by spreading false information and contributing to infodemics.[40] Worse still, distrust of published information and lack of public support may be the result of inconsistencies between government recommendations, social media, and recommendations of global health organizations.[8] For instance, misinformation on social networks and lack of informative and beneficial programs based on government strategies affect the acceptance of use of masks in the society.[7] COVID-19 highlighted the need for contextual-based mechanisms that governments need to put in place to deal with misinformation and decrease their destructive effects during pandemics.[40]

16.2.5 Digital technology

The global emphasis on social distancing and the need for telecommunication and distance education, has made the provision and access to digital technology one of the main challenges for many countries in this particular era. Undoubtedly, COVID-19 pandemic will lead to structural changes in the relationship between technology and society, with visible consequences in many areas. For example, teleworking, which was the format of working in some contexts partially, has become the normal way of doing jobs in many countries depending on their communications infrastructure.[41] Some religious places have made it possible for citizens to remotely perform their duties instead of attending places of worship. With hundreds of millions of students out of school due to the outbreak of COVID-19, many governments have enabled virtual and distance education system through the internet. Most of gatherings and events, from small company meeting, mid and big size conferences, up to the highest level political meetings of heads of states, went online and through video conferencing. Most notably, for the first time in their entire 70-year history, the United Nations held its annual general assembly (UNGA) online in September 2020,[42] and the WHO held its World Health Assembly (WHA) virtually in May 2020.[4] Hundreds of thousands, or millions, of webinars and virtual meetings have been held by different universities and research centers around the world on various dimensions of this pandemic. Virtual medical consultation is becoming a new normal to overcome or minimize the harms for both patients and service providers.

Despite their many benefits in this situation, the need to use digital platforms have brought some fundamental challenges to many governments in the LMICs, i.e., the weak information and communication technology (ICT) infrastructure, particularly in many rural and slum areas, which might hinder the use of cyberspace capacity and increase the already existing inequalities in communities' access to the services.[15] In addition, the prolongation of the

COVID-19 crisis could lead to a reduction in the capacity of the internet, even in more industrialized countries with strong digital infrastructures.[12]

16.2.6 Psychosocial perspective

While more attention has been paid to the issues related to the physical complications of COVID-19, evidence shows that psychosocial effects of the outbreak, from internal strife to racist attacks and growing nationalist behavior, have increased.[12] Citizens' mental health conditions might deteriorate in affected communities, with disorders such as anxiety, depression, and panic disorders being at stake.[43] The global concern is that the psychosocial effects of this epidemic may have long-term consequences that are more dangerous than the disease itself. Enhancing family disputes are among the negative psychological effects of this pandemic. As members of a household have been living together for a long period of time and in the case of many citizens in small-sized places, tensions within the family structure have increased. Unfortunately, there are some data stating that divorce rate, family disputes, marital problems, child abuse, and rape will have augmented during and after this period. In addition, due to economic problems and rising unemployment rates globally, more psychological pressure is placed on those who have lost their jobs. As people seek to regulate their negative emotions, they may use a wide range of coping strategies including drug abuse, alcohol consumption, suicide, etc., which often have adverse consequences for their mental and physical health. Hence, to prevent these complications, it is necessary to implement public health education programs and campaigns to raise public awareness to prevent and combat adverse psychosocial effects of the crisis.[12]

16.3 Main challenges of environmental health engineering during prevention of COVID-19 pandemic

COVID-19, similar to other two coronaviruses, which had caused major epidemics in recent years (namely, SARS and the Middle Eastern Respiratory Syndrome [MERS]), transmits through direct human-to-human contact with symptomatic and asymptomatically infectious persons, contaminated objects or surfaces, or by inhalation of respiratory droplets. Environmental factors such as air temperature, humidity, air pollution, water and sewage, inanimate surface, insects, etc., are also important influential factors in the spread of microbial threats, including the COVID-19 virus.[44]

Despite the influence of environmental factors on the transmission of the COVID-19 virus, the type of this influence has not yet been properly identified. For instance, regarding the effects of temperature on the transfer of COVID-19, different studies have reached different and sometimes contradictory results. A study of 122 cities in China found that rising temperatures had no effect on reducing the incidence of COVID-19.[45] In another study, rising temperatures and declining humidity were determined as favorable factors for virus survival.[46] In contrast, another study found that with the onset of spring and summer, virus transmission would decrease.[47] A study of 429 cities around the world found an increase in disease transmission in colder cities.[48] The differences in the results of different studies are due to several factors, including the impact of other variables on the spread of the virus and different measures and policies by the governments in dealing with COVID-19, which highlights the need for further investigation, while controlling other variables. Another study revealed that changes in temperature and humidity alone, without severe public health interventions, will not necessarily reduce the spread of COVID-19.[49]

Although there is currently no evidence that the virus is transmitted through air diffusion, the virus can be transmitted from asymptomatic-infected people through sneezing, coughing, talking to other people at close distance, and polluted airflow.[44,50] Therefore, it is recommended to use viral filters and PPEs such as standard mask, gown, face shield, gloves, etc. Further, closed suction, air flow changes, and negative pressure air pollution in closed environments are where the virus is likely to spread,[51,52] requiring use of standard masks.[53] Also, people who are suspected to be infected with COVID-19 virus and patients with clinical symptoms should be quarantined, while others should use preventive methods such as frequently hand washing, using disinfectants, wearing PPEs, and considering social distance.[54]

So far, there is no safe and effective medication for COVID-19 and current vaccines are not widely distributed. Therefore, nonpharmacological measures such as quarantining patients with clinical symptoms; environmental health activities and promoting hygiene activities such as wearing PPEs; social distancing; frequently disinfecting hands and surfaces; and avoiding touching one's face including mouth, eyes, or nose with unwashed hands are currently the most effective methods to tackle the crisis and to prevent the virus spread in the societies.[44] The IHME predicts the impact of social distance with and without the mask on the global prevalence of COVID-19, as shown in Fig. 16.5.

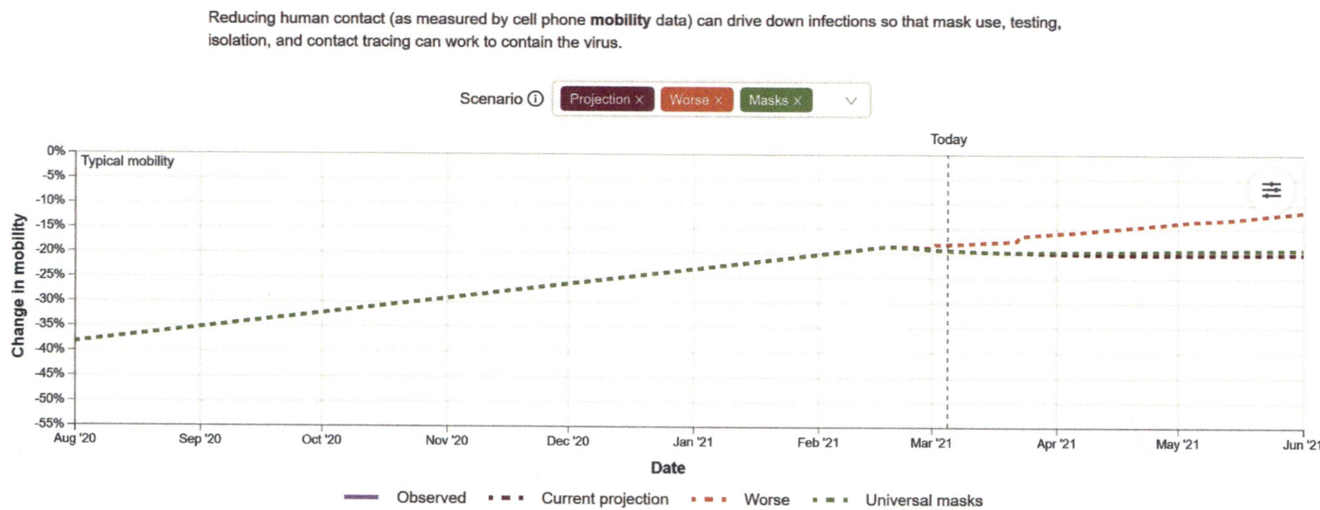

Figure 16.5 Forecasting the effect of social distancing with and without universal masks use on the global prevalence of COVID-19 until March 5, 2021. *(Source from: Health Data, the Institute for Health Metrics and Evaluation. License: Creative Commons Attribution-NonCommercial-NoDerivatives 4.0 International License (https://creativecommons.org/licenses/by-nc-nd/4.0/). Institute for Health Metrics & Evaluation.* COVID-19 projections; *2020. https://covid19.healthdata.org/. [Accessed 11 December 2020].)*[55]

Despite these recommendations, most countries have faced a shortage of PPE and restrictions on the use of this equipment for the public and health staff.[56] In addition, most vulnerable groups in the community, including street sleepers, the homeless, people who live in overcrowded households or in unventilated homes, migrants, and slum dwellers do not have access to PPE and cannot observe health protocols properly.[57]

For the first time in modern history, the world observed only recently that some countries in the southern hemisphere that finished winter in October 2020 witnessed dramatic decrease in the incidence of seasonal influenza compared to previous years.[58] It is speculated that protective public health measures for COVID-19, i.e., hand washing, mask wearing, and social distancing has been the key in reducing the number of cases with seasonal flu, which also play an essential role in controlling COVID-19 pandemic.[59] This can decrease the immediate demand for health care services and reduce the consequences of the disease, including mortality and morbidity.[60]

Ensuring access to safe water as well as providing sanitation and waste management services are essential in preventing the transmission of COVID-19.[61] There is no evidence regarding the virus transmission through the stool of an infected person. Nevertheless, reports confirmed that the COVID-19 virus can survive in sewage for days or weeks,[62] findings which could be useful to monitor and track the rate of virus rotation in communities.[63] In spite of the virus survival in the wastewater, its presence in water resources highly depends on factors such as temperature, sunlight, the presence of organic compounds, as well as other antagonistic microorganisms. Despite the survival of the COVID-19 virus in sewage, the risk of virus transmission through contaminated drinking water has not been confirmed.[61,64]

16.4 The mental health issues associated with the COVID-19 pandemic

The COVID-19 outbreak has affected almost all significant economic, political, and social aspects of human life around the world.[65] It is not surprising, therefore, that the crisis has been described as a "mental health catastrophe," which is among the most serious and worrying difficulty that human beings has experienced in the recent history.[66] Raising fear and anxiety among population, especially in vulnerable groups, i.e., elderly, children, medical staff and care providers, and people with underlying special health problems, are reported as being heightened during the pandemic.[67] Because health is a complete state of physical and mental conditions, governments have to be aware about the most notable impacts of COVID-19 on human beings' mental health during this outbreak and afterward.[65]

The significant amount of unknowns about various aspects of COVID-19 has imposed additional stress and anxiety on both people and officials.[65] Worse still, the loss of loved ones, including friends, relatives, and colleagues due to COVID-19,[69] self-isolation, implementation of lock-down and quarantines, restricted face-to-face socialization, obligation to teleworking, children's home-schooling, temporary unemployment, job losses, and economic insecurity have exacerbated the burden of mental health problems during this pandemic. Additionally, it has been forecasted that loneliness, child abuse, and divorce would be increased in this pandemic, since people have to stay at home and must tolerate each other for a long time period. Substance abuse may also increase during the pandemic as some people may use opioids for excitement, filling their leisure time or overcoming their anxiety and stress.[65,68]

Even before the beginning of COVID-19 pandemic, mental health diseases were prevalent and accounted for 13% of the global burden of disease. On October 6, 2020, for the occasion of World Mental Health day, WHO published the result of a current study on the effects of COVID-19 on mental, neurological, and substance use (MNS) services among 130 WHO member states. The results show that most countries did not be succeed in providing required mental health services for their citizens, which made them more vulnerable in the sensitive and critical time of crisis.[69] This also reminds us urgent the need for urgent strategies to reduce the burden of psychological impact of pandemic and preventing the ongoing difficulties to get worse.[67]

16.5 COVID-19 global consequences and the future of the sustainable development goals

In 2015, the United Nations (UN) launched the sustainable development goals (SDGs) with the aim of ending poverty and moving the world toward greater equity, peace, resilience, prosperity, and opportunities for all people in a healthy planet by 2030. Despite the uneven and sluggish progress in achieving some SDGs until 2020 and the probability of not meeting the goals by 2030, some gains have been achieved. These include reducing the share of children and youth out of school, declining the incidence of many communicable diseases, improving access to safe and secure drinking water, and increasing the appointment of women to leadership roles.[70]

Today, going through one-third of the path into 2030 agenda, the advent of COVID-19 has adversely affected all aspects of SDGs, i.e., health, social, economic, cultural, and political. As of March 5, 2021, the total number of cases in 223 countries and territories exceeded 114 million, while 2,554,694 people died from the disease.[4,71] Many SDGs are threatened by the pandemic and its vast consequences. The health systems in many countries have been on the verge of collapse, over 1.6 billion children have been out of school, many households have lost their incomes, many workers have been forced to turn to informal jobs for a living, and millions of people have been suffering from extreme poverty and hunger (Table 16.1 and Fig. 16.6).[70] The outbreak has also increased inequalities globally, where vulnerable groups have been increasingly exposed to direct and indirect effects of the pandemic.[15,72–74] Poor families are getting poorer, marginalized individuals and groups are more likely to get infected and die due to COVID-19 or even other chronic or acute diseases, while women and children have been exposed to more domestic violence than ever before.[70] COVID-19 has endangered achieving SDGs significantly, and even in some of 169 targets the world has fallen behind.[38,70] These show that policy makers and researchers need to explore multiple dimensions beyond population health and well-being and create a sustainable and strengthened context for further progress toward SDGs.[29] In the following section, we will describe some vital consequences of the pandemic on selected SDGs.

16.5.1 SDG 1: No poverty

Prior to COVID-19, the world was looking to end poverty by 2030, yet it was predicted that 6% of the world's population would remain in extreme poverty by 2030. Now, with the emergence of COVID-19, it is estimated that an additional 71 million people will live in extreme poverty. For the first time in decades, COVID-19 has increased global poverty.[70] Although income inequality is diminishing in some countries, the global economic downturn following the epidemic could plunge millions into poverty and exacerbate inequalities.[75]

16.5.2 SDG 2: Zero hunger

Despite substantial progress since 1990 to reduce hunger, the world is still struggling with a growing hunger crisis. In 2019, the number of serious starving people suffering from crisis level hunger rose from 113 million to

... main effects of some of the sustainable development goals, before and after the outbreak of COVID-19.

	Targets	Before covid-19	Current forecast after covid-19 (2020)	2030
SDG 1: No poverty	Proportion of people living below $1 90 a day	8.2% (2019)	8.8%	6%
SDG 2: Zero hunger	Prevalence of moderate or severe food insecurity	25.9% (2019)		N/A
	The proportion of children under 5 years of age suffering from stunting	21% (2019)		
	The proportion of children under 5 were affected by wasting, or acute undernutrition	6.9% (2019)		3%
SDG 3: Good health and well-being	Under-5 mortality rates	17.7% (2018)	Hundreds of thousands of additional under-5 deaths are expected	N/A
	Coverage of the required three doses of diphtheria	86% (2018)	It was reported moderate-to-severe disruptions or a total suspension of vaccination services	
	Coverage of the second dose of measles-containing vaccine	69% (2018)		
SDG 4: Quality education	The proportion of children and youth out of primary and secondary school	17% (2018)	About 90% of all students (1.57 billion) were out of school	Over 200 million children will still be out of school
	Prolonged absence from school is associated with lower retention and graduation rates and worse learning outcomes	85% (2019)	Prolonged absence from school, accompanied by reduced retention and graduation rates and worse learning outcomes	N/A

Continued

Table 16.1 Comparison between main effects of some of the sustainable development goals, before and after the outbreak of COVID-19.—cont'd

	Targets	Before covid-19	Current forecast after covid-19 (2020)	2030
	Only 65% of primary schools have basic handwashing facilities			
SDG 5: Gender equality	Despite improvements, full gender equality remains unreached (2019)		up to 30% increase in domestic violence in some countries/ Women now spend about three times as much as men in unpaid domestic and care work as men.	N/A
SDG 6: Clean water and sanitation	The proportion of the global population using safely managed drinking water services	2.2 billion people lack safely managed drinking water/4.2 billion people lack safely managed sanitation (2017)	3 billion people worldwide lack basic handwashing facilities at home	Water scarcity could displace 700 million people
	Some countries experience a funding gap of 61% for achieving water and sanitation targets.			
SDG 8: Decent work and economic growth	GDP per capita growth rate	1.5% (2019)	Real GDP per capita is expected to decline by 4.2%	N/A
	The growth rate in labor productivity	1.4% (2019)	Changes in labor productivity growth, influenced by global reductions in working hours and declining economic growth	
	The percentage workers in the informal economy	61% (2016)	up to 60% drop in informal workers income	
	Global unemployment rate	5% (2019)	up to 14% drop in global working hours	
		400 million job losses in 2nd quarter of 2020.		

SDG 9: Industry, innovation, and infrastructure	51% decrease in the number of air passengers from January to May 2020 compared to the same period in 2019			
SDG 11: Sustainable cities and communities	Share of urban and rural population	Living in slums rose to 24% (2018)	Over 90% of COVID-19 cases are in urban areas	N/A
SDG 16: Peace, justice, and strong institutions	Change in intentional homicide rates	5.8 per 100,000 population	Rising cases of violence and murder in Latin American countries/reducing cases of violence and murder in European countries	5.2 per 100,000 population
	Increased risk of COVID-19 with overcrowding in prisons in 60% of countries			
SDG 17: Partnerships for the goals	Global foreign direct investment (FDI)	Increase by 2% (2018)	Decrease by up to 40%	N/A
	Global merchandise trade	2.1% (2018)	Declined by 13%–32%	

Figure 16.6 How SDGs have been influenced by COVID-19. *(Based on this is an adaptation of an original work by United Nations.* The sustainable development goals report 2020. *New York, NY, USA: United Nations Publications; 2020.)*

135 million, indicating that the world has not only failed to achieve SDG 2, but has also fallen behind. Regrettably, in 2020, with the emergence of COVID-19, achieving this goal has become even more challenging. The UN World Food Program (WFP) has estimated that this number is likely to double to 265 million in 2020.[76]

All four dimensions of food security, including availability (is the supply of food adequate?), accessibility (can people obtain the food they need?), utilization (do people have enough intake of nutrients?), and stability (can people access food at all times?), are directly affected by COVID-19.[77] This can be mainly due to individuals' income loss and their inability to pay in addition to rising food prices. Because of travel restrictions, seasonal farmworkers cannot travel to harvest crops, the possibility of food exportations is also limited, that in turn will constrain food availability and accessibility by affecting supply chain.[38] The absence of international cooperation, inadequate protection by governments, and disruption of supply chains may also lead to food shortages and rising food prices around the world.[29] Consequently, these could affect all four aspects of food security and may lead to rising the number of people starving.

The COVID-19 pandemic showed that sustainable use of land and water resources, plus planning for resilient growing of basic nutrients, is essential. To do this, we need to reduce food waste to a minimum. Annually, near $400 billion worth of food is wasted worldwide. This amount can feed approximately 1.26 billion people annually.[78] Moreover, panic buying across the world following the pandemic has led to more food wastage and an impact on the quality of diets due to lack of fresh food.[78]

16.5.3 SDG 3: Good health and well-being for all in all ages

Despite progress in SDG 3, COVID-19 has raised growing concerns about the regression of all that has been achieved so far. COVID-19 accounts for the bulk of the health system's limited resources, as a result of which many other health needs have been underconsidered. The UN Children's Fund (UNICEF), for example, has notified that 116 million newborns and mothers are at risk of inadequate health care due to the COVID-19 pandemic.[79] A study published in The Lancet Global Health predicts that more than 2 million under-five deaths and more than 1 million mothers' death could occur within 12 months in 118 LMICs due to limited access to basic care, food, and vaccinations.[80] During this pandemic, 70 countries are at the risk of interrupting childhood vaccination programs.[70,75] Moreover, many other health care services, i.e., cancer screening, family planning, or other non-COVID-19 infectious diseases, have been discontinued or neglected in many countries.[75] Consequently, any disruption in the continuation of other patient treatment and provision of health care services may influence inversely the years of progress in this area and affect the population health in the future.[75]

16.5.4 SDG 4: Quality education

COVID-19 is the most unprecedented challenge the world educational systems have ever encountered. In an attempt to maintain social distancing and prevent the virus transmission, many countries were forced to shut down their educational institutions, at least temporarily, and switch to distance education, which in turn has posed new challenges to both learners and educators. The UN Educational, Scientific, and Cultural Organization (UNESCO) declared that more than 1.6 billion students are out of school due to school closure.[81] Lack of access to education, in addition to the short-term effects of decreasing literacy, increases the likelihood of dropping out and not returning to school in the long-term, which might in turn further threaten future socioeconomic opportunities.

It is also predicted that the negative effects of school closures will be disproportionately greater on the most vulnerable groups, which will increase existing inequalities, in particular, gender inequality.[29] Before the COVID-19 crisis, near 130 million girls were out of school. Shutting down the schools will lead to additional 11 million girls not going back to school. Depriving girls from going to school, in addition to jeopardizing years of striving for progress toward gender equality, will raise other concerns, including the risk of forced early marriage, adolescence pregnancy, and domestic violence.[82]

In many rural or marginalized areas, there is no adequate infrastructure and facilities for remote learning. Students in these areas suffer from digital inequality, meaning lack of access to the internet and computers. These vulnerable children, who have less opportunity to learn at home, are at greater risk of exploitation.[29,70] Furthermore, using the internet for distance learning may rise the exposure of children to cyberbullying, high-risk online behaviors, and online hunters.[70]

School closures are also a serious threat to the health and safety of children, especially vulnerable children. For millions of children, school is not only a place to learn, but also a safe place away from violence. Children at school can receive health services such as vaccinations, as well as nutritional services, including free meals and iron supplements. As a result of shutting down schools, approximately 379 million children may face food insecurity because they do not have access to free or subsidized school meals, which is a threat to the their immune system and their ability to fight diseases.[70]

With the end of the summer holidays, the reopening of schools has created a new challenge, the so-called "duality in prioritizing health or education" for both students and teachers around the world. On the one hand, to realize the right of students to enjoy education and due to the challenges of distance education, there is an urgent need to reopen the schools. On the other hand, the threat of COVID-19, especially as it coincides with the beginning of autumn and the second wave of the disease, has raised concerns among students and their families.[83] Considering some of the positive effects of school reopening on students' health, especially vulnerable children, some, including the director of school-based programs at the WFP, believe that the effects of school closures on students' health are greater than its reopening. Therefore, many believe that instead of shutting down schools, reopening should be considered along with some protective strategies such as reducing class sizes, improving hygiene, alternating days at school, etc.[84]

To prevent inequalities escalation, the educational systems' responses to COVID-19 should be adjusted according to the contextual factors of communities and taking into account cultural, socioeconomic, geographical, and gender differences.[29]

16.5.5 SDG 8: Decent work and economic growth

The global economy, including many national and international industries, e.g., tourism, international trade, and travel, have been tremendously affected by the COVID-19 pandemic. For instance, the tourism industry has dropped down dramatically due to travel restrictions and border closures. The UN World Tourism Organization (UNWTO) announced a 22% reduction in international tourism during the first quarter of 2020 as the result of COVID-19 pandemic. It estimates the number of international tourism will also experience 60%–80% fall during 2020. This amount is responsible for $910 billion to $1.2 trillion loss from tourism revenues, which will impose a major impact on the global economy.[85] In particular, countries whose tourist attractions accounted for the bulk of their gross domestic product (GDP) will face severe recession.[38] This endangers millions of livelihoods and may reverse the progress made so far in advancing the SDGs.[85]

Due to the dependence of other SDGs on financing and extra investment, the economic downturn caused by the COVID-19 pandemic, in addition to challenging the economic growth goal and targets, has indirectly posed challenges to other SDGs as well. Even before COVID-19, SDG financing was underfunded by $2.5 trillion.[38]

16.5.6 SDG 11: Sustainable cities and communities

In an attempt to fight COVID-19, many governments have encouraged people to stay home to prevent virus transmission. Such a situation is even more devastating for the 79.5 million displaced people who have neither a safe house nor a formal job. Globally, almost 1 billion people live and work in informal, low-service, and insecure urban areas, most of whom live in crowded conditions with limited access to safe tap water, electricity, health, and other services. The COVID-19 pandemic poses a serious threat to the access of displaced people (refugees, migrants, or internal displaced people) to health care services and economic safety nets. This situation causes them to be considered as a threat for the settled citizens, which in turn increases their isolation, deprivation, and vulnerability.[29]

16.5.7 SDG 13: Climate action

COVID-19 is not the only global threat to SDGs. The world is predicted to encounter more stressful conditions including more catastrophic pandemics or epidemics, floods, earthquake, storm, wildfires, droughts, and so forth. The frequency and severity of all of these may be intensified as a result of climate change. Indeed, climate change is a greater challenge to national security than the COVID-19 pandemic.[38,86,87] The UN SDG progress report 2020 stated that as a result of the pandemic, restrictions and extreme decline in people activity, greenhouse gas emissions has fall down by 6% that is less than 7.6% target in the Paris Agreement.[70]

16.5.8 SDG 15: Life and land

In addition to disrupting ecosystems, wildlife trafficking is responsible for the spread of infectious diseases.[70] Therefore, reducing smuggling as well as the supply and demand of illegal wildlife products, which is considered in SDG 15, would reduce the likelihood of transmitting new viruses to humans.[88] Pangolins, for example, which are suspected to be the intermediary animals in the transmission of COVID-19 from bats to humans, are being trafficked more than any other animal in the world.[89] Pangolin meat is sold in wet markets, which are known to be the possible sites of early transmission of the coronavirus to humans. Between 2014 and 2018, 370,000 pangolins were seized worldwide.[70] Focusing these goal and targets may be helpful to reduce probability of further outbreaks.[38]

16.5.9 SDG 16: Peace, justice, and strong institutions

Even in the absence of COVID-19, conflict is recognized as a social determinant of health, which creates a wide range of health problems, from the lack of access to quality health care services to inadequate infrastructures. Measures to mitigate virus spread may increase domestic violence and threaten citizens' mental health. In order to properly respond to and prevent COVID-19 within vulnerable groups, the UN Security Council called upon the whole world for peace and a cease fire. Due to the increasing focus on COVID-19, attention to international agreements, treaties, and peace agreements has been also decreased. In addition, according to the UN High Commissioner for Human Rights, under exceptional or emergency measures, there has been a dramatic increase in police harshness and civil rights violations.[29]

16.5.10 SDG 17: Partnerships for the goals

As a result of the economic crisis caused by COVID-19, many donor countries prefer to spend most of their capital on their own citizens, which has led to a meaningful reduction of many official development assistances. With the withdrawal of the USA from the WHO, a significant proportion of the organization's funding was cut. The UN predicts that global foreign direct investment will decline by 40% in 2020.[70] Up to 16% of GDP across African countries comes from remittances.[29] Due to lockdown and the economic challenges posed by COVID-19, these remittances, which supported millions of households in the LMICs, are expected to decline dramatically from \$554 billion in 2019 to \$445 billion in 2020.[29,70] Ten months into the crisis, the real consequences of COVID-19 is not clear yet. Until safe vaccines are widely distributed or a certain effective treatment become available to control the outbreak, we cannot determine the precise consequences of the pandemic on sustainable development.[6] The world needs more strict surveillance and precise monitoring system to accurately track the virus and predict its future behavior, viral evolutions, transmissibility, and pathogenicity, all of which will influence on the virus mortality rates and prognosis as well as its global consequences.[90,91]

16.6 Lessons learned from COVID-19 outbreak

In the fight against COVID-19, an expanded spectrum of policies has been adopted by different countries, each of which led to its own consequences. Meanwhile, health policy learning tailored to each country's context can be helpful in increasing countries' preparedness to make more appropriate and timely decisions in the face of possible future waves of the COVID-19 as well as probable future crises, which in turn will have more favorable consequences for promoting global health. Countries like South Korea, China, and Singapore, which had learned lessons from the previous outbreaks such as MERS, SARS, etc. have often taken the COVID-19 threat more seriously, performed more effective in this battle, and therefore, achieved more desirable consequences. Evidence shows that countries with decisive and effective leadership, which mostly have appropriate coordination, were significantly less affected by COVID-19 than other countries.[92–94]

Our study to compare 10 selected countries' policies during the first wave of outbreak, we discovered the importance of three main strategies for better control of the outbreak, including timely measures, extensive open public testing for early detection of the cases, and comprehensive contact tracing. Adopting these strategies in addition to the strong political support played key role in response efficiently to the crisis.[6]

Despite the many threats and challenges, pandemics can be an opportunity to increase creativity and innovation. Increasing the use of novel technologies in tracking, tracing, testing, diagnosis, and treatment, will directly help control the outbreak and can be also effective to prevent the disease by reducing exposure. Also, by replacing humans with new technologies, such as telemedicine, governments can protect the economic cycle while maintaining social distancing and reducing contact.[78]

COVID-19 reminded us of the global fragile preparedness and the countries' weak ability to face large-scale crises. It also taught us that countries need to build and steadily improve their capacity for sustainable prevention, detection, and rapid response to health emergencies.[95] Global public health security needs to be invested, prioritized, and funded more efficiently. To increase global emergency preparedness, all countries need to have long-term, predictable, flexible, and sustainable financing on a large scale and based on global solidarity. Relying on a small number of HIC aid, foundations, and institutions is not enough to deal with public health security threats. Formulating more efficient policies and strategies to improve global health security along with implementing more preparedness interventions is also required globally.[96,97]

According to the International Health Regulations (IHR) (2005), all member states are obliged to develop minimum core public health capacities for the effective implementation of the IHR (2005). In 2008, the WHA adopted a resolution under Article 54 of the IHR (2005) requiring member states and the WHO to report progress on IHR implementation. In this regard, since 2010, WHO developed a self-assessment questionnaire for countries' progress self-assessment toward implementing IHR capacities. In 2018, the WHO introduced a new State Parties Self-Assessment Annual Reporting tool (SPAR). The SPAR is a quantitative tool and the only mandatory component of the IHR monitoring and evaluation framework.[98] The 2019 data showed progress in all 13 IHR capacities worldwide, particularly in the area of surveillance, laboratory, and IHR Coordination and National IHR Focal Point, but more effective and sustainable efforts are needed with regard to the chemical events, points of entry, and radiation emergencies. Based on this report, the average score of the member states was 64%, showing more work is needed to reach an acceptable global status.[99]

The 2019 global health security (GHS) index, jointly published by Johns Hopkins Center for Health Security, Nuclear Threat Initiative, and The Economist Intelligence Unit, reported 195 countries preparedness in the face of outbreaks. The GHS measures countries capability to prevent, detect and report, and rapid respond, in addition to the country's health system to protect people, their compliance with global norms, and the countries environmental risks and vulnerability to microbial threats. According to the GHS index, global average score was 40.2 out of 100, which highlighted the weak preparedness of the world.[100] The index also showed that the USA is the most prepared country in this regard; nevertheless, the COVID-19 experience violated the result and showed the need to revised the GHS indicators and develop more comprehensive and realistic indices.[28] Actually, none of the previous indices could precisely predict countries' preparedness to deal with the coronavirus threat and save lives. Also, the COVID-19 crisis showed that to measure the countries' preparedness in the face of public health threats, in addition to indices related to the health sector, other indices from other sectors, i.e., social protection and political support should also be considered and interventions should be made to promote these indices before the crises occur.

The COVID-19 crisis also highlighted the importance of investing on prevention rather than treatment. Estimations resealed that ending COVID-19 crisis costs between \$8.1 and \$15.8 trillion globally (obtained by calculating the reduction in gross domestic product plus the cost of economic loss due to hundreds of thousands of workforces death from the disease worldwide); while, globally \$22.2–\$30.7 billion per annum is required to prevent or remarkably reduce new diseases transmission that is about 500 times less than ending the pandemic up.[101]

In addition, the high prevalence of COVID-19 even in countries with a strong health system, once more, proved the increasingly need for universal health coverage (UHC) more than ever. Governments should spend more on health systems strengthening to improve its resiliency and maintain health security in the face of crises. This requires a strong political commitment with a view to achieving sustainable health development.[13,102]

The COVID-19 crisis affects all aspects of societies and needs inter-sectoral collaborations and multilateral interventions within not only the health system but also the governments and societies, so-called whole-of-government and whole-of-society approaches.[103] Until widespread distribution of the vaccine or detection of a definite treatment, in addition to the government, people are equally responsible to preserve each other from being infected. Ethically, it is everybody's duty to be committed to the personal hygiene, not only to protect his-self/her-self but also other people in the society.[97]

The important of one-health approach become more apparent than ever through COVID-19 outbreak.[104] Prevention, detection, and controlling zoonosis requires the integration of both humans and animals surveillance health systems.[90] Prior to COVID-19, previous outbreaks revealed that the occurrence of zoonotic diseases are increasing and might be continued[105] due to the increasing human-animal cohabitation. More research is required to build reproducible animal models to prevent the occurrence or minimize the consequences of any other emerging zoonotic infectious diseases.[90] Toward utilizing one-health approach, improving zoonotic infection prevention and control measures could be an applicable solution that has evidently been implemented in camel farms following the MERS outbreak in 2012.[103]

16.7 Main points

- The COVID-19 pandemic revealed global health systems' fragility.
- A pandemic is a global event that affects human beings' livelihoods and does not respect boundaries.
- The complexity of fighting pandemics requires integrated and coordinated global actions, beyond the whole-of-government and whole-of-society approaches, and in line with a whole-of-world approach.
- Although evidence has not confirmed airborne diffusion of COVID-19 thus far, the virus can be transmitted from person to person through sneezing, coughing, talking to other people at close distance, and polluted airflow.
- Environmental factors such as air temperature, humidity, air pollution, water and sewage, inanimate surface, insects, etc., are effective in spreading the COVID-19 virus.
- COVID-19 pandemic has made the world upside-down by creating uncertain conditions. This has led many people to experience fear, worry, and anxiety during this period.
- Some vulnerable groups of population, i.e., health care providers, refugees, elderly, informal workers, and people who suffer from underlying disease, might be more exposed to mental problems, due to their special characteristics.
- Given the effects of COVID-19, updating SDGs and their targets should be taken seriously, taking into account three points: (1) will SDG goals and targets still remain priority after COVID-19? (2) If so, will be we witnessing development or growth in these indicators? (3) Is the path to achieving SDGs resilient enough to overcome sudden disruptions, including the ongoing and probable upcoming global crises?

- The COVID-19 pandemic highlighted the necessity of global alliance, solidarity, and partnership to achieve good health and well-being for all (SDG 3).
- Imposing problems for national, local and communicable levels of almost all countries, COVID-19 does not respect international boundaries.
- Future pandemics look inevitable, rendering all countries, irrespective of their level of development, to become more resilient and strong enough to face the future long-scale crisis, especially in the field of health.
- Ever since COVID-19 outbreak initiated, one-health approach has attracted global attention. It has become more explicit and clearer, now more than ever, that human beings and animals' health systems are highly interconnected.

16.8 Conclusion

Historically, the most efficient reforms in the health systems have usually occurred in response to major crises. For example, health care systems in Europe and Japan improved dramatically after World War II. Similarly, high incidence of HIV/AIDS and financial crises led to robust plans to reach UHC in Thailand. Thus, governments can use the experiences and policy lessons learned from the current pandemic to establish more resilient health care systems that are strong enough to overcome any possible future crisis. Current response to COVID-19 needs to be strengthened urgently; otherwise, the world is likely to face chaos and devastating catastrophes for years and perhaps decades to come.

Unfortunately, although current vaccines are not distributed widely and there is no definitive treatment for COVID-19, nonpharmacological interventions and environmental health activities such as quarantining patients with clinical symptoms; wearing PPEs; social distancing; frequently disinfecting hands and surfaces; and avoiding touching one's face including mouth, eyes, or nose with unwashed hands have been suggested as effective methods to control the disease. However, given that implementing social distancing, lockdown, quarantine, and obligation to stay at home might intensify the psychological burden of this outbreak, it is necessary that health authorities prioritize people mental health and prepare appropriate strategies for people to cope with psychological consequences of the outbreak.

COVID-19 revealed that SDG targets are not resilient enough in the face of such a powerful global threat. Some reports indicate that not only two-thirds of SDG targets are under threat as a result of COVID-19 pandemic, but 10% of goals may also intensify problems and reinforce the effects of further outbreaks. Nevertheless, some of the targets including Target 15.7 (end poaching and trafficking of protected species and address demand and supply of illegal wildlife products), are critical to reducing the risk of outbreaks such as COVID-19. In addition, SDG 3, in particular achieving UHC (SDG 3.8), is vital to contain the impact of outbreaks, especially in the LMICs. Therefore, SDG revisit might be necessary to replace some SDG priorities with a set of more realistic targets. It should be considered which goals are achievable in a less connected world with a slow global economic growth.

Despite its devastating challenges, the pandemic and its lessons learned can provide an opportunity for many governments to rebuild their health system toward greater surveillance and preparedness and plan and implement necessary measures to reach UHC and rebuild global trust. The ongoing pandemic reminds all countries and their governments the importance of intersectoral and multisectoral collaboration inside and outsides of their borders. COVID-19 reminds authorities across the globe that reaching UHC in all countries is necessary, now more than ever.

References

1. Liu Y, Lee JM, Lee C. The challenges and opportunities of a global health crisis: the management and business implications of COVID-19 from an Asian perspective. *Asian Bus Manag* 2020;**19**(3):277–97.
2. Hens N, Vranck P, Molenberghs G. The COVID-19 epidemic, its mortality, and the role of non-pharmaceutical interventions. *Eur Heart J Acute Cardiovasc Care* 2020;**9**(3):204–8.
3. World Health Organization. *Statement on the second meeting of the International Health Regulations (2005) Emergency Committee regarding the outbreak of novel coronavirus (2019-nCoV)*. 2020. www.who.int/news-room/detail/30-01-2020-statement-on-the-second-meeting-of-the-international-health-regulations-(2005)-emergency-committee-regarding-the-outbreak-of-novel-coronavirus-(2019-ncov). [Accessed 30 January 2020].
4. World Health Organization. *WHO coronavirus disease (COVID-19) dashboard*. WHO; 2020. https://covid19.who.int/. [Accessed 5 March 2021].
5. Singer D. Clinical and health policy challenges in responding to the COVID-19 pandemic. *Postgrad Med* 2020;**96**(1137):373.
6. Raoofi A, Takian A, Haghighi H, et al. COVID-19 and comparative health policy learning; the experiences of 10 countries. *Arch Iran Med* 2021;**24**(3) [Epub ahead of print].
7. Seale H, Dyer CEF, Abdi I, et al. Improving the impact of non-pharmaceutical interventions during COVID-19: examining the factors that influence engagement and the impact on individuals. *BMC Infect Dis* 2020;**20**(1):607.
8. Doogan C, Buntine W, Linger H, Brunt S. Public perceptions and attitudes toward COVID-19 nonpharmaceutical interventions across six countries: a topic modeling analysis of Twitter data. *J Med Internet Res* 2020;**22**(9):e21419.

9. Davis M, Stephenson N, Flowers P. Compliant, complacent or panicked? Investigating the problematisation of the Australian general public in pandemic influenza control. *Soc Sci Med* 2011;**72**(6):912–8.
10. Bish A, Michie S. Demographic and attitudinal determinants of protective behaviours during a pandemic: a review. *Br J Health Psychol* 2010;**15**(Pt 4):797–824.
11. Ferguson NM, Laydon D, Nedjati-Gilani G, et al. *Report 9 – impact of non-pharmaceutical interventions (NPIs) to reduce COVID-19 mortality and healthcare demand.* 2020. https://www.imperial.ac.uk/mrc-global-infectious-disease-analysis/covid-19/report-9-impact-of-npis-on-covid-19/. [Accessed 11 December 2020].
12. World Economic Forum. *Challenges and opportunities in the post-COVID-19 world.* Switzerland: World Economic Forum; May 19, 2020.
13. Takian A, Aarabi M, Haghighi H. *The role of universal health coverage in overcoming the covid-19 pandemic.* The BMJ Opinion; 2020. Accessed April 20, https://blogs.bmj.com/bmj/2020/04/20/the-role-of-universal-health-coverage-in-overcoming-the-covid-19-pandemic/.
14. Anwar S, Nasrullah M, Hosen MJ. COVID-19 and Bangladesh: challenges and how to address them. *Front Public Health* 2020;**8**. 154–154.
15. Takian A, Kiani MM, Khanjankhani K. COVID-19 and the need to prioritize health equity and social determinants of health. *Int J Public Health* 2020;**65**(5):521–3.
16. Takian A, Bakhtiari A, Ostovar A. Universal health coverage for strengthening prevention and control of noncommunicable diseases in COVID-19 era. *Med J Islam Repub Iran* 2020;**34**(1):1050–3.
17. World Health Organization. *Maintaining essential health services: operational guidance for the COVID-19 context: interim guidance.* Geneva, Switzerland: World Health Organization; 2020.
18. Bell JS, Reynolds L, Freeman C, Jackson JK. Strategies to promote access to medications during the COVID-19 pandemic. *Aust J Gen Pract* 2020;**49**(8):530–2.
19. Cavallo JJ, Donoho DA, Forman HP. Hospital capacity and operations in the coronavirus disease 2019 (covid-19) pandemic—planning for the nth patient. *JAMA Health Forum* 2020;**1**(3):e200345.
20. Rosenbaum L. The untold toll – the pandemic's effects on patients without COVID-19. *N Engl J Med* 2020;**382**(24):2368–71.
21. Nicholson S, Asch DA. Hospitals need cash. Health insurers have it. *Harv Bus Rev* 2020. https://hbr.org/2020/03/hospitals-need-cash-health-insurers-have-it. [Accessed 13 September 2020].
22. Anoushiravani AA, O'Connor CM, DiCaprio MR, Iorio R. Economic impacts of the COVID-19 crisis: an orthopaedic perspective. *J Bone Joint Surg Am* 2020;**102**(11):937–41.
23. Bedingfield W. *What the world can learn from South Korea's coronavirus strategy? Wired.* 2020. Accessed March 21. https://www.wired.co.uk/article/south-korea-coronavirus.
24. Mareiniss DP. The impending storm: COVID-19, pandemics and our overwhelmed emergency departments. *Am J Emerg Med* 2020;**38**(6):1293–4.
25. Zhao S, Lin Q, Ran J, et al. Preliminary estimation of the basic reproduction number of novel coronavirus (2019-nCoV) in China, from 2019 to 2020: a data-driven analysis in the early phase of the outbreak. *Int J Infect Dis* 2020;**92**:214–7.
26. Wang FS, Zhang C. What to do next to control the 2019-nCoV epidemic? *Lancet* 2020;**395**(10222):391–3.
27. Buss PM, Tobar S. COVID-19 and opportunities for international cooperation in health. *Cad Saude Publica* 2020;**36**(4):e00066920.
28. Raoofi A, Takian A, Akbari Sari A, Olyaeemanesh A, Haghighi H, Aarabi M. COVID-19 pandemic and comparative health policy learning in Iran. *Arch Iran Med* 2020;**23**(4):220–34.

29. Lambert H, Gupte J, Fletcher H, et al. COVID-19 as a global challenge: towards an inclusive and sustainable future. *Lancet Planet Health* 2020;**4**(8):e312–4.
30. Shammi M, Bodrud-Doza M, Towfiqul Islam ARM, Rahman MM. COVID-19 pandemic, socioeconomic crisis and human stress in resource-limited settings: a case from Bangladesh. *Heliyon* 2020;**6**(5):e04063.
31. The World Bank. *The global economic outlook during the COVID-19 pandemic: a changed world*. 2020. https://www.worldbank.org/en/news/feature/2020/06/08/the-global-economic-outlook-during-the-covid-19-pandemic-a-changed-world. [Accessed 24 October 2020].
32. Maliszewska, M., Mattoo, A., & van der Mensbrugghe, D. (2020). *The Potential Impact of COVID-19 on GDP and Trade: A Preliminary Assessment*.
33. Poudel PB, Poudel MR, Gautam A, et al. COVID-19 and its global impact on food and agriculture. *J Biol Today's World* 2020;**9**(5):221.
34. Baldwin R, Tomiura E. Thinking ahead about the trade impact of COVID-19. In: Baldwin R, di Mauro BW, editors. *Economics in the time of COVID-19*. London, UK: CEPR Press; 2020. p. 59–73.
35. Gössling S, Scott D, Hall CM. Pandemics, tourism and global change: a rapid assessment of COVID-19. *J Sustain Tourism* 2021;**29**(1):1–20.
36. Shah SGS, Farrow A. A commentary on "World Health Organization declares global emergency: a review of the 2019 novel Coronavirus (COVID-19)". *Int J Surg* 2020;**76**:128–9.
37. World Bank. *Global economic prospects*. Washington, DC: World Bank; June 2020.
38. Naidoo R, Fisher B. Reset sustainable development goals for a pandemic world. *Nature* 2020;**583**(7815):198–201.
39. Pandemic Response Accountability Committee. *Pandemic oversight*. 2020. https://www.pandemicoversight.gov/track-the-money/funding-overview. [Accessed 13 September 2020].
40. The Lancet Infectious Disease. Challenges of coronavirus disease 2019. *Lancet Infect Dis* 2020;**20**(3):261.
41. International Labour Organisation. *Teleworking during the COVID-19 pandemic and beyond: a practical guide*. Geneva, Switzerland: International Labour Office; 2020.
42. United Nations. *Workings of the general assembly during the COVID-19 pandemic*. 2020. https://www.worldbank.org/en/news/feature/2020/06/08/the-global-economic-outlook-during-the-covid-19-pandemic-a-changed-world. [Accessed 24 October 2020].
43. Qiu J, Shen B, Zhao M, Wang Z, Xie B, Xu Y. A nationwide survey of psychological distress among Chinese people in the COVID-19 epidemic: implications and policy recommendations. *Gen Psychiatr* 2020;**33**(2):e100213.
44. Eslami H, Jalili M. The role of environmental factors to transmission of SARS-CoV-2 (COVID-19). *Amb Express* 2020;**10**(1):92.
45. Xie J, Zhu Y. Association between ambient temperature and COVID-19 infection in 122 cities from China. *Sci Total Environ* 2020;**724**:138201.
46. Bu J, Peng D-D, Xiao H, et al. *Analysis of meteorological conditions and prediction of epidemic trend of 2019-nCoV infection in 2020*. medRxiv; 2020. 2020.2002.2013.20022715.
47. Oliveiros B, Caramelo L, Ferreira NC, Caramelo F. *Role of temperature and humidity in the modulation of the doubling time of COVID-19 cases*. medRxiv; 2020. 2020.2003.2005.20031872.
48. Wang M, Jiang A, Gong L, et al. *Temperature significant change COVID-19 transmission in 429 cities*. medRxiv; 2020. 2020.2002.2022.20025791.
49. Poirier C, Luo W, Majumder MS, et al. The role of environmental factors on transmission rates of the COVID-19 outbreak: an initial assessment in two spatial scales. *Sci Rep* 2020;**10**(1):17002.

50. Li J-PO, Lam DSC, Chen Y, Ting DSW. Novel Coronavirus disease 2019 (COVID-19): the importance of recognising possible early ocular manifestation and using protective eyewear. *Br J Ophthalmol* 2020;**104**(3):297–8.
51. Bowdle A, Munoz-Price LS. Preventing infection of patients and healthcare workers should Be the new normal in the era of novel coronavirus epidemics. *Anesthesiology* 2020;**132**(6):1292–5.
52. Malhotra N, Gupta N, Ish S, Ish P. COVID-19 in intensive care. Some necessary steps for health care workers. *Monaldi Arch Chest Dis* 2020;**90**(1):161–2.
53. Jones RM, Xia Y. Annual burden of occupationally-acquired influenza infections in hospitals and emergency departments in the United States. *Risk Anal* 2018;**38**(3):442–53.
54. Ng Y, Li Z, Chua Y-X, et al. Evaluation of the effectiveness of surveillance and containment measures for the first 100 patients with COVID-19 in Singapore – January 2–February 29, 2020. *Morb Mortal Wkly Rep (MMWR)* 2020;**69**(11):307–11.
55. Institute for Health Metrics & Evaluation. *COVID-19 projections*. IHME; 2020. https://covid19.healthdata.org/. [Accessed 5 March 2021].
56. Pung R, Chiew CJ, Young BE, et al. Investigation of three clusters of COVID-19 in Singapore: implications for surveillance and response measures. *Lancet* 2020;**395**(10229):1039–46.
57. Garcia LP, Duarte E. Nonpharmaceutical interventions for tackling the COVID-19 epidemic in Brazil. *Epidemiol Serv Saude* 2020;**29**(2):e2020222.
58. Olsen SJ, Azziz-Baumgartner E, Budd AP, et al. Decreased influenza activity during the COVID-19 pandemic – United States, Australia, Chile, and South Africa, 2020. *Morb Mortal Wkly Rep (MMWR)* 2020;**69**(37):1305–9.
59. Ngonghala CN, Iboi E, Eikenberry S, et al. Mathematical assessment of the impact of non-pharmaceutical interventions on curtailing the 2019 novel Coronavirus. *Math Biosci* 2020;**325**:108364.
60. Kantor BN, Kantor J. Non-pharmaceutical interventions for pandemic COVID-19: a cross-sectional investigation of US general public beliefs, attitudes, and actions. *Front Med* 2020;**7**:384.
61. World Health Organization. *Water, sanitation, hygiene, and waste management for SARS-CoV-2, the virus that causes COVID-19*. COVID-19*: Infection prevention and control/ WASH*. 2020. https://www.who.int/publications/i/item/WHO-2019-nCoV-IPC-WASH-2020.4. [Accessed 25 October 2020].
62. Qu G, Li X, Hu L, Jiang G. An imperative need for research on the role of environmental factors in transmission of novel coronavirus (COVID-19). *Environ Sci Technol* 2020;**54**(7):3730–2.
63. Lodder W, de Roda Husman AM. SARS-CoV-2 in wastewater: potential health risk, but also data source. *Lancet Gastroenterol Hepatol* 2020;**5**(6):533–4.
64. Naddeo V, Liu H. Editorial Perspectives: 2019 novel coronavirus (SARS-CoV-2): what is its fate in urban water cycle and how can the water research community respond? *Environ Sci* 2020;**6**(5):1213–6.
65. Haider II , Tiwana F, Tahir SM. Impact of the COVID-19 pandemic on adult mental health. *Pak J Med Sci* 2020;**36**(COVID19-S4):S90–4.
66. Russell MC, Figley CR. Generational wartime behavioral health crises: part One of a preliminary analysis. *Psychol Inj Law* 2015;**8**(2):106–31.
67. Serafini G, Parmigiani B, Amerio A, Aguglia A, Sher L, Amore M. The psychological impact of COVID-19 on the mental health in the general population. *QJM* 2020;**113**(8):531–7.
68. Panchal N, Kamal R, Orgera K, et al. *The implications of COVID-19 for mental health and substance use*. Kaiser Fam Found; 2020.
69. The Lancet Infectious Diseases. The intersection of COVID-19 and mental health. *Lancet Infect Dis* 2020;**20**(11):1217.

70. United Nations. *The sustainable development goals report 2020.* New York, NY, USA: United Nations Publications; 2020.
71. World Health Organization. *Coronavirus disease (COVID-19) pandemic.* WHO; 2020. https://www.who.int/emergencies/diseases/novel-coronavirus-2019. [Accessed 5 March 2021].
72. Zaharieva R. *Lessening the impact of COVID-19 on vulnerable groups. COVID-19.* Health Inequalities; 2020. Accessed Jul 13, https://epha.org/lessening-the-impact-of-covid-19-on-vulnerable-groups/.
73. Ataguba OA, Ataguba JE. Social determinants of health: the role of effective communication in the COVID-19 pandemic in developing countries. *Glob Health Action* 2020;**13**(1):1788263.
74. Molina-Betancur JC, Martinez-Herrera E, Pericas JM, Benach J. Coronavirus disease 2019 and slums in the global South: lessons from Medellin (Colombia). *Glob Health Promot* 2020. 1757975920962797.
75. The Lancet Public Health. Will the COVID-19 pandemic threaten the SDGs? *Lancet Public Health* 2020;**5**(9):e460.
76. Global Network Against Food Crises, Food Security Information Network. *Global report on food crises. Joint analysis for better decisions.* Food Security Information Network (FSIN); 2020.
77. Laborde D, Martin W, Swinnen J, Vos R. COVID-19 risks to global food security. *Science* 2020;**369**(6503):500−2.
78. Torero M. Without food, there can be no exit from the pandemic. *Nature* 2020;**580**(7805):588−9.
79. *Pregnant mothers and babies born during COVID-19 pandemic threatened by strained health systems and disruptions in services.* New York, USA: UNICEF; May 7 2020. [press release].
80. Roberton T, Carter ED, Chou VB, et al. Early estimates of the indirect effects of the COVID-19 pandemic on maternal and child mortality in low-income and middle-income countries: a modelling study. *Lancet Glob Health* 2020;**8**(7):e901−8.
81. *UN educational, scientific and cultural organisation. Global education coalition. COVID-19 education response.* 2020. https://en.unesco.org/covid19/educationresponse/globalcoalition. [Accessed 21 September 2020].
82. UN Educational, Scientific and Cultural Organisation. *Keeping girls in the picture.* 2020. https://en.unesco.org/covid19/educationresponse/girlseducation. [Accessed 21 September 2020].
83. Simon A, Huebner J, Berner R, et al. Measures to maintain regular operations and prevent outbreaks of SARS-CoV-2 in childcare facilities or schools under pandemic conditions and co-circulation of other respiratory pathogens. *GMS Hyg Infect Control* 2020;**15**. Doc22.
84. UN Educational, Scientific and Cultural Organisation. *Reopening schools safely: the challenges for health and education.* 2020. https://en.unesco.org/news/reopening-schools-safely-challenges-health-and-education. [Accessed 25 October 2020].
85. World Tourism Organization. *International tourist numbers could fall 60-80% in 2020.* 2020. https://www.unwto.org/news/covid-19-international-tourist-numbers-could-fall-60-80-in-2020. [Accessed 20 September 2020].
86. Institute for Economics & Peace. *COVID-19 and peace.* Sydney: IEP; 2020.
87. Institute for Economics & Peace. *Cological threat register 2020: understanding ecological threats, resilience and peace.* Sydney: IEP; 2020.
88. Johnson CK, Hitchens PL, Pandit PS, et al. Global shifts in mammalian population trends reveal key predictors of virus spillover risk. *Proc Biol Sci* 2020;**287**(1924):20192736.
89. Hitchens P, Johnson CK. *Don't blame the pangolin (or any other animal) for COVID-19. Animals, Food & You.* 2020. Accessed April 8, https://pursuit.unimelb.edu.au/articles/don-t-blame-the-pangolin-or-any-other-animal-for-covid-19.

90. Contini C, Di Nuzzo M, Barp N, et al. The novel zoonotic COVID-19 pandemic: an expected global health concern. *J Infect Dev Ctries* 2020;**14**(3):254–64.
91. Sohrabi C, Alsafi Z, O'Neill N, et al. World Health Organization declares global emergency: a review of the 2019 novel coronavirus (COVID-19). *Int J Surg* 2020;**76**:71–6.
92. Al Saidi AMO, Nur FA, Al-Mandhari AS, El Rabbat M, Hafeez A, Abubakar A. Decisive leadership is a necessity in the COVID-19 response. *Lancet* 2020;**396**(10247):295–8.
93. Forman R, Atun R, McKee M, Mossialos E. 12 Lessons learned from the management of the coronavirus pandemic. *Health Pol* 2020;**124**(6):577–80.
94. Sergent K, Stajkovic AD. Women's leadership is associated with fewer deaths during the COVID-19 crisis: quantitative and qualitative analyses of United States governors. *J Appl Psychol* 2020;**105**(8):771–83.
95. World Health Organization. *Strategic Partnership for International Health Regulations (2005) and Health Security (SPH)*; 2020. https://extranet.who.int/sph/covid-19. [Accessed 16 September 2020].
96. Khanna RC, Cicinelli MV, Gilbert SS, Honavar SG, Murthy GSV. COVID-19 pandemic: lessons learned and future directions. *Indian J Ophthalmol* 2020;**68**(5):703–10.
97. Global Preparedness Monitoring Board. *A world in disorder. Global preparedness monitoring board annual report 2020.* Geneva: World Health Organization; 2020.
98. World Health Organization. Global Health Observatory (GHO) data. *International health regulations (2005) monitoring framework.* 2020. https://www.who.int/gho/ihr/en/. [Accessed 16 September 2020].
99. World Health Organization. *WHO coronavirus disease (COVID-19) dashboard; Iran (Islamic republic of).* 2020. https://covid19.who.int/region/emro/country/ir. [Accessed 10 May 2020].
100. The Global Health Security Index. *Global health security index: building collective action and accountability.* 2019. https://www.ghsindex.org/wp-content/uploads/2019/10/GHS-Index-Report_FINAL_Oct2019.pdf. [Accessed 15 March 2020].
101. Schwab J. *Fighting COVID-19 could cost 500 times as much as pandemic prevention measures.* 2020. https://www.weforum.org/agenda/2020/08/pandemic-fight-costs-500x-more-than-preventing-one-futurity?fbclid=IwAR0MdKIwh1_xrq7a28n-0oiDu23lXY52_-ksMSbKQxB0DQJfMXim4Gi2H9E. [Accessed 17 September 2020].
102. Armocida B, Formenti B, Palestra F, Ussai S, Missoni E. COVID-19: universal health coverage now more than ever. *J Glob Health* 2020;**10**(1):010350.
103. Kock RA, Karesh WB, Veas F, et al. 2019-nCoV in context: lessons learned? *Lancet Planet Health* 2020;**4**(3):e87–8.
104. Sleeman JM, DeLiberto T, Nguyen N. Optimization of human, animal, and environmental health by using the One Health approach. *J Vet Sci* 2017;**18**(S1):263–8.
105. Kelvin DJ, Rubino S. Fear of the novel coronavirus. *J Infect Dev Ctries* 2020;**14**(1):1–2.

Index

Note: 'Page numbers followed by "f" indicate figures and "t" indicate tables.'

Q

R

S

Z

FSC
www.fsc.org
MIX
Papier | Fördert
gute Waldnutzung
FSC® C083411

Kolibri 360 GmbH
Ferdinand-Jühlke-Straße 7
99095 Erfurt, Deutschland
produktsicherheit@kolibri360.de